The United States Merchant Marine in World War I

Also by Greg H. Williams
and from McFarland

The Liberty Ships of World War II: A Record of the 2,710 Vessels and Their Builders, Operators and Namesakes, with a History of the Jeremiah O'Brien (2014)

World War II U.S. Navy Vessels in Private Hands: The Boats and Ships Sold and Registered for Commercial and Recreational Purposes Under the American Flag (2013)

The French Assault on American Shipping, 1793–1813: A History and Comprehensive Record of Merchant Marine Losses (2009)

World War II Naval and Maritime Claims Against the United States: Cases in the Federal Court of Claims, 1937–1948 (2006)

Civil War Suits in the U.S. Court of Claims: Cases Involving Compensation to Northerners and Southerners for Wartime Losses (2006)

Civil and Merchant Vessel Encounters with United States Navy Ships, 1800–2000 (2002)

The United States Merchant Marine in World War I

Ships, Crews, Shipbuilders and Operators

Greg H. Williams

McFarland & Company, Inc., Publishers
Jefferson, North Carolina

Library of Congress Cataloguing-in-Publication Data

Names: Williams, Greg H., author.
Title: The United States Merchant Marine in World War I : ships, crews, shipbuilders and operators / Greg H. Williams.
Description: Jefferson, North Carolina : McFarland & Company, Inc., Publishers, 2017 | Includes bibliographical references and index.
Identifiers: LCCN 2017006103 | ISBN 9781476667034 (softcover : acid free paper) ♾
Subjects: LCSH: World War, 1914–1918—Transportation—United States. | Merchant marine—United States—History—20th century. | Shipping—United States—History—20th century. | Shipbuilding—United States—History—20th century. | Merchant ships—United States—History—20th century. | World War, 1914–1918—Naval operations, American. | World War, 1914–1918—United States.
Classification: LCC D639.T8 W55 2017 | DDC 940.4/5973—dc23
LC record available at https://lccn.loc.gov/2017006103

British Library cataloguing data are available

ISBN (book) 978-1-4766-6703-4
ISBN (ebook) 978-1-4766-2672-7

Front cover *top* Merchant Marine seal; photograph of the steamer *Hawaiian* built in 1919 for the American-Hawaiian Steamship Company, a typical freighter of the day (from the author's collection)

Printed in the United States of America

McFarland & Company, Inc., Publishers
Box 611, Jefferson, North Carolina 28640
www.mcfarlandpub.com

Table of Contents

Preface

When the general history of America's wars is recounted, the sea services—and in particular the merchant marine—are usually mentioned as an afterthought, if they are mentioned at all. The focus is always on the great land battles—Bunker Hill, Saratoga, the Doughboys at the Meuse-Argonne, the Leathernecks on Guadalcanal, and Joe and Willie on their way to Berlin.

The battles at sea are often mentioned in passing and the loss of merchant shipping, if it is mentioned at all, is usually expressed in terms of anonymous "tonnage."

The World War I Centennial Commission is currently building a $30 million monument in Washington to the American soldiers who lost their lives in the Great War. A film clip on its website shows soldiers disembarking from a troop transport but there is no mention of the ship, the crew who brought the ship "over there," or the gunners who protected it.

A merchant seaman, especially an engineer, who signed articles for a North Atlantic voyage in the winter of 1917–18 or 1941–42 might just as well have signed his own death certificate. Unless survivors from a torpedoed ship were rescued fairly quickly, the odds of dying from exposure were very high. No ship in a convoy, even one very close by, was permitted to stop and pick up survivors and the closest escort might be off chasing the attacking submarine. In heavy winter weather it would be very difficult to locate anyone still alive, much less get him aboard a rescue vessel. Even so, very few ships left port without full crews although during the Great War an all–American crew was almost unheard of.

Information on this subject is widely scattered and available only within short chapters in books, newspaper and magazine articles, and notices in trade publications of the day. This book is an attempt to produce a reasonably comprehensive account of the American merchant marine during the Great War.

Thanks to Kim E. Demory at the National Vessel Documentation Center for help with ship information; Louis Hough, grandson of wood freighter designer Edward S. Hough for information on those ships; and Gina Bardi, reference librarian at the San Francisco National Historical Park.

Introduction

In 1895, despite the recent dramatic change from sail to steam, it was noted that more than 1,700 American sailing vessels entered New York Harbor. Ship building costs in the United States were about 30 percent higher than in England and American law did not allow American registry for a foreign-built ship.

Between 1900 and 1910 the nation's foreign commerce increased by more than a billion dollars yet only 8 percent of the trade was carried in American ships.

When Uncle Sam's Great White Fleet made its grand tour from 1907 to 1909, foreign-flagged colliers and merchant ships furnished about 90 percent of the supplies the big warships needed because there weren't enough ocean-going American ships for the job. All but a relative handful of the nation's biggest ships were either on the Great Lakes or in passenger service.

By 1913, America's foreign trade amounted to $4,268,892,000, yet the nation had only 85 ocean-going ships and most of them went only to Caribbean and Central American ports. The Treasury Department was paying British shipping companies thousands of dollars a year to carry U.S. mail. In 1914, American importers paid a total of $52,131,000 in freight charges and of that, $46,818,000 went to foreign steamship companies. Maritime historian Alfred D. Rathbone stated in 1945: "Had we subsidized or otherwise brought about the construction and maintenance of even a modest American merchant marine with some of the money paid to foreign operators, our plight in 1914 could not possibly have been so serious." After the war started in August 1914, much of the foreign shipping the United States had depended on, mostly British, to carry American imports or exports was no longer available and goods and merchandise began piling up in warehouses and on piers.

The nature and effectiveness of Germany's submarine warfare caught everyone by surprise so there was no anti-submarine naval force to protect shipping. There weren't enough destroyers to go around and the new 110-foot wooden subchaser was designed and built after the submarine war was already well underway.

Despite the dramatic increase in ship values and freight charges—along with marine insurance rates—most private capital was reluctant to chance huge losses and held back on investing in shipbuilding until government war risk insurance, contracts, and subsidies became available.

The pent-up demand for ships then created a frenzy of shipbuilding. Idled shipyards retooled, new ones sprang up, and the building of sailing vessels—an all but vanished occupation—began anew. Retired shipwrights sharpened up their rusty tools and went back to work. As new businesses opened and existing companies increased their wartime production, workers became much more selective in their choice of employer and generally went after the best deal. The sea had not been all that popular of a calling for American boys before the war and very few, if any, American ships left port with all–American crews.

During the war, the Emergency Fleet Corporation requisitioned all domestic and foreign contracts for ocean-going steel ships then building, often completing them to government specifications. Various financial arrangements were made with the owners for the use or purchase of their vessels and the operation of all ocean-going ships came under control of the Shipping Board.

The Great War lasted four years, three months, and 16 days. The United States was in it for one year, seven months, and five days. Considering mainly the maritime aspects, the war went through six distinct phases:

1. In November 1914 Great Britain blockaded the North Sea to starve Germany, which it nearly did. All neutral shipping was inspected and contraband seized.
2. In February 1915 Germany declared the waters around the British Isles a war zone. All belligerent and neutral shipping was subject to attack without warning.
3. In March 1915 Great Britain halted all neutral shipping to Germany, causing worldwide protest.
4. In April 1917 the United States entered the war and overwhelmed the German army with 2,200,000 troops out of a total of 3,764,677 in the service. Many went over on foreign ships.
5. The German people became frustrated with a war they knew they couldn't win. While devastating to Allied shipping, the U-boat war had ultimately failed and with it Germany's only hope for victory. Increasing air attacks, labor strikes, and mutinies in the navy plagued the Kaiser's war effort. Better anti-submarine warfare tactics resulted in heavy U-boat losses.
6. The Northern Mine Barrage was the final straw. Very serious submarine losses caused mutinies in what was left of the U-boat fleet and the end came swiftly.

On October 21, 1918, the German government sent a note to President Wilson advising him that orders had been issued to submarine commanders to cease torpedoing passenger ships and that steps toward an armistice would be considered. On October 26 the U-boat crews mutinied and the war officially ended at 11 a.m. on Monday, November 11, 1918, French time.

The survivors from American vessels who were brought aboard U-boats reported that the German crews seemed to have no idea of how many American troops had actually gone overseas, grossly underestimating the real numbers.

On the homefront, the relatively short period of frenzied merchant ship building resulted in three generally unpleasant outcomes:

1. Most of the ships contracted for by the Shipping Board Emergency Fleet Corporation were delivered long after the war ended.

2. Almost every hastily-organized shipyard had contracts cancelled, leaving the builders holding the bag for capital improvements, equipment purchased, labor costs, and large loans. A huge amount of public money was spent compensating owners and settling claims for breach of contract.
3. The nation was left with a glut of no longer economical sailing vessels, wooden freighters nobody wanted, and big, coal-burning steel freighters that rapidly became obsolete.

The oil-burning steel freighters served well through the post-war years until the Great Depression when the merchant marine survived only through government mail-carrying subsidies. By 1941 these ships were tired, too small for the needs of the day, and all but obsolete.

The position the United States found itself in by April 1917, with respect to the merchant marine and the Navy, would be repeated almost down to the last detail in 1941 with the additional consequences of the disastrous naval arms limitation treaties, which the United States embraced, Japan renounced, and Germany took steps to evade.

Part One

Prelude to War

1. The Ascension of Germany

In November 1870, after decades of intrigue and a late Prussian war against France lead by Otto Eduard Leopold, Prince of Bismarck, Duke of Lauenburg, the North German Confederation concluded treaties with the independent southern states and on December 10 they united under Prussian rule.

Bismarck became chancellor, in reality a dictator, and on January 18, 1871, Wilhelm Friedrich Ludwig von Preußen, the King of Prussia, became Emperor of Germany. On March 9, 1888, the aged—some said senile—Wilhelm died at age 90 and was succeeded by his son Friedrich Wilhelm Nikolaus Karl, who then became Deutscher Kaiser und König von Preußen as Friedrich III. He had married Queen Victoria's oldest daughter, Princess Victoria "Vicky" Adelaide Mary Louisa, on May 19, 1857, and had he lived world events might have been very different. Unfortunately, he died of throat cancer on June 15, 1888, and was succeeded by his son, Friedrich Wilhelm Viktor Albrecht von Preußen—Kaiser Wilhelm II—who had the opposite temperament of his father. He shook his "mailed fist" at Europe and declared, "Germany must be supreme on sea as well as land."

Efforts had been afoot for decades to build a 60-mile channel connecting the Baltic Sea with the North Sea. The canal would save ships a 240-mile trip from Kiel up north and around Denmark to get to the North Sea. No action was taken until 1886 when plans were finalized and on June 3, 1887, Kaiser Wilhelm II presided over the start of construction of the Nord-Ostsee-Kanal from the new naval base at Kiel to Brunsbüttel on the Elbe River, 17 miles east of Cuxhaven at the mouth of the river on the North Sea.

In 1814 England had acquired Heligoland Island in the North Sea just off the coast of what would become Germany. The Admiralty found it unsuitable for a naval base and it eventually became a tourist attraction. With Germany's naval expansion and work on the Nord-Ostsee-Kanal going well, Heligoland Island was seen as an ideal naval base for protecting the canal and nearby naval installations. On July 1, 1890, Germany and England concluded a deal whereby Germany traded land in East Africa for the island. Germany eventually transformed the island into the "Gibraltar of the North." The new canal opened on June 21, 1895, and at the same time Germany began building fast passenger liners to challenge British trans-Atlantic supremacy.

Europe was then seething with old and new resentments, jealousies, broken alliances, new alliances, and a host of troubles. Czar Nicholas II of Russia became concerned about the situation and in 1899 called for a conference to discuss arms limitations and to try and find some way to settle disputes without resorting to war. Delegates from many nations, including Germany, met at The Hague and while the conference failed to set a limit on arms, it did produce a voluntary court for settling disputes.

In 1903, England formed a defensive alliance with France and trouble was brewing between Russia and Japan. Theodore Roosevelt proposed a second Hague conference to convene in 1904 to expand on the first but it had to be postponed when Russia and Japan went to war. Roosevelt was instrumental in ending that conflict but then England concluded another defensive alliance with Russia and the three, France, England, and Russia, all leery of Germany, became known as the "Triple Entente" and the "Circle of Steel." "Entente" derives from the French as applied to the parties of an agreement. This annoyed the Kaiser, but Germany attended the Second Hague Conference, held between June and October 1907. The relevant maritime matters discussed were:

Convention VI—The Status of Enemy Merchant Ships at the Outbreak of Hostilities.
Convention VII—Conversion of Merchant Ships into War Ships.
Convention VIII—The Laying of Automatic Submarine Contact Mines.
Convention XI—Restrictions with Regard to the Exercise of the Right of Capture in Naval War.
Convention XII—Creation of an International Prize Court.

All of these Conventions were ratified by everyone, including Germany, but no agreement was ever reached on setting up an international prize court.

The legalities of privateering, prize law, naval blockade, the duties and responsibilities of a neutral government toward a belligerent in their harbor, and the doctrine of visitation and search by naval commanders on the high seas had long been settled by World War I. English courts of admiralty, being those of the major sea power, had tried many of these cases over the decades and the most important of the decisions handed down were generally accepted by all the other maritime nations in the interest of stability, expediency, and continuity.

However, as usually happens when push comes to shove, the generally accepted laws of nations were often interpreted by belligerents to suit their own purposes, diplomacy, treaties, and conventions going only so far as self-preservation obtains.

2. *The Arming of Merchant Ships*

The arming of merchant ships had pretty much waned along with the age of sail and the advent of bigger, more powerful, rifled naval guns against which a lightly armed merchant vessel had no defense. There were some notable exceptions to carrying armaments and one occurred during the Yukon gold rush.

The North American Transportation & Trading Co. steamer *Portland* ferried gold seekers and goods from Seattle up to St. Michael's via Dutch Harbor. In August 1897 the company was preparing for its last voyage of the season and anticipated bringing back a number of prospectors who had struck it rich and would be coming down with a large quantity of gold. The company felt some protection might be needed so traffic manager Charles H. Hamilton sent their agent, Portus B. Weare, to see Secretary of the Treasury Lyman J. Gage about borrowing the type of Gatling gun used by the Navy, "to protect patrons of this company against pillage or robbery by outlaws who had not been successful in the gold fields of Alaska." Weare was told there were none available for use by private parties and to go out and purchase two of Hiram Maxim's machineguns, one for the bow and one for the stern. Weare purchased one gun and obtained the services of an expert "who will attend to fitting it up and putting it in working order" and the two left Washington on August 24. The mounting of the gun and its highly effective performance was well advertised in advance of the steamer's sailing and the *Portland* arrived in Seattle without incident with $2 million worth of gold.

One way to rapidly increase a naval fleet is to put guns on existing merchant ships and, with the requisite modifications, transform it into a ship of war. By the turn of the century, a merchant ship could not be effectively armed without the assistance of the government under whose flag the ship sailed. Suitable guns and ammunition were manufactured only under naval contracts and would have to be sold or loaned to a shipowner. In addition, commercial ships were not designed or built for the mounting and firing of a modern 3-inch, 4-inch, or 5-inch gun. There were no magazines for the safe storage of shells and powder or living space for a naval gun crew.

First Lord of the Admiralty Winston Churchill, ever mindful of England's status in relation to other navies, watched Germany's naval expansion program with concern, especially their plans to convert existing merchant ships into warships in order to rapidly increase its naval fleet. Kaiser Wilhelm II's aspirations of overtaking Great Britain as the major sea power had also made considerable progress with trans–Atlantic passenger service.

On March 26, 1913, right after President Woodrow Wilson took office, Churchill addressed Parliament on several subjects and then stated:

> I turn to one aspect of trade protection which requires special reference. It was made clear at the second Hague Conference and the London Conference, that certain of the Great Powers have reserved to themselves the right to convert merchant steamers into cruisers, not merely in national harbours, but if necessary on the high seas. There is now good reason to believe that a considerable number of foreign merchant steamers may be rapidly converted into armed ships by the mounting of guns. The sea-borne trade of the world follows well-marked routes upon nearly all of which tonnage of the British mercantile marine largely predominates. Our food-carrying liners and vessels carrying raw materials following these trade routes would in certain contingencies meet foreign vessels armed and equipped in the manner described. If the British ships had no armament, they would be at the mercy of any foreign liner carrying one effective gun and a few rounds of ammunition. It would be obviously absurd to meet the contingency of considerable numbers of foreign armed merchant cruisers on the high seas by building an equal number of cruisers. That would expose this country to an expenditure of money to meet a particular danger, altogether disproportionate to the expense caused by any foreign Power in creating that danger. Hostile cruisers, wherever they are found, will be covered and met by British ships of war, but the proper reply to an armed merchantman is another merchantman armed in her own defense.

The Great Powers at that time were England, the United States, Germany, France, Japan, Russia, Italy, and Austria-Hungary. When Germany allied with Austria-Hungary they became known as the Central Powers owing to their central geographical location within Europe. When Italy joined they were known as the Triple Alliance. During the coming war, Germany, Austria-Hungary, Turkey, and Bulgaria were referred to as the Central Powers.

Churchill went on to say that the Admiralty has elected to loan guns and to provide ammunition and training to merchant crews. The shipping companies were to cover the nominal cost of structural work needed before guns were put aboard and concluded by saying:

> No one can pretend to view these measures without regret, or without hoping that the period of retrogression all over the world which has rendered them necessary, may be succeeded by days of broader international confidence and agreement, than those through which we are now passing.

That May, Sir Owen Cosby Philipps, 1st Baron Kyslant and president of the Royal Mail Steam Packet Co., informed his stockholders that the company's larger liners had been armed. By January 1914, 29 British steamers had guns on their fantails—to fire while fleeing, if necessary.

British ships were given only one stern gun and provided with a trained Royal Navy gun pointer and a sight setter. Merchant seamen made up the rest of the gun crew. When the war started, 39 British ships were defensively armed with two 4.7" guns.

Germany's position was that the arming of a ship that remained in merchant service, as opposed to converting an existing merchant ship into a warship, was illegal under international law and that "a merchantman assumes a warlike character by armament with guns."

Part Two

The War in Europe

3. *Breakout of War*

On June 12, 1914, Benjamin Rush, vice-president of the Insurance Company of North America, arrived at New York on the Cunard liner *Mauretania* with his 18-year-old daughter Charlotte after vacationing in Europe. He recalled later that if someone had said to him at that moment, "Mr. Rush, there's going to be a war," he would have replied, "Why, has Pancho Villa finally gone too far?" War with Mexico seemed far more likely than any conflagration in Europe.

Seventeen days later, on Monday, June 29, a one-column, non–headline making Associated Press bulletin appeared on the front pages of the nation's daily newspapers. The heir to the Austrian throne, Archduke Franz Ferdinand and his wife Sophie had escaped a bomb thrown at them in Sarajevo but a few hours later they were shot and killed by a "student assassin"—identified as a Bosnian Serb—and riots had caused authorities to declare martial law.

Given the long and bloody history of the parties involved, including a recent war, the event did not engender any great anxiety outside the region. But then Austria accused Serbia of official complicity and on July 23 gave Serbia 48 hours to admit it, repudiate anti–Austrian propaganda, and submit to a formal investigation of the shooting by Austrian officials.

This ultimatum alarmed Sir Edward Grey, the British foreign minister, and Czar Nicholas II of Russia as they were all too aware of the existing defensive alliances that would come into play should a war actually be declared. They immediately proposed a conference in London to mediate the dispute. France and Italy quickly agreed but Germany declined. As one observer later stated, "because their hearts had been set on war and their preparations completed down to the last detail."

On July 25, prior to the actual ultimatum deadline, Austria began mobilizing troops and when Serbia's reply came it was deemed unsatisfactory and the world held its breath.

On Tuesday, July 28, Austria declared war on Serbia. This declaration caused immediate panic in the European financial markets and resulted in heavy losses to Americans who held securities there. The next day, July 29, was a little calmer and hopeful as British and French diplomats got to work trying to restore calm in the region while Germany, Russia, and France began mobilizing troops and the Royal Navy's Home Fleet was put on alert. Very aggressive diplomatic efforts ensued but on Thursday, July 30, hopes began to dim as German and Russian talk became more strident and the New York Stock Exchange had its biggest drop since the Panic of 1907. The marine insurance world turned upside down. Most policies on German and Austrian ships were cancelled and some companies refused to underwrite any trans–Atlantic voyage.

German and Austrian ships stayed in port, their sailings cancelled. French and British passenger service stayed on schedule but with many army reservists on board reporting for duty. Much of the world's marine traffic stayed in port, unable to obtain insurance. Grain shipments stopped.

On Saturday, August 1, the world blew up, overtaking the heat of summer, when Germany declared war on Russia over Russia's mobilization of land and sea forces Germany considered a threat to their survival. The next day, Russia invaded Germany and German troops entered France, Luxembourg, and Switzerland. On the 4th, Germany invaded Belgium and England declared war on Germany. The United States declared neutrality on the 5th and the next day Austria declared war on Russia and Serbia declared war on Germany.

The Shipping Crisis in America

In 1914 the American merchant marine was second only to Great Britain in total gross tonnage at around 8,319,000 but only about 1,500,000 tons of that was in foreign trade. The bulk was running coastwise or on the Great Lakes, which put the United States behind Germany, France, Norway, and Japan in overseas trade.

Because of the higher cost to borrow money and build and operate a ship in the United States, most American operators realized higher profits at home. An English operator had much lower shipbuilding and labor costs and was content with a three or four percent return on his overseas shipping investment.

In addition, a very old American law prohibited a foreign-built ship from obtaining American registry so American merchants were unable to take advantage of lower building costs in England. As a result of all this, American shippers had been depending largely on British ships to get their goods to Europe but when the war started, the British needed their ships to transport supplies for their troops going to France. Almost all shipping from Atlantic ports ceased, a situation not seen since President Jefferson's disastrous commercial embargo of 1807.

Lloyds of London, "that lodestar of marine rates," closed its doors and American underwriters were quoting 25 to 30 percent higher rates for North Sea voyages. Lloyds reopened on August 4 with rates that were close to what American underwriters had settled on while Lloyds was closed. When reports started coming in about German raiders in the South Atlantic and Indian Oceans, rates to South America went up 20 percent and 15 to 20 percent to India and the Far East. Hundreds of fully-loaded ships were held up in port and rail traffic backed up all the way to Midwest rail centers. Prices of commodities jumped overnight. The value of 6,000 tons of sugar aboard a Norwegian freighter at Wilmington, Delaware, increased $50,000 in one day.

England was the leading marine insurance carrier of the day, followed by Germany—both now belligerents—then the United States. The marine insurance sector in America was much smaller than fire and life at the time so the ability of underwriters to cover all ordinary marine, and now war risks, was beyond the financial capacity of the industry, especially as more ships entered trans–Atlantic service when freight rates began to rise, creating a long-awaited, renewed opportunity for profit.

Rates soon began to drop, especially to the British Isles, as the Royal Navy gained some control of the sea lanes and the German fleet was kept in check. But the U-boats were essentially unchecked and on August 10, a group of prominent shipping men including California shipping pioneer Robert Dollar and James A. Farrell, president of U.S. Steel, went to Washington to ask for help the shipping crisis and with marine insurance rates.

Congress responded and on August 18 allowed foreign-built vessels to be admitted to American registry. By June 30, 1915, 147 foreign-built ships were under the American flag.

The problem with marine insurance was a little more complicated, but on September 2, 1914, after considerable opposition, the Bureau of War Risk Insurance was established within the Treasury Department. Secretary of the Treasury William G. McAdoo appointed well-known New York marine underwriter William C. De Lanoy as director. Marine insurance notables William Wallace, of Boston, Hendon Chubb, of New Jersey, and William N. Davey, of New York served as the advisory board for $1.00 a year. The purpose of the Bureau was stated:

> That the said Bureau of War Risk Insurance, subject to the general direction of the Secretary of the Treasury, shall as soon as practicable, make provisions for the insurance by the United States of American vessels, their freight and passage moneys, and cargoes shipped therein, against loss or damage by the risks of war, whenever it shall appear to the Secretary that American vessels, shippers or importers in American vessels are unable in any trade to secure adequate war risk insurance on reasonable terms.

The Bureau was to determine a specific rate for each voyage based on the particular ship, the route proposed, and cargo carried, much like a private underwriter would do.

In the meantime, the State Department had serious concerns about Britain's arming of merchant vessels against submarines. Official inquiries were made and on August 25 the British ambassador in Washington, James Bryce, assured the State Department that the guns on British merchant ships were for defensive purposes only and would never be used in an offensive manner. If offensively armed British merchant ships were allowed to enter and leave American ports, it would subject the United States to violations of neutrality and in turn subject American shipping to possible destruction.

Imports and Exports

On August 13, 1914, Dr. Heinrich F. Albert, 40, left Copenhagen, Denmark, on the *Oscar II* and arrived at New York on the 26th. Albert was the agent for the Zentral Einkaufs Gesellschaft (Central Purchasing Agency) in Berlin, whose role was to purchase and ship American goods and commodities on the account of the Imperial German Government. Albert was also to act as the commercial attaché and financial advisor to Count Johann Heinrich von Bernstorff, the German ambassador in Washington. His other job was to do everything he could to keep supplies from reaching the Allies.

In 1910, Karl and Eugen Neumond were brewers and grain manufacturers and exporters at Frankfort-am-Main, Hesse, Germany. They had an office at 2218 Decatur in New Orleans run by Ludwig Eisemann. In 1913, Karl ran the K. & E. Neumond exporting business at 15 Broad St. in New York. In 1914 Eugen Neumond was in New Orleans. They were the main exporter of grains, fodder, and foodstuffs from the United States to Germany and they were the only exporter in the United States to have an office in Germany. The company got large contracts with the German government after the war started and worked closely with Dr. Albert.

In 1914, one million typewriter ribbons were made from imported cotton cloth and 69 million tin cans were manufactured from imported tinplate and lead. American industry used huge amounts of imported raw materials for the manufacture of items that were then exported abroad.

Germany was second only to England in the dollar value of goods traded by Americans. For the 12 months ending June 30, 1914, the United States exported $344,794,276 worth of merchandise to Germany and imported $189,919,136. During the same period the United States exported $548,641,399 to England and imported $248,089,913. Next in volume was France and all three were now at war.

Several important commodities were not manufactured in the United States in enough quantity to meet demands including optical glass made in Germany, England, and France and textile dyes made in Germany.

For some time prior to the outbreak of war, the Department of Commerce's Bureau of Standards had become concerned about optical glass since it was of vital military importance for binoculars, telescopes, rangefinders, periscopes, bombsights, cameras, and gunsights of all types.

Virtually no optical glass was produced in the United States and foreign producers kept a very tight lid on their manufacturing processes.

France and England had no glass to spare for export so in July 1914 the Bureau hired Heber D. Curtis, an astronomer at the Lick Observatory, to begin experimenting with optical glass production in Pittsburgh. His work continued throughout the war and the plant was eventually moved to Washington.

At the beginning of the war, only six American companies made aniline dyes, amounting to about 800 tons a year. It was costing $20 million a year for imported dyes. By November 1917, 150 companies were producing 25,000 tons but the price had gone up to $50 million. It was estimated that a $30 suit of clothes had 20¢ worth of dyes.

Cotton has always been a major export item that is traditionally put up in 500-pound bales. Between 1900 and 1913, around 65 percent of raw cotton was exported. In 1913, 14,795,000 bales were produced and 62.5 percent was exported, amounting to 22.5 percent of the total value of all exports. After 1913, domestic consumption began to increase.

The banner crop of 1914 produced 16,992,000 bales, of which 52.5 percent was exported, accounting for 26.2 percent of the total value of all exports.

To ease the shipping crisis and keep exports moving through the war zone, Secretary of the Treasury William G. McAdoo suggested the government purchase and operate merchant vessels. President Wilson concurred but there was strong opposition to the "Ship Purchase" measure and he wanted evidence it was necessary. Secretary McAdoo then solicited facts and figures from the collectors of customs at various Atlantic ports and in February 1915 several collectors telegraphed McAdoo. The Collectors reported goods were piling up "owing to a tie-up of shipping due to the European war." Grain was piling up at New York and Baltimore and farmers were becoming very nervous. Collector Norman Hamilton at Baltimore reported:

> There is much congestion of merchandise here on the docks and in warehouses awaiting shipment, both foreign and coastwise, the coastwise freight destined ultimately for export. There are in warehouse and

> under shed at Norfolk at the present time 85,565 bales of cotton, valued in Germany at $5,476,160. There is great congestion in warehouses, under shed and on cars of general export merchandise freight, this consisting principally of tobacco, lumber, flour, cottonseed oil, &c. With warehouses crowded, lumber is still laden on several hundred cars waiting at local deep-water terminals for available ships to carry it to foreign destination.

Between 1915 and 1919 the cotton crop remained in the 12–12.9 million bale range but in 1916 only 46.6 percent was exported, accounting for only 8.76 percent of the total value of all exports. In 1917, only 36.9 percent was exported. In 1918, 43.7, and in 1919 it was back up to 56.2.

The 1921 crop was very low at 8,351,000 bales, of which 77.5 percent was exported, not nearly enough to keep a large number of brand-new ships busy and accounting for only 12.2 percent of the total value of all exports. The Great War had a devastating economic affect on Europe and that contributed to "a rapid shrinking of the value of raw cotton."

The Declaration of London

The subject of what constitutes contraband during war had always been debated and disputed as technology and merchandise evolved. England, being the major maritime power, took the lead in these discussions and on February 26, 1909, promulgated the Declaration of London "relating to neutral rights and commerce." Contraband was defined as either Absolute or Conditional and the ships carrying them were liable to destruction or adjudication as prize, depending on the circumstances.

The original list included certain ores which was objected to by Sweden and were subsequently removed. On August 11, 1914, balloons and flying machines and their accessories were added to the Absolute Contraband list.

On August 20, an Order in Council was published containing the expanded list and the same day the German government relayed a message via Copenhagen that they would apply the current Declaration of London "provided its provisions are not disregarded by other belligerents." German commanders would play by the rules and for the most part they did.

Absolute Contraband, as defined by Great Britain, France, Russia, and Italy included 43 items directly related to the prosecution of war. Item No. 1 was "Arms of all kinds, including arms for sporting purposes, and their component parts." By the end of the war many more items had been added.

Conditional Contraband consisted of foodstuffs, fuel, forage, animal feeds, money in all forms, glues, and such—all commodities that could be used for military purposes but which had equal or greater value in civilian use. The list of 16 Conditional Contraband items remained about the same during the war.

The Declaration was amended during the war. One provision allowed a ship to be seized on its return voyage if it was found to have transported contraband using false papers. By late 1914 the list of Absolute Contraband included unwrought copper, nickel, iron pyrites, ferro chrome, lead, motor vehicles, rubber, mineral oils, resinous products, camphor, turpentine, various chemicals, aluminum, antimony, raw wool, worsted yarns, tin, castor oil, paraffin, wax, and ammonia.

The selling of arms and ammunition to belligerents by neutrals was an age-old practice that by World War I was still generally accepted but was being hotly debated. When is a neutral no longer a neutral? When has the line been crossed? Many fine points and details entered the debate and there was never any accepted consensus.

In February 1793, France declared war on England and American citizens started selling munitions to France. When England complained, Secretary of State Thomas Jefferson replied:

> Our citizens have always been free to vend and export arms; it is the constant occupation and livelihood of some of them; to suppress their callings, the only means of their subsistence because a war exists in foreign and distant countries in which we have no concern would scarcely be expected; it would be hard in principle and impossible in practice.

That is not to say there were no rules. The firmly established legal principle in the export of arms, or any contraband, was the difference between a government selling goods to a belligerent as opposed to a private individual doing so at his own risk. There was no law prohibiting an American citizen from shipping contraband in an American ship and from the beginning of the war President Wilson made no attempt to restrict American arms manufacturers from shipping their goods.

It was a long-standing violation of the neutrality laws of the United States and of the laws of nations for a citizen or resident alien to supply, fit out, or load contraband onto a vessel belonging to a belligerent in an American port and on October 15, 1914, this was widely publicized by the State Department. This was unpopular with the Congressmen who wanted unrestricted aid made available to France and England.

On February 16, 1915, Germany stated that neutral nations had a right to sell arms and ammunition to belligerents and said, "it is the right of neutrals to carry on their accustomed trade upon the breaking out of a war." The German scholar and jurist, Friedrich H. Geffcken, concluded that a belligerent "had only the right to demand that a neutral maintain its pre-war relations with a belligerent but go no further if neutrality would not be violated."

On March 18, 1915, England added cotton as a Conditional item but Sir Edward Grey assured Washington that American cotton shipments would not be interfered with although several cargoes were seized "having contraband concealed in or under the bales."

On April 4, 1915, Count von Bernstorff accused the United States of violating the spirit of neutrality by greatly enlarging its arms and munitions industries and should, if truly neutral, prohibit all arms and munitions from leaving the country.

In early 1915, Britain added corn, wheat, and flour to the list of Absolute Contraband owing to "the German Government's taking over the control of the supply of these articles throughout the Empire." Germany added them as well and later included pig lead in blocks, sheets, or pipes; all rough lumber except certain exotic woods; cylinder tar, crude or refined sulphur; and sulphuric acid.

A Ship from San Francisco

The ink on Germany's declaration of war had hardly dried when Germans in the United States began plotting ways to aid the Vaterland.

Germany had a substantial merchant marine, which the Royal Navy soon rendered useless, and the owners of freighters and liners whose ships were caught in foreign ports after the war started chose to keep them there for the duration, safely out of the reach of Allied warships.

When Congress repealed the law forbidding American registration to foreign-built ships, some owners saw an opportunity to keep their ships sailing and maintained and aid Germany at the same time.

The Kosmos Line steamer *Alexandria* left Hamburg in late April 1914 under Capt. Martin Schulz. On July 29, they left Champerico, Guatemala, for San Francisco. From San Francisco they were going to Victoria, B.C., but war was declared when the ship was at sea. They arrived at San Francisco on August 7th and there they stayed.

Franz Bopp, 51, was the German consul general at San Francisco. Bopp came to the United States in 1893 and married an Illinois girl, Olivia Voche, two years later. They lived in Berkeley at 2313 Hearst Ave. with their son William, 17, Franz H., 9, Olivia's mother Eliza, 65, and two servants.

Bopp sat in his office at 201 Sansome St. and began plotting a way to use the *Alexandria*. He, or someone he knew, was acquainted with a marine surveyor named Philip R. Thayer.

Thayer lived at 414 Fairmount Ave. in Oakland with his wife Hazel. He was born in California on December 17, 1875, and was an agent for his father, Ignatius E. Thayer, a Civil War veteran from Massachusetts and a manufacturer at 110 Market St.

Bopp and Thayer enlisted a group of local investors and established the Northern & Southern Steamship Co. with Philip as president. The firm purchased the *Alexandria*, renamed it the *Sacramento*, and declared their intention to engage in the South American trade as formerly done under Kosmos operation. Kosmos was the Hamburg-American affiliate for the company's South American trade. In October 1914 the company obtained American registry.

Rumors had swirled around the waterfront for weeks about a strange deal in the works and the British Consul General in San Francisco, A. Carnegie Ross, C.B., had kept a close watch on developments. Ross said the change of flag was a violation of the Hague Convention and the ship could be subject to attack outside the three-mile limit. When London learned of the situation, the *Sacramento* was "blacklisted" and was liable to seizure on the high seas as enemy property.

The ship was loaded with coal, engine room supplies, sauerkraut, beer, and other provisions and applied for clearance to Valparaiso, Chile, from Collector of Customs John O. Davis. On the cargo manifest, the coal was consigned to the Compañia Tranvias Electricas, an electricity generating plant in Valparaiso.

Collector Davis cleared Capt. Thomas A. Anderson, a former Kosmos employee, to sail on the 14th. Anderson left the next day. His chief mate was Edmund Johnson.

When the ship arrived at Valparaiso on November 20 in ballast, eyebrows went up and after examining the ship's papers Chilean customs officials immediately became suspicious. While Capt. Anderson reported to the Kosmos office, officials contacted the electric company. The company said they had no knowledge of a coal shipment consigned to them.

Naval authorities were then notified and Capt. Anderson was brought in for questioning. He told them he had been stopped and boarded by officers from the German cruiser *Dresden* and escorted to the northeast bay on Mas Afuera Island and robbed of supplies for use by the German Pacific Squadron. He said he was permitted to leave on the 18th after embarking the crew of the plundered French sailing vessel *Elue*.

Chilean authorities knew the German warships *Scharnhorst*, *Gneisenau*, *Dresden*, *Leipzig*, *Titania*, and *Prinz Eitel Friedrich* had sailed from China, bombarded French Tahiti, and had been at Mas Afuera, a remote island in the Juan Fernandez group, for about a week taking on coal and stores from the seized French ships *Elue* and *Valentine* and the Norwegian *Helicon*.

Meanwhile, Collector Davis at San Francisco had been informed that in all probability the *Sacramento* had obtained American registry under false pretenses and had loaded the coal for the German warships. No one on the San Francisco waterfront was surprised.

On December 16, after a thorough investigation, Chile declared the *Sacramento* a German auxiliary and Capt. Anderson was given 24 hours to leave Valparaiso or have his ship arrested. Anderson didn't leave and the U.S. State Department advised Chile it was fine with them if Capt. Anderson and the crew were detained until brought back to face charges. The Justice Department's Bureau of Investigation also got to work at home.

The owners of the ship hired a Washington law firm, Britton & Gray, to represent their case to the Secretary of State in order to clear them of any wrongdoing but the Bureau of Investigation had obtained evidence that Franz Bopp organized the whole thing and the ship's purser, Benno Klocke, 36, the former chief mate on the *Alexandria*, was the real leader of the expedition. Capt. Anderson had prior knowledge of the scheme and it was proved the ship had communicated with the *Dresden* two days before reaching Mas Afuera.

On January 27, 1915, Chile decided to seize the ship and the American vice-consul at Valparaiso, Alfred H. West, was informed that wireless operator Gustave Traub's unlicensed wireless set was being dismantled and packed away over the strong objections of Capt. Anderson.

In February 1916, after extensive investigations in the United States and Canada, a federal grand jury indicted 32 conspirators on various charges including murder and arson. In the *Alexandria* case, those indicted were Franz Bopp; Philip Thayer; Heinrich W.E. Kauffman, chancellor of the German consulate-general; Robert Cappelle, the San Francisco agent for North German Lloyd; Maurice Hall, consul-general for Turkey; wholesale grocers John and Julius Rothschild; Baron Eckhardt H. von Schack, vice-consul general; ship brokers George and James Flood; ship brokers Robert H. Swayne and John G. Hoyt; Capt. Anderson, Benno Klocke, and Charles D. Bunker.

In December 1916, after investigating other California cases, Baron George William von Brincken, the German vice-consul general in San Francisco, Charles C. Crowley, "a detective in the employ of the German consulate," and Crowley's secretary, Margaret Cornell, were indicted for "conspiracy to interfere with and destroy commerce of the Allies, and use of the mails to incite arson, murder and assassination by burning certain buildings of munition making firms at Ishpeming, Michigan, and Pinole, California." On January 27, 1917, all were sentenced to a year in jail and a $5,000 fine. Margaret Cornell was released from San Quentin at 7 a.m. on November 20 for good behavior after 9 months and 20 days. She hid from reporters but three hours later went out a side door and was taken by friends to Napa.

In December 1918, Special War Attorney John W. Preston and U.S. Attorney Annette Adams filed suit for $250,000 against the Northern & Southern Steamship Co. for obtaining American registry under false pretenses and violating American neutrality. Swayne, Bunker, Capt. Anderson, Joseph H. Bley, and the officers of Northern & Southern pleaded guilty to conspiracy, aiding the enemy, and violations of American neutrality. The defendants admitted engaging the *Sacramento, Mazatlan, Maverick, Olson, Mahony,* and the schooner *Annie Larsen* to supply the German Pacific fleet with provisions.

Lloyds of London, still the premier marine insurance underwriter, said that if any more German sympathizers were allowed American registry for ex–German ships they would raise rates high enough to cripple the American merchant marine.

4. *Submarines*

Lessons from the recent Russo-Japanese War were at hand where the "automobile torpedo" fired from surface ships had seriously damaged, but not sunk, Russian capital ships at Port Arthur. Mines had still been far more effective.

Every major navy had submarines, whose threat and deadly potential was well known, even as their ultimate use was being debated. In mid–1913, England had 70; France 60; the United States and Russia 30 each; Germany 23; Italy 20; Japan 13; and Austria-Hungary 6.

England had been the world's greatest sea power for 200 years and some higher ups in the Royal Navy had not taken the new submarine very seriously believing they were incapable of sinking a much faster capital ship at sea, speed being the major factor. Some still saw mines as the more effective weapon while other old salts held it was a sneaky, ungentlemanly way to conduct a good, old-fashioned war at sea.

In 1906, Sir William White, in discussing the Russo-Japanese war, stated:

> The operations at Port Arthur and the battle in the Korean Straits have demonstrated conclusively the power of swift torpedo vessels, even in daylight, when handled by gallant and skillful seamen, and have given fresh illustrations of the tactical advantages attaching to high speed. It cannot be overlooked that while our latest submarines have a surface speed of 13 knots, torpedo vessels of not much greater displacement and about the same cost have more than twice that speed, associated with a greater radius of action, superior sea-keeping qualities, an equal torpedo equipment, and an armament of quick-firing guns. Under circumstances when submarines would be of little or no fighting value—as, for example, against ships proceeding at good speed in a considerable sea—surface torpedo vessels would remain efficient. The true conclusion, therefore, is that submarines are not and cannot be substituted for torpedo-boats and destroyers, but are complimentary thereto. It may be doubted whether, on the whole, swift surface torpedo-boats may not prove more effective than slow submarines in getting torpedo attacks "home" and injuring an enemy even in daylight. Great risks must be taken, no doubt, but the chances of escape seem to be greater than would be anticipated from *a priori* consideration, especially when attacks of torpedo craft are supported by other vessels of the fleets to which they are attached.

It was noted then that no submarine constructed up to that time had attained a surface speed of 15 knots, which seems to have been the threshold speed. The next year, 1907, Mason S. Chace, Esq., a member of the Society of Naval Architects and Marine Engineers, stated:

> Existing types of submarines used in conjunction with mines tend to limit the role of the battleship, in wars of the future, to fighting on the high seas. If submarines are to fight battleships at sea and to threaten their

existence, they must be of a type which possesses a surface speed at least equal to that of the battleship, combined with sufficient endurance to enable them to get within striking distance. Without such speed they will not be able to place themselves as to utilize their ability to fight as submarines.

With technological advances constantly being made, Germany in 1913 commissioned the 210'-6" unterseeboot *U-19* with a surface speed of 15 knots and a submerged speed of 9 knots. Armament was four torpedo tubes, six torpedoes, one 3.7-cm Hotchkiss gun, one 8.8-cm, and one 10.5-cm deck gun. The sub had a range of 11,000 miles and carried enough food for an average 27-day patrol.

The early German subs were extremely uncomfortable to live and work in while submerged so they stayed on the surface as much as possible, often disguised as all manner of small craft to get within range of an unarmed merchant ship traveling alone. The basic tactic was to lie in wait off a port or submerged on a shallow coastal shelf. Almost all attacks were made along the approaches to ports of entry and fairly close to shore. Very few attacks were made far out to sea.

The later U-boats carried 8 or 10 torpedoes, 200–300 rounds of deck gun ammunition, and 500–600 "bombs" that were used to blow a hole in the side of merchant vessels to save torpedoes and deck gun ammunition. In the early days, when merchant vessels were stopped and boarded, the submarine could replenish its food supply from the ship's stores.

The nature of submarine warfare against maritime commerce changed the rules. Accepted prize law was immaterial and irrelevant since no submarine could carry enough sailors to put a prize crew aboard a captured ship and take it into a port for adjudication before a recognized prize court. They simply took off whatever stores and supplies they wanted, got the crew and passengers into lifeboats, and sank the ship. It was recognized early on that a submarine commander had no other choice since one shot from an armed merchant vessel could fatally injure his boat.

In addition to submarines, Germany also had a number of innocent-looking merchant ships fitted out with concealed guns that operated in the Atlantic, Caribbean, the Gulf of Mexico, and the Pacific. The majority of these surface raiders operated out of Tsingtau, China, against British shipping in the Pacific.

The *Pathfinder*

It wasn't long after the war started that a lightning bolt hit the Royal Navy. On Saturday, September 5, 1914, Kapitänleutnant Otto Hersing in *U-21* spotted Capt. Francis Martin-Peake's British cruiser *Pathfinder* off Eyemouth, Scotland. Martin-Peake was steaming at only 5 knots owing to a shortage of coal and, as fate would have it, Hersing's torpedo hit a magazine. The ship sank very quickly leaving only 18 survivors out of about 268 aboard.

Hersing was the first submarine commander to sink a ship with a self-propelled torpedo. The notion that the attack was by submarine torpedo sent shockwaves through the Royal Navy. Some old hands refused to believe it was a torpedo, choosing instead to believe the ship must have hit a mine.

In discussing countermeasures, the Royal Navy's most promising methods involved somehow locking the subs up in their bases, possibly with mines; sending ships out in groups "convoyed" with a naval escort to force the subs to come to the ships; and developing small, fast vessels equipped with underwater listening devices to detect a sub's presence underwater and bombs to blow them up. The term "convoy" came from the French "convoyer," to accompany or escort.

The Board of Invention and Research stepped up their efforts to develop more effective underwater locating devices. While the idea of an underwater explosive device for use against submarines was not new, it took the British until January of 1916 to develop the first practical depth charge. They were initially set to explode at 80 feet but this was later changed to 100 feet.

The *Glitra*

In October 1914, Christian Salvesen's 215' British freighter *Glitra* left Grangemouth, Scotland, with coal, oil, and iron plate. On the 20th, Kapitänleutnant Johannes Feldkirchener in the *U-17* stopped the ship just outside Norwegian territorial waters. After searching the ship Feldkirchener ordered everyone into the lifeboats and the Germans blew up the ship and cargo.

The *Glitra* was the first merchant vessel sunk with bombs placed aboard. By the end of the first year of war, the U-boats had sunk ten merchant ships and eight warships but five subs had also been lost in the effort.

Part Three

America's Early Role in the War

The British North Sea Blockade

On November 3, 1914, Great Britain declared the entire North Sea a British military area and blockaded all German ports. All neutral merchant ships bound for Germany were required to put into a British port for inspection. Any contraband cargo, including food, was removed and the ship escorted through British minefields toward their destination. They also seized contraband on neutral ships going to neutral countries, which they believed would end up in Germany. Most shipmasters readily complied.

The "Starvation Blockade" and its overall effectiveness has been debated ever since. After the blockade was set up, Dr. Heinrich Albert's plan was to flood the Scandinavian market with food in hope the surplus would eventually make its way to Germany.

On November 18, 1914, German naval authorities issued a directive to agents overseas to begin sabotaging shipments to England and France. Thus began an undeclared war on America's homefront and on neutral commerce in general.

5. *1915—American Ships Face the Dangers*

As freight rates on cargo began to rise because of war risks, more risks were taken by adventuress shipowners. Besides the danger of having any given cargo declared contraband by a submarine or raider commander, England and Germany both laid minefields.

The *William P. Frye*

The first American ship lost to enemy action prior to the United States formally entering the war on April 6, 1917, was Arthur Sewall & Co.'s 4-masted, ship-rigged, steel sailing vessel *William P. Frye.*

Capt. Herman H. Kiehne, Jr., 31, left Seattle on November 4, 1914, for Queenstown, Ireland, with 186,950 bushels of wheat under charter to Max H. Houser, a grain wholesaler in Portland, Oregon. Capt. Kiehne, of 1510 Thames St., Baltimore, had his wife Mamie Elizabeth, and son, Herman B., 9, on board.

On January 27, 1915, they were stopped southeast of Brazil by Korvettenkapitän Wilhelm Therichens's armed merchant cruiser *Prinz Eitel Friedrich,* an ex–Norddeutscher Lloyd passenger liner. The ship's pre-war master, Capt. Carl Mundt, became Therichens's first officer when the Kaiser took the ship.

An armed boarding party was sent over to examine Capt. Kiehne's papers and when the wheat was discovered, Therichens declared it contraband. He intended to jettison the wheat and let the ship go but the unloading was slow and the next day he decided to just sink the ship and remaining cargo. Everyone was ordered into the lifeboats and taken aboard the raider while a charge of dynamite was put aboard.

Therichens then sank two more French and three British ships and took their crews aboard. On February 20, he sank the British freighter *Willerby* but by then he was almost out of coal. His options were extremely limited so he decided to head for Newport News, Virginia, and land the survivors. He must have known he would never leave again but at least his ship would be kept out of British hands.

Collector of Customs Norman Hamilton notified the State and Treasury Departments and when word got out the Royal Navy sent three cruisers toward Newport News. The raider arrived early on the morning of Wednesday, March 10. Hamilton went aboard, told Capt. Therichens to retain his crew on board, and presented him with a letter:

> Sir—I have the honor to advise you that under instructions received from the Secretary of the Treasury, formulated with the advice of the State Department, your stay in United States waters will be governed as follows:
>
> You will be expected to leave United States jurisdiction within twenty-four hours in accordance with Article 12, Convention 13, The Hague, 1907, unless a further period of time is proved to be necessary to put on board fuel or stores, or to make repairs.
>
> If you so desire, you may have fuel and lubricants sufficient, with those already on board, to take your ship to the nearest German port.
>
> You may have provisions in the quantity not more than necessary to bring your supplies to the usual peace standard for the complement of the ship.
>
> If necessary, you may have repairs made, but only to the extent that is necessary to make the vessel seaworthy, the amount of repairs and the time necessary for their execution to be determined by designated authority of the United States.
>
> You may land the captured passengers, officers and seamen after having passed by the United States immigration authorities as being eligible aliens for landing in the United States.
>
> You may land and forward your mail after arrangements have been completed with the United States Post-office authorities for the proper forwarding of the same.
>
> You may land any Americans among the officers, seamen, or passengers from captured vessels.
>
> You are required not to use your radio equipment in port for purposes of transmission.
>
> In conclusion, I have the honor to request that you furnish me today a signed statement of the circumstances of the destruction of the ship *William P. Frye*, or any other American ship that has been destroyed by the vessel in your command.

The ship's arrival caused a sensation. The Atlantic Fleet was ordered up from Guantanamo Bay and two more British cruisers joined the others waiting outside the Virginia Capes. The battleship *Alabama* steamed off the James and Elizabeth Rivers. British merchant ships in port were detained and local pilots were ordered not to take any belligerent vessel out until further notice, which the British promptly protested.

The Neutrality Board in Washington declared that all repairs the Germans might want to have done be made under the supervision of a naval officer.

On March 12, Capt. Therichens requested permission to allow his crew to go ashore within the limits of Newport News, which was granted.

The State Department filed a formal protest over the *Frye* sinking as a violation of the treaties of amity and commerce of September 19, 1785, and July 11, 1790, between Prussia and the United States. German ambassador von Bernstorff replied that the wheat was contraband going to an enemy port and "that its treaty with this country merely obliges it to make compensation." He also added that if anything illegal took place it was Therichens's fault, not Germany's.

The *Frye* was valued at $150,000, the wheat $39,759.54, and the freight $37,500. A claim for $228,059.54 was forwarded by secretary of state William J. Bryan and presented to the German government by James W. Gerard, the American ambassador at Berlin.

It was expected that Therichens would declare that he intended to remain in port and around April 7 he did that. The raider was promptly seized, moved to Philadelphia, and in 1917 became the Navy transport *DeKalb*.

It was later reported that 14 German officers and sailors escaped from interned ships.

Two were caught and returned, two from the *Eitel Friedrich* were known to have reached Germany. One sailor, Richard Wagner, made his way to California. On September 28, 1917, Roy O. Samson, of the Bureau of Investigation, filed a report at Los Angeles:

> It was reported yesterday by Otto C. Fritze, German spy under order of internment, that one Richard Wagner, formerly among the crew of the *Prince Eitel Friedrich*, is the said German Agent in this section of the Country, and in authority over all other German Agents; that under Wagner's instructions he prepared plans for the destruction of the irrigation system at Sharp's Heading; that Sharpe's Heading is located at Mexicali, about two miles South of the border town of Calexico, California; that Sharp's Heading furnishes all of the water for the Imperial Valley, and the destruction of this Heading would prevent a drop of water coming into the Imperial Valley, and would deprive the people of water for all purposes, drinking included, and ruin a million or a million and a half dollar cotton crop. That the destruction of the cotton crop was primarily the object of their conspiracy, as they felt it could not then be used for Military purposes.
>
> Wagner is said to drive a Hudson car, a roadster of the speedster type, red wire wheels, black radiator pipes with red background, two extra tires, left side, one or two bumpers, victoria top, black body, California and Mexico licenses.
>
> Wagner's apprehension will be sought.

Germany Declares a War Zone

In September 1914, Germany's land offensive toward Paris was quickly stopped at the Battle of the Marne. Both sides dug in and constructed trenches and bunkers all the way from the North Sea to France's border with Switzerland. The Kaiser's hopes for taking Paris and all of France and Belgium without risking his High Seas Fleet in a confrontation with the Royal Navy were thus dashed on the Western Front.

With this stalemate, Germany realized they could not defeat the Allied forces on land and promptly offered peace terms. Unfortunately, no agreement was forthcoming because both sides still believed they could force an unconditional surrender on the other and there was no reason to compromise. Germany believed England could be starved and forced to sue for peace and even if the United States came in on Britain's side, all the provisions needed by England and the supplies needed by the Allied forces could be stopped with U-boats.

After the Marne victory, marine insurance rates stabilized and American export consignments to Denmark, Norway, and Italy leaped far above pre-war levels. Most of these shipments were copper, aluminum, and petroleum, which ultimately found its way to Germany.

On February 4, 1915, Adm. Hugo von Pohl announced the Kaiser's new policy. There are various translations, but all say essentially the same thing:

> Proclamation.
>
> 1. The waters surrounding Great Britain and Ireland including the whole English Channel are hereby declared to be a war zone. On and after the 18th of February, 1915, every enemy merchant ship found in said war zone will be destroyed without its being always possible to avert the dangers threatening the crews and passengers on that account.
>
> 2. Even neutral ships are exposed to danger in the war zone as in view of the misuse of neutral flags ordered on January 31 by the British Government and of the accidents of naval war, it cannot always be avoided to strike even neutral ships in attacks that are directed at enemy ships.
>
> 3. Northward navigation around the Shetland Islands, in the eastern waters of the North Sea and in a strip of not less than 30 miles width along the Netherlands coast is in no danger.
>
> VON POHL,
> *Chief of the Admiral Staff of the Navy.*
>
> Berlin, February 4, 1915.

The policy applied to any British, Russian, or French ship and Germany warned neutral nations, including the United States, that no ship would be safe from attack in the designated waters. The United States and other nations protested the measure as illegal but to no avail.

In an effort to stay safe, some British shipmasters flew the American flag on their ships and on February 10, President Wilson protested this practice.

At first, the U-boat commanders observed the traditional, internationally accepted rules for visit and search on the high seas that any surface ship commander would do on encountering a merchant vessel. The ship was hailed, usually by a shot, and a boarding party sent aboard. On finding the ship to be a belligerent or a neutral vessel carrying contraband cargo, the crew and passengers were ordered off and into lifeboats.

Once the submarine war started, certain rules of etiquette became customary. In return for allowing the crew to escape, the merchant ship was not allowed to take any offensive or defensive action, flee, or to send out a radio message. Some U-boat commanders even gave the survivors supplies or directions to the nearest land and some even towed lifeboats in toward land.

The *Evelyn*

The 252' ship *Badsworth* was built at Southampton, England, and in 1899 was wrecked in American waters. The ship was purchased by an American, repaired under provisions of the "Wreck Act," which then allowed American registry, and was renamed *Evelyn*. Around 1908, A.H. Bull purchased the ship, sold it to the Philadelphia & Gulf Steamship Co., who sold it to O.C. Hempstead & Co., of Philadelphia, and they sold it to the big New York cotton merchants Harriss, Irby & Vose. The high demand for cotton in Germany in 1914 could make the company a lot of money so a voyage to Bremen was planned.

Young Capt. David T. Smith, 28, of Milton, New York, was hired for the trip. He was understandably nervous, having heard of the various minefields, so he visited the German consul at 11 Broadway for sailing instructions. The Kaiserliches Deutsches Generalkonsulat, Dr. Horst Paul Falcke, showed him a German Admiralty publication from November 4, 1914, which stated, "All merchant ships bound for the Eider, Elbe, Weser and Jade must first steer towards Listertief-buoy. Ships bound for the Ems must steer directly toward the Ems. Ships, to insure their own safety, are obliged, after leaving the Listertief-buoy, to be piloted." The approximate location of the Listertief-buoy was stated to be "55° 3–¾' N. and 8° 17–½' E." but there were no instructions on how to get to the buoy itself.

The *Evelyn* loaded around 5,500 bales of cotton worth $350,000. The fair market value of the ship was $210,000 and the owners carried $100,000 in government war risk insurance on the cotton.

Capt. Smith left on January 29, 1915, and arrived at the Downs in the English Channel on February 15 where a British officer came aboard to inspect the cargo. Smith asked for specific sailing instructions and he was shown two routes, one up the coast of England via Farn Island and one across the Channel to the coast of Holland. The officer declined to recommend a course but did say that most American ships went via Holland. Smith also visited other ship captains at the Downs who were going to Germany and all said they were going via

Holland. Smith then decided his best option was to go to Rotterdam where he could get the latest information and pick up a Dutch pilot who could take him to Bremen "or at least to the German line."

At Rotterdam he hired a Dutch pilot "who held himself out as an experienced North Sea pilot." The pilot assured him they would encounter a German patrol boat and could obtain specific instructions to Bremen. Smith then concluded that the German instructions he had reviewed at the consul's office were for the purpose of obtaining a pilot at the Listertief buoy and decided he could simply head due south for the Weser since he had already secured the services of a competent pilot. The pilot concurred with Smith's conclusion.

Early on February 19, the ship was off the Dutch coast about 10 miles from the Borkum Lightship between Norderney and the Isle of Borkum. Capt. Smith, nervous and pacing, was in the wheelhouse, having been up all night. At 4:05 a.m. an enormous explosion lifted the bow of the ship out of the water. Smith told everyone off while he gathered up his passage money, papers, and some blankets. At 4:20 another explosion blew out the starboard side and the ship began to settle while the crew was still lowering three boats. Capt. Smith and 28 sailors hung around but 13 others took off on their own and headed for Holland. The ship sank seven hours later.

The American consul at Bremen, William Thomas Fee, reported the loss on Saturday, February 21, and on the 22nd reported that Capt. Smith, the Dutch pilot, and 13 others had been rescued at 10 a.m. on Saturday morning but their location was unknown. Chief mate Senson and 13 others were picked up by the German "scoutship" *Mars* and taken to Heligoland. Thirteen were subsequently removed to a sailors' home in Bremerhaven. The only known casualty was a Spanish oiler who died of hypothermia and was buried at sea.

Secretary of State Bryan cabled London, Berlin, and the American minister at The Hague, Henry Van Dyke, to request aid be given to all survivors and an investigation of the "nationality of the mine." Allegations were quickly made the mine was German and Berlin just as quickly said the steamer must have been off course.

Count von Bernstorff said the Dutch pilot was not qualified to navigate in German waters and had been put aboard by the Rotterdam office of the Furness Steamship Co., a British firm who had no interest in seeing valuable goods reach Germany.

The Germans had an extensive mine barrage west of Heligoland Island that extended in a southerly and southwesterly direction toward, but not reaching the mouth of the Ems River where Borkum Island lies. During January and early February, the area had experienced heavy weather and it was supposed during a subsequent investigation that mines had been torn from their moorings from this minefield and drifted into the sea lanes.

Capt. Smith boarded Capt. Frederick G. Avery's Mallory Line steamer *Denver* for passage home. They ran into heavy weather, hull plates sprung, the boiler fires went out, and the ship wallowed in the water. Wireless calls went out and when Capt. George Sheafe's *Megantic* arrived on March 23 the *Denver* was listing 42 degrees. The *St. Louis* arrived and spread oil to calm the seas. Captain Smith, Capt. Avery, and his wife Olive got aboard the *Megantic*. They left Liverpool on March 17 and arrived at New York on the 27th. Capt. Smith later stated he thought the *Denver* would not have lasted more than two hours.

The *Carib*

The Clyde Steamship freighter *Carib* was another cotton ship. It left the States earlier than the *Evelyn* but got into trouble at a later date.

The freighter was chartered by Strachan & Co. to take Harriss, Irby & Vose cotton to Bremen. The loading of 4,600 bales at Charleston, South Carolina, was finished on January 25, 1915, and Captain Edgar L. Cole and 29 crew cleared for Bremen on the 27th. Capt. Cole found out he was going to Bremen about a week before sailing. He had 200 tons of pig iron aboard for ballast and 550 tons of coal. Extensive repairs were made to the ship before sailing. Capt. Cole had never sailed to the Continent or to any German port and the only sailing instructions he got were from Joseph Whitsit, Strachan's agent at Charleston.

Clyde and Strachan believed the Channel route would be the safest, "as the English government claimed it would allow all American vessels going to Germany with cotton free passage." They also knew that "parts of the northern route were mined." Capt. Cole was instructed to sail for the Hook of Holland and contact the Furness Shipping Co. at Rotterdam and they would arrange for Nicolaus Haye & Co., of Bremen, to have a German pilot meet him there.

On February 14 they reached the Isle of Wight and anchored. A uniformed Trinity House pilot came aboard to take them to the Downs where they anchored at 9:30 p.m. Bad weather delayed their sailing until the 19th when they were cleared to leave. A pilot told Capt. Cole to go by Galloper Buoy "as Holland-American steamers were constantly going back and forth in that track."

They saw quite a few mines on the passage and arrived at the Maas Lightship at 1:30 p.m. on the 20th where they were stopped by Dutch patrol boats. Capt. Cole requested a pilot to take him to the Weser River and he was told the pilots were in the Hook of Holland. He went there and anchored near the gas buoy with the pilot flag hoisted. A pilot boat came out thinking the ship was bound for Rotterdam. Capt. Cole was told no pilot would take him to the Weser but he could go ashore and telegraph Bremen to send one out. While ashore he met the chief mate from the Atlantic Gulf & West Indies steamer *Ocmulgee* who said they had a pilot and the Furness office at Rotterdam had suggested the *Carib* follow the *Ocmulgee* but Capt. Cole declined the offer. He contacted Furness himself and a pilot named Kruize came aboard at 1 p.m. on the 21st and they sailed at 4:15 that afternoon. Capt. Cole later stated, "The pilot told me he was a North Sea pilot and had been sent by the Furness Shipping Company. He was sober when he came on board and remained so the entire time." The pilot had his own chart with him.

At 10:58 a.m. on February 23 they hit a mine ten miles west of Norderney. Captain Cole stated later:

> It seemed to strike the vessel just about amidships and directly from below as the ship seemed to lift. The explosion broke the ship in two and burst one of the boilers, which blew coal and steam up into the air and into the Captain's room and almost instantly the engine room was full of water up to the cylinders. The rush of water and steam is, in my opinion, what caused the death of Third Assistant Engineer William Bazzell, Fireman Antonio Martinez, and Coal Passer Segunda Blaz as no trace of these men could be found, although a search for them was made as thoroughly as possible.

Capt. Cole, 3rd Mate Charles H. Winnett, the pilot, and 15 sailors were in one boat and Chief Mate George H. Gifford had 12 crew in his boat. After rowing for about 40 minutes in heavy fog they spotted a boat, toward which they steered, and around 12:30 p.m. they were taken aboard the German patrol boat *Annie Busse*. Capt. Cole said they were treated "most courteously and kindly." The fog delayed their entry into Wilhelmshaven until 1 p.m. of the 24th. From there they were taken to Bremerhaven, arriving at 10 p.m. that night, and reported to Consul William T. Fee at Bremen.

William Bazzell lived at 76 W. Garden St., Pensacola, Florida. Martinez and Blaz were from Coruña, Spain. Watch oiler Frank Redmond survived. Jesse Boyd was the 2nd Mate, George S. Keowen, chief engineer; Frank Hubel, 1st engineer; and Thomas P. Tracy, 2nd engineer. The survivors returned home on the steamer *Southerner*.

The loss of the *Evelyn* and *Carib* pretty much ended any further attempts to ship goods directly to Germany. When asked by a reporter if his company would make any further cotton shipments, William L. Harriss didn't have much to say except, "This is a risk—a war risk—that we take."

The *Dacia*

Edward Breitung was born in Schalkan, Germany, on November 10, 1831, came to America in 1849, and settled in Richland, Michigan. He attended school to learn English and worked in a grocery store for a couple of years then went to Detroit as a book keeper. He moved to Marquette in 1855 and ran a general store and then took an interest in the iron business. In 1859 he moved to Negaunee where he and Israel B. Case operated the Pioneer iron blast furnace. He soon began investing in mining lands and is credited with developing the iron mining industry in the Upper Peninsula in the Negaunee Range.

He married a Belgian girl, Mary Pahlon, and on November 1, 1871, they had a son, Edward Nicholas Breitung. In 1880 Edward was 46 and the mayor of Negauee. Mary was 27 and their son Edward N. was 8. In 1883 Edward was elected to Congress and served in the House of Representatives for two years.

Edward Breitung died on March 3, 1887, and Edward N. and his widow Mary inherited a substantial estate, which Edward N. managed. Six years after Edward's death, Mary married Nathan Kaufman, a business associate of her late husband. Nathan had a sister named Charlotte who married Edward N. in 1890 and they had a daughter, Juliet.

Edward N.'s company, E.N. Brietung & Co., Ltd., was headquartered in Marquette and had an office at 11 Pine St. in New York. He had interests in railroads in Canada and Peru and after the war started he found the high freight rates in shipping to be a very attractive investment.

Edward N. had a nephew named Max Breitung, who became a business associate. Max made the acquaintance of Hans O. Schundler, a lumber dealer in Brooklyn who had a similar interest in shipping. The Breitungs enlisted his aid in purchasing ships and in August 1914 sought to purchase a number of lakers but no sales were realized. Efforts to purchase British ships from broker Samuel Holmes in New York failed as they were for sale to British buyers only.

On December 3, 1914, Schundler submitted a report to Max Breitung on ships available and freight rates on grain and cotton to Gothenburg, Copenhagen, Rotterdam, Bremen, and Liverpool. The freight on a standard 500-lb. bale to Gothenburg or Copenhagen pre-war was $1.25 per 100 pounds and had gone to $1.50 or higher and to Rotterdam from $1.75 to $2.50. Bremen and Hamburg was upwards of $3.00 and to Liverpool $1.00 and up.

Egon Von Novelly was born in Germany on November 8, 1885, and lived at 200 W. 58th St. in New York as a resident alien. He was general manager of the Liberty Navigation Co. at 66 Broadway and apparently was acquainted with Edward N. Breitung.

William G. Sickel was vice-director of freight and traffic for the Hamburg-American Line in New York. On December 7, 1914, Von Novelly visited Sickel and made an offer to purchase the company's freighter *Dacia* for $165,000, then laid up at Port Arthur, Texas. The 335' steamer left Hamburg on June 17, 1914, for Newport News, Virginia, and on the 28th left for Port Arthur. When the war started, Hamburg-American left the ship there out of reach of the Royal Navy with plans to sell their ships. Hamburg-American was controlled by the German government.

A sales contract was drawn up and on December 9 Von Novelly signed a charter-party agreement with L.A. Wright & Co., the agent for Tom B. Owens & Co., of Forth Worth, Texas, to carry 11,000 bales of cotton from Galveston to Bremen. Owens was a cotton exporter and vice-president of the Fort Worth Grain & Cotton Exchange.

Contracts for the sale and transportation of the cotton were made at Bremen on December 12 and 14 while the ship was still under the German flag. The cotton was consigned to Harold von Linstow of Bremen who agreed to pay 16 cents a pound or $80 a bale and freight at $3.50 a bale, the highest freight rate paid to date.

Von Novelly made the first payment on the ship, $16,500, on December 16 and the same day transferred the title to the ship to Edward N. Brietung. Von Novelly made the final payment of $148,500 for the ship on December 21. The checks were drawn on the Guaranty Trust Co. in New York City.

On January 4, 1915, the ship was issued a temporary register at Port Arthur. Capt. George MacDonald arrived at Galveston on January 7 and commenced loading $880,000 worth of cotton, which was completed on January 21. A war risk policy for $715,000 was taken out in the name of Guaranty Trust and the ship departed for Norfolk on the 23rd and then left for Bremen. Breitung told port officials the ship was going Rotterdam and the cotton re-shipped from there to Bremen.

On February 27, 1915, the *Dacia* was seized by the French auxiliary cruiser *Europe* at the entrance to the English Channel and taken to Brest for adjudication as prize of war. At trial, Edward Breitung held himself out as sole purchaser of the ship but the court found copies of all the relevant documents in New York. William Sickel apparently denied selling the ship to Von Novelly and said he did not know of the title transfer, never met Breitung, and never spoke to him, but told Von Novelly that Breitung was "a promoter of fictitious ventures" and was insolvent "and one with whom no ship-owner could seriously enter into a contract."

On August 4, 1915, the court ordered the ship sold on the grounds that French law did not recognize the transfer of title from a belligerent after the outbreak of hostilities. The ship was renamed *Yser* and was torpedoed on November 8, 1915, while in a convoy to Bizerte, Tunisia. All aboard were rescued by the Italian steamer *Elisa Francesca*.

Max Breitung was arrested in November 1915 along with five others and indicted on conspiracy charges to damage ships and cargoes. It was known at the time he was related to Edward N., "the wealthy shipping and mining man of acknowledged German sympathy."

The Seamen's Act of 1915

For literally hundreds of years the relationship between a sailor and the captain or owner of a merchant ship was governed by laws universally adopted by the maritime nations that favored the master or the owner. The profession, in a legal sense, was regarded by sailors as an almost involuntary servitude.

Over the decades, several acts of Congress were passed to address various issues but the first known attempt to organize sailors was made in San Francisco in January 1866. On the evening of the 11th a meeting resulted in the formation of the Seamen's Friendly Union & Protective Society. Alfred A. Enquist was president. Enquist, oddly enough, owned a wine and liquor store on the corner of Davis and Clark. In 1867, the Seaman's Boarding Master Association had an office at 603 Front St. but the next year both groups were gone and Enquist had quit the liquor business to run the Oriental Hotel at Washington & Drumm.

In 1872 the Shipping Commissioners Act required that a sailor sign on in the presence of a federal shipping commissioner to avoid being forced on by a crimp, but even this was viewed by sailors as tightening their legal bonds to a vessel. A seafarer, like anyone ashore, chose his occupation—no one was forced to go to sea—but any serious infringement like insubordination, simple assault, mutiny, or desertion could be punished in ways not seen on land.

A factory worker who struck a supervisor would not be flogged. Walking off a job would not result in arrest and imprisonment by the authorities, a loss of all earned pay, or the forfeiture of personal property left at work. Add to that bad or meager food, unsafe ships, brutal shipmasters or "bucko mates," and the sailors' life could be miserable.

Sailors generally got little sympathy from the public since they were often seen as brawling, whoring, syphilitic drunks who spent all their money in waterfront dives, exposing themselves to crimps and who gave the captain and the shipowner—decent men who were just trying to make an honest living—no end of trouble.

In 1875 the United Seamen's Association was organized in New York but this group and the Seamen's Friendly Union & Protective Society had little influence.

The December 21, 1898, White Act allowed a sailor to leave a ship in any port of the United States without arrest or penalty other than the forfeiture of any pay that was due. That was a beginning but abuses continued.

In 1906, Standard Oil of California's 4-masted steel bark *Atlas* took on a cargo of case oil and left New York under Capt. John C. Amberman for Yokohama, Japan. On the way back they would stop at Honolulu and load sugar.

What started the trouble is unknown, but it was alleged that by the time they arrived at Yokohama, three sailors were so badly injured from beatings they couldn't get out of their bunks and even on arrival in port, seaman Francis McKay said he was being beaten so badly he climbed up the rigging and jumped overboard to escape. He swam to a sampan and was taken ashore where met up with the entire crew of 20 to report to Henry B. Miller, the American consul. The sailors asked for protection and said the entire voyage had been dreadful. They said they were beaten by Amberman or his bucko mate and the only food they got was two ounces of beef and one pint of water a day supplemented by whatever birds or fish they could catch.

The White Act also made seamen who were discharged from service in a foreign port as the result of injury or illness a ward of the American consul and their care in port and transportation back to the United States was paid for out of a special fund.

Francis McKay signed on the steamer *Montara* at Yokohama, left there on June 9, and arrived at Seattle on July 3. Other crewmembers were apparently put on the Army transport *Sherman* by the consul and arrived at San Francisco on August 17.

Meanwhile, Capt. Amberman signed on a new crew of 20 Japanese sailors and headed for Honolulu where they arrived at noon on July 15. Third mate Henry Hackney immediately filed charges that Amberman ran a "hell ship." Capt. Amberman responded:

> He was a sick man during the voyage. He had diabetes. I nursed and cared for the man during the entire trip. He was practically worthless as a third mate for he was unable to go aloft. He certainly had no cause for complaint. I disrated some of the men in the crew for I considered that they were not worth the wages that they would have received under the rating they signed under on the articles. This matter was taken before the American consul and he decided in my favor. The sensational stories about my knocking men down with belaying pins and their not getting enough food are false.

Normally, a sailor who was sick and unable to perform his duties would have been left ashore in a hospital under charge of the American consul and the captain would have signed on a new third mate, but he didn't.

When the *Atlas* got to Honolulu the entire crew deserted and headed for Waimanalo to the home of one of the crewmen. None of them intended to go back to the ship, but they didn't have permission from the immigration inspector to be ashore so they were arrested. One of the sailors told authorities of a stowaway on board who disappeared during the voyage and said he saw the boatswain throwing his personal affects overboard. Capt. Amberman was subpoenaed by assistant district attorney Joseph J. Dunne and the matter was investigated. Capt. Amberman denied knowing anything about a stowaway or a death. The failure to record a death in the log book, if a death occurred, was a serious offense. One witness said the work was much harder than on other ships and the food was poor—mostly slices of bread and boiled beans. Apparently there wasn't any evidence of a murder or death so Capt. Amberman signed on another crew and the *Atlas* left for the Delaware Breakwater on August 13.

In December 1913, Robert Dollar's British-flagged steamer *Bessie Dollar* was at Shanghai when Capt. Abernethy signed on Dung Pau and 18 other Chinese nationals for the voyage back to San Francisco. Dollar's intention was to transfer them to the *Mackinaw* at San Francisco for $8 a month. The American sailors who were discharged were then earning $45 a month.

At that time, the average worker in a Chicago slaughterhouse made about $25 a month for a 37.5-hour work week. When the Chinese were discovered it was alleged that Dollar had concocted a scheme to eventually replace American seamen with cheap foreign labor.

The U.S. Commissioner of Navigation reported that during the fiscal year ending June 30, 1914, only 31.5 percent of seamen on American ships were native-born Americans and only 47.4 percent were American citizens.

Robert Dollar was sued for violating contract labor law and the case went to court. In the meantime, efforts to improve the lot of American seamen were slowly moving forward but it was the *Titanic* disaster that shook the maritime world to its core.

The International Conference on Safety of Life at Sea was convened in London and was signed by the delegates on January 20, 1914. Watertight construction, radiotelegraphy, life-saving appliances, and fire protection, and safety certificates were addressed for ships carrying 12 passengers or more.

In the United States, Sen. Robert M. La Follette, "a landlubber from Wisconsin," and Andrew Furuseth, president of the International Seamen's Union, saw an opportunity to expand these safety measures into a general improvement of the sailor's lot and to employ more Americans on American ships. With war on the horizon they felt their chances were greatly improved.

Around that time, the desertion rate was 1.39 percent. A foreign sailor who deserted his ship in an American port could be arrested and prevented from signing on an American ship.

When the Great War started in August 1914, American shippers began to see an opportunity to greatly expand foreign trade. They fiercely opposed anything that would increase their costs and hamper their efforts. Their views were generally supported by economists and after two years of very heated debate, President Wilson signed the Seamen's Act on March 4, 1915. The Act was set to go into effect on November 4, 1915. The full title was "An Act to promote the welfare of American seamen in the merchant marine of the United States; to abolish arrest and imprisonment as a penalty for desertion and to secure the abrogation of treaty provisions in relation thereto; and to promote safety at sea." Desertion was no longer punishable by arrest and imprisonment. Prior to the Act it was customary for a master to withhold crew wages in foreign ports to prevent desertion but a sailor could now demand one-half the wages earned up to the arrival at any given port. If a sailor took half his pay and then deserted, he could not be prosecuted for desertion but he did forfeit all personal effects left on board as well as the balance of accrued wages. This provision was to apply to all foreign vessels in American ports after March 4, 1916, "or at such later date as existing treaties may be abrogated."

After that date, if a low-wage sailor deserted a foreign ship the master was required to secure a substitute "of the same or higher rating" and at the prevailing wage at the port.

For fiscal year 1916 the percentage of native-born or naturalized American citizens shipping out still held at around 43 percent.

On July 2, 1917, Albert Brannon signed on the British White Star freighter *Bovic* at Manchester as a fireman and trimmer for a voyage to New York and return at £9 a month. Prior to arriving at New York he drew an advance on his wages and while at New York he deserted his ship. He then signed on the American steamer *Byron* at the exchange rate of £18 1s a month. The captain of the *Bovic* then had to find a replacement at £15 a month.

In England, desertion was still desertion, and he was brought up on charges. He was still owed £19 4d for his time on the *Bovic* and his wages for the *Byron*, £11 14s 11d had not yet been paid. The court ordered Brannon to forfeit his unpaid wages and reimburse White Star for the expenses they incurred in replacing him.

Because of the new American law, it was becoming almost impossible to find pierhead jumps in American ports to replace sick or injured crewmembers and "the temptation of men to desert and resign was very great." It was said that White Star had lost over 400 sailors as of May 1, 1917, and 30 sailors had deserted from another ship at New York. Some British sailors were even jailed on arriving home and more than a few foreign sailors stayed on American ships and never went home again.

It was anticipated by proponents of the Act that the "freedom clause" in the Act would benefit all sailors worldwide by forcing low-wage foreign owners to match American salaries. It never quite worked out that way and a foreign sailor who arrived back home could find himself in trouble for desertion and be prosecuted.

Section 2 of the Act provided for two deck and three engine room watches and sailors either shipped out for deck or engine and could no longer be forced to work alternate shifts unless it was necessary for the safety of the vessel or cargo.

A day's work in port was 9 hours with Sunday, New Year's Day, July 4, Labor Day, Thanksgiving, and Christmas off, but a ship could depart on any of those days to maintain a schedule.

The space allotted to each sailor in the forecastle was increased. Each sailor was entitled to 120 cubic feet for sleep and 16 square feet of deck space. Suitable wash rooms were also required. Welding Ring, a partner with Mailler I. Quereau, New York commercial merchants at 35 William St., stated:

> This is two-thirds more cubic space and one-third more floor space than was formerly required. A certain 10,000-ton ship [deadweight] owned by a New York company carried 39 men in the crew. It must now devote 2,808 cubic feet to the crew. It will hereafter have to devote 4,680 cubic feet, an increase of 1,872 cubic feet. This will take 47 tons from the cargo space. Hospital quarters will displace 9 tons more.

Flogging for insubordination or simple assault was replaced with fines and all forms of corporal punishment were prohibited although an out-of-control or mutinous sailor could always be confined.

No court could attach or arrest the wages of a seaman or apprentice while under articles. Prior to the Act, a sailor in New York made from $25 to $30 a month. In February 1916, the pay went up by $5 a month. On April 15, 1916, pay went up to $45 a month. Firemen got raises from $5 to $10 a month. In the Spring of 1917 the wages went to $60 a month for sailors and firemen on both coasts and $72 a month on the Great Lakes.

The Act created the official certification for the rating of able seaman and initially required 40 percent of the crew to be able seamen. An AB had to be at least 19 and have 3 years' service on deck at sea or on the Great Lakes. Candidates had to undergo a physical examination and take a test administered by the Steamboat Inspection Service. At New York, between November 4 and 11, physical exams were given to 913 sailors and of those 101 failed mostly due to vision problems or overall weakness and physical vitality.

Of 100 seamen issued AB certificates in New York on November 9 and 10, only 22 were Americans. Norway was second at 16, the rest Russian, Scandinavian, and various Europeans including 6 Germans. There was no provision in the Act requiring American citizenship, it pertained only to sailors shipping out on American-flag vessels regardless of their nationality.

It was reported that a large number of sailors on the Pacific Coast were failing the tests, resulting in protests to Washington to repeal the law.

It was required that no vessel covered by the Act could leave port unless 75 percent of the crew could speak the language spoken by the ship's officers, but not necessarily English. Tests for proficiency were given by the ship's officers and not by the customs collectors who had to clear the ship to sail. The language provision was intended to "drive Chinese crews off American vessels in the Pacific" but turned out to be a "body blow" to Pacific Coast shippers. On September 18, 1915, Secretary of Commerce William C. Redfield stated the law "is in no sense intended to be used to create embarrassment or to cause unnecessary expense or delay."

Membership in the maritime unions burgeoned during the Great War and the formation of the War Labor Board facilitated negotiations between management and labor. As a result, there were no major strikes by merchant seamen during the war.

There were quite a few national and regional maritime unions in the United States. Some of the regional unions also enlisted fishermen and inland tugboat and workboat crews. The major unions during the Great War were:

Sailors' Union of the Pacific, 59 Clay St., San Francisco.
Marine Cooks & Stewards' Association of the Pacific Coast, 42 Market St., San Francisco.
Marine Firemen, Oilers, & Watertenders' Union of the Pacific, 58 Commercial St., San Francisco.
International Seamen's Union of America, 355 N. Clark St., Chicago. The Union had branches throughout the country.
International Seafarer's Federation, 355 N. Clark St., Chicago.
Marine Firemen, Oilers, Watertenders & Coal Passers of the Great Lakes, 71 Main St., Buffalo, New York.
Sailors' Union of the Great Lakes, 355 N. Clark St., Chicago.
Seafarers International Union—Atlantic, Gulf, Great Lakes, and Inland.
Eastern & Gulf Sailors' Association, 67–69 Front St., New York.
Marine Cooks & Stewards' Union, 35 W. Eagle St., Buffalo, New York.
Marine Cooks & Stewards' Association of the Atlantic & Gulf, 12 South St., New York.
Marine Firemen's, Oilers' & Watertenders' Union of the Atlantic & Gulf, 70 South St., New York.

All of the major maritime powers had deepwater sailors' unions, including Germany's Internationale Transportarbeiter-Föderation in Berlin established in 1896.

It was supposed by supporters of the Act that the added costs to shippers would be offset by greater labor efficiency provided by a "better class of men to the seamen's trade." Shipping executives never saw it that way and continued to complain.

England Halts All Waterborne Commerce to Germany

England's response to Germany's declaration of a war zone around the British Isles was the Order in Council of March 11, 1915, which sought an end to all neutral trade with Germany. The Order was published on March 15 and was met with worldwide howls of protest as entirely illegal. Even though any losses suffered by neutrals would be reimbursed, the Order was thoroughly resented by all of Germany's neutral trading partners and created as much ill will toward England as there was toward Germany for starting the war in the first place.

On March 30, 1915, Secretary of State William Bryan sent a note to Great Britain:

> The order in council of the 15th of March would constitute, were the provisions to be actually carried into effect as they stand, a practical assertion of unlimited belligerent rights over neutral commerce within the whole European area and an almost unqualified denial of the sovereign rights of the nations now at peace.
>
> This government takes it for granted that there can be no question what those rights are. A nation's sovereignty over its own ships and citizens under its own flag on the high seas in time of peace is, of course, unlimited, and that sovereignty suffers no diminution in time of war, except in so far as the practice and

consent of civilized nations has limited it by the recognition of certain now clearly determined rights, which it is conceded may be exercised by nations which are at war.

A belligerent nation has been conceded this right of visit and search, and the right of capture and condemnation if upon examination a neutral vessel is found to be engaged in unneutral service or to be carrying contraband of war intended for the enemy's government or armed forces. It has been conceded the right to establish and maintain a blockade or an enemy's ports and coasts and to capture and condemn any vessel taken in trying to break the blockade. It is even conceded the right to detain and take to its own ports for judicial examination all vessels which it suspects for substantial reasons to be engaged in unneutral or contraband service and to condemn them if the suspicion is sustained. But such rights, long clearly defined, both in doctrine and practice, have hitherto been held to be the only permissible exceptions to the principle of universal equality of sovereignty on the high seas as between belligerents and nations not engaged in war.

It is confidently assumed that His Majesty's government will not deny that it is a rule sanctioned by general practice that, even though a blockade should exist and the doctrine of contraband as to unblockaded territory be rigidly enforced, innocent shipments may freely be transported to and from the United States through neutral countries to belligerent territory without being subject to penalties of contraband traffic or breach of blockade, much less to detention, requisition or confiscation.

Moreover, the rules of the Declaration of Paris of 1856—among them that free ships make free goods—will hardly at this day be disputed by signatories of that solemn agreement.

The note went on to cite judicial precedents and acknowledged the unique set of circumstances faced by Great Britain and "acquitted Britain of illegal purposes."

Leon Thrasher

Leon Thrasher was an early American casualty of the war and the reaction to his death reflected the mood of the times: 1) exasperation with President Wilson's inaction and a just cause for severing diplomatic relations with Germany and, 2) an example of someone being where he shouldn't have been.

Leon Chester Thrasher was born on September 19, 1883, in Hardwick, Maryland, and became a machinist then a mechanical engineer. He had worked with the Panama Canal Railroad at Colon and in early 1915 he went to London to accept a job with the Broomassie Mining Co. in West Africa. Horace Secombe, an employee of the company, was going there and booked passage on the unarmed Edler Dempster cargoliner *Falaba* and Thrasher went with him.

Capt. Frederick Davies left London with a crew of 95, 147 passengers, and 13 tons of gunpowder and cartridges. On March 28 they were stopped south of Ireland by Korvettenkapitän Freiherr von Forstner in the *U-28*. Forstner ordered Capt. Davies to heave to and get everyone off but Davies "took flight" and fired distress rockets. An angry Forstner then gave Capt. Davies ten minutes to get everyone off but when that deadline passed he took no action. Twenty-three minutes later he saw "suspicious steamers hurrying to the aid of the *Falaba*" and fired torpedoes. Not everyone had time to get off and 104 people were lost, including Capt. Davies.

There were reports of the submarine crew "laughing and jeering while the *Falaba*'s crew were struggling for their lives." Germany blamed England for allowing ships in the war zone. England regarded the submarine captain as a pirate.

Leon's parents were farmers in Hardwick and he had two brothers and a sister.

The *Greenbrier*

The Coast Steamship Co. freighter *Greenbrier* loaded cotton at New Orleans and left New York for Bremen on March 29 under Capt. James Dalton and a crew of 38. They were detained at Kirkwall, Orkney Islands, for three days for investigation by British customs authorities then released. It was reported there were 14 Germans aboard and the chief engineer, who was English, was arrested at Bremen and jailed. The ship left Bremen on April 1 for New York and the next day hit a British mine in the North Sea near Amrum, North Fresian Islands, and sank very quickly. There were no casualties and the survivors reached the German coast in 15 hours in two lifeboats. They well treated and sent to Rotterdam by train. On April 23, 25 survivors arrived at New York on the *Rotterdam*. Capt. Dalton lived at 721 49th St. in Brooklyn.

The *Cushing*

In April 1915 Capt. Lars Larsen Herland's Standard Oil of New Jersey tanker *Cushing* was en route to Rotterdam with 9,000 tons of crude consigned to the government of Holland. At about 7 p.m. on the 28th the ship was 25 miles off Antwerp and 8 miles from the North Hinder lightship when a lookout spotted what he thought was a British airplane several thousand feet up and apparently coming from the coast of England. The biplane swooped down to 500 feet, then 400, then to about 300 feet and "a second later there was a tremendous explosion and a solid wave of water flooded the stern deck." The crew on deck got below in a hurry just before a second "missile" missed the port quarter by a foot or so but sending more water "cascading over the lower deck." The plane then swung up into the wind and dropped a third bomb, "which just grazed the starboard rail and shot into the sea." Capt. Herland said it was a "dastardly act, a deliberate attempt to sink an unarmed vessel and murder the members of the crew."

The bridge watch said it appeared the pilot was attempting to drop the bombs down the smokestack and wreck the boilers. No one was hurt. The plane had "a white flag with a black cross in the center." The ship had "CUSHING NEW YORK USA" painted on the hull in letters eight feet high and two American flags, one on the masthead and one from the taffrail.

En route home, first mate Charles Christopher was on watch when a lookout spotted a floating mine off Deal, England. The ship "grazed" the mine but his skillful ship handling averted disaster. Germany admitted attacking the ship but said no flag was showing and offered compensation.

The *Gulflight*

Gulf Refining's tanker was the first American ship attacked by a German submarine. Capt. Alfred Gunter left Port Arthur, Texas, for Rouen, France, with 56,000 barrels of gasoline. On the morning of Saturday, May 1, 1915, they were met by two British patrol boats off the Isles of Scilly. At noon they were 15 miles southwest of Bishop Rock when 2nd mate Paul

Bower, of Chicago, and the quartermaster on watch saw a submarine about 1½ miles off the port bow. No flag was seen on the sub, which then disappeared. About 25 minutes later a torpedo hit the ship forward on the starboard side.

Wireless operator Charles C. Short, of Chicago, and a Spanish able seaman, Eugene A. Choponto, of Port Arthur, Texas, panicked and jumped off the ship and were never seen again. The water all around the ship was covered with oil but there was no fire or explosion. Two lifeboats got away with 34 survivors and were soon taken aboard the patrol boat *Iago*.

Very early the next morning, between 1 and 2 a.m., Capt. Gunter became "fearfully upset" and died at 3:30, supposedly of "heart failure due to shock." Capt. Gunter's body and 33 survivors were brought to Penzance on the steamer *Lyonnesse*. The tanker was towed into Crow Sound, Isles of Scilly, by patrol boats and repaired and returned to service.

Second engineer Huster B. Crist later stated: "I was on watch in the engine room when we were torpedoed and so terrific was the blow that the *Gulflight* seemed to be tumbling to pieces. It appeared to be lifted high in the air and then to descend rapidly. I told the boys to beat it as quickly as possible and shut the engine down."

Germany said the ship had no flag showing and offered compensation. Bower said the flag was flying as usual and fishermen on a British trawler who saw the whole thing said the American flag on the stern was plainly visible. The commander of *U-30*, said he thought the ship was British but patrol boats prevented him from getting in for a closer look.

Capt. Gunter was born in Germany in 1863, came to America in 1882, and was naturalized in 1890. He married Elizabeth "Lizzie" Law, 19, of New Jersey, on January 19, 1906, in Manhattan and the next year their daughter Louise was born. The family lived at 96 W. 16th St. in Bayonne, New Jersey.

The attack caused great anger in America. Col. Theodore Roosevelt was apoplectic and condemned the act as piracy.

The *Lusitania*

This incident is important because of the impact it had on American neutrality and submarine warfare. The fact that 159 Americans chose to sail on a ship belonging to a belligerent that was destroyed in a war zone caused a furor between the two opposing sides of American society: one side wanted an immediate declaration of war while the other side, still hoping to avoid war, said Americans and American ships shouldn't go where there was trouble. And there was trouble.

In the 1890s, Germany set out to enter trans-Atlantic passenger service with the specific purpose of taking the Blue Riband from England, which they did in 1897 with the 22-knot *Kaiser Wilhelm Der Gross*. Cunard built the turbine-powered *Lusitania* and triple-expansion *Mauretania* to reclaim it, which in 1907 the 24-knot *Lusitania* did, followed by the 26-knot *Mauretania*.

When war came, trans-Atlantic passenger traffic naturally dwindled since few vacationers considered being blown to pieces by a torpedo a fun summer activity. The *Mauretania* became a hospital ship and Cunard ran the *Lusitania* at 75 percent of boiler power to reduce coal and labor costs, reducing the speed to 21 knots, still very respectable, and still an irritation to the Kaiser. Still, more than a few people believed Germany would never attack a passenger ship.

German agents in the United States knew everything there was to know about the ship and its operations and all was communicated to Berlin. Count von Bernstorff was reportedly against any idea of deliberately sinking the ship but would do his part if ordered.

On January 30, 1915, the *Lusitania* left New York and on approaching the war zone declared by Germany all the lifeboats were made ready and the American flag was hoisted in place of the red duster, apparently at the request of the American passengers on board. The ship altered course "further off from the south coast of Ireland than formerly" to lessen their time in the war zone and ship arrived without incident. On February 10 the American ambassador, Walter Hines Page, protested the use of the American flag but Sir Edward Grey, Secretary of State for Foreign Affairs, replied that he understood the American passengers had requested it.

The ship made another uneventful passage to England and on April 17, 1915, Capt. William Thomas Turner left Liverpool and arrived at New York on the 24th. The ship was scheduled to sail from New York for Liverpool on May 1. On orders from Berlin, von Bernstorff purchased ads in the *New York Times, New York Tribune, New York Sun, New York Herald,* and the *New York World* that stated:

NOTICE!

Travelers intending to embark on the Atlantic voyage are reminded that a state of war exists between Germany and her allies and Great Britain and her allies. That the zone of war includes the waters adjacent to the British Isles. That in accordance with formal notice given by the Imperial German Government, vessels flying the flag of Great Britain or of any of her allies are liable to destruction in those waters, and that travelers sailing in the war zone on ships of Great Britain or her allies do so at their own risk.

IMPERIAL GERMAN EMBASSY
April 22, 1915. Washington, D.C.

Some say the idea was to divert attention from a plot to sink the ship, which was well-known to Germans in New York, but not believed by Americans, and to absolve Berlin.

The ship left New York on Saturday, May 1, 1915, and on approaching the war zone on May 6, Captain Turner ordered the lifeboats swung out, all bulkhead doors and portholes closed, and he rang up Full Ahead. Lookouts were doubled on the bow, on the bridge wings, and in the crow's nest. At 7:50 p.m. Capt. Turner received a warning, which he believed was from Queenstown: "Submarines active off south coast of Ireland" and directed his course to southwest of Old Head of Kinsale. where he believed he would pick up his escorts.

On April 30, Kapitänleutnant Walther von Schwieger got underway in the *U-20* with orders to patrol southwest of Ireland and to strictly enforce Germany's war zone policy around the British Isles, which meant sinking any vessel he encountered belonging to or flying an enemy flag. On May 5 he sank an English sailing vessel and the next morning an English steamer. At noon on the 6th he sighted a passenger liner but it was too far away. That evening he sank another English steamer before heavy fog began forming. Von Schwieger patrolled for two more days but visibility was too low to effectively operate. By the morning of the 7th, fuel oil was getting low and he had only two torpedoes left so he decided to head for Wilhelmshaven. By early afternoon the fog had lifted enough to see. Von Schwieger's log stated:

2.20 p.m. Directly in front of us I sighted four funnels and masts of steamer at right angles to our course, coming from south-southwest and going toward Galley Head. It is recognized as a passenger steamer.

2.25 Have advanced eleven meters toward steamer, in hope it will change its course along the Irish coast.

> 2.35 Steamer turns, takes direction to Queenstown, and thereby makes it possible for us to approach it for shot. We proceed at high speed in order to reach correct position.
>
> 3.10. Torpedo shot at distance of 700 meters, going 3 meters below the surface. Hits steering centre behind bridge. Unusually great detonation with large cloud of smoke and debris shot above the funnels. In addition to torpedo, a second explosion must have taken place. (Boiler, coal, or powder?) Bridge and part of the ship where the torpedo hit are torn apart, and fire follows.
>
> The ship stops very quickly leans over to starboard, at the same time sinking at the bow. It looks as though it would capsize in a short time. There is great confusion on board. Boats are cleared and many of them lowered into the water. Many boats, fully loaded, drop down into the water bow- or stern-first and capsize. The boats on the port side cannot be made clear because of the slanting position. At the front of the ship the name *Lusitania* in gold letters can be seen. The chimneys are painted black. The stern flag is not hoisted. The ship was going about twenty miles an hour.

Von Schwieger implied that he thought his second and last torpedo might be needed to sink the ship but stated, "I could not have sent a second torpedo into the crowd." The ship sank in less than 20 minutes.

Germany said the ship was carrying artillery shells. Why would a passenger liner with holds full of general merchandise and passengers' luggage explode and sink so fast?

Many Americans called for an immediate declaration of war. Secretary of the Navy Josephus Daniels asked the General Board—the Navy's planners—to "prepare a comprehensive naval building program," which was done and submitted to Congress, but noting much came of it. When a senator later asked Admiral Sims why more destroyers, in particular, weren't built sooner for escort work he replied:

> If we could have imagined that the Germans would do what they did do we could have prepared for it and built destroyers galore, if we could have persuaded Congress to give us the money. Nobody had any experience with this kind of war at all, and nobody could be savage enough in his disposition to know what the Germans would do, and therefore to prepare for it. So I guess that I would advise you to be a little gentle in criticisms of naval officers in general because they were not prepared for this war because we are a more civilized people.

The sinking caused worldwide outrage. On May 13 President Wilson sent Germany a very strong note saying "that this Government would omit no word or act to protect its citizens against murder on the seas." It was known and admitted at the time that the ship was carrying munitions in defiance of American law. Secretary of State Bryan demanded reparations.

Between June 15 and July 1, 1915, the British Wreck Commissioner's Court in London investigated the sinking and took testimony and depositions from passengers and crew into the "inexpressibly cowardly attack upon an unarmed passenger liner."

Lawsuits seeking $6,000,000 in damages were filed in the federal District Court of New York against the Cunard Steamship Co., Ltd., by 67 plaintiffs on the grounds that the *Lusitania* carried ammunition and high explosives and had been improperly navigated through a submarine zone. On August 24, 1918, Judge Julius M. Mayer found the ship had no explosives of any kind on board but was carrying 18 fuse cases, 125 empty shrapnel cases, 4,200 cases of safety cartridges, and 189 cases of infantry equipment "such as leather fittings, pouches, and the like" but nothing of the sort that would "explode" and that appears to be the defining factor. Item No. 5 on the list of Absolute Contraband was "Projectiles, charges, and cartridges of all kinds, and their component parts." No. 14 was "Clothing and equipment of a distinctive military character."

Almost every witness said there were two explosions, but whether they were from two torpedoes or a high order internal explosion remained a major dispute. Despite popular belief to the contrary, von Schwieger denied he was sent out specifically to get the *Lusitania*.

The *Nebraskan*

In 1915, Capt. John S. Greene took American-Hawaiian's freighter *Nebraskan* out of San Francisco with California produce for New York and was one of the first ships through the Panama Canal. At New York the ship was chartered by the White Star Line on May 1 for the return trip in ballast to load another cargo and on May 24 Greene left Liverpool for the Delaware Breakwater. The ship had "NEBRASKAN OF NEW YORK" painted on both sides of the hull in white letters six feet high.

At 4:33 p.m. on the 25th they passed the Fastnet Rock. At 8 p.m. when Third Mate Wilfred Harold, Sr., 22, of 197 Athens St., Boston, came on watch, Chief Engineer Frank Chester Yandell, of 1445 76th St., Brooklyn, was out on deck getting some fresh air. At 8:24 he noticed a "white streak in the water about fifty feet away"—the only one to actually see it—and, according to Capt. Greene, "the ship met with a violent shock, followed instantly by a terrific explosion, bursting No. 1 hatch and deck, throwing hatch beams and cargo derricks 30 feet in the air and filling No. 1 lower hold completely with water." An S.O.S. went out and Chief Engineer Yandell immediately went below and found the 3rd engineer had shut the engines down. The crew got off but when it appeared the ship would remain afloat they returned on board and headed for Liverpool. The ship was drydocked and examined by naval attaché Lt. Cdr. John H. Towers and naval constructor Cdr. Lewis B. McBride. Both concluded the ship had been torpedoed and not hit a mine.

The explosion was witnessed by several fishing boats and freight schooners from Crookhaven, Ireland, who all said they saw "a steamer flying the American flag going west" and reported hearing two loud reports. The steamer then sounded its whistle to summon help.

The sub commander later said no flag or identification as a neutral was showing and the German government offered compensation. Residents of Liverpool were anxious to see the American response to the incident, firmly hoping it would be the last straw and America would enter the war against Germany.

A German's Response to America

President Wilson sent notes of carefully worded protest to Germany on May 13 and May 21 regarding the *Falaba*, *Cushing*, *Gulflight*, and *Lusitania*. The notes referred to the acts as violations of the treaty of 1828 between the United States and Germany, addressed the inhumane nature of Germany's submarine warfare, and asserted the right of neutrals to travel where they please. On May 28, 1915, Count Ernst Graf zu Reventlow, a German naval officer and author from Rotterdam, offered a response to Wilson's notes:

> Trade between Germany and America has shrunk to microscopic dimensions. What they receive from us is more valuable and necessary than what we receive from them. Complete cutting off of negotiations

will not affect us. America would only be able to damage us by confiscating the ships left in her harbors and such other German property. Further dangerous deeds of war from America are not to be feared because they are not possible.

Also we must not forget certain interior difficulties in America. That is another side of the business. On the other hand, any stopping of the submarine war, if only for a time, would have the most important results. Any orders to submarine commanders to conform to any formal conditions laid down by international law would mean hindering their actions and make the submarine war an empty farce.

There is no drawing back. There cannot be even a fraction of withdrawal. The empire cannot compromise itself nor its way of carrying on this war. That would be to commit hari-kari amidst the jeers of the world.

Because Italy has now become our enemy, America does not seem any more terrifying to us than before. Germany is not to be frightened with big words and threats. We believe every German expects the German empire to answer the American note by maintaining its position regarding the submarine war with all its means and without consideration.

America may then do or not do what it likes. We neither care for her bluff nor her deeds.

On June 6, 1915, the Kaiser stated that passenger liners would no longer be attacked.

The *Leelanaw*

Harriss, Irby & Vose loaded 6,000 bales of cotton worth $400,000 on Harby's steamer at Galveston. The cotton was consigned to Russia but was being shipped to Gothenburg, Sweden, for trans-shipment to Moscow. The ship's papers were examined and cleared by the Russian and British consuls in New York.

Capt. Eugene B. Delk left New York on May 17 and on June 1 he was stopped by the British and ordered to Kirkwall for examination. They arrived the next day and Capt. Delk was informed that Sweden prohibited the export of cotton. William Harriss told British authorities they would land the cotton at a Norwegian port instead but the British would not agree to that and ordered Capt. Delk to proceed directly to Archangel. The British were certain the cotton would find its way to Germany and if the cotton was consigned to Russia, why not go to Russia?

The ship was released on June 26, the cotton was discharged at Archangel, and a cargo of flax was loaded for Belfast. On the afternoon of June 25 they were about 60 miles northwest of the Orkney Islands when a lookout spotted a submarine. Capt. Delk attempted to escape but was fired on from about two miles away. The shots fell short but Capt. Delk hove to and at 2:10 the sub came up and the commander asked for his papers. He declared the flax absolute contraband and gave the crew of seven officers and 32 sailors time to gather a few possessions and get into their boats. Capt. Delk requested his papers back but was refused. Five shots were fired at the ship, which had little effect, and then a torpedo was fired. The crew was taken aboard the sub and two more shots were fired, one starting a fire on the ship.

The sub towed the lifeboats toward shore and about 90 minutes later the *Leelanaw* was seen sinking. At about 8:30 p.m., when they had travelled about 50 miles, another steamer was sighted and the survivors were ordered back into their boats. All made landfall at Kirkwall at 6:30 the next morning. E. Haldeman Dennison, the American consul at Dundee, ordered the crew "sent to Dundee on tomorrow's boat." The crew left Liverpool on July 31 on the American Line steamer *St. Paul* and arrived at New York on August 8.

The ship's papers were sent to the prize court at Hamburg. Germany had declared flax absolute contraband on April 18 but there was no way for Capt. Delk to know that. The incident created a diplomatic crisis and sharp disagreements over the wording of the Prussian-American Treaty of 1828. It was fully expected to produce an immediate declaration of war as the State Department said the cargo could have been removed and the ship spared.

Theodore Roosevelt was livid, loudly calling the incident a "damnable outrage" and describing the Administration's "peace at any price" policy as "perfectly hellish."

On September 23, 1915, Germany announced that American ships "carrying conditional contraband would not be sunk by submarines under any conditions" and agreed to make full reparation for the loss of American ships carrying absolute contraband pending an arbitrator's interpretation of the 1765, 1799, and 1828 treaties involved. They also implied they would cease torpedoing merchant ships without warning.

The *Pass of Balmaha* and the *Vincent*

In early 1915, Americans purchased the two big, full-rigged sailing ships *Pass of Balmaha* and *Vincent* in England for $30,000 each. The dramatic increase in freight rates to Europe had made sailing vessels popular and economical again and the length of their passages often compared favorably with steamers, minus the expense of larger crews and fuel. The sailing vessels *Dirigo* and *John Ena* were engaged to carry case oil from Philadelphia to Japan at $50,000 in freight charges for the outbound voyage. An editorial in *The Outlook* magazine in March 1915 stated, in part:

> This happy state of affairs may be but temporary. The merchant marine may be still hampered by senseless legislation after the war and unable to maintain the strategic position it now holds. The fact remains that it is now experiencing such a boom as has not been seen since the halcyon days when merchants retired on the profits of a single spicy cargo brought in some snowy-pinioned clipper from the Orient.

The editors were right. The 245', steel-hulled, 3-masted *Pass of Balmaha* was owned by the Harby Steamship Co., of New York and homeported at Boston under Capt. William Scott.

To celebrate this happy, but temporary, state of affairs, a race between the *Pass of Balmaha* and the *Vincent* was planned in January 1915 to carry cotton to Bremen, Germany. Old salts were beside themselves with excitement and thousands of dollars in wagers were put on the two captains. A trans-Atlantic race between two big clipper ships had not been seen in 50 years. The big New York house of cotton merchants, Harriss, Irby & Vose, chartered the *Pass of Balmaha* and loaded 6,000 bales aboard.

The *Vincent* had carried cotton to Cuxhaven, Germany, soon after the war started. More excitement was created, almost a frenzy, when Capt. Scott said he would go through the English Channel "by way of the Downs" while Capt. John C. Amberman said his course in the *Vincent* would be to the north of Scotland to get to the North Sea. Most money was on the *Vincent*.

They left New York on January 25. The *Pass of Balmaha* was seized by the British on or about March 9 and sent in to Kirkwall for inspection. The bark and the crew of 28 were released around March 19 but went aground near Sylt in the North Fresian Islands in heavy fog. Two-thirds of the cargo was removed and the vessel refloated about April 5. Both vessels were subsequently sold to new owners in New York for $75,000 each.

In June 1915 the *Pass of Balmaha* departed New York under Capt. Scott for Arkhanglesk, Russia, with 5,000 bales of cotton consigned to Russia. On August 4 they were stopped by

HMS *Victorian* off the coast of Norway and boarded. The cargo was declared contraband and a prize crew of one officer and six marines was put aboard with orders to sail toward Kirkwall and Scapa Flow. Capt. Scott was incredulous that a cargo destined for an ally of Great Britain was being delayed. Perhaps they wanted it for their own textile mills.

Over Capt. Scott's vehement objection, the British prize officer ordered the American flag taken down and replaced with a British flag. That was a mistake.

On August 5, the flag was sighted by Kapitänleutnant Ernst Graff in *U-36*. When Graff surfaced about a half a mile away, and the British officer saw the sub, he immediately took the British flag down—which Graff did not see—and put the American flag back up, but it was too late. Capt. Scott could barely contain himself imagining the British officer in a Prussian prison, but at the same time he didn't want to lose his ship so he ordered the prize crew below.

A German officer came aboard and asked where the Union Jack was. Capt. Scott said there was no British flag, he must be mistaken. Scott produced his papers and was ordered to sail for Cuxhaven. When they arrived 11 days later, Capt. Scott brought the British officer and marines up and they were promptly arrested along with the ship. The American crew was allowed to leave Germany in a neutral vessel. Capt. Scott left Liverpool on November 20 on the steamer *Orduna* and arrived at New York on November 29.

On June 9 the *Vincent* left New York for Archangel, arrived on July 30, and departed on September 4 for London. On September 27 the ship hit a mine off Cape Orloff in the White Sea and caught fire and sank. Those injured, all Americans, were Capt. Amberman, 226 Baltic St., Brooklyn; 1st mate William Leary, of St. John, New Brunswick, Canada; 2nd mate Samuel Latham, 716 N. 6th St., Philadelphia; and the carpenter, who was not identified.

The rest of the crew of 18 returned to New York on October 29 on the steamer *Dwinsk*. Capt. Amberman returned to New York on the steamer *Czar* from Archangel on November 12.

The 1915 cotton crop was one of the biggest in history and the growers, brokers, and shippers were very nervous. Fully two-thirds of American cotton was exported as the domestic market could not absorb the amount of cotton grown.

On December 12 the *Pass of Balmaha* and the cargo were condemned by a prize court at Hamburg. The American vice-consul, Henry H. Morgan, attended the hearing. The cotton was sold for 2,500,000 marks ($625,000). The ship was declared an enemy vessel on the grounds that it was British until the end of 1914 then sold to an American buyer after the war started and its pre-war Canadian registry was illegally changed after the war started. American ambassador Gerard filed a formal protest on the grounds the ship was American-owned.

After this, about 2,000 merchants who traded overseas retained a law firm to represent them in Washington and hired attorney William W. Bride, formerly employed in the Foreign Trade Office of the State Department, to represent their interests in England, The Netherlands, and France. Bride left on the steamer *Rotterdam* around February 1, 1916.

* * *

Helen W. Martin—Percy & Small's big 5-masted schooner left New York on July 27 for Archangel, Russia, and arrived on September 13. They hit a mine at 4 p.m. on November 18 in the North Sea three miles off Orfordness with four reported injured. A tug came out and took the vessel to London where it was repaired.

Petrolite—Capt. Thompson Thompson's Standard Oil tanker left New York on November 7, 1915, for Alexandria, Egypt. On December 5, en route home about 120 miles off Crete, the commander of the Austrian submarine *U-39* came up and demanded the ship stop for search. Capt. Thompson decided to flee instead and the sub commenced firing. One shell entered the engine room and injured a Danish sailor, Frederick Larsen. Capt. Thompson stopped after the 14th shot and the sub came alongside. One sailor was taken off as a hostage while sub's crew looted the tanker's stores and provisions. The cruiser *Des Moines* came out from Canea, Crete, and took a report from Capt. Thompson. The incident caused a sensation and a diplomatic crisis with Austria. In response to Capt. Thompson's protest, the State Department acknowledged that a warship has a right to seize supplies from a merchant ship under certain circumstances but the legality of taking stores from a neutral by force was unsettled.

Orleanian—the Universal Transportation Co. freighter *Orleanian*, under Capt. Charles R. Bergman, left New York on December 23, 1915, for Malta with case oil. Neither ship nor crew was ever seen or heard from again and it was supposed all were lost off Gibraltar. Capt. Bergman's wife Alma was on board as a stewardess. They lived at 105 Summer St., Malden, Massachusetts.

Ship Values Increase

The sinking of ships, and the *Gulflight* and *Lusitania* especially, sent the value of ships up by 50 to 100 percent and also caused great concern to shipowners. Coastwise steamers and ships that had been laid up due to low freight rates were refitted and sent out. In August 1914, one charterer had a chance to buy an old steamer for $90,000. By May 1915 he was chartering the same ship for $90,000 a month. Freight rates went up 500 percent and cargoes 50 percent in value.

A ship that used to be worth $400,000 with freight charges of $20,000 and a cargo worth around $1,000,000 on a typical voyage was now worth $600,000–$800,000, the freight $120,000, and the cargo $1,500,000. Marquis James, an insurance man, stated:

> This 70-percent increase in the underwriting load was for one sailing only. Double and treble the number of sailings and one gets nearer the true dimensions of the problem in dollars. Nor does the mere matter of dollars tell the whole story. In addition to the extraordinary risks of war—submarines, mines and surface raiders—ordinary risks of the sea were greatly enhanced by the suppression of navigation lights, by hasty inspections, inadequately trained officers and crews, want of time for proper repair or overhaul, and other factors. Additional sources of cargo loss were thefts from warehouses and piers, damage due to speed or carelessness in handling and to insecure packing. In the emergency there was not time to see to those things.

The steamer *Caroline* was reportedly purchased in October 1915 and left New York on December 4 for Europe. The freight earned not only paid for the ship but put extra cash in the owner's bank account. In November 1915 Merchants & Miners sold their 30-year-old steamer *Frederic* for $45,000. It was chartered in less than a week for a voyage to Europe. The freight paid for the ship with money left over.

Talk of building a national merchant marine that would also serve as a naval auxiliary began with Secretary McAdoo's "Ship Purchase" idea in October 1915 owing to the lack of ships to move goods and the reluctance of enough private capital to take the necessary risks.

6. 1916—A Year of Tension for America

Several attempts to arrive at a peace deal were made by both sides but failed. President Wilson nevertheless decided to run for president again. His opponent was Charles Evans Hughes, an associate justice on the Supreme Court who resigned his position to run for president. Hughes was very critical of Wilson's policies, opposed an arms embargo, and felt the United States should strongly assert the right of its citizens to travel and trade feely without fear. Wilson's campaign slogan, "He kept us out of war" won him a very narrow victory, thanks to California voters.

The arming of merchant ships became more complicated when the armed Transatlantica Italiana liner *Guiseppe Verde* arrived at New York from Naples on January 6, 1916. After the ship left, the Austro-Hungarian government objected to the United States allowing an armed ship to enter and leave a neutral port. The Austrian ambassador, Baron Erich Zwiedinek, called on Secretary Lansing to discuss the matter as a violation of neutrality. Lansing told him the ship was allowed to leave after assurances were given that the guns were for defensive purposes only. The Austrian government said an official protest would follow if a second armed ship was allowed to enter.

The State Department then considered revising its rules by either not allowing armed vessels to enter and leave or by keeping American citizens off armed ships. At the beginning of the war, English and French ships informally agreed to disarm their ships before leaving a U.S. port.

International rules regarding submarines and armed merchant ships had been established. Merchant ships were required to stop when hailed by a submarine, as they would be by any other warship under the old doctrine of visit and search. It was recognized that a submarine could be sunk or completely disabled by one shot from another vessel and because of this, the Central Powers held that a submarine "not having an opportunity to warn armed merchant ships, has no other course left but to sink such vessels without warning."

The United States at that time held that after being warned by a belligerent vessel, a merchant ship has a right to defend itself from capture by returning fire and fleeing. However, once a shot is fired, the merchant ship assumed the character of a warship and remained so until coming to a stop and firing ceased. Once the ship stops, international law required the submarine commander to allow passengers and crew to "reach a place of safety" before destroying the ship. Passengers and crew aboard an armed ship could not hold the government whose flag the submarine flies liable for lives lost or property damaged during a fight.

Regulating Shipping

Talk of a federal shipping board got the immediate attention of William H. Douglas, president of the Chamber of Commerce of the United States. Douglas first decided to take a poll among his shipping industry members. The shipping men were staunchly against the purchase, lease, operation, or construction of any ships by the government, 690 to 89. As to subsidies "to offset the cost of operation between sailing under the American flag and under foreign flags" the members approved 713 to 52. The members also approved the creation of a federal shipping board but without government officials in ex-officio positions.

In January 1916, New York attorney Stuart G. Gibboney addressed the congestion of goods waiting for shipment in New York Harbor in the January issue of Marine Engineering magazine and endorsed Secretary McAdoo's idea of a 500,000-ton government-operated merchant marine due to "The utter failure of private capital to come forward and back enterprises which would tend toward the rehabilitation of American shipping, despite the inducement offered by rising freight rates."

A bill, H.R. 10500 was introduced and, to formally explore the issue, the Committee on the Merchant Marine and Fisheries under Chairman Joshua W. Alexander, of Missouri, held hearings in Washington from February 10 to March 16, 1916. The topic stated:

> A bill to establish a United States Shipping Board for the purpose of encouraging, developing, and creating a naval auxiliary and naval reserve and a merchant marine to meet the requirements of the commerce of the United States with its territories and possessions, and with foreign countries, and for other purposes.

The bill essentially nationalized all waterborne commerce after January 1, 1917. No vessel would be allowed to carry freight or passengers anywhere between states, coastwise, or foreign without first obtaining a license from a shipping board. William Douglas was the first witness called to testify and his first statement to the Committee summarized the Chamber's position, stating in part:

> We believe that regulation should be adopted. We do not want drastic legislation that is going to interfere in any way with shipping measures or shipping companies or the proper regulation, by those who may run lines, of our shipping, but we do think that a board should have the same rights of supervision which, under the railroad system, is now working quite successfully in this country, that is, if shippers feel that they are in any way injured by the rates or by anything else which a company which is running regular lines, carries out as a policy, that they should have the right to go to the board and demand a fair and proper hearing.

Douglas was very concerned about the difference between "regulating" and "supervising" and how those two words were used in the language of the bill. Chairman Alexander replied:

> The Interstate Commerce Commission was created a great many years ago and it has the power of supervision, but it was only within the last decade that it was given the power to "regulate" and it is only within the last decade that they have had real power over the railroads in the matter of transportation rates. Now, if you study this bill you will find that the power is vested in this board to regulate. It does not say that they shall do it, but that they shall supervise, and, whenever necessary, they shall have power to step in and compel the observance of reasonable regulations.

Steamship company executives, naval officers, shipbuilders, insurance men, and other expert witnesses gave testimony and opinions on every aspect of maritime conditions and trade. Government officials and bean counters submitted reports regarding available registered tonnage, ships then building, and all manner of statistics on foreign trade. Letters and affidavits were introduced from others not present whose opinions were solicited. These talks continued formally and informally for months.

The *Sussex*

While the Shipping Board bill was being debated, events occurred that naturally affected the course and shape of these discussions and helped sway them in one direction or another. One involved the French cross–Channel ferry *Sussex.*

On March 24, 1916, the *U-29* sank the 275' French cross-Channel ferry *Sussex* en route from Folkestone to Dieppe with considerable loss of life.

According to Secretary of the Navy Daniels, "All preparations were made to mobilize the Fleet..." and on April 18, President Wilson sent a message to Germany:

> Unless the Imperial Government should now immediately declare and effect an abandonment of its present methods of submarine warfare against passenger and freight-carrying vessels, the Government of the United States can have no choice but to sever diplomatic relations with the German Empire altogether.

On the 27th a Mobilization Plan was sent out to all commands by Adm. William S. Benson. On May 4, a note was received from Germany:

> In accordance with the general principles of visit and search and destruction of merchant vessels recognized by international law, such vessels, both within and without the area declared as naval war zone, shall not be sunk without warning and without saving human lives, unless such ships attempt to escape or offer resistance.

The *Sussex* attack created more worldwide outrage and another diplomatic crisis. On May 4, Germany offered reparations and said they would no longer target passenger vessels or unarmed merchant ships and would make sure everyone got off safely. This was generally known as the "Sussex Pledge."

From May 6 to 8 the Navy conducted an extensive, extraordinary communications drill that linked up all radio, telegraph, and telephone systems into one seamless network for rapid communication between yards, stations, and ships at sea.

In anticipation for war, Count von Bernstorff secretly ordered the crews on all German ships stuck in American harbors to disable their engines and do as much mechanical damage as possible.

During this time several American ships were damaged by mines without any casualties. A few were stopped by submarine commanders who examined their papers and released them.

A Blockade Runner Visits Baltimore

The British cut Germany's trans–Atlantic cables and blockaded their ports. In order to transfer uncensored mail, obtain products German industry needed, and gold to bolster the treasury, and in return supply American manufacturers with vitally needed goods, someone suggested using unterseeboots as commercial freighters. Designers came up with a 214' submarine with a cargo capacity of 750 tons in two holds. It was planned to build seven boats but only two were completed, the *Deutschland* and the *Bremen*.

The plan was to send the *Deutschland* to Baltimore and a Norddeutscher Lloyd shipmaster, Kapitän Paul Lebrecht König, was chosen to command the boat. At the other end, Kapitän Hans F. Hinsch, master of the German liner *Necktar*, laid up in Baltimore, was appointed overall coordinator of the venture.

The agent for North German Lloyd at Baltimore was A. Schumacher & Co., at Hansa Haus. Henry G. Hilken, of 4 Bishop Rd., was an officer with the company. By March 1916 the *Deutschland* was almost ready and in May he incorporated the Eastern Forwarding Co. at 2 E. German St. with a capital stock of $50,000 to act as forwarding agents for the cargoes. Hilken was president.

The sub was painted a "steel gray" and two small-caliber guns were mounted on the conning tower for defensive purposes. Since the freighters were defensively armed, Germany's argument against the United States supplying England in the same manner was negated.

The *Deutschland* took aboard $1,000,000 worth of chemicals and aniline dyes while König obtained a health certificate from the American consul at Bremen, William T. Fee, who notified the authorities at Baltimore that a submarine freighter was on its way. The news spread quickly and Royal Navy commanders got orders to sink the sub on sight.

König left on June 14 and stopped at the Heligoland naval base to throw off the British. He departed on the 23rd traveling on the surface unless an aircraft or another vessel was sighted. König took the shorter route through the Straits of Dover, made the trans–Atlantic passage without incident, and arrived between the Virginia Capes at about 1:45 a.m. on Sunday, July 9. At about 2 a.m. König surfaced off Cape Henry about 50 yards from Capt. Fred Cocke's pilot boat *Relief*, which had been anxiously waiting, along with the tugboat *Thomas F. Simmons*. An especially excited Capt. Hinsch was aboard the tug as the official welcoming committee.

Kapt. König told Capt. Cocke he wished to abide by all American laws. Hinsch was aboard for about two hours, apparently discussing strategy, and the sub arrived in town "under the very noses of the British cruisers just outside." The *Deutschland* was the first submarine to cross the Atlantic.

As expected, England and France vehemently objected to this development on the grounds that visitation and search was impossible if the ship submerged. The State Department replied there was nothing in the law of nations or the conventions and treaties then in force to prevent a submersible cargo ship from sailing as long as all the proper clearances and papers were obtained and all the proper navigation lights required under the rules were displayed.

The supercargo on the *Deutschland* was Gotthold Prusse, who stayed ashore with a widow, Mary Blanck, of 3204 Winfield Ave. in the Forrest Park area. Prusse was on the payroll of North German Lloyd and was to act as port captain, arrange for any repairs the *Deutschland* might need, and await the arrival of the *Bremen*.

No visitors were allowed on board the sub. Only two people, Guy W. Steele, the U.S. Surveyor of Customs at the Port of Baltimore, and a friend of his, Ary Johannes Lamme, a stock broker who lived at 82 58th St. in New York, went aboard. Lamme was an acquaintance of Steele and a native of Holland who spoke German and Steele asked him to accompany him on his official business. The two were besieged by reporters wanting details of the boat. Lamme drew a sketch that was later used in the papers and in *Popular Mechanics* magazine.

The cargo was discharged and over the next 23 days took aboard 802,037 pounds of crude rubber worth $568,855; 752,674 pounds of nickel worth $376,337; 181,049 pounds of tin worth $108,630, and reportedly a consignment of gold. No consignee was stated on the cargo manifest, just the destination of Bremen.

While in port, the sub was repainted in a camouflage pattern of sea green and "dirty white" irregular stripes across the whaleback. An observer stated, "Where the waters are deep and green it will be impossible to distinguish her once she submerges." The sub cast off lines at 5:38 p.m. on August 1 to the cheers of a small crowd that had gathered, including a contingent of German sailors who had been stranded in Baltimore since the war began. The boat was towed out past Locust Point, evaded Allied warships off Hampton Roads, and arrived at the mouth of the Weser River at Bremen on the 23rd.

The sub immediately prepared for another voyage to New London, Connecticut, with chemicals, dyestuffs, and securities worth 1.8 million pounds sterling. Eastern Forwarding was notified and the boat sailed on October 10. The sub arrived at New London on November 1 and moored at the State Pier, hidden from public view. The cargo was discharged and then took aboard nickel, copper, and $150,000 in silver bullion from the San Francisco mint. Count von Bernstorff and his wife Jeanne visited the sub and the crew were the guests of the Frohsinn Club at Mystic. Bernstorff denied any mail was being put aboard but said it would be carried on the next trip.

The sub departed on the 17th, escorted by the T.A. Scott Co. tugs *T.A. Scott, Jr.*, alongside the sub, and *Cassie*, about a mile astern. Thomas A. and John A. Scott were submarine engineers and wreckers at 292 Pequot Ave.

About an hour out, near Race Rock, off the end of Fisher's Island in Long Island Sound, the *Scott* and sub the collided. The tug was under the command Capt. John Gurney and Hans Hinsch was a passenger aboard to see his charge off. The tug sank in three minutes with the loss of Capt. Gurney, engineer William A. Caton, fireman Edward Stone, cook Clarence B. Davidson, and deckhand Eugene Duzant. Hinsch jumped overboard and was the only survivor. He was saved by Alber Mix, a Scott Co. superintendent aboard the *Cassie*, who threw him a life preserver and hauled him aboard. Hinsch couldn't swim.

The sub returned to New London for repairs, departed on the 21st, and arrived back home on December 10, 1916. The accident was ascribed to the *Scott* steaming too close to the sub and being caught in strong currents. Eastern Forwarding said all claims for damages would be settled.

While the *Deutschland* was in port, a second sub with a deck gun had been sighted, causing wild excitement and rumors in New London. Commander Yates Stirling, of the Submarine Flotilla, conducted an inquiry and assured everyone it was an American boat on training maneuvers.

It was supposed by naval authorities and the Justice Department that these voyages were of a strictly commercial nature and nothing was observed that would suggest otherwise.

On April 6, 1917, Gotthold Prusse became an enemy alien and was contacted by the Justice Department's Bureau of Investigation. His movements were restricted to one-half mile of his residence and he was contacted on numerous occasions thereafter. His quarters were searched regularly and a "cover" was placed on his mail.

On May 15, 1917, Prusse applied for a permit to reside in the United States as a resident alien naming Henry G. Hilken as his sponsor. Henry and Paul Hilken, Heinrich Ruhstrat, and Carl A. Luederitz were A. Schumacher & Co. officers.

The permit was denied but on July 13, 1917, he was admitted to the United States as a "ship building engineer for the purpose of superintending the mechanics of the German submarines in this country." Prusse was also on the payroll of the Inter-Ocean Steamship Company. The *Deutschland* and *Bremen* were expected to make several commercial voyages but the *Deutschland* made just the two trips and the *Bremen* disappeared and was never seen again.

In August 1917, Baltimore City detective George Armstrong contacted the Bureau of Investigation regarding a "mysterious German" living in the home of Mary Blanck. Agent Billups Harris was notified and had the subject followed. When he was spotted in a prohibited area at Lexington and Park Avenues on August 24 he was arrested. After Prusse was jailed on orders from Washington, Henry Hilken and Paul Johannsen, a "decidedly pro–German" Police Judge, "got busy at once over the phone trying to get to see him." Agent Harris wrote, "I repeat that it seems to me Prusse might be extremely dangerous as one trip along the neighborhood of forts, ships, etc., to a man of his skill would be equivalent to a photograph."

Prusse had no incriminating papers on him when he was arrested, none were ever found in his quarters, and no evidence of any anti–American activity was ever discovered. Early on Tuesday morning, September 18, 1917, Gotthold Prusse hung himself by his belt in the city jail.

Another person on board the *Deutschland* was Adelbert A. Fischer, 46, president of the Schutte & Koerting Co. and the General Condenser Co. at 12th & Thomas in Philadelphia. He lived at 6906 Wissahickon Ave. in Baltimore with his wife Helen, 37, daughters Ruth, 16, Francis, 7, and sons Kermit, 12, John, 10, and two servants.

Fischer was the son-in-law of Ernest Koerting, who was known as the "Carnegie of Germany." Schutte & Koerting held patents for automatic ejector and injector boiler valves that were used on every American naval vessel and the company owned $1.7 million worth of real estate, factory equipment, and other assets in the United States. Fischer came to America in 1903 to oversee the business.

On March 17, 1917, Fischer transferred all of the common stock Ernest Koerting owned to his wife, Helen. Helen paid for the stock with a note due and payable in 20 years. When the sub came in, Helen attempted to get papers to her husband on board the sub but was stopped. Adelbert was arrested and interned at Fort Oglethorpe. Adelbert was found to be good friends with Capt. Thierichens, of the raider *Prinz Eitel Friederick* and both of them worked closely with the German sailors who were being detained at Philadelphia on propaganda projects.

The arrival of the *Deutschland* stunned the banking world. Several days before the ship arrived, Julius Pirnitzer, president of the Transatlantic Trust Co. in New York, began telling people they would soon be able to transmit messages and money by submarine "to the other side, despite the British blockade." The thousands of Hungarian immigrants in America had been cheated repeatedly by scores of brokers, ticket agents, and bankers entrusted to hold or transmit money home for them. Pirnitzer was a Hungarian national sent over from Budapest to organize Transatlantic Trust and to enlist "Americans of good financial and business standing" to join his organization and to conduct an honest and efficient bank, which they did. Pirnitzer told his associates he intended to become an American citizen but actually had no intention of doing so.

Transatlantic Trust was organized by three large banks in Hungary in collusion with the Royal Hungarian Postal Savings Bank, which allowed Transatlantic to make exclusive money transmissions through the Hungarian postal system, which other banks were not allowed to, over fierce protest.

The bank was actually building a huge web of informers across the country through its depositors, "to keep the home Government informed." Pirnitzer felt confident enough to advertise—in foreign languages—the bank as "the official agent of the German and Austro-Hungarian Governments." The bank was at the service of the Central Powers and, according investigators, "vigorously pushed the sale of war bonds of the Central Powers."

The bank was seized by the Department of Justice and Pirnitzer was arrested as a dangerous enemy alien and imprisoned. At that time, there were around 13,000 immigrant depositors and over 1,000 agents providing all sorts of information.

The German firm of Crossman & Sieleken (Herman Sieleken) was a large coffee importing business and when the company was seized by the Alien Property Custodian they had 200 pounds of rubber ready to ship on the *Deutschland*.

A third voyage was planned but with the worsening of U.S.–German relations by late 1916 the *Deutschland* was rebuilt as a warship and in February 1917 became the *U-155* under Kapitänleutnant Karl Meusel. The *151* to *157* were of the *Deutschland* type.

The *U-155* operated in Canadian waters and out in the Atlantic, never coming any closer than 200 or 300 miles from the East Coast.

The United States Shipping Board

On August 16, 1916, it was announced in Britain that 218 neutral vessels had been lost to date. On the 20th, the London Times castigated the United States for remaining neutral and it was insinuated there would be an economic price to pay after hostilities ended.

On September 7, 1916, Congress passed the Federal Shipping Act, which created the U.S. Shipping Board (39 Stat. L. 729) "for the purpose of encouraging, developing, and creating a naval auxiliary and naval reserve and a merchant marine to meet the requirements of the commerce of the United States with its territories and possessions and with foreign countries; to regulate carriers by water engaged in the foreign and interstate commerce of the United States." (Public Law No. 260, 64th Congress). The Shipping Board was authorized to "form, with a capital stock not exceeding $50,000,000, one or more corporations for the purchase, construction, equipment, lease, charter, maintenance, and operation of merchant vessels in the commerce of the United States."

On December 22, 1916, President Wilson nominated a chairman and four board members and they were confirmed by the Senate of January 19, 1917: Bernard N. Baker, of Baltimore, founder of the Atlantic Transport Co., was appointed chairman for a five-year term. William Denman, of California; John White, of Kansas City, Missouri; Theodore Brent, of Louisiana, and John A. Donald, of New York, were members.

Politics immediately entered the picture. Since California voters had put Wilson back in the White House, Treasury Secretary William McAdoo—Wilson's son-in-law—thought a Californian should chair the board. Wilson agreed and when Baker heard Denman was to be chairman he resigned and stated publicly the only way to make the Board work was to replace Denman with someone who knew about ships and could work with Gen. Goethals. Raymond B. Stevens, of New Hampshire, took Denman's place.

The *U-53* Brings War to American Waters

In the fall of 1916, "Kaiser Bill" thought it would be nice if one of his submarines paid a friendly call on an American naval base. Plans were then made for the *U-53* to visit the naval station at Newport, Rhode Island.

Around September 20, the *U-53* left Wilhelmshaven for Newfoundland. At about 2 p.m. on Saturday, October 7, Lt. George C. Fuller's training submarine *D-2* was cruising on the surface about three miles off Point Judith when a lookout spotted what appeared to be another submarine "standing toward Newport flying the German man-of-war ensign and the commission pennant and carrying two guns in a conspicuous position."

Fuller sent in a coded report and altered course to steam alongside the sub. When they reached the Brenton Reef Lightship, the German commander requested permission to enter the naval station at Newport, which was granted, and the *D-2* escorted the boat in. On arrival Lt. Fuller went over to the cruiser *Birmingham* to confer with Rear Adm. Albert Gleaves. The sub was directed to Berth No. 1 in the anchorage and an hour or so later a launch from the *Birmingham* was sent over and brought the sub's commander ashore to call on Rear Adm. Austin M. Knight, commandant of the Narragansett Bay Naval Station. He introduced himself as Kapitänleutnant Hans Rose, said he didn't require any assistance or supplies, and that he intended to leave by 6 p.m. that evening. His only purpose was to visit and invited anyone who wished to come aboard his boat.

After Rose left and went back aboard his sub, the commandant sent his aide, Lt. Cdr. William D. Puleston, to return the call, observing all the normal peacetime rules of naval etiquette. Puleston was joined by the commander of the destroyer force staff and a few other destroyer officers.

They learned the craft was the *U-53* and asked Rose not to use any of his radio equipment while in port. Under Rose were executive, navigating, ordnance, engineer, electrical, and radio officers and 33 sailors. Cdr. Puleston later stated, "The officers were in the regular uniform, new and natty in appearance. The crew wore heavy blue woolen knit sweaters, coats and trousers of soft, thin black leather lined with thin cloth, top boots and the regular blue flat cap. They were freshly shaved or with neatly close-trimmed beard or hair, and presenting a very neat appearance."

Rose and his crew were hospitable, friendly, and very forthcoming with technical information about the boat. Crew members showed their guests around every space, explaining anything they wished to know except their names, which they would not divulge.

It was learned the sub was 214' overall, had a cruising radius of 10,000 miles, and a speed of 18 knots on the surface. Propulsion was a six-cylinder, four-cycle, 1,200-hp Diesel engine. Armament was two 18" torpedo tubes forward and two aft, and ten torpedoes with a range of 2,000 yards. They also had 600 "bombs" to place aboard merchant craft. They had left

Wilhelmshaven with a six-week supply of provisions and had used two-thirds of that supply already.

The *U-53* was the first submersible warship to cross the Atlantic. The visit was considered by naval authorities to be a "path-finding" expedition and everyone knew Rose didn't come all that way just to host afternoon teas. Rose departed at 5:17 and submerged at 7 p.m. just outside the three-mile limit. All the available destroyers and Coast Guard cutters were already out on Neutrality Patrol and kept anxious watch waiting for the other shoe to drop.

It didn't take long. At 5:30 the next morning, Sunday, October 8, Rose stopped Capt. Smith's American-Hawaiian steamer *Kansan* en route from New York to Rotterdam via Boston with a cargo of steel consigned to the Italian government at Genoa. Rose allowed the ship to proceed and Smith arrived at Boston late that night amid frantic wireless traffic. Rose then really got to work.

Capt. Wilson's 375' British freighter *Strathdene*, under French charter, left New York for Bordeaux early Sunday morning, October 8, and was torpedoed at 6 a.m. off Nantucket Shoals Lightship. The first reports of Roses' activities were sent in by the lightship crew.

Capt. Harden's British freighter *West Point*, London to Newport News, was sunk by gunfire ten miles off Nantucket at 10:45 a.m.

Capt. Smith's Red Cross Line British passenger liner *Stephano*, en route from St. Johns, Newfoundland, to New York, with 75 crew, 44 first class and 39 second class passengers, and 3,000 tons of general cargo was sunk at 4:30 p.m. The crew of destroyer *Ericsson* witnessed the sub fire three shots at the ship. All aboard were rescued by destroyers *Balch* and transferred to the *Drayton*. Many Americans, most from New York, were on board returning from Halifax and St. Johns. The ship had recently been sold to Russia.

At 7:50 p.m. the British freighter *Kingston* was torpedoed southeast of Nantucket.

The Dutch freighter *Bloomersdijk*, New York to Rotterdam, with a crew of 50, and a cargo of grain valued at $500,000 consigned to the Netherlands government was torpedoed at 8 p.m. south of Nantucket. The ship was valued at $1.5 million.

The Norwegian freighter *Christian Knudson*, New York to London, was torpedoed at 8:30 p.m. south of Nantucket.

Destroyers *Fanning, Ericcson, Drayton*, and *Benham* left Newport to search for the survivors of the ships attacked off Nantucket Light.

At 2 a.m. on October 9, the *Ericcson* brought 81 survivors in to Newport, including ten children. The *Drayton* had 68, *Benham* 36, and the *Jenkins* 31. *Fanning* brought back six people later on the 9th.

These attacks led to widespread speculation that there might be a secret sub base somewhere around Long Island or Block Island.

Rose sank six ships within 100 miles of Boston without loss of life and then sank another two ships. In all, 16 destroyers rescued 216 passengers and crew. At that time there were three neutral and 27 British or French ships at sea within striking distance of the *U-53*.

On the night of Wednesday, October 11, 1916, all merchant craft were warned by wireless that a U-boat was off the coast. Capt. J.L. Jones, of the armed White Star freighter *Bovic*, was en route from Liverpool to New York when he got the message. He had his lifeboats swung out, the 4" gun on an upper deck loaded and ready, and his speed increased to "every pound of steam the firemen and engineers could giver her" and got up to 13½ knots.

At 8:15 on Friday morning, October 13, he had just passed the deeply-laden, unarmed

Scandinavian-American liner *Hellig Olav* when a lookout spotted a periscope about 2½ miles away. The liner was seen to increase speed and change course to the south, apparently having sighted the sub. The sub was then seen to change course to follow the liner, apparently finding a heavily-loaded, unarmed ship a better prospect than an armed ship. Jones sent a wireless message saying he had spotted a sub 28 miles southeast of the Nantucket Lightship, very near where the *U-53* sank the *West Point* on October 8 but none of the incoming ships received it. The *Bovic* arrived at New York on the morning of October 14 without incident.

When Halvor Jacobsen, Scandinavian-American's general manager in New York, heard of Capt. Jones' report, he wired the *Hellig Olav* and at 2:53 that afternoon received a reply saying they were 653 miles east of the Ambrose Channel and were on schedule. He came to believe the crew of the *Bovic* had been mistaken and might have taken a spar from a sunken ship sitting upright in the water for a periscope. The captains of several other ships that arrived without incident tended to agree. All believed the *U-53* was "well on her way back to Bremen." The consensus was that Rose might have hung around Cape Race lying in wait for a transport but he was nowhere near Nantucket. In any case, the sub's activities created widespread panic.

The French liner *Espagne* was due to sail for Bordeaux but 80 first class, 10 second class, and 5 steerage passengers cancelled their trip. Capt. Emile Laurent also believed the *U-53* was over 1,000 miles east of Sandy Hook. Besides, he could make 19 knots and had a 3.5" gun. The ship had arrived safely with 85 first class, 65 second class, and 60 third class (steerage) passengers.

On October 14 a communication was received from Berlin via Sayville, Long Island, asserting Germany's right to operate submarines off the American coast, "so long as American territorial rights are respected and pledges to the United States kept." It was denied a blockade was in place and that, "no sensible American citizen will believe that a secret depot for supplying German submarines could be established on the American coast." It was pointed out that British cruisers "approach the coast of New York so closely that they can be seen from the roofs of tall buildings with the naked eye."

Reports from survivors indicated two other subs, *U-48* and *U-61* were in the area but when Rose arrived back in Germany on November 1, he had *U-61* painted on his conning tower.

On February 3, 1917, Kapitänleutnant Otto von Schrader's *U-53* sank the *Housatonic* and at 4:21 p.m. on December 6, Capt. David Worth Bagley's destroyer *Jacob Jones* off the Isles of Scilly with the loss of two officers and 64 sailors. The ship sank in eight minutes and was the first American destroyer to be sunk by enemy action.

The *Jones* was at Newport, Rhode Island, when the *U-53* called at the Naval Station in 1916. Von Schrader came alongside and took two sailors prisoner: Albert De Mello, S2c, of 121 Hathaway St., New Bedford, Massachusetts, and John Francis Murphy, SC1c, of 53 Hall Ave., Newport, Rhode Island.

The first group of survivors was rescued at 8 p.m. that night by the steamer *Catalina* and another group was picked up at 8:50 a.m. the next morning when a British patrol boat came by. The executive officer on the *Jones* was Lt. Norman Scott. On a dark, sweltering night in November 1942 off Guadalcanal, Admiral Scott was lost aboard his flagship *San Francisco*.

Von Schrader was aboard the *U-53* at Newport in 1916 and reportedly recognized Bagley and sent in radio report of the location of the sinking to Queensland.

The *Lanao*

On October 28, 1916, the small Findlay Miller Steamship Co.'s unarmed Philippine-flagged steamer was en route from Saigon to Le Havre with rice under Capt. Henry Mainland and was sunk with bombs from *U-63* about 30 miles off Cape Vincent, Portugal. Although Philippine-registered, the ship flew the American flag as was customary being considered an American vessel.

A protest was filed by the State Department and on December 9, 1916, a telegram was sent to the American embassy at Berlin from Arthur Zimmermann at the German Foreign Office in Berlin:

> The undersigned has the honor to reply as follows to the note of Mr. Grew, Charge d'Affaires of the United States of America, dated the 21st ultimo: Foreign Office Number 14420. Relative to the sinking of the steamer *Lanao* by a German submarine.
>
> On October 28th last a German submarine encountered the steamer *Lanao* on the way to Le Havre and captured it as a prize because it was carrying contraband to the enemy. Since the submarine was unable to conduct the *Lanao* into a German or Allied port without exposing itself to the danger of sinking, it sank the steamer after it had taken the crew on board. The crew was then sent on board the Norwegian steamer *Tromp*, which was just at the time in sight. The steamer *Lanao* belonged originally to the Findlay Millar Steamship Company, of Manila; it was, however, sold in July, 1916, to the shipping firm of Hannevig Brothers, of London, and sailed since under the English flag. The owners of the firm of Hannevig Brothers are the sons of the shipowner Christian Hannevig of Borre, Norway, and are, like him, Norwegian subjects.

There were no casualties among the 30 crew and all landed at Barry, Wales.

The *Columbian*

Capt. Frederick Curtis's American-Hawaiian passenger liner *Columbian* with a crew of 109 was en route from Boston to Brest via England under foreign charter at 11 knots in rough seas. At about 2:30 p.m. on November 7, 1916, Capt. Curtis was eating dinner when he heard a shot that sounded "remote." After he arrived on the bridge, second shot went over the ship and he could just make out a submarine about four miles off. He ordered the American flag hoisted and the *U-49* came up. They were about 50 miles northwest of Cabo Ortegal, Spain. Visibility was about three miles with heavy wind and sea. The sub signaled "Send your papers on board." Capt. Curtis replied, "I cannot make out your flag." The sub then came closer and said to follow them but not approach or try to escape.

The crew put on lifebelts and the ship followed the sub through the night "guided by rockets" and the *Columbian* was ordered to fire rockets from time to time.

At about 6 a.m. the next morning six crewmen rowed a petty officer over to the sub, picked up a German officer and two sailors, and rowed back to the ship. The sailors placed two bombs aboard the ship while the crew was searched for weapons "electric pocket lamps." They were allowed to retrieve a small amount of personal items and ordered into lifeboats while foodstuffs were removed to the sub. The Norwegian steamer *Balto* was standing by and the crew went aboard. Capt. Curtis was taken aboard the sub. The bomb fuses were ignited and 12 minutes later they exploded sinking the ship. The *Balto* and the sub left and when they encountered the Swedish ship *Väringen* the crew was transferred off and the *Balto* sunk. Shortly afterward the Norwegian steamer *Fordalen* was sighted, the crew transferred and the *Väringen* was sunk. The *Fordalen* took the survivors to Coruña, Spain, greeted by a large crowd, and then went on to London.

Captain Curtis was held prisoner for six days on the sub along with Capt. Arthur Patterson of the British steamer *Seatonia*. The survivors reached Coruña, Spain, Capt. Curtis later stated:

> Our cabin was very small. It contained a little folding table, a folding chair, and three bunks. Everything was permeated with the odor of benzene. There was no communication with the exterior and the cabin was absolutely dark night and day.
>
> We were fed in the morning with a few morsels of black bread, a cup of cocoa, and a small portion of bad butter. At noon with a stew of canned meat and soup and at supper at 10 o'clock with coffee or tea and black bread with butter or marmalade. The hours spent in this narrow prison were long and disagreeable.

The American crew of the *Columbian* was augmented by 5 Mexicans, 5 Spaniards, 1 Chilean, 1 Puerto Rican, 5 Chinese, 3 Russians, 3 Swiss, and 1 Italian.

* * *

Villamer—the Tabacalera Steamship Co.'s Philippine-flagged freighter was sunk on November 7, 1916, west of the Madeira Islands with the loss of two.

The *Chemung*

On February 11, 1916, Charles W. Morse executed a contract to purchase the 325' laker *Chemung* from the Erie Railroad for $400,000 cash. On March 7, Morse assigned the contract to the Harby Steamship Co. Harby completed the purchase on April 22 and the 28-year-old ship was taken to New York and registered for trans–Atlantic service.

Capt. John J. Duffy and a crew of 35 left New York for Genoa and Naples with iron and steel products, electrical and other machinery, machine tools, blasting caps, chemicals, dried fish, wooden staves, 5,146 pieces of copper, and 200 bales of Harby cotton.

On November 28, 1916, the ship was sunk by the Austrian-manned *U-38* 14 miles off Cabo de Gata, Spain. The crew was taken aboard a Spanish merchant vessel and landed at Valencia. Capt. Duffy later stated:

> The *Chemung* went down with the American flag flying but I had no argument over the flag with the Austrian submarine commander. We were fired on after I had hoisted a signal that I was stopping. We had been on the lookout for submarines and had determined to stop as soon as we were signaled. Therefore, when the first gun was fired our signal that we were stopping went up. Nevertheless, a shell fired afterward struck our bows.
>
> When the ship was struck by the torpedo our boats were not out of range of the debris, which shot up into the air for a thousand feet and fell all around us. It was lucky no one was hurt.
>
> Third officer Jacobsen, who went to the submarine with the ship's papers, says the Austrians first intended to make me a prisoner. But he said to them, "Don't, the captain is a very good sort." I was not taken aboard the submarine, but Jacobsen said she was a very large type, being 170 feet long. He said she had a high bow like a yacht, not like the bow of an ordinary submarine.

The sub commander towed the lifeboats to within five miles of the coast of Valencia. Germany admitted liability and said the contraband cargo would be paid for pending "diplomatic negotiations." The company said the ship was worth $814,650 on the date it was sunk. Harby paid $131 a gross ton for the ship, which was "slightly above the market at the time of the contract of purchase" according to the post-war claims officer Edwin B. Parker. After the loss, Harby collected $584,650 in war risk insurance from American and French underwriters. That figure represented a market value of $191 per gross ton. Harby owned the ship for 8 months and 19 days. On May 14, 1926, Parker ruled that Germany owed Harby $1,000 under the Treaty of Berlin of August 25, 1921.

* * *

Kansan—In 1916, the France & Canada Steamship Co. chartered American-Hawaiian's big freighter *Kansan* for $47,000 a month to carry supplies to St. Nazaire. On December 18, Capt. Enoch A. Forsyth's ship hit a mine in the Bay of Biscay. Six were injured but Capt. Forsyth managed to beach the ship in shallow water. Salvage efforts were successful and they returned to New York in April 1917. When the ship got back, the charter rate went up $72,000 a month, one of the highest rates for the day.

Germany's New Submarine Policy

Germany's "Sussex Pledge" of May 4, 1916, curbed its submarine campaign but Berlin continued to watch Allied developments very closely. They knew their current submarine campaign had depleted worldwide shipping but that shipyards were blossoming like the desert after a rain and the American merchant marine had grown to 2,191,000 tons in 1916, double what it was in 1914, and American exports to Britain and France amounted to $3,012,000,000 of which $1,290,000,000 were munitions. British food supplies were getting short and prices for everything were rising.

On December 18, 1916, President Wilson transmitted a peace proposal to all belligerents. Germany was the only nation that responded but it was "brief and evasive." On December 30, 1916, the Allied powers, referring to Germany's various peace offers that were rejected, stated, "that peace could not be secured without reparation, restitution, and the assurance of the impossibility of future wars on the part of Germany." With the rejection of its offers of peace, including the "Sussex Pledge" of May 4, 1916. German planners kept Kaiser Wilhelm II informed and the Kaiser was finally persuaded to start using U-boats against all merchant shipping in all waters.

On December 22, 1916, Adm. Henning von Holtzendorff, head of the German Admiralty, sent out the Strictly Secret memo B-35840-L detailing a new policy for the U-boats and stated, in part:

> The beginning and the declaration of the unrestricted U-boat war must follow so quickly one upon the other that there is no time for negotiations, especially between England and the neutrals. The wholesome terror will exercise in this case upon enemy and neutral alike.

The memo went on to say the new program was to begin on February 1, 1917, with the intention of starving England out in five months and forcing an Allied surrender by August 1st. Adm. Holtzendorff was well aware that this would cause America to enter the war and stated:

> As regards tonnage this influence would be negligible. It is not to be expected that more than a small fraction of the tonnage of the Central Powers lying in America and many other neutral harbors could then

be enlisted for the traffic to England. For the far greatest part of this shipping can be damaged in such a way that it cannot sail in the decisive time of the first months. Preparations to this effect have been made. There would also be no crews to be found for them.

Just as little decisive effect can be ascribed to any considerable extent to American troops, which, in the first place, cannot be brought over through lack of tonnage.

7. 1917—Ties Are Broken

On January 6, 1917, the Bay Steamship Company of America's freighter *Sacramento*, the ex–*Alexandria*, was on a passage from Le Havre to Cardiff under Capt. Charles Edward Slater. At 11 a.m. they were off the coast of France 2½ miles north of the Casquets lighthouse when they were fired at by a sub. Capt. Slater ordered the flag hoisted and he started toward the sub to have his papers examined but the U-boat submerged and left. The ship arrived at Cardiff on the 8th.

On the late afternoon of January 31, as instructed, von Bersnstorff called on Secretary of State Bryan to explain Germany's position on submarine warfare subsequent to the failure of peace negotiations. One memo stated:

> After bluntly refusing Germany's peace offer, the Entente Powers stated in their note addressed to the American government, that they are determined to continue the war in order to deprive Germany of German provinces in the west and east, to destroy Austria-Hungary, and to annihilate Turkey. In waging war with such aims the Entente Allies are violating all the rules of international law, as they prevent the legitimate trade of neutrals with the Central Powers, and of the neutrals themselves.
>
> Germany has, so far, not made unrestricted use of the weapons which she possesses in her submarines. Since the Entente Powers, however, have made it impossible to come to an understanding based on equality of rights of all nations, as proposed by the Central Powers, and have instead declared only such a peace to be impossible which shall be dictated by the Entente Allies and shall result in the destruction and the humiliation of the Central Powers, Germany is unable further to forego the full use of her submarines.
>
> The Imperial Government therefore does not doubt that the Government of the United States will understand the situation thus forced upon Germany by the Entente Allies' brutal methods of war and by their determination to destroy the Central Powers, and that the Government of the United States will further realize that the now openly disclosed intention of the Entente Allies gives back to Germany the freedom of action which she reserved in her note addressed to the Government of the United States on May 4, 1916.
>
> Under these circumstances, Germany will meet the illegal measures of her enemies by forcibly preventing, after February 1, 1917, in a zone around Great Britain, France, Italy and in the Eastern Mediterranean, all navigation, that of neutrals included, from and to England and from and to France, etc. All ships met within the zone will be sunk.

On Friday, February 2, President Wilson met with his Cabinet. Wilson, according Navy Secretary Daniels, was usually "genial and smiling at the gatherings of his official family" but that day he was "grave and serious." All agreed that diplomatic relations with Germany must be terminated. Wilson agreed it was time to act, but exactly what to do caused bitter debate.

Essentially, there were three main choices: keep American ships out of Germany's way;

convoy ships under naval escort; or put guns on the ships. Opponents of arming or convoying, still hoping to avoid war, said that taking any military-like action would be seen by Germany as an act of aggression. Everyone knew that arming merchant ships for offensive action would immediately lead to war.

When President Wilson took the position that the United States had a right to arm merchant ships for defensive purposes, as England had done, the effort went in that direction. He believed he had the authority to authorize arming but he wanted the support of Congress. Secretary Daniels assured Wilson that, "We can arm them as fast as the ships are ready." The legal grounds for the idea went back to the old 1780 League of Russia, Prussia, Sweden, and Denmark's theory of "Armed Neutrality."

On February 3 the United States severed diplomatic relations with Germany. Ambassador von Bernstorff, in Washington since 1908, was given his passport by the State Department and the American ambassador in Berlin, James W. Gerard, was recalled.

American Prisoners

The severing of relations caused great anxiety among Americans taken prisoner by U-boat and raider commanders and incarcerated in Germany. Four American veterinarians were on ships captured and sunk in December 1916 by Kapitänleutnant Hans Berg's raider *Möwe*.

On the 2nd, Berg captured the British freighter *Voltaire*, en route to New York from France. On board was Dr. Herbert D. Snyder, 23, of 305 Westover Ave., Norfolk, Virginia, a member of the New York City health department.

On December 6, the Canadian horse transport *Mount Temple* was seized and three people on the freighter died in shelling from the raider. Dr. Richard Zabriskie, 30, of 8 Cottage Place, Englewood, New Jersey, was employed on board.

The White Star liner *Georgic* sailed from Philadelphia on August 12, 1916, and was seized on December 10. Dr. John E. Davis, 49, of 612 Main St., Columbus, Mississippi, and Dr. Orville Ernest McKim, 32, of Watertown, New York, were passengers.

The next day, December 11, Berg captured the British freighter *Yarrowdale* and a prize crew brought the ship to Pernambuco, Brazil.

On January 5, 1917, he captured the Japanese freighter *Hudson Maru* and put 287 survivors from 26 ships on board and on January 16, 1917, a prize crew brought the *Maru* to Pernambuco.

Several hundred survivors, including 72 Americans, were then put aboard the *Yarrowdale* and taken to Germany where they were imprisoned at Swinemuende, Karlsruhe, and Dulken. Professional men and officers were taken to Karlsruhe. Enlisted sailors were sent to Dulken, where conditions were much worse. The American veterinarians were sent to Karlsruhe.

When word was received that the German ships in American ports had not been seized and their crews jailed, prison officials were ordered to release the vets. On February 27 they were escorted to Warnemunde and put aboard a Dutch ferryboat in the custody of Dutch police. The police took them to a hotel in Gjedser where they enjoyed their first real meal and a hot shower bath since their capture. They arrived at Copenhagen by train on March 2.

Zabriskie, Snyder, and Davis left Bergen, Norway, on April 15, 1917, on the *Bergensfjord* and arrived at New York on April 30. In August, McKim was a 2nd Lt. in the Army Veterinary Corps and was stationed in the United States.

They reported that airplanes attacked the railroad station at Karisruhe four times and bombs "rained on all sides of their camp" near the station and they felt they were lucky to be alive.

According to Dr. Paul Ritter, the Swiss minister in Washington, D.C., who acted as the spokesman for the German government, the rest of the American prisoners were to be released on March 7 and 56 were ultimately released. Germany said "as paid members of the crews of armed merchantmen they were due no immunity."

The *Housatonic*

Ironically, an ex–German ship was the first American ship sunk under Germany's new policy on the very day relations were severed.

Capt. Thomas A. Ensor and his crew of 25 had left Galveston on January 6 for Liverpool with 144,200 bushels of winter wheat and flour provided by the Texas Transport & Terminal Co. On February 3, they were 20 miles southwest of Bishop Rock when a sub appeared. The sub's commander, Hans Rose, declared the wheat contraband and expressed regret at having to sink the ship. The crew got off in three boats and bombs were placed aboard and at 12:30 the ship blew up. Rose took the lifeboats in tow toward shore at 15 knots. About two hours later a boat was sighted on the horizon but its crew did not see the sub or the lifeboats. Rose said, "That fellow is asleep but I will wake him up for you" and he fired a shot. The boat then headed toward the sub at full speed "with thick black smoke belching out of her tall funnel." Rose then cut the boats loose and left. The boat was the British trawler *Salvator*. The crew was taken ashore where the American consul at Plymouth, John G. Stevens, took charge of them.

Capt. Thomas arrived in New York on February 21 on the liner *Orduna* from Liverpool with a souvenir he took from the sub—a lifebelt marked "*U-53*." The crew arrived at New York on the *Philadelphia* the next day.

The owners received a payment of $273,353.30 on their war risk insurance policy and filed a claim for another $839,000, the replacement value of the ship. They were awarded $4,500 in 1926.

At the time, Germany had about 100 subs that were sinking 15 British ships a day. During January 1916, the last month of "restricted" submarine warfare, U-boats sank 368,000 tons of shipping. In February it was 540,000, and in March 593,000.

The Atlantic Fleet Mobilizes

Vice Adm. Henry T. Mayo was commander-in-chief of the Atlantic Fleet. It was customary in those days for the Atlantic Fleet to repair south to Guantanamo Bay, Cuba, for training exercises after the holidays.

At 11:45 a.m. on February 3, the commandant of the naval station at Guantanamo Bay informed Mayo that "unofficial information has been received to the effect that diplomatic relations between the United States and Germany have been broken off."

In regard to Germany's new submarine policy, Mayo later stated: "In view of the previously received press reports of the activity of German submarines in the Gulf of Mexico and in South American waters, a consideration of the defense of the fleet in Guantanamo Bay against surprise attack by German submarines was immediately taken up."

The Navy Department had no idea what Germany would do next and it was fully expected that U-boats would be encountered off the Atlantic Coast in very short order.

At around 2:50 p.m. on February 4, Chief of Naval Operations Adm. William S. Benson, telegraphed Adm. Mayo: "Position of fleet well known to everybody. If considered advisable on account of submarines, shift base to Gulf of Guacanayabo or elsewhere at discretion. Inform Department confidentially".

At 3:59 p.m. Mayo sent his own message and both crossed in transit: "Unless instructions are received to the contrary, propose to shift fleet base to Gulf of Guacanabayo after spotting practice February 5th; then proceed with schedule of all gunnery exercises." On February 5, Mayo took the fleet from Guantanamo Bay to the Gulf of Guacanayabo but left the armored cruiser *Montana* behind to guard the naval station.

President Wilson Declares a National Emergency

On February 5 President Wilson declared a national emergency and issued Proclamation 1354 subsequent to Section 9 of the United States Shipping Act of September 7, 1916:

Emergency Water Transportation of the United States

Whereas, Congress did by "An Act to Establish a United States Shipping Board for the purpose of encouraging, developing, and creating a naval auxiliary and naval reserve and a merchant marine to meet the requirements of the commerce of the United States with its Territories and possessions and with foreign countries; to regulate carriers by water engaged in the foreign and interstate commerce of the United States; and for other purposes," approved September 7, 1916, provide that "during any national emergency the existence of which is declared by proclamation of the President, no vessel registered or enrolled and licensed under the laws of the United States shall, without the approval of the board, be sold, leased, or chartered to any person not a citizen of the United States, or transferred to a foreign registry or flag";

And Whereas, many shipowners of the United States are permitting their ships to pass to alien registers and to foreign trades in which we do not participate, and from which they cannot be brought back to serve the needs of our water-borne commerce without the permission of governments of foreign nations;

Now Therefore, I, Woodrow Wilson, President of the United States of America, acting under and by virtue of the authority conferred in me by said Act of Congress, do hereby declare and proclaim that I have found that there exists a national emergency arising from the insufficiency of maritime tonnage to carry the products of the farms, forests, mines and manufacturing industries of the United States, to their consumers abroad and within the United States, and I do hereby admonish all citizens of the United States and every person to abstain from every violation of the provisions of said Act of Congress, and I do hereby warn them that all violations of such provisions will be rigorously prosecuted, and I do hereby enjoin upon all officers of the United States, charged with the execution of the laws thereof, the utmost diligence in preventing violations of said Act, and this my proclamation issued thereunder, and in bringing to trial and punishment any offenders against the same.

In Witness Whereof, I have hereunto set my hand and caused the seal of the United States to be affixed.

Done at the City of Washington this 5th day of February in the year of our Lord one thousand nine hundred and seventeen and of the Independence of the United States of America the one hundred and forty-first.

On February 10 the Navy's General Board issued a plan for preparations for war. Item 2a was an "Armed guard plan" and Item 14 said "To indicate the steps tending to make life, ships, property, of vessels carrying armed guards, self-protecting." Item 15 said: "To outline the general plan whereby naval vessels may be requisitioned for guns to place on merchant shipping."

The General Board assigned the armed guard problem to Cdr. William Pitt Scott.

The *Lyman M. Law*

Capt. Stephen McDonough and a crew of 12 were en route to Italy on the Benedict-Mason Marine Co. schooner with a cargo of "thin wood" from Maine used to make crates for lemons. They were stopped on February 12 by *U-35* in the Mediterranean about 25 miles off Cagliari, Sardinia. The crew was taken off and the schooner's stores looted before the boat was sunk with gunfire. There were no casualties and the crew reached Caligari safely. Capt. McDonough later said he believed the sub commander just wanted the seven months' supply of stores and provisions.

The *Laconia*

On February 17, 1917, Capt. William R.D. Irvine's Cunard liner *Laconia* left New York for Liverpool with a crew of 216, including 14 Americans, 75 passengers including six Americans, and a cargo of 2,000 cases of brass cartridges, 1,000 boxes of bar silver, 42,000 bushels of wheat, 3,600 tons of structural steel, 200 tons of iron plate, 2,843 bales of cotton, 900 tons of general foodstuffs, 1,408 boxes of apples and oranges, 150 tons of machinery, and 1,300 sacks of mail transferred from the American ship *St. Louis*. The ship was armed with two 4.7" guns.

At 10:30 p.m., Sunday, February 25, the ship was torpedoed without warning by *U-50* about 150 miles from Ireland in a choppy, ice cold sea. Twenty minutes later a second torpedo hit the ship. Thirteen lifeboats got away. The sub surfaced and told the second officer help was on the way but 12 perished, six crew and six passengers.

At 4 a.m. the next morning patrol boats arrived and took the survivors to Queenstown. An official report from Wesley Frost, the American consul there, was received in Washington that afternoon.

Two Americans were lost, Mary E. Hoy and her daughter Elizabeth, of Chicago. The sinking illustrated the fate survivors often faced in open boats, as told by the Rev. Dunstan Sargent:

> Mrs. Hoy died in the arms of her daughter. Her body slipped off into the sea out of her daughter's weakened arms. The heart-broken daughter succumbed a few minutes afterward, and her body fell over the side of the boat as we were tossed by the icy waves.
>
> In icy water, up to her knees for two hours, the daughter all the time bravely supported her aged mother, uttering words of encouragement to her. From the start both were violently seasick, which, coupled with the cold and exposure, gradually wore down their courage. Both were brave women.
>
> The first to die in our boat was Irvine Robinson, of Toronto. After his body had been consigned to the sea we tossed about for an hour getting more and more water until the gunwales were almost level with the sea. Then Cedric P. Ivatt, of London, who was not physically strong, succumbed in the arms of his fiancée, who was close to him, trying in vain to keep him warm by throwing her wealth of hair about his neck.

Albert Harris Hoy was born in Ohio in 1846, became a physician, and married Mary Young, of Chicago. They had a daughter, Elizabeth, in 1878, and a son, Austin, in 1881. The family lived at 461 N. State St. in Chicago with two servants.

Austin became a representative for the Sullivan Machinery Co. and moved to London. Around 1913, the rest of family decided to join him there and Dr. Hoy set up a medical practice. The Hoys lived at 13 Sloane Ave. in London.

In late 1916, Mary and Elizabeth decided to return home to visit friends and spend Christmas in Chicago. They had twice cancelled plans to return to England due to fear of submarine attack and they spent two weeks in Washington, D.C., before finally leaving on the *Laconia*.

Floyd Gibbons, a correspondent for the Chicago Tribune, was on the ship and was initially reported missing. He later wrote a graphic account of the incident. He went ashore with an English friend of his who asked him if this would be the last straw for the Wilson Administration. It was not and as one writer later stated, "Another act in the comedy of delay and vacillation had yet to be played." Austin Hoy lost no time in sending a heated telegram to President Wilson.

Germany's new policy began shaking the pacifism out of Congress and the American people and strident voices were beginning to demand action.

President Wilson Takes Action

By February 19, the British had obtained copies of the coded Zimmermann cable through commercial avenues and they brought a translation to the American embassy in London where Edward Bell, the secretary saw it. The ambassador, Walter Hines Page, saw it the next day and on the 23rd met with British foreign minister Arthur Balfour to get a confirmation of its authenticity. Once convinced, Page reported to President Wilson.

On February 26, President Wilson took the extraordinary step of addressing Congress in person to ask for a policy of "armed neutrality."

The House promptly passed a bill to arm merchant ships but Sen. Robert M. La Follette stopped the effort. Before adjourning, a large group of senators signed a paper saying they would have voted for it given the opportunity. Wilson then asked Attorney General Thomas E. Gregory if he had the authority to authorize the arming of merchant ships and Gregory said he did.

Navy Secretary Daniels later said a group of Congressmen, "acting with reckless disregard of the country's needs, prevented important legislation."

The *Orleans*

On February 11 the Oriental Navigation Corp. armed freighter *Orleans* left New York under Capt. Allen Tucker arrived at Bordeaux, France, on the 27th. The *Orleans* was the first American ship to reach France after Germany declared the submarine zone. The ship was greeted by a cheering crowd of thousands waving handkerchiefs and throwing hats into the air. The crew was paraded through the streets where flowers were thrown from balconies and

Capt. Tucker was presented with a huge bouquet tied with red, white, and blue ribbons when they reached the prefecture. At a luncheon held in Capt. Tucker's honor, Mayor Gruetz stated:

> It is with a feeling of legitimate pride that I salute the sailors of the great American republic. Disregarding the insolent provocations of a nation of prey, they crossed the so-called blockade zone despite the perils with which they were menaced and thus affirmed the right of neutrals to unrestricted navigation. This successful voyage is something besides a defiance. It is much better still. It is an attraction that a great people, proud and free, are firmly resolved to defend the laws of humanity and to have respected the most sacred principles of civilization against the action of a nation which has disregarded them all.

The ship's arrival was delayed by fog, which also hindered the submarines. Germany said the ship's arrival was due to luck and not to any lack of effort on their part. Foreign Secretary Zimmermann stated, "We make absolutely no distinction. Our determination is unshakable since that is the only way to finish the war this summer, a desire in which we all join."

The Zimmermann Cable

In hopes of creating a distraction that would divert, or at least slow down, war materials going to the Allies, someone in Berlin got the bright idea of having Mexico declare war on the United States on Germany's side. When the British cut Germany's trans-Atlantic cables early in the war, Arthur Zimmermann, Germany's foreign secretary in Berlin, prevailed upon the American ambassador, James Gerard, to use his system. Gerard consented on condition only uncoded messages were sent, but on January 16, 1917, Zimmermann sent a coded telegram to von Bernstorff in Washington. The message stated:

> We intend to begin on the first of February unrestricted submarine warfare. We shall endeavor in spite of this to keep the United States of America neutral. In the event of this not succeeding, we make Mexico a proposal of alliance on the following basis: make war together, make peace together, generous financial support and an understanding on our part that Mexico is to reconquer the lost territory in Texas, New Mexico, and Arizona. The settlement in detail is left you. You will inform the President of the above most secretly as soon as the outbreak of war with the United States of America is certain and add the suggestion that he should, on his own initiative, invite Japan to immediate adherence and at the same time mediate between Japan and ourselves. Please call the President's attention to the fact that the ruthless employment of our submarines now offers the prospect of compelling England in a few months to make peace.
>
> Zimmermnn

The British Naval Intelligence Service had been monitoring all communications they could and had worked on deciphering codes. Room 40 intercepted the cable and quickly deciphered part of it. The contents stunned the British but they were then over the proverbial barrel since they had obtained the message by eavesdropping on American diplomatic correspondence.

Everyone knew Great Britain desperately wanted the United States to enter the war on their side and they knew the message would inflame passions in America and probably result in a declaration of war. They surmised the pacifist element would immediately declare the message fraudulent and they would be accused of ulterior motives. That, in turn, would require them to prove in some way they had intercepted the cable from a legitimate American source. And lastly, to make matters worse, Germany would learn their codes had been broken. The only thing to do was to keep it quiet and try to find the message somewhere else in the transmitting system so they could release it without revealing its true source, which they eventually did.

On January 19 von Bernstorff relayed the message to Heinrich von Eckhardt, the German ambassador in Mexico City. Mexico dismissed the idea of starting a war they couldn't win and they didn't entirely trust the Germans either.

At that time, Maj. Gen. John J. "Black Jack" Pershing was in Mexico trying to capture Pancho Villa and things were not going well. On January 28 President Wilson ordered Pershing to bring his troops home. Pershing would soon have other fish to fry.

President Wilson released the Zimmermann cable to the press on February 28 and on March 1 it was made public.

The Navy Takes Civilian Vessels

Destroyers were recognized as the major anti-submarine weapon but on January 1, 1914, the U.S. Navy had only 62 greyhounds. Of those, 29 were on active duty with the Atlantic Fleet and two were in reserve at Newport and Annapolis.

On June 30, 1914, just prior to the outbreak of war in August, Congress authorized the construction of six torpedo-boat destroyers. As the Germans waged their sea war against Great Britain, the extent and effectiveness of Germany's U-boat campaign startled naval planners everywhere and on March 3, 1915, Congress authorized six more destroyers.

On March 10, 1916, as things really began to heat up, Congress authorized the President "to undertake the construction of or to acquire and convert, not to exceed 500,000 tons of modern naval vessels," of which 315,000 tons could be combatant ships. Fifteen destroyers were built under this act, but new, improved destroyers could not be built overnight and when America finally entered the war on April 6, 1917, the Navy had only 67 destroyers in commission. The last pre-war destroyer completed was the *Wilkes*, commissioned on November 10, 1916.

The Act of August 29, 1916, provided for 50 more destroyers but only 14 of these were delivered during the war. The last was the *Robinson*, commissioned on October 10, 1918.

Recognizing the urgency of the situation, the Act authorized the Secretary of the Navy to acquire privately owned vessels suitable for coastal patrol and ocean escort work to supplement the destroyer force. They could be acquired by outright purchase, bareboat, or time charter to suit the Navy's needs.

The Naval Appropriation Act of March 4, 1917, was "construed as giving the department the right to commandeer vessels needed for naval use in war" according to Secretary Daniels. After the United States entered the war, there was an immediate need for auxiliary vessels of all sorts and the Navy Department set its sights on the larger fishing boats and the big steam yachts owned by wealthy individuals for ocean escort work. This Act would prove to be inadequate.

The Navy Armed Guard

On March 1 Cdr. Pitt Scott submitted an Armed Guard Plan which stated, "It has been decided to arm merchant vessels with naval guns for defense. Our decision is made. The problem that now confronts us is to how we are to accomplish this: in other words, to prepare a plan." Pitt's document laid out the "why" for the decision. The problem of "how" was turned over to Capt. Frank H. Schofield in the plans section of the Navy Department and by March 8, 1917, he and Capt. William V. Pratt, assistant to Adm. Benson, the Chief of Naval Operation, had the basic scheme and policies worked out. It was anticipated that arming merchant vessels would reduce losses by at least 80 percent.

Their research revealed the total number of guns available for merchant vessels to be 32 6-inch; 84 5-inch; 176 4-inch; 88 3-inch; 136 6-pounders; and 188 3-pounders. It was decided to distribute and mount guns "so as to give the greatest degree of protection to each vessel, no matter what direction from which it is attacked." It was also noted that, "The actual assignment of guns to vessels can not be made until a list of vessels to engage in trade in or near danger zones is obtained, and until the desires of owners are ascertained."

The idea was to furnish each ship with two guns and a full Navy gun crew of around 20 sailors and also include radio operators and signalmen. The commander would be "a chief petty officer of great experience" or a regular commissioned officer.

On March 9 the Armed Guard was officially authorized by President Wilson and on the 11th, Admiral Mayo received an order on board his flagship *Pennsylvania* from the Bureau of Navigation:

> Detail immediately 30 trained torpedo defense gun crews consisting of two gun pointers, sightsetters, plugmen, gun captain, each crew. Above crews to be divided into groups, each group to consist of 39-men crews. Detail immediately 10 lieutenants (senior grade) competent to command each group three guns. Detail in addition one gunner's mate, one petty officer for each group of three guns. Above personnel to be placed aboard United States merchant vessels as armed guard for defense against German submarines. The above personnel to be detailed from 13 vessels battleship force now with flag. Send above personnel to receiving ship *New York* via destroyer to Key West and rail to New York. Radio probable date arrival party at New York. Be prepared for additional details. Acknowledge.

This message came from the Bureau of Navigation, which had clear authority regarding personnel matters, but there was no endorsement from Adm. Benson, the Chief of Naval Operations. Adm. Mayo assumed Benson had knowledge and concurred but he was somewhat alarmed by the loss of so many essential, trained ratings.

On March 13, 1917, Secretary of the Navy Josephus Daniels published his instructions:

> REGULATIONS GOVERNING THE CONDUCT OF AMERICAN MERCHANT VESSELS ON WHICH ARMED GUARDS HAVE BEEN PLACED.
>
> 1. Armed guards on American merchant vessels are for the sole purpose of defense against the unlawful acts of the submarines of Germany or of any nation following the policy announced by Germany in her note of January 31, 1917. Neither the armed guards nor their arms can be used for any other purpose.
>
> 2. The announced policy of Germany in her note of January 31, 1917, to sink all vessels that enter certain areas of the high seas, has led the Government of the United States to authorize armed guards on merchant vessels to resist any and all attempts of the submarine of Germany or of any nation following the policy announced by Germany in her note of January 31, 1917, to put that policy into practice.
>
> 3. It shall be lawful for the armed guard on any American merchant vessel to fire upon any submarine of Germany or of any nation following the policy announced by Germany in her note of January 31, 1917, that attempts to approach or lies within 4,000 yards of the commercial route of the vessel sighting the submarine, if the submarine is sighted within the zone proscribed by Germany.
>
> 4. No armed guard on any American merchant vessel shall fire at any submarine that lies more than 4,000 yards from the commercial route of the vessel sighting the submarine, except that the submarine shall have fired first.

5. No armed guard on any American merchant vessel shall take any offensive action against any submarine of Germany or of any nation following the policy announced by Germany in her note of January 31, 1917, on the high seas outside of the zones proscribed by Germany, unless the submarine is guilty of an unlawful act that jeopardizes the vessel, her passengers, or crew, or unless the submarine is submerged.

6. No armed guard on an American merchant vessel shall attack a submarine that is retiring or attempting to retire either within or without the zone proscribed by Germany, unless it may be reasonably presumed to be maneuvering for renewal of attack.

7. In all cases not herein specifically excepted the armed guard on American merchant vessels shall be governed by the principles of established international law and the treaties and conventions to which the Government of the United States is a party.

8. American merchant vessels are forbidden to pursue or search out the submarine of any nation or to engage in any aggressive warfare against them.

9. American merchant vessels shall make every effort compatible with the safety of the merchant vessel to save the lives of the crew of any submarine that may be sunk, or that submits, or is in distress.

10. (This paragraph stricken out.)

11. (This paragraph stricken out.)

12. American merchant vessels should communicate with the commandant of the naval district before leaving a United States port to make sure of the latest information.

13. The safety of American merchant vessels requires that they obey all instructions of the vessels of war of the United States.

ON SIGHTING A SUBMARINE IN THE PROSCRIBED ZONES.

14. If a submarine is sighted beyond torpedo range, bring submarine abaft the beam and keep her there. If submarine attempts to close, bring her astern and proceed at highest possible speed.

15. If submarine is sighted close aboard forward of the beam, the greatest safety lies in changing course directly toward the submarine.

16. If submarine is sighted close aboard abaft the beam, the greatest safety lies in turning away from the submarine and proceeding at highest speed.

ON OPENING FIRE IN DEFENSE AGAINST
THE UNLAWFUL ACTS OF SUBMARINES.

17. Hoist national colors before first shot is fired.

18. Once it has been decided to open fire, do not submit to the gunfire of a submarine so long as the armed guard can continue to fire.

19. Send all persons except the bridge force and the armed guard below decks while vessel is under fire.

20. Watch out for torpedoes and maneuver to avoid them. If unable to avoid them, maneuver so that they will strike a glancing blow.

21. The armed guard is commanded by the Senior Naval Officer on board. He shall have exclusive control over the military functions of the armed guard and shall be responsible for the execution of all regulations given herein governing the employment of the armed guard.

22. The military discipline of the armed guard shall be administered by the naval officer commanding the armed guard.

23. The armed guard shall be subject to the orders of the master of the merchant vessel as to matters of nonmilitary character, but the members of the armed guard shall not be required to perform any ship duties except their military duty, and these shall be performed invariably under the direction of the officer commanding the armed guard.

24. The decision as to opening fire or ceasing fire upon any submarine shall reside exclusively with the naval officer commanding the armed guard.

25. The enlisted personnel of the armed guard shall be quartered and messed together on both in port and at sea, at the expense of the owners of the vessel on which the armed guard is serving, in a manner satisfactory to the naval officer commanding the armed guard.

26. The naval officer commanding the armed guard shall take precedence next after the master, except that he shall not be eligible for succession to the command of the ship. He shall be quartered and messed

on board both at sea and in port, at the expense of the owners of the vessel on which he is serving, in a manner appropriate to his precedence next in line after the master.

27. The master of the merchant vessel shall, on request of the commander of the armed guard, detail members of the crew to handle ammunition, clear decks, and otherwise supplement the service of the gun.

28. The naval officer commanding the armed guard shall be responsible for—

(*a*) The condition of the battery and its appurtenances.

(*b*) The training of the guns' crews and spotters, including members of the ship's force detailed by the master to assist in the service of the guns.

(*c*) The readiness of the ship's battery at night.

(*d*) The readiness of the armed guard to perform its duties at all times.

(*e*) The continuous lookout near each gun by a member of the armed guard.

(*f*) The making of all reports required by the Navy Department.

JOSEPHUS DANIELS.
Secretary of the Navy.

MARCH 13, 1917.

On March 14, 30 more gun crews were ordered north.

According to the agreement signed between the steamship owner and the government, the owner agreed "to quarter and subsist the armed guard both at sea and in port in a manner satisfactory to the commander of the armed guard without expense to the government or to the members of the armed guard." For the most part relations between the merchant crew and the armed guard were cordial. Occasionally a shipmaster would object to practice firing on the grounds that "it would shake loose the rivets on his decrepit ship." In one case, a wad of hair was found in a batch of peach stew and the master accused the gunners. In another case when a submarine was detected, the black gang had to be "driven below at the point of a pistol" but these instances were rare and thus were noted for posterity.

It was generally accepted that when a ship was armed and a Navy crew brought aboard, as was stated, "the ship's company comes under naval discipline," a state of affairs which has become controversial ever since as regards the status of merchant seaman during both world wars.

Item 18 required gun crews not to surrender while they still had ammunition.

Item 23 implies that the Armed Guard commander took physical control of the ship during "military" matters.

Item 24 gives the Armed Guard commander the authority to determine when a "military" situation exists.

Item 27 required the master to order civilian crewmembers to assist the gun crews and to train with them, which they did in both world wars. When, as stated, a merchant ship assumes a warlike character upon firing a gun at the enemy, the merchant ship becomes a warship. It then follows that any civilian who is ordered by the Secretary of the Navy to take part in combative action is under military orders and therefore became a member of the armed forces for the duration of the action. If Item 27 had the force of law, a shipmaster or crewmember who refused could be prosecuted for dereliction of duty under military law and if a civilian crewmember suffered any injury he would be eligible for medical and other care through the Veterans Administration.

The only way to resolve these administrative conflicts would have been in a court of law or for the Navy to purchase or bareboat charter the ships and put all–Navy crews aboard and this was done when the Naval Overseas Transportation Service was established on January 9, 1918.

Subs usually attacked on the surface at night and could be repulsed by heavy rifle and machinegun fire. Masters were instructed to report any attack by radio in plain language but to "Make sure of the loyalty of your radio operator."

It was decided that troopships, carrying the most valuable cargo, should have first priority in arming. The International Mercantile Marine Co. consented and over two days guns and Armed Guard crews were put aboard their passenger liners *Manchuria, St. Louis,* and *Aztec.* Four days later the *St. Paul* and *New York* were armed. *Philadelphia, Kroonland,* and *Finland* were next.

The *Manchuria* was the first ship to be armed and sailed from New York on March 16, 1917. Sailing vessels were armed as well. A typical schooner would have two 6-pounders and an Armed Guard under a warrant officer and six sailors.

Chief Boatswain's Mate John I. Eopolucci, of Washington, D.C., was the first American serviceman lost in the war and the Armed Guard's first casualty. A total of 384 ships had Armed Guard sailors aboard and had 227 reported encounters with U-boats. Two armed ships were sunk by gunfire from enemy submarines, one was sunk by a bomb placed on board, 33 were sunk by torpedo, and 113 attacking subs were driven off by gunfire. A total of 347 attacks by subs were reported by wars' end. Armed ships made 1,832 trans–Atlantic trips.

The *Algonquin*

In December 1916, the American Star Line purchased the Canadian laker *Algonquin* from Robert O. MacKay and Adam Brown MacKay, in business as the Hamilton & Montreal Navigation Co. at Hamilton, Ontario, for trans–Atlantic service.

The unarmed freighter left New York on February 20, 1917, under Capt. Aager Nordberg for London with crew of 27 (ten American citizens) and $1.7 million worth of foodstuffs consigned to Swift & Co. The ship was one of the first American vessels to sail for the United Kingdom after Germany's announced blockade of the British Isles. Everything was fine until March 12 when the ship was about 65 miles off Bishop Rock, Isles of Scilly. According to a crewman:

> Suddenly at 6 o'clock on Sunday morning a shell screamed and passed across our bow. This was the first warning of the presence of the submarine, which kept on shelling, the shots falling all around. It was a hellish thing to do. We lowered two boats and we believe the submarine deliberately tried to smash them. A hole was made in one and the shelling only stopped when we were well away from the side of the vessel. The submarine six times circled the *Algonquin* with only the periscope showing, apparently fearing danger but finally several of the crew boarded the ship and placed bombs on it causing it to sink in ten minutes. Then the submarine came up to our boats, questioned the crew, and laughed in our faces as the vessel sank. They refused to tow our boats, saying they expected to sink other ships. We pulled throughout Monday and Monday night and made land on Tuesday after 27 hours of incessant toil. The *Algonquin* flew the American flag and the American colors were painted in large stripes on the side. We counted 22 men on the deck of the submarine, which carried two guns, and was named *U-39*.

Chief mate Frank Kerney and chief engineer James Finnerty identified the sub as the *U-38*. Capt. Norberg said when he asked the sub commander for a tow toward land he was told, "I am too busy." There were no casualties but "the crew suffered severely from exposure to cold in lifeboats during the night." The ship was valued at $400,000.

John D. Stephanidis, an attorney and apparently part-owner of the ship, stated:

> If this does not constitute an overt act I do not know what does. I expect to go to Washington tomorrow to take up the matter with President Wilson and Secretary Lansing. I am sorry if the destruction of my ship must be a *casus belli* but I believe that this occurrence must be the subject of action of some kind against the German government.
>
> The *Algonquin* carried no munitions but was laden with foodstuffs. It flew the American flag and the flag was also painted on the side.

John Demetrin Stephanidis, 38, was born in Constantinople, Turkey, on March 26, 1876, and came to America in November 1902. He spoke Greek and worked as an official interpreter. He was a lawyer in 1917 when he was naturalized. His 18-year-old wife Flora was born in Greece, came to America in 1915, and was naturalized in 1920. They lived at 740 West End Ave. in Manhattan with Mary Raplewitz, their 18-year-old Polish servant.

On March 15 a report was received from Copenhagen that merchant seaman James Moore, a crewman on the unarmed Norwegian steamer *Storskog* was being detained in Germany along with August Henning, of Chicago, a bookkeeper with the National Radiator Co. after the ship was sunk by a U-boat.

The *Vigilancia*

Capt. Frank A. Middleton's Gaston, Williams & Wigmore's freighter *Vigilancia* left New York on February 28 for Le Havre with general merchandise and 43 on board. The clearly marked ship was attacked on March 16 without warning at 10 a.m. by *U-70* about 145 miles west of Bishop Rock, Isles of Scilly, and sank in seven minutes. The weather was clear with a moderate swell. Fifteen were reported lost. The survivors made landfall after 54 hours in the boats "and suffered greatly from injuries and exposure." Capt. Middleton stated:

> After rescuing as many of the crew as possible in the boats we had biscuits and water. At night I fired distress signals. Several times, by the glare of the lights, I saw a submarine following fifty yards from the boats between 10 o'clock Friday night and 2:50 o'clock Saturday morning, but it made no attempt to help us. We suffered great hardships in the boats. One man of the engine room staff is paralyzed as the result of exposure.

Those lost were drowned while launching boats: Neils P. North, 3rd mate, naturalized American, father in Denmark; Joseph Sibera, quartermaster, American; T. Roudon, boatswain, Venezuelan; F. Brown, carpenter, American, Virginia; C.F. Aderhold, 3rd engineer, American, father D.E., Attalla, Alabama; Alexander Rodriguez, oiler, Puerto Rican; E. Dimitrios, fireman, Greek; A. Galitos, fireman, Greek; Joe Livios, watertender, Spanish; G.S. Sparro, watertender, Spanish; R. Gonzales, coal passer, Spanish; Juan Nesz, coal passer, Spanish; M. Vasquez, coal passer, Spanish; Esteban Lopez, messboy, naturalized American. Porto Rico; Julio P. Montario, messboy, Peruvian.

The *City of Memphis*

On March 16, the Ocean Steamship Co.'s unarmed passenger ship *City of Memphis* left Cardiff, Wales, for New York under Capt. Louis P. Borum, 34, and a crew of 58, of which 29 were Americans.

At 3:55 p.m. on the 17th they were about 35 miles south of Fastnet in heavy seas when they encountered *UC-66*. The sub commander fired a shot from about 3 miles away and signaled the ship to stop. This signal was not seen and when the sub was about a mile away another shot was fired and shell fragments hit the ship. Capt. Borum sounded four short whistle blasts for abandon ship and stopped. The wireless operator, Thomas J. Welch, did not send any message.

The sub commander gave everyone 15 minutes to get off and when the entire crew was in five boats by 4:15 he fired eight shots into the ship. There was huge explosion, which 3rd engineer William Thompson thought might have been a torpedo, and in 20 minutes the ship sank stern first. The sub commander refused to tow the boats toward shore and they became separated in a southwest swell and moderate breeze. At 3:45 the next morning, Chief Mate Charles Gilbert Laird's boat was found by the crew of a British patrol vessel and were brought aboard. Chief Engineer Walter I. Piercy's boat was found at 6:30 a.m. by the same vessel and in all 34 survivors were landed at Queenstown at 4:30 p.m.

Initial reports said 15 survivors landed at Schull at 7 a.m. on the 18th, 34 were on a British naval vessel, and Capt. Borum and 8 sailors were still missing. They were believed to have been taken aboard the sub as prisoners. On the 20th, the American Consul at Queenstown, Wesley Frost, reported that Capt. Borum and 8 survivors had landed safely at Glasgow.

All were eventually accounted for and the survivors left Liverpool on March 31, 1917, on the *St. Louis* and arrived at New York on April 9, 1917.

The incident caused a furor and when questioned, Secretary of State Lansing replied it was "Too serious to discuss." Everyone fully expected President Wilson to call on Congress.

The *Illinois*

Capt. Hjelmar Iversen's Texas Steamship Co.'s tanker was en route from London to Port Arthur, Texas, in ballast. At 8 a.m. on March 17 the ship was shelled without warning in the English Channel about 20 miles north of Alderney. Firing continued while the crew was getting into the boats and all aid and towing toward shore was refused. The ship was sunk with bombs. The crew of 34, with 16 Americans, landed at Guernsey at 2:40 a.m. on the 18th and headed for Southampton, where they arrived on the 23rd. One sailor was slightly injured. The survivors left Liverpool on March 31, 1917, on the *St. Louis* and arrived at New York on April 9, 1917.

The Day of Decision

Secretary of the Navy Daniels said March 20 was the "Day of Decision" for President Wilson's cabinet. The unprovoked sinking of American ships had become too much and it was decided to ask Congress for a declaration of war. In response to the still strong pacifist block, Wilson stated, "We are not governed by public opinion. I want to do right whether it is popular or not." The next morning Wilson called on Congress to convene on April 2 instead of the 16th "to receive a communication by the Executive on grave questions of national policy which should be taken under consideration."

The *Healdton*

The unarmed Standard Oil of New Jersey tanker left Philadelphia for Rotterdam on January 26, 1917, with 42,754 barrels of refined oil worth $106,886. Capt. Charles Christopher arrived at Halifax on February 9, departed on March 2, and arrived at Bergen, Norway, on the 17th. On March 20 they departed Bergen for Rotterdam. Vice-Consul Gerhard H. Krogh at Rotterdam cabled the State Department:

> Voyage uneventful until 8:15 p.m., evening March 21, when ship lay 23 miles north by east of Terschelling Lightship; torpedoed twice without warning. First torpedo hit amidships; all lights went out. Second torpedo hit further aft, under flag painted on port side.
>
> Bunkers were ablaze and ship began to settle. After first torpedo shock Captain ordered engine stopped and crew took to boats. Vessel began list aport. Captain made farewell search for missing crew members. One boat tried to pull away from sinking vessel and capsized. Captain saw submarine approaching when he was lying off *Healdton* watching vessel sink. Nationally submarine impossible to determine because it possessed no distinguishing marks and was seen from a distance of 100 feet. No words were exchanged with submarine. From action, Captain and crew believed submarine undoubtedly German. No other vessels were present or within sight when *Healdton* torpedoed.
>
> Submarine made no effort to assist crew or officers to save their lives but disappeared immediately after *Healdton* sank. Ship's papers were not demanded. Captain and 12 men rowed and sailed in open boat for Terschelling Lightship. At 8 o'clock, morning March 22, boat was picked up by Dutch trawler *Java 164* about 10 miles from Dutch coast. Another boat containing first mate and 7 men picked up by Dutch torpedo boat *G-13* near Terschelling after 17 hours' exposure. Rescued at 2 o'clock, afternoon March 22.
>
> No details at hand concerning those drowned in capsized boat or on ship except one Norwegian with arms and legs frozen, brought Ymuiden by steam trawler *Ocean*. Perils and hardships suffered by all survivors extraordinary. Some possess no clothes, others had no shoes. Two naked stokers, Jose Jacinto and Jose Gonzales, died from exposure, the former in mate's boat, latter in hospital today.
>
> Stokers had clothes burned off by fire in boiler rooms. All members of the crew partially clothed. One man in Captain's boat, G. W. Embry, home New Orleans, first assistant engineer, would have died of exposure and fatigue without clothes if trawler had picked up small boat one hour later. All suffered from fatigue.
>
> Second assistant engineer only wore underclothes and barefooted. Every member of crew states decisively that *Healdton* torpedoed without warning. Ship made no effort to resist. Captain ordered engines stopped after first torpedo. No flight attempted because ship crippled. Vessel going ordinary speed on peaceful voyage unarmed when torpedoed.
>
> *Healdton* was torpedoed when all members crew, 41 in all, were on vessel. Both torpedoes struck before anyone could leave ship. Casualties were as follows: 19 men either died in capsized boat by ship or from suffocation in bunkers. Captain believes two members crew were never able to reach lifeboats because on watch below or in engine room. Two other men, as previously mentioned, died from exposure and burns. Engineer Embry badly frozen. Other survivors with minor afflictions as bruised and frozen ears or fingers.
>
> When vessel torpedoed there was moderate northwest breeze on choppy sea. Heavy snow squalls were encountered by survivors in open boats and weather during whole night was below freezing. All survivors have now reached Rotterdam. Of crew, number 41, 13 were Americans and 28 other nationalities; 13 members od crew Spanish, Portuguese, Swedish, and Norwegian were drowned.

The consul went on to report that those saved were Capt. Christopher, 147 Nicholas Ave., Brooklyn; 1st mate Otto Willerup, 2274 7th Ave., New York; chief engineer John Caldwell, 160 Manhattan Ave., New York; 1st engineer G. Swanson, Brooklyn; and wireless operator Herman Parker, 2246 N. Van Peldt St., Philadelphia. The casualties were:

W.C. Chandler, 2nd mate, American, father C.S. Chandler, 510 W. 124th St., New York. Drowned or suffocated.

C.F. Hudgins, 3rd mate, American, father Charles, Laban, Virginia.

Southward Whittington Smith, able seaman, American, 6347 Ellis Ave., Chicago. Drowned.

Abrhamson Fridjof, seaman, Norwegian. Wounded and frozen.

J. Mathewson/Matheson, seaman, Norwegian, father Otto, Bergen, Norway. Drowned.

George W. Embry, 2nd engineer, American. 2034 Felicity St., New Orleans. Badly frozen in lifeboat but survived.

Walter C. Johnson, 3rd engineer, American, wife, 1038 Park Ave., Hoboken, New Jersey. Drowned or suffocated.

Francisco Martinez, pumpman, Spanish, wife Pela, Mura, Spain. Drowned.

Emery le Veaux, oiler, American, father John, RFD 2, Indington, Michigan. Drowned.

Rebovido Beningo (aka Benny Rebordo), fireman, Spanish, Coruña, Spain. Drowned.

Antonio Gonzales, fireman, Spanish. Drowned.

Juan Gonzales, fireman, Spanish, Coruña, Spain. Died from exposure.

Eugenio Lamar, fireman, Spanish. Drowned.

Jacinto Cocho, coal passer, Cuban, wife Antova, Malaga, Spain. Died of burns.

Antonio Cordoza, coal passer, Portuguese. Drowned.

Pablo Herera (aka Piable Tnewer), coal passer, Spanish, father Mari, Santander, Spain. Drowned.

Jose Soto (aka Josi Sato), coal passer, Spanish. Malaga, Spain. Died of burns.

Ludwig van der Dose, chief steward, Dutch, wife Aggie, Rotterdam, Holland. Drowned.

George Bealey, 2nd steward, American, father George, 1221 Ellis St., San Francisco. Drowned.

J.P. Lenevelt, 2nd cook, Dutch, mother Florence Johnson, Rotterdam, Holland. Drowned.

John W. Steiner, messman, American, mother Louisa, 520 Avery St., Pittsburgh. Drowned.

Christopher, Caldwell, Willerup, Embry, and Parker arrived at New York from London on April 13, 1917, on the liner *Manchuria*.

Germany said the ship was being escorted by British torpedo boats, which the Admiralty emphatically denied. This sinking sparked outrage throughout the land. Sen. Elihu Root addressed an angry crowd of 15,000 at Madison Square Garden on March 22 where a demand was made for a declaration of war and a force sent immediately to France under command of Col. Roosevelt, who was apoplectic, and announced he was ready to go.

The *Aztec*

The Oriental Navigation Co. armed freighter under Capt. Walter O'Brien and a crew of 47 (16 native-born Americans) left New York on March 18, 1917, for Le Havre, France, with foodstuffs and general merchandise worth $500,000. The 11-member Armed Guard crew under Lt. Fuller Gresham had two 5" guns. The *Aztec* was the first ship with an Armed Guard crew to sail from an American port.

At about 9:30 p.m. on April 1 they were off Ushant Island on the northwest coast of France in heavy seas and blacked out. Chief engineer Charles Erickson was out on deck and noticed a light showing through a port in the wireless room and went to tell the operator, Watson Sluney. As they were talking outside, a huge explosion occurred right below them. The chief was thrown into the sea and was never seen again. Sluney blacked out and was thrown about 30' but stayed on the ship. When he came to he went to send an S.O.S. but his room and set were wrecked.

While the black gang was struggling to get valves lined up and the pumps going, Capt. O'Brien ordered the engine secured and everyone off.

The first lifeboat smashed against the ship's side and the 11 occupants were lost. The second boat got away and the third boat with Capt. O'Brien and five of the Armed Guard crew also got away. A lifeboat with 19 survivors was found and brought aboard the French patrol boat *Jeanne d'Arc* and landed at Brest. Capt. O'Brien's boat was spotted three hours later and put aboard the French patrol boat *Sirius*. The third boat with 2nd mate Larkin and 18 sailors was not accounted for.

Chief Boatswain's Mate John Isadore Eopolucci was lost and became the first serviceman lost in the war and the Armed Guard's first fatality.

John enlisted in the Navy on May 7, 1915, at Rosebank, New York. He was survived by his mother, Annie M. Eopolucci, of 649 L St., Washington, D.C. He was buried in the Brookwood American Military Cemetery at Brookwood, Surrey, England.

Herbert Collins, third engineer, of 510 E. 5th St., Wilmington, Delaware, survived. Those reported lost, presumably by drowning, and their next of kin, if known, were:

Glendall C. Larkin, 2nd mate, naturalized American, sister Eugenia and brother Herbert M., 54 South St., Reading, Massachusetts. The family was from Yarmouth, Nova Scotia. Dad, Capt. Anthony Larkin, and mom, Miriam, both died in 1913. Herbert and Eugenia were living with an uncle in Reading.

Thomas Louse, carpenter, Finnish, mother Jemeina, Nusekishka, Finland.

Timari Pain, carpenter.

Werner Grip, seaman. Finnish, mother Elli, Helsingfors, Finland.

Kino Helmroos, seaman. Finnish.

Einar Kimsaas, seaman, Norwegian, father Edward, Norvik, Norway. Body washed ashore and recovered.

Hennivig F. Lundberg, seaman, Swedish, father Paul, Sturnstad, Sweden.

Werener Orpana, seaman, Finnish.

E. Panude, seaman, Chilean, father John, Antafagasta, Chile.

Arne Schlan, seaman, Finnish, father Ajlmer, St. Michael, Finland.

Harold Turvik, seaman, Norwegian, father Albert, Tuesurk, Norway. Body washed ashore and recovered.

Charles Erickson, chief engineer, American, wife Isabella, 839 55th St., Brooklyn.

Bernhard Borjesson, 1st engineer, American, wife A.M., 245 Forbel Ave. Brooklyn.

Harry Lewis Larkin, 2nd engineer, naturalized American, Mrs. Eva Larkin, 85 Garfield St., East Lynn, Massachusetts. Harry was 2nd mate Glendall Larkin's older brother.

Sverrt Ekenes, oiler, Norwegian, brother Samuel, 323 44th St., Brooklyn.
Carl Johanssen, oiler, Swedish.
Nils Martenson, oiler, Swedish, brother Otto Landskivno, Sweden.
Luis Franco, fireman. Mexican.
Julian Ray Macomber, 27, fireman, Hawaiian. His father, Charles, was a carpenter from Massachusetts who moved to Kau, Hawaii, and married a local girl. They had 3 sons and 5 daughters.
Charles Pinnopu/Pineoppu, fireman, Hawaiian, mother Mary Koloa, Kauai, Hawaii.
John Davis, wiper, Hawaiian.
Ekela Kaohi, wiper, Hawaiian, 1323 Fort St., Honolulu.
Charles Kelly, wiper, American.
Chester L. Lee, chief steward, American, sister C. Hue, 111 Sacramento St., San Francisco.
J. Leon, cook, Malay.
A. Sammis, 2nd cook, Malay, wife Allie, Singapore.
John Maar, messboy, Russian.
Hiel Kapu Rice, messboy, Hawaiian, father William Boardman Rice, 2316 N. King St., Honolulu. Born in Honolulu on July 8, 1890. In December 1916 he was the storekeeper on the steamer *Rockingham* at Baltimore.
Henry L. Hong, cabin boy, Hawaiian.

It was later determined the ship was torpedoed by *U-46*.

By the spring of 1917, the situation in Germany since the war began in 1914 had become increasingly hostile to the monarchy and the war and a committee had been organized to establish a German Republic. The chairman was a Berlin newsman, Dr. Roese Meyer. Bread was being rationed, 250,000 workers in Berlin were on strike, and agitations to force the Kaiser to seek a peace agreement were being made. Reports were received in Washington of sailors on surface ships at Kiel being forced at gunpoint to man submarines.

PART FOUR

America Goes to War

8. A Declaration of War

On April 3, President Wilson asked Congress to declare war. The next day the Senate passed a war resolution 82 to 6.

Continuing Losses

Marguerite—Universal Transportation Co.'s 4-masted schooner *Marguerite*, of New York, was en route from Caglieri to Algiers in ballast under Capt. Charles W. Willard, of South Portland, Maine, and 11 foreign sailors. At 8 p.m. on April 4, 1917, they were about 35 miles southwest of Sardinia when the *U-35* came up. Capt. Willard was given 10 minutes to evacuate. He was offered a tow toward land but refused. There were no casualties and sub commander kept the ship's papers. They were put aboard a French torpedo boat 46 hours later and landed at Bizerta. Capt. Willard signed on his crew at Marseilles.

Missourian—Capt. William Lyon's American-Hawaiian freighter left Genoa on April 4 in ballast for home. Sunk with gunfire by *U-52* in the Mediterranean near Porto Maurizio, Italy, around 4:30 p.m. There were no casualties. The total crew was 53, including 32 Americans.

War Resolution

On April 5 the House passed a war resolution 373 to 50. The United States had about 550 ocean-going merchant vessels, not counting those plying the Great Lakes, and when the Shipping Board and Navy Department began a survey of all available craft, it was estimated the number of large, seagoing yachts available for arming against submarines to be something over 100.

On the night of April 6, 1917, Wilson addressed Congress and in closing stated:

> It is a distressing and oppressive duty, gentleman of the Congress, which I have performed in thus addressing you. There are, it may be, many months of fiery trial and sacrifice ahead of us. It is a fearful thing to lead this great, peaceful people into war, into the most terrible and disastrous of all wars, civilization itself seeming to be in the balance. But the right is more precious than peace and we shall fight for the things which we have always carried nearest our hearts—for democracy, for the right of those who submit to authority to have a voice in their own governments, for the rights and liberties of small nations, for a universal dominion of right by such concert of free peoples as shall bring peace and safety to all nations and make the world itself at last free.

> To such a task we can dedicate our lives and our fortunes, everything that we are and everything that we have, with the pride of those who know the day has come when America is privileged to spend her blood and her might for the principles that gave her birth and happiness and the peace which she has treasured. God helping her she can do no other.

With but a few dissenting votes in the House and Senate, Congress declared that a state of war then existed, 373–50. At that time, the Chief Executive and Cabinet consisted of:

President—Thomas Woodrow Wilson of New Jersey
Vice-President—Thomas R. Marshall of Indiana
Speaker of the House—Champ Clark of Missouri
Secretary of State—Robert Lansing of New York
Secretary of the Treasury—William G. McAdoo of New York
Secretary of War—Newton D. Baker of Ohio
Secretary of the Navy—Josephus Daniels of North Carolina
Attorney General—Thomas E. Gregory of Texas
Postmaster General—Albert S. Burleson of Texas
Secretary of the Interior—Franklin K. Lane of California
Secretary of Commerce—William C. Redfield of New York
Secretary of Agriculture—David F. Houston of Missouri
Secretary of Labor—William B. Wilson of Pennsylvania

Strictly speaking, the United States was never a member of the 26 "Allied" nations because no treaty of alliance was ever concluded with any of them. The United States was an "Associate" of those nations fighting to defeat the Central Powers and joined groups like the Supreme War Council in November 1917 "to establish unity of purpose among the great powers in the prosecution of the war."

9. *Setting Up Shop in Europe*

President Wilson sent Rear Admiral William Snowden Sims, president of the Naval War College, to London as Commander of Naval Forces in Europe.

Sims and his aide, Cdr. John V. Babcock, went over incognito on the *New York* as "S.W. Davidson" and "V.J. Richardson" respectively, their fellow passengers having no idea who they really were. The ship hit a mine on April 9 just outside Liverpool and the passengers had to be transferred off to get ashore.

On arrival Sims met with Admiral of the Fleet John Rushworth Jellicoe, 1st Earl Jellicoe, who told him bluntly that England was going under from losses due to submarines and if starvation set in, which was Germany's plan, they would be forced to surrender. The British had about 200 destroyers dispersed throughout the realm, but about half of those were permanently assigned to the Grand Fleet.

Sims then met his British counterpart, Adm. Sir Lewis Bayly, who said their experience showed that destroyers were the best tool against submarines and on April 16 Sims requested that every available destroyer be sent over. At this time, there wasn't much of anything else in the fleet for anti-submarine work. A new type of 110' wooden "submarine chaser" was in the works. The Russians also urgently requested four patrol vessels and four destroyers for the Arctic coast.

On April 18, the Navy Department asked Sims why the U-boats couldn't be "blocked up in their bases." Why the British weren't already doing it wasn't asked. The British actually devised a strategy that had been contemplated and then abandoned as impractical. The plan would resurface later.

Ports in France

American freighters, tankers, and troopships had to have somewhere to go in Europe. After almost three years of war, most Channel ports in England and France were bursting at the seams with Allied traffic and there wasn't much space left to land American troops and supplies.

Getting the troops and supplies to Europe was the Navy's job and they naturally wanted the shortest trans–Atlantic route possible. Even if the strategically closer English Channel ports were available, they were much farther away and therefore presented a more dangerous route of travel. The same held for Mediterranean ports. The only real choice for the United States was the Bay of Biscay on the west coast of France. Although furthest from the fronts, the area did have the advantage of a mild year-round climate. The ports considered were Brest, Saint-Nazaire, Nantes, La Pallice, and Bordeaux.

Bordeaux was the principal port at that time but Brest is the northernmost port. The harbor at Brest was large and deep, had little current to interfere with navigation and anchoring, but facilities there were limited. Saint-Nazaire is about 140 miles south of Brest and could accommodate ships with a draft of 30 feet but had room in the harbor for only 29 ships at a time. Nantes is just slightly north of Saint-Nazaire but is 31 miles inland and limited ships to 24 feet of draft.

La Pallice is about 200 miles south of Brest but was determined to be suitable only for limited troop disembarkations and the discharging of limited types of cargoes.

Bordeaux is 240 miles south of Brest, lies 60 miles "up" the River Gironde, in a southerly direction, and is also 150 miles further to the west of Brest longitudinally. In addition, the bar at the entrance to the River Gironde would allow only ships drawing less than 25 feet to pass over and the current in the river was strong enough to interfere with anchoring.

Brest, being a reasonably good harbor and closest to the United States, was the logical choice for the Navy. Rear Admiral William Bartlett Fletcher, Sr., was then appointed Commander of U.S. Naval Forces on the French Coast.

When Brest was chosen, a minor conflict arose between the Army and the Navy. The Army's primary concern was the distance from the port of debarkation to their planned inland bases and training centers. In their planning, the Army chose Bordeaux and Saint-Nazaire and anticipated that fully two-thirds of all American Expeditionary Forces would pass through Bordeaux.

When all the facts were presented, everyone realized that harbor accommodations had to be the priority so Army engineers got to work making Brest suitable for troopships. On July 1, 1918, Maj. Frederick William Green, vice president of the St. Louis Southwestern Railway, became superintendent of the Army Transport Service and was responsible for unloading the ships. Maj. Harry O'Neill, head of stevedores for the White Star Line at Boston, was second in command.

A total of 41,544 American doughboys went over on American merchant ships; 911,047 on Navy transports; 1,006,987 on British ships; 68,246 on ships under charter to the British; and 52,066 on French, Italian, or other Allied ship.

Contrary to the Kaiser's plan, no American transport was ever sunk en route to France but American lives were lost on the British-operated *Tuscania, Moldavia,* and *Otranto*. The *Tuscania* had 2,179 National Guard soldiers from Michigan and Wisconsin aboard and of those 204 were lost. Of 480 American soldiers on the *Moldavia,* 53 members of Co. B, 58th Infantry, Fourth Division were lost.

10. Finding the Path to Victory

The Emergency Fleet Corporation

In the month of April 1917, nearly 1,000,000 tons of shipping was lost to U-boats, but the replacement tonnage during this same time amounted to about 650,000 tons, a loss ratio of four to one. Something had to be done, and done fast. British Premier Lloyd George stated in April, "The road to victory, the guarantee of victory, the absolute assurance of victory, is to be found in one word, ships, in a second word, ships, and in a third word, ships."

That inspired the cry of "Ships, ships, and more ships!" throughout the land and a sort of frenzy akin to the California Gold Rush of 1849 began. The early reluctance of private capital to expose itself to risk in 1915 suddenly evaporated when government contracts were made available. It was recognized early on that to get ships built as quickly and efficiently as possible, the federal bureaucracy had to have a minimal role. The job needed to be done by experts in the field who had the business acumen to see it through.

On April 16, 1917, the Shipping Board organized the United States Shipping Board Emergency Fleet Corporation under the laws of the District of Columbia "and delegated to that Corporation the execution of its construction program."

President Wilson asked Gen. George W. Goethals, builder of the Panama Canal and president of the Panama Steamship Co., to lead the company and he accepted. Congress appropriated $50,000,000 for the project.

In order to save steel and get ships built quickly, a proposal had been made to build wooden steam freighters. Theodore E. Ferris, chief naval architect for the Shipping Board, got to work on a design while William Denman got busy contacting builders.

Gen. Goethals strongly opposed the idea of wooden ships and said, "birds are still nesting in the trees." He thought it was ludicrous to put a steam engine in a 3,500-deadweight ton ship that required the same size crew when the same engine and crew could go into a 10,000-ton steel ship. Naturally, the wood shipbuilders, particularly on the Pacific Coast in Denman's California and Wilson's voting bloc, disliked such talk and made their feelings known in Washington.

The resulting battle caused Goethals, Denman, White, and Brent to eventually resign. Edward N. Hurley, of Illinois, became chairman of the Shipping Board and Bainbridge Colby, of New York; and Charles N. Page, of California joined. Lester Sisler, of Florida, was secretary.

Admiral Washington Lee Capps became General Manager of the Emergency Fleet Corporation but Goethals' argument was taken to heart and the original appropriation of $50,000,000 was increased to $700,000,000 for steel ships and shipyards.

Admiral Capps subsequently resigned out of frustration at being subordinate to the chairman of the Shipping Board.

On July 11, 1917, President Wilson delegated to the Emergency Fleet Corporation the authority to requisition and complete ships then building and to construct new ships.

By the end of 1917, the Corporation had over 1,000 employees, including many technical experts who took significant cuts in pay to lend their expertise to the war effort. The Corporation had 16 offices that were supervising the building of 1,118 contract vessels in 116 shipyards around the country. The training of shipyard workers and building housing for them were other significant endeavors. The Shipping Board even had an official yell used at launchings: "Ships! Ships! Ships! Away! Away! Away!"

The First Shots at Sea

The first shots fired from an American merchant ship in the World War were reported on April 19 by Lt. Bruce R. Ware, Jr., the Armed Guard commander on Capt. Emery Rice's steamer *Mongolia* off Beachy Point in the English Channel. He said they blew the plating off a U-boat's conning tower at about 500 yards range. April 19 was noted to be the 142nd anniversary of the Battle of Lexington. Lt. Ware reported:

> Immediately the entire spot became covered with a cloud of blue-gray smoke that did not dissipate readily but hung like a pall over the spot where the periscope was last seen. A ragged splash, presumably the periscope knocked off on shell contact, was seen immediately behind the point of aim and a further rugged splash was seen about fifty yards behind.

After that, a large oil patch was seen. In the meantime, the Navy was frantically trying to beef up its very low inventory.

The Submarine Signal Company

The case of the Submarine Signal Co. in Boston illustrates the entanglement of German and American technologies and business that existed in days of peace.

Submarine Signal was started in the early days of radio research and incorporated in 1910. By the mid–1920s the company held 20 patents in the United States, Germany, and other countries for various transmitting and receiving devices and by the 1940s was a leading maker of sonar, radar, and fathometer equipment.

On January 1, 1908, the company signed licensing contracts with the German firm of Norddeutscher Maschinen und Armatuern Fabrik Gesellschaft Mit Beschrankter Haftung and at a later date licensed the Atlas-Werke Co., of Bremen, to "manufacture, sell to governments, and lease to shipowners submarine receiving apparatus within a certain prescribed district or territory consisting of Germany, Belgium, Holland, Austria-Hungary, and Russia."

The agreement with Atlas-Werke called for a joint account in which manufacturing costs were debited, rents and leases were credited, and the difference split 50–50. And, since ships move around a lot, the company in whose territory a vessel was located became responsible for collecting rents and performing maintenance. Submarine Signal was responsible if the ship was in neither company's recognized territory. This worked well until World War I.

The joint resolution of May 12, 1917, authorized President Wilson to seize all belligerent merchant vessels then in U.S. ports and on June 30 he issued an executive order directing the United States Shipping Board to seize the vessels. Consequently, 53 German vessels with Submarine Signal gear on board manufactured and installed by Atlas-Werke, were seized.

The equipment was an early form of sonar. Listening devices were installed in the bottom of a ship, port and starboard, and wired to an "A" and a "B" receiver in the wheelhouse, pilothouse, or chart room. Most lighthouses and lightships had a submerged signal sending apparatus whose sound could be picked up and identified by a merchant or naval vessel irrespective of the weather.

Beginning in April, and at various times thereafter, the company notified Secretary of the Treasury William G. McAdoo, the Shipping Board, and the collectors at various ports, that many of the seized German ships had equipment belonging to Submarine Signal. At the same time the U.S. government requested the company to look after and maintain the equipment and they did this from time to time on various ships. Submarine Signal also sought payment for their services from the Navy Department, Shipping Board, and the Treasury Department and to find out exactly which vessels had been seized and who controlled them. Because the vessels were transferred between the Army, Navy, and Shipping Board, the company could never be certain who was operating any given ship at any given time during the war.

The company did manage to secure rental agreements from the government on some of the equipment and were paid $27,262.70. In January 1923 the company obtained a full list of the seized German vessels and the times during which each department operated them. From this they attempted to compile a statement of rental and service charges which amounted to $109,671.69. They placed a value of $52,500 on the equipment itself. The invoices were not honored and Submarine Signal had to go to court to press their claim.

The case dealt with Fifth Amendment and war powers law. U.S. citizens owned the property but it was on enemy vessels seized by the president by authorization of Congress. The court decided:

> If the war power of the Government is legitimately exercised and the property taken thereunder found on vessels of the enemy belongs to a citizen of the United States the injury done is incidental to the exercise of the war power, and there is no taking of the property under the limitations or restrictions of the Fifth Amendment, and there is no right to compensation on account of such taking. If the property taken is so situated that it can not be separated from the enemy property its taking is an incident to the exercise of the war power, and it is not the taking of private property for public use for which compensation can be demanded. And in the exercise of the war power it is immaterial whether the owner of the property taken is a citizen of the United States or not. The apparatus was on vessels of the enemy, was placed on said vessels by the licensee of the plaintiff, which licensee was an enemy, was being used by the enemy, and was being devoted to the enemy's use by the agent of the plaintiff, and is a proper subject for confiscation.
>
> The plaintiff might have some claim for compensation for repairs and inspections actually made of the submarine apparatus attached to these vessels which were rendered after they were taken and which were done at the request of authorized officials of the Government. But the plaintiff has failed to prove such a claim, and has stated that there was no practical way of preparing a statement showing the value of such services. Therefore the claim must be disallowed for failure of proof.

Crews for Merchant Ships

Henry Howard, 48, was a chemical engineer and vice-president of the Merrimac Chemical Co. at 33 Broad St., Room 104, in Boston. The company was incorporated in 1863 and manufactured acids, oils, sulfate, tin crystals, wood and denatured alcohol, and other assorted products. Howard lived at 36 Amory St. with his wife Alice, 38, daughter Catherine, 18, and sons Henry, 17, Thomas, 16, and John, 15, and three servants.

Howard was a member of the Council of National Defense and had big ideas for training merchant seamen. The Shipping Board was favorably impressed and on May 29, 1917, he was appointed director of the Recruiting and Training Service, which was officially established on June 1 on the 12th floor of the Custom House in Boston.

In July it was reported that 4,239 seamen were working on the Great Lakes, where most of the biggest ships were operating, far too few to man existing ships, the seized German ships, and the new ships expected to be coming off the ways. It was anticipated that around 25,000 merchant seamen would be needed to man 1,600 ships. The problem of recruiting and training needed to be immediately addressed.

Beginning on August 1, 1917, a major conference was held in Washington to discuss the manning of ships. It was chaired by Secretary of Labor William B. Wilson and attended by cabinet members, labor leaders, steamship company officials, ship captains, sailors, and anyone who had an interest in the subject. Every aspect of crewing ships was discussed. It was agreed that the primary problem would be a shortage of qualified deck and engineering officers since it was anticipated that 5,000 new licensed deck officers and engineers would be needed. The solution was to offer qualified seamen, who had at least two years at sea, past or present experience, free schooling for a 3rd mate or 3rd assistant engineer's ticket.

There was one immediate problem that had to be addressed—the question of loyalty and sabotage. On August 2, Andrew Furuseth stated, in part:

> According to a census which we have taken there are about 5,000 Germans sailing under the American flag in different capacities. It may be said that they are among the highest skilled men, whether they sail in the forecastle, in the fire-room or in the galley. Of these men, 3,751 are in the organizations of the seamen.... It is a very serious matter to displace them and I want to suggest the serious thought, based upon the psychology of the seamen as I know it, whether it would not be well to use those men where they can be used in order to release other men for the trans-Atlantic trade. The importance of the matter is to get highly skilled men. A fast vessel and a highly skilled crew is the real defense against a U-boat.
>
> When a seaman has begun to sail in different vessels and, in a sense, divorced himself from his own country, his primary loyalty is to the ship he is on and to his shipmates and I do not think that there is any danger in making use of those men in the purely coastwise trade and in the Lakes trade. It may be said that they can destroy something in the Lake trade, but then anybody can do if they particularly want to do it they are for sale and they disregard the shipmates they are sailing with.

In mid-September 1917 the Shipping Board published a 22-page pamphlet outlining the course of instruction. Applicants had to be at least 19, a native-born or naturalized American citizen, in good health and proficient in English and ordinary arithmetic. Deck officer candidates had to pass a "color sense and visual acuity" test and engineers a hearing test to hear bell orders. Recruiting for these schools was handled by the Steamboat Inspection Service. There was no pay while attending school but tuition, books, and instruments were provided free of charge. Deck school was six weeks, engineering four weeks. Upon graduation and passing the examination, a student was required to ship out on an ocean-going steamer as a junior officer for at least two months at a wage of $75 a month. Many schools for deck training were set up all over the country in the major and minor ports.

Marine engineer schools were established at the Massachusetts Institute of Technology at Cambridge, the Stevens Institute of Technology, the Case School of Applied Science, Johns Hopkins University, Tulane University, the University of Washington at Seattle, the Armour Institute of Technology, and a school aboard the steamer *John Weaver* at Philadelphia. Oilers and watertenders on the Great Lakes and stationary and locomotive engineers were encouraged to apply.

To recruit apprentice seamen, it was decided to put literature and recruiting officers in 6,000 Rexall drug stores around the country. Any male between the ages of 21 and 30 who was reasonably fit could apply. Each candidate was examined by a volunteer physician and his application was sent to Boston. If accepted, the Shipping Board covered railroad fare to the nearest training center for another examination by a Shipping Board doctor and if approved he was outfitted for recruit training in deck, engine, or steward's school. Sailors otherwise fit for military service could be exempted by their district draft board for industrial reasons. No compensation was received and no allowance was made for expenses while in school. In the first five months, around 15,000 applications were received.

The Robert Dollar Case

While the crewing situation was being resolved, an item of intense interest was winding its way through the courts. On January 3, 1917, an item had appeared in California newspapers regarding a lawsuit against Robert Dollar:

> An industrial case of particularly keen interest to Pacific Coast shipping interests and sailors shortly will come before the supreme court of the United States. In general the case involves the contract labor law and specifically the right of American shipowners to sign on oriental crews on vessels in far eastern ports and then, after the ships have reached an American port, assign the foreign crews to other vessels.
>
> The case comes up from the federal district court of California on a writ of certiorari just issued by the highest court. That the case will be reviewed is largely due to the efforts of Paul Scharrenberg, secretary of the California State Federation of Labor, who was the moving spirit of the prosecution in the lower court.
>
> It appears that the Pacific Mail Steamship company brought forty-two Chinese to America to man their steamer *Venezuela*. Mr. Scharrenberg alleged in the importation of the orientals to be in violation of the contract labor law and appealed to Commissioner General Caminetti of the immigration service. While the commissioner was investigating the charges, the case got into the California court and a decision was rendered in favor of the defendant, mainly on the precedent of a similar case of a few years ago in which the Dollar Steamship company was held not to have violated the contract labor law in signing oriental crews in much the same manner.
>
> Pacific coast labor organizations fear, it is understood, that unless the decision of the federal district court of California is reversed, it will be but a short time until Japanese and Chinese crews will replace American sailors and practically all ships engaged in the Pacific trade because of the considerable difference in the wages for which they will work.

In Scharrenberg v. Dollar Steamship the Supreme Court held on November 5, 1917, that a merchant seaman was not a "laborer" within the meaning of contract labor law, and that sailors, while working in international waters, were not inhabitants of the flag of the ship they were on. Court stated, in part:

A ship of American registry engaged in foreign commerce is not a part of the territory of the United States in such a sense that seamen employed in it can be said to be "laboring in the United States" or "performing labor in this country" within the inhibition of the alien contract labor law.

That seamen employed on an American vessel engaged in foreign commerce are Chinese is without significance in determining whether they were imported into the United States in violation of the alien contract labor law, as the act applies to all alien contract laborers without regard to their origin or nationality [245 U.S. 122].

The decision sent shockwaves through the maritime labor unions. A November 1917 editorial in the San Francisco Chronicle, viewed by the Sailors' Union of the Pacific as hostile to the Seamen's Act of 1915, stated:

Nineteen Chinese sailors were brought to this port and transferred to an American ship not in the coastwise trade. It was sought to punish the company responsible for the transfer, but the Supreme Court holds that the transfer was not in violation of the Exclusion Act or the law forbidding the importation of contract labor, since they were not landed and the home of the sailor is the sea. The Seamen's Act does not seem to be involved in the case.

Those who would exclude Asiatics say that if Chinese can be employed on American ships, Americans will not become merchant sailors. They do not now, and have not in recent years. The number of native Americans serving in crews of merchant ships is negligible. Three-fourths of the crews are aliens, and of the citizens most of them are naturalized.

What present conditions and wages are doing does not appear, but with opportunities on land equal to or better than opportunities at sea, there is no reason to suppose that more Americans than usual are becoming merchant sailors. Those who attracted will enter the Navy.

What will be the conditions after the war no one can tell. While urgent necessities remain to be filled, freights and wages will be high when conditions again become normal, freight will be carried by those who can do it cheapest. And if that should be those who employ Asiatic crews, we should say that so far as labor is concerned it makes no difference what flag the ship flies, while it is a national advantage to have a large mercantile marine flying the American flag.

John E. Tennison, acting secretary of the Sailors' Union of the Pacific, wrote a response on November 8, 1917, that was printed in the paper:

With reference to your recent editorial on the Supreme Court decision permitting the importation of Chinese for service on American ships, I beg to call your attention to a possible aftermath of the most serious portent. If a court can say that this law was not intended to apply to seamen, it can certainly apply the same contention to any other calling, with the result that a law passed to prevent competition of pauper labor with American labor will soon be entirely nullified by decisions of courts. The claim that a seaman's home is upon the sea is mere twaddle. Carry that to its logical conclusion and the right to franchise can be denied a seaman. Go still further and it must be held that he has no right to the opportunity of marrying, acquiring a home and rearing a family. A seaman's home is no more upon the sea than a railroad operative upon a train.

In the past few decades American boys have not gone to sea in great numbers. Just at present, however, there is a concerted nationwide effort to induce the American boy to return to the sea. Can any sane person conceive that American boys will go to sea in competition with Chinese?

On the face of your editorial it seems that at this time when the dangers of a seaman's life are multiplied by passages through the war zone, etc., they are expected to do the work, as no one would dream of sending a vessel out now with a crew of Chinese. When the dangers have passed, however, the men that are relied upon now must again compete with Chinese. If this is to be the national policy, the logical thing at this time would seem to be to start in right now and man all our offshore ships with Chinese. Then we shall see how long it will take to bring all our war affairs to a standstill.

* * *

President Wilson authorized licensed officers from neutral nations to take out American papers and the first applications were received in September 1917 from the Dutch steamer *Ombilin* at San Francisco.

Between June 1 and October 31, 1917, 604 deck recruits graduated and 416 engineers graduated. By the end of December 1917 there were 27 free navigation schools and eight free engineering schools. To liven things up, Stanton H. King, of Boston, an old salt, was hired to lead and teach sea chanties to the recruits.

Big plans were made for 13 training ships to handle 4,000 students a month but that never materialized, due in part to a less than enthusiastic response to seaman's training.

On January 1, 1918, the coastwise passenger steamer *Calvin Austin* was put in service as a training ship for unlicensed seamen at east Boston. Sister ship *Governor Dingley* was soon put into service and then the *Governor Cobb*. An ex–Army transport, the *Meade* was also added. In June 1918 the submarine tender *Iris* was put into service on the West Coast. In September 1918 there were 3,125 apprentice seamen on board for the 6-weeks' program. Around 156 applications a day were received for this training from every state and every walk of life. After graduating, four students are assigned to each ship as ordinary seamen along with six able-bodied seamen. Ordinary seamen earned $60 a month with an overseas bonus of 50 percent. Firemen earned $75 a month with a 50 percent war-zone bonus. Engineering positions were always in demand. Cooks were also in demand and earned $60 to $90 a month depending on experience. The Board was well aware that bad food on a ship rapidly led to discontent and low morale and good food had the opposite effect whether in peacetime or war.

By September 1918 there were 36 free, government officer training schools for those with two years at sea. Of those, 24 trained deck officers and 12 trained engineers, with a total of 1,139 students. Of those, 631 were in the deck program and 508 in engineering. The largest schools were in Philadelphia, New York, Boston, San Francisco, and Chicago.

From August 1917, 8,069 students had graduated and passed examinations conducted by the Steamboat Inspection Service as third or second mates or third or second assistant engineers.

Shipyard Labor

One item on the agenda at a 1902 meeting of the American Foundrymen's Association was "How Many Hours Shall Constitute a Day's Work?"—a hot topic of the day. It was generally accepted that the work day was ten hours, but the "so-called laboring element" had petitioned Congress for an 8-hour day and on May 20, 1902, that became the law for federal, state, and city employees and any worker employed on a government contract. Some, but not all, in the Foundrymen's Association believed a 9-hour day was a reasonable compromise. One member stated in a paper presented at the meeting:

> The average man is by nature lazy and would rather not work at all if he could get his necessities and comforts without it. But without work we cannot secure our wants, much less our comforts or pleasures by which life is made worth living. Accumulated wealth, so called, is the surplus product of toil and measures our prosperity and happiness and just as wealth accumulates the possibilities of prosperity and happiness increase. Therefore, it follows that the more work we do or the longer the working day, the more wealth, the greater the prosperity, the higher the wages and the greater the happiness of all our people.

In 1905, the Industrial Workers of the World—the infamous "Wobblies"—was established to concentrate worldwide labor against capital. They were fiercely opposed to the war, seeing only great war profits for the capitalist and his son and death or a lifetime of maiming, debilitating injuries for the common man's son. This sentiment was shared by many Americans before the country actually entered the war, but once war was declared and seen as justified, a patriotic fervor swept the land, the likes of which has not been seen since.

In the summer of 1917, James Phillips, secretary of the marine transport branch of the I.W.W., was detailed to cause as much interference with shipping activities on the Atlantic and Gulf Coasts as he could in retaliation for public opposition to I.W.W. members. While most Americans never believed Germany was a threat to the survival of the United States, they did believe the war had to be won. But at the same time, workers refused to be exploited to achieve that goal. In addressing labor unrest and the wages of shipyard workers during the war, the editor of *Shipping* magazine stated, in part:

> Do any of the "friends" of the workingman ever give a thought to the fact that the men who create all the trouble are not those who earn little, but those who earn the most? With every advance in wages the quality of the work performed suffers a decline and when, as at present, semi-skilled workmen manage to pile up from $150 to $200 in a week, it is simply impossible to enforce any semblance of discipline among them. Cojointly with the loss of discipline there arises a craving for all sorts of useless articles that cost money, and the net result is that after a few months of affluence the men are just as poor as they were before, while they have lost much of their old time energy, and the families have been demoralized by the sudden accretion of riches which found concrete expression in the purchase of thousand-dollar mechanical pianos and talking machines, along with extravagant clothes and flashy jewelry purchased at several times its actual value.

The Russian Revolution had a profound effect on the more strident members of organized labor, especially the Wobblies, and some ships, especially Italian, were seen flying the red flag on entering port after the war.

In February 1918, the ship carpenters union threatened to strike but cancelled the walkout after Wilson sent a strong protest and rebuke. In May 1918, 112 Wobbly leaders were put on trial.

On September 18, 1917, around 3,500 longshoremen went on strike in San Francisco. Union organizer and A.F. of L. founder Samuel Gompers arrived to help resolve disputes. That week, approximately 50,000 shipyard workers belonging to 25 unions walked off the job. Most were skilled mechanics on the West Coast objecting to low wages. On September 23, about 30,000 shipyard workers in San Francisco, Portland, and Seattle went back to work after receiving a telegram from President Wilson, who promised the newly-established Shipbuilding Labor Adjustment Board would work to solve wage issues. In New York, 6,500 longshoremen agreed to arbitration.

In 1920, the Bureau of Labor Statistics Division reported there were 4,324 strikes in 1917, the first year of the war in all industries, and 3,232 in 1918.

The First Victory at Sea

On May 30, 1917, Capt. John Charlton's armed tanker *Silver Shell* was given credit for being the first American merchant ship to achieve a confirmed sinking of a sub in the war. The episode occurred off Balearic Island in the Mediterranean. Chief Turret Captain William J. Clark was in charge of the Armed Guard detachment. Clark reported:

> May 30th, at 5:35 p.m. a submarine was sighted off starboard beam, flying no flag or marks of identification. Manned and loaded both guns. Hoisted colors and waited about ten minutes. Fired after gun with sights set at 4,000 yards, scale 49, and fired purposely short to see what the submarine would do, as she was closing in on the *Silver Shell*. At the same instant of our flash the submarine fired a shot, the shell dropping 100 yards short amidships. The ship was swung to port to bring the submarine astern. Twenty-five rounds were fired at the submarine, the last two of which appeared to be hits. As the last shot landed the submarine's bow raised up and went down suddenly. The crew of the submarine who were on deck did not have time to get inside so it is believed there is not much doubt her being hit.
>
> The submarine fired in all 32 rounds, the last four of which were shrapnel and exploded overhead. The second shot from the submarine dropped 100 yards off our starboard beam, the range perfect. The third shot off the port beam with the same range. As there was a heavy swell running, it is believed that caused the submarine to miss hitting the *Silver Shell*. The fourth shot passed over the bridge and fired purposely short to see what she would do. The submarine all the time was gaining on the *Silver Shell*. The submarine appeared to be 250 to 300 feet long. There were two guns mounted on the deck, a six inch (apparently) gun forward of the conning tower and a smaller gun aft. The forward gun was the only one used. The first shot fired at the submarine only reached half way, so the submarine was about 7,000 yards off.
>
> The engagement lasted from 6 p.m. to 7:30 p.m. There were no casualties. As soon as the submarine fired the first shot an S.O.S. call was sent out giving the position and course which was answered by F.U.T. (Toulon) stating assistance was under way. A few seconds after a Spanish ship commenced sending with her wireless with apparent intent to block our message.

Capt. Charlton later reported:

> As the submarine displayed no flag and was coming nearer, we fired a shot. The submarine, which was then about 7,000 yards away, replied immediately with what seemed to me, on account of the range, a six-inch gun. Her first shot fell amidships about 100 yards short. We changed our course due west, increasing speed. The submarine followed us, keeping up fire. The fight lasted from 6 o'clock until about 7:30 p.m. We fired 25 shots. The submarine fired over 30, including some shrapnel toward the end, which exploded astern of our ship too high to do any damage.
>
> Our last two shells seem to have hit the submarine in the forward part. A few seconds later her bow jumped up, and then she went down, stern upward at an angle of about 45 degrees. The submarine did not come up again, and I believe it was sunk then and there. No damage was done to our ship, and there were no casualties.

The *Silver Shell* had two more close calls during the war with torpedoes passing very close by while in convoys under escort.

The *Norlina*

In May 1917 Capt. Jesse Forster left Baltimore and in June headed home from Liverpool in company with two British steamers. One was the *Manchester Port* with a cargo of pyrites from Huelva for New York. At about 4:50 p.m. on June 4 they were off Cape Wrath, Scotland, when the bridge watch saw one of the armed British ships come under attack by a U-boat. After about 30 minutes of gunfire the sub submerged but a short time later the *Manchester Port* sent out an S.O.S. reporting a sub just off their stern. After another 20 minutes of firing in plain sight of the *Norlina*, the sub submerged again and the British steamer reported they were undamaged. Things were quiet until the crew was at supper around 6:30. A lookout on the bow gun yelled, "Here she comes, torpedo port side!" Wireless operator Oscar Foy reported what happened next:

The torpedo had a red head about 16 inches in diameter. She was about 10 feet long and tapered off at the end. Her propeller left a wake of long whitecaps after her. The torpedo struck us on the port side a glancing blow amidships right near the engine room. The torpedo must have weighed a ton and shook our ship from end to end. Our ship was empty and we all thought she had exploded from the terrific noise she made when she hit. Simultaneously the ship's whistle blew short and successive blasts, which was the signal to abandon ship and man the life boats.

I sent out a signal of distress which was picked up by the *Manchester Port, T.L.O.* (a British ship) and Malin Head, the latter a British shore station. I was then ordered to my position in the starboard boat. There were about 20 men assigned to this boat.

The Captain and Mr. Lloyd, Second Mate, lowered our boat to the water.

Mr. Baker and Mr. Robinson were in charge of the life boat on the port side. They had difficulty in lowering the life boat on account of boiling hot water coming from the ship's overboard exhaust which partly filled the boat. The crew of this boat were mostly foreigners and were greatly excited. Mr. Baker ordered them at the point of a gun to be quiet and sit down. Otherwise both port and starboard boats were launched clear of the falls without incident. It was as bright as day. The sea was perfectly calm, weather clear and cool. After we had cleared the falls and ship's side, the Captain ordered us to stand by.

The Captain, First Mate, Second Mate, Chief Boatswain's Mate Gullickson and Mr. P Savage, Boatswain's Mate First Class, remained on the ship. While the German commander was a game one, he had nothing on our captain. While apparently our ship was in a sinking condition, our Captain inspected the engine room and holds and took soundings in the bilges. He found that the torpedo had failed to explode.

All hands were then ordered back on the ship. We were laying perfectly still for at least an hour. The submarine which fired at us apparently took it for granted we were sinking and started in pursuit of the two British merchant ships, which were trying to get away. When the commander of the submarine saw our crew coming back from the life boats and climbing up on deck, he immediately gave up chasing the British vessels and started for us again. The submarine was about 2,000 yards off our starboard beam. We were helpless. We were also in danger of our boilers exploding as the engineer's force were still in the boats.

Suddenly in tones loud and clear, "Man the guns!" The guns' crew scrambled to the deck. They ran fore and aft to their respective positions. The chief gunner gave them their ranges from the bridge. In a few seconds the four-inch shells were screeching, popping, and breaking all around the on-coming submarine. She was hard to hit on account of her rapid speed. Although our ship was perfectly still, our gunners could not get her range. The submarine would come full speed and then come to a full stop. When about 600 yards off our starboard quarter, a shell from our forward gun hit her and she submerged. Again she appeared and our after gun hit her and blew away her periscope. Another shot from our forward gun fell right on top of her. A shower of black specks rose high in the air, followed by a great commotion of bubbles of water and a light blue smoke arising from the stern of the submarine which a second before had been the eyes of our enemy.

Our crew, which were lined up against the starboard rail watching the battle, gave a hearty American cheer when the submarine disappeared.

Nineteen shots in all were fired, which is the Secretary of the Navy military salute. Our gunners afterwards said we ought to have given them two more and made it twenty-one, the Presidential or National salute.

The *Manchester Port* arrived safely at New York. The Armed Guard commander was Chief Boatswain's Mate Olaf J. Gullickson. Gullickson worked his way up and was a commander in the Navy during World War II. The sub was later reported to be the *U-88*.

The First Convoy

In the absence of destroyers, it was recognized early on that the best defense against a submarine was speed and alternating course by "zigzagging" for not more than 15 minutes on each leg. Since every ship, fast or slow, was needed, the idea was to group ships together and travel at the speed of the slowest ship. The submarine would have to come within range of the naval escorts and if an attack was made, the sub's location would be immediately apparent.

This idea had its advocates and detractors. As Secretary of the Navy Daniels said to Adm. Sims, "In regard to convoy, I consider that American vessels having armed guard are safer when sailing independently." A good many shipmasters felt the same way.

From the beginning, the Admiralty had considerable difficulty with some of their merchant ship captains, whose attitude was, "give us a gun and let us look out for ourselves." They did not like being in convoys. Having to keep station in blackout conditions, often in rough weather, with the constant danger of collision or attack was extremely stressful.

The first trans-Atlantic convoy left Quebec with Canadian troops for Southampton on Saturday, October 3, 1914, at 10 knots. On October 8, German agents in New York reported to their naval authorities that a 24-ship convoy escorted by eight warships left Quebec on the 2nd, presumably going to Boulogne. *U-8* and *U-20* were dispatched but they were spotted on the 12th by the French off Cap Gris Nez and the convoy "took shelter" at Plymouth. The subs never got close enough to do any damage. The greatest danger from submarine attack was close to shore.

It was decided to separate the faster troopships from freighters. Troops were put only on ships that could make at least 12 knots, normally the big ocean liners, and the troopship convoys were routed to avoid lanes used by freighters and had greater escort protection. The normal speed for a freighter convoy was 6 to 9 knots. Any Allied or neutral ship could join a convoy. All convoys zigzagged except in darkness or fog and information regarding submarine activity or mines was sent by blinker light.

The general policy was to send convoys out at regular intervals, usually every four or eight days, so escort movements could be coordinated. Escorts stayed with a convoy through the submarine danger zones and then turned around and joined a convoy going in the opposite direction. This required a high degree of coordination. Ship movements were never made public and no ship was referred to by its real name if information had to be relayed by telephone.

As of May 31, 1917, 1,308 A.E.F. personnel had arrived in Europe.

The first American Expeditionary Force convoy carried Maj. Gen. John J. "Black Jack" Pershing's troops and supplies. It was decided to make up the convoy in four sections that sailed six hours apart. When the first group left New York on June 14 at 15 knots in heavy fog they had no final destination. All the merchant ships were under Army control with civilian crews.

Group 1:
U.S.S. *DeKalb* (troop transport)—Cdr. Walter R. Gherardi.
Havana—passenger ship, New York & Cuba Mail Steamship Co. (Ward Line).
Saratoga—passenger ship, New York & Cuba Mail Steamship Co. (Ward Line).
Tenadores—passenger ship, United Fruit Co.
Pastores—passenger ship, United Fruit Co.
Escorted by cruiser *Seattle* (flagship for Adm. Gleaves), destroyers *Roe, Terry,* and *Wilkes,* and converted yacht *Corsair*. The *Corsair* had to turn back and was replaced by destroyer *Fanning*.

Group 2:

U.S.S. *Henderson* (troop transport)—Lt. Cdr. George W. Steele, Jr., with 1,500 soldiers and 24 mules.
Momus—passenger ship, Southern Pacific Steamship Co.
Antilles—passenger ship, Southern Pacific Steamship Co.
Lenape—passenger ship, Clyde Steamship Co.
Escorted by cruiser *Birmingham,* destroyers *Burrows* and *Lamson,* and converted yacht *Aphrodite,* Lt. Cdr. Ralph D. Craft.

Group 3:
U.S.S. *Cyclops* (fuel ship)—Lt. Cdr. George W. Worley.
Henry R. Mallory—passenger ship, Mallory Steamship Co.
Finland—passenger ship, International Mercantile Marine.
San Jacinto—passenger ship, Mallory Steamship Co.
This section was escorted by cruiser *Charleston* and destroyers *Allen, McCall,* and *Preston.*

Group 4:
U.S.S. *Hancock* (transport)—Cdr. Edwin T. Pollock.
U.S.S. *Kanawha* (fuel ship).
Montanan—passenger ship, American-Hawaiian Steamship Co.
Dakotan—freighter, American-Hawaiian Steamship Co.
El Occidente—freighter, Southern Pacific Co.
Edward Luckenbach—freighter, Luckenbach Co.
Escorted by the cruiser *St. Louis* and destroyers *Shaw, Ammen, Flusser,* and *Parker.*

On June 20, while the convoy was at sea fighting submarines, Adm. Sims received a message from Secretary of the Navy Daniels:

> There will be no additional movements before August of troops. You will be furnished fully with information as to sailing of Army supply ships as far as possible in advance, and the actual sailing intended route and probably dates of arrival will be reported. We hope to sail four Army supply ships now fitting out in about 10 days' time. The 32 destroyers, which are all that are available, have sailed; 110-feet chasers, which are to be sent to France, should begin to deliver in August. Fishing vessels, 12 in number, will sail in August for France. There are no other small craft available at present, although work on yachts is being pushed, probably ready 15th July. In regard to convoy, I consider that American vessels having armed guard are safer when sailing independently.

On June 21, Saint-Nazaire was designated Base Port No. 1 and on June 26 the *Tenadores,* fresh from a sunny, relaxing Caribbean cruise to Cuba, became the first American merchant ship to reach France. The next day the second group arrived and on the 29th the fourth and final section of the convoy arrived and anchored out. In all, 346 officers, 11,607 enlisted men, and 46,700 tons of supplies were brought over.

From July 1917 to November 1918 the War Department sent 628 ships of its own carrying 5,706,551 tons of cargo, totally independent of the Navy, to supply the A.E.F. Only a part of this cargo came across in convoys.

After the war, Adm. Henry Wilson pointed out, quite proudly:

> No troopship coming to the coast of France escorted by an American escort was ever successfully attacked, and only five troopships west-bound without troops were successfully attacked.... If we had destroyers enough to have let us have no other thought than to send our west-bound ships through an area of maximum safety, it is firmly believed that no troop transport, whether east-bound or west-bound, would have been successfully attacked or have been lost.

Convoy routing through the approaches to the various ports was perfected to reduce losses. After the war Adm. Sims said to a Senate committee:

> I might state in this connection that my opinion as to the ability of merchant seamen to handle their ships in convoy came from the fact that I had been associated at one time for three years in instructing young men in navigation and merchant seamanship and I thought I knew those men better than the average officer has an opportunity to know them. At any rate, that opinion turned out, in the end, to be correct. Merchant skippers found no difficulty, under the guidance of a convoy commander, in handling their ships even up to the number of 40, in close formation, 500 yards apart, 1,000 yards between the lines, or maneuvering those ships simultaneously like a school of fish, zigzagging back and forth, in order to get out of the way of submarines.

Sims also stated in 1920:

> The [Navy] department caused serious embarrassment and delays in putting into effect the convoy system which was the most important of all the measures used in defeating the submarine war against allied shipping.
>
> The figures which I have before me show clearly that in April, 1917, the allied cause seemed doomed on account of the losses of tonnage. In the first four months of the year there had been a net loss of over 2,000,000 tons, or 7 per cent of the total Allied and neutral shipping, and the rate of losses had been increasing every month. In the month of April alone the net loss amounted to 800,000 tons, or twice as much as in the whole period of the war before January 1, 1917. It was apparent that these losses, if continued, would soon reduce the tonnage to such an extent that military requirements, and requirements of the population of the allied countries, could not longer be maintained. The imports had already been reduced by 40 per cent from the prewar figures. They could not be reduced further without starving the armies or the civil population of the allied countries. Any further reduction at the rate then existing would have made it impossible to have transported an American army or to maintain it when once across.
>
> With the adoption of the convoy system, and the antisubmarine measures put into effect by the Allies with our assistance in 1917, the losses were gradually reduced until in October, 1918, they amounted to only 100,000 tons. The period between the beginning of the German unrestricted submarine campaign and the armistice can be divided into three phases so far as losses of merchant tonnage are concerned:
>
> First. The period from February 1, 1917, to the end of July, 1917, when American aid was lacking and when shipping was not being convoyed. Average losses, 640,000 tons per month.
>
> Second. The period from August 1, 1917, to February 1, 1918, when there was a partial employment of the convoy system and moderate assistance from America. Losses, 390,000 tons per month.
>
> Third. The period from February 1, 1918, to the armistice, when full cooperation was given by America, and consequently full use could be made of the convoy system. Losses, 250,000 tons per month.

After the war, Rear Admiral Albert P. Niblack, then Director of Naval Intelligence, had this to say:

> I think we are apt to exaggerate the effect of the convoy system in lessening the number of sinkings, because I think we should take into consideration, as Admiral Mayo also points out, the employment of new and offensive measures through the use of depth charges, mystery ships, airships, kite balloons, the laying of mine barrages, the firing of torpedoes from Allied submarines, combined with the use of organized patrols fitted with listening devices and hunting the submarine systematically.

Trade Restrictions

German commercial presence in the United States was very heavy. There were individual investors manufacturers, importers, and almost all of the big commercial houses in Germany had subsidiaries and offices not only in the United States but all over the world.

The investors generally fell into two classes: private investments by individuals in real estate, mortgages, and industrial and transportation stocks, and the major institutional investments "made by combined capital in Germany having close affiliations with the great political and financial powers of the Empire."

On June 15, 1917, Congress passed the Espionage Act. Section 1 of Title VII of the Espionage Act authorized the President to prohibit the export of any article, and Section 3 authorized any collector of customs who had probable cause to suspect a prohibited article was about to be shipped out of the United Sates on any foreign or domestic vessel to hold the vessel until the matter was reviewed by the secretary of commerce.

On June 22 President Wilson created the Exports Council to oversee the administering of the Espionage Act and any proclamations related to it.

On July 9 President Wilson proclaimed that certain categories of goods "shall not on and after the 15th day of July, 1917, be carried out or exported from the United States or its territorial possessions." These banned goods included coal, coke, fuel oils, kerosene, gasoline, bunker fuels, food grains, flour, meal, fodder and feeds, meat, and fats. Fats were used to make explosives.

Prohibited destination countries were listed in the proclamation and included Russia.

On August 21 Wilson created the Export Administration Board to oversee all provisions of Title VII of the Espionage Act. On August 27 the number of contraband items was substantially increased to include "all contrivances for or means of transportation on land or in the water or air" and on the 30th all goods taken out of the United States had to be approved by the Export Administration Board.

Vessels for the Navy

From the beginning, ships and boats were needed for naval and civilian service and early efforts were made to take control of vessels in service and those on the builders' ways.

The first effort was made on March 4, 1917, when Congress passed the Naval Appropriation Act, which gave the Secretary of the Navy limited power to seize civilian vessels for naval use. The early convoy escorts were appropriated under this Act but it proved to be inadequate.

The Act of May 12, 1917, was a resolution authorizing the President "to take over for the United States the possession and title of any vessel within its jurisdiction, which at the time of coming therein was owned in whole or in part by any corporation."

The provisions of The Naval Appropriation Act of March 4, 1917, "were not sufficient to fully meet the situation" and on June 15, 1917, Congress passed another act, "granting broader powers under which possession of property needed by the Navy for prosecution of the war might be taken possession of by the department subject to the payment of just compensation to the owners, which compensation the act in question provided should be fixed by the President." Congress allocated $34,500,000 to requisition ships.

The Act also gave President Wilson "far-reaching authority to requisition, construct, and operate ships without limitations or conditions" within the amount of money appropriated by Congress.

President Wilson appointed a board of experts to which he delegated this duty and appointed Capt. Alexander F. Halstead as senior member of the Board of Appraisal of Merchant and Private Vessels, "To safeguard the interests of the Government and avoid as far as possible the payment of excessive and unreasonable prices for the vessels so acquired...." Other members were Lt. Cdr. Edward C. Kalbfus, and naval constructor Horatio G. Gillmor. Acting Navy Secretary Franklin Roosevelt requested the services of three civilians to serve without compensation. All accepted and were subsequently commissioned lieutenant commanders in the Naval Reserve Force: Stevenson Taylor, president of the American Bureau of Shipping; naval architect A. Loring Swasey, designer of the new 110' wooden subchaser; and shipping man Daniel Bacon of New York. When Capt. Halstead was subsequently ordered to France he was succeeded by Capt. Robert L. Russell. Secretary of the Navy Daniels later stated:

> The board was directed by the department to appraise and fix the fair value of each piece of vessel property which the department considered acquiring, either by purchase or by charter, and the values so fixed were used by the department in its negotiations with the owners of the property. In no case was any vessel property acquired by the Navy if the demands of the owner were materially at variance with the findings of the board as to the fair value of the property. The department found a general wish to cooperate on the part of owners and acquired a number of vessels in this way. The spirit of cooperation was especially noticeable among the owners of yachts and other pleasure vessels, a large number of which were required almost immediately after the entry of the United States as a belligerent.

The Board set up shop at 26 Cortlandt Street, New York, later at 280 Broadway. Marine surveyors were engaged and all the big yachts and fishing boats thought suitable for patrol and escort service were hauled out and inspected. If found adequate they were given a Scout Patrol (SP) number. Not all vessels inspected and numbered were commissioned into naval service. A number of menhaden fishing boats were requisitioned along with 238 yachts for inshore and offshore patrol and escort work. All were Navy-manned.

The designation "SP" engendered considerable interest among the residents of coastal yachting communities like Newport, Rhode Island, who saw fancy, gleaming, polished yachts turned into dark gray boats full of strangers painting and oiling big guns. In the warm days of 1917's spring and summer, the "SP" was said to stand for "Slacker's Paradise," but when the gales of winter arrived, that changed to "Suicide Packet," according to author Frank Hunter Potter in 1919.

All the converted yachts and fishing boats were commanded by regular Navy, Naval Reserve Force, or National Naval Volunteer officers with a few chiefs riding herd over mostly new recruits, many of whom had never been to sea before.

The honor of owning the first S.P. went to Charles Longstreth, of Philadelphia, who owned two wooden steam yachts, the 71' *Arawan II* and the 73' *Zenith*. The *Arawan II* was taken on March 27, 1917, inspected and designated S.P. 1. The boat was armed with one 1-pounder, one .30 cal. machinegun, and commissioned on April 9 under Lt. Cdr. Longstreth. He was a member of the U.S. Power Squadron and patrolled the 3rd Naval District out of Philadelphia until his boat was formally returned to him on November 29, 1918.

The *Zenith* was likewise inspected, acquired on April 21, 1917, numbered S.P. 21, and commissioned on April 23. The boat patrolled the 4th Naval District and was decommissioned on November 21, 1918.

Longstreth was born on April 11, 1868, in Philadelphia, became a machinist and then president of the U.S. Metallic Packing Co. at 429 N. 13th. The company manufactured metallic packing material for stationary and marine engine piston rods and valve stems. He lived with his wife Kathryn at 1681 Locust St. After the war they retired to Coronado, California, and enjoyed traveling. Among the early conversions for offshore work and the first to arrive in France were:

Carola IV—167' steel yacht owned by Leonard Richards, president of Richards & Co., a chemical manufacturer in Stamford, Connecticut. Commissioned on July 7 as *Carola IV* under Lt. (jg) Harold R. Keller. Arrived at Brest on August 29 and commenced patrol duties. In October the boat was moored on the French coast and used for berthing and as a makeshift receiving ship.

Remlik—199' yacht owned by huckster and newspaperman Willis Sharpe Kilmer, of Binghampton, New York. Commissioned on July 11 as *Remlik* under Lt. Cdr. Isaac C. Johnson, Jr.

Emeline—163' yacht owned by Robert T. Graves, of New York. Commissioned on July 14 as *Emeline* under Lt. Cdr. John P. Miller.

Wanderer—197' yacht owned by banker and social gadfly Henry Augustus Coit Taylor, of Newport, Rhode Island. Commissioned on July 14 as *Wanderer* under Lt. Pierre L. Wilson.

Corona—172' yacht owned by physician Henry A. Laughlin, of Philadelphia. Commissioned on July 20 as *Corona* under Lt. Cdr. Lemuel M. Stevens. Departed New York on July 31 and arrived at Brest via Newfoundland and the Azores on August 29.

Guinevere—198' yacht owned by Edgar Palmer, president of the N.J. Zinc Co. Commissioned on July 20 as *Guinevere* under Lt. Cdr. Guy Davis. Departed Newport, Rhode Island, on August 1 and arrived at Brest via Newfoundland and the Azores on the 29th.

Alcedo—275' yacht owned by Philadelphia financier and newspaper publisher George William Childs Drexel. Commissioned on July 28 as *Alcedo* under Lt. Cdr. William T. Conn, Jr. Departed Newport, Rhode Island, on August 4 and arrived at Brest via Newfoundland and the Azores on the 29th.

Wakiva—239' yacht owned by Harry Stephen Harkness of New York. Son of Standard Oil officer Lamon V. Harkness. Auto racer and aviation pioneer. Commissioned on August 6 as *Wakiva* under Cdr. Thomas R. Kurtz. Converted to a minesweeper and left Provincetown, Massachusetts, on August 26 in Capt. Thomas P. Magruder's squadron and arrived at Brest via the Azores, on September 18 as Capt. Magruder's flagship. Reported to the Rochefort District on January 23, 1918.

Bauman—158' freighter *P.K. Bauman* operated by M.M. Davis & Sons at Solomons, Maryland. Commissioned on August 10 as *Bauman* under Lt. Charles F. Chambers. Converted to a minesweeper and left Provincetown, Massachusetts, on August 26 in Capt. Thomas P. Magruder's squadron and arrived at Brest via the Azores on September 22.

Cahill—150' wooden Menhaden steam trawler *Winfield S. Cahill* operated by the Eubank Tankard Co., Kilmarnock, Virginia. Commissioned on August 10 as *Cahill* under Lt. (jg) James B. Will. Converted to a minesweeper and left Provincetown, Massachusetts, on August 26 in Capt. Thomas P. Magruder's squadron and arrived at Brest via the Azores on September 22.

Courtney—155' wooden Menhaden steam trawler *William J. Courtney* operated by the C.E. Davis Packing Co., Reedville, Virginia. Commissioned on August 10 as *Courtney*.

Converted to a minesweeper and left Provincetown, Massachusetts, on August 26 in Capt. Thomas P. Magruder's squadron and arrived at Brest via the Azores on September 22.

Douglas—158' wooden Menhaden steam trawler *Otis W. Douglas* operated by the Douglas Co., Inc., Reedville, Virginia. Commissioned *Douglas* on August 10 under Lt. Grant T. Stephenson. Converted to a minesweeper and left Provincetown, Massachusetts, on August 26 in Capt. Thomas P. Magruder's squadron and arrived at Brest via the Azores on September 22 apparently under Lt. Archibald McGlasson.

Hinton—160' wooden Menhaden steam trawler *John B. Hinton* operated by the Seaboard Oil & Guano Co., Reedville, Virginia. Commissioned on August 10 as *Hinton*. Converted to a minesweeper and left Provincetown, Massachusetts, on August 26 in Capt. Thomas P. Magruder's squadron and arrived at Brest via the Azores on September 22.

Hubbard—155' wooden Menhaden steam trawler *B.H.B. Hubbard, Jr.* operated by the Taft Fish Co., Reedville, Virginia. Commissioned on August 10 as *Hubbard* under Lt. (jg) Edgar S. Husband. Converted to a minesweeper and left Provincetown, Massachusetts, on August 26 in Capt. Thomas P. Magruder's squadron and arrived at Brest via the Azores on September 22.

James—150' wooden Menhaden steam trawler *W.T. James* operated by the Taft Fish Co., Tappahannock, Virginia. Commissioned on August 10 as *James* under Ens. Alexander M. Burr. Converted to a minesweeper and left Provincetown, Massachusetts, on August 26 in Capt. Thomas P. Magruder's squadron and arrived at Brest via the Azores on September 22.

Anderton—the 139' wooden Menhaden steam trawler *Raymond J. Anderton* operated by the Atlantic Fertilizer & Oil Co. Commissioned on August 18 as *Anderton* under Chief Boatswain Frederick L. Muller. Converted to a minesweeper and left Provincetown, Massachusetts, on August 26 in Capt. Thomas P. Magruder's squadron and arrived at Brest via the Azores on September 22.

Rambler—176' yacht owned by Kenneth B. Van Riper, president of the Carbondale Oil Co. Commissioned on October 18 as *Rambler* under Lt. Cdr. Reuben K. Dyer. Departed November 4.

May—239' yacht owned by Joseph R. DeLamar, 43 Exchange Place, New York. Commissioned on October 7 as *May* under Cdr. Frank T. Evans. Arrived at Rochefort on January 6, 1918.

Lydonia—181' yacht owned by William A. Lydon, president of the Great Lakes Dredge & Dock Co. in Chicago. Commissioned on October 27 as *Lydonia* under Lt. Richard P. McCullough.

Galatea—192' yacht owned by Edsel L. Ford, of Grosse Point, Michigan. Commissioned on November 16 as *Galatea* under Lt. Harold D. Hinckley, USCG.

Nokomis—243' yacht owned by Horace E. Dodge, of Detroit, Michigan. Commissioned on December 3 as *Nokomis* under Lt. Cdr. Raleigh C. Williams. Reported to the Rochefort District on March 14, 1918.

When the converted Menhaden fishing boats arrived it became evident almost immediately that they were unsuitable for ocean escort work. The addition of deckhouses, guns, ammunition, and magazines reduced their already low freeboard to the point that they could not ship water fast enough in a seaway. They had made the trip from Boston to Brest without incident under ideal weather conditions but they simply were not fast enough or seaworthy enough for ocean service. The *Rehoboth,* foundered off Ushant Island shortly after arriving. All the fishing boats were subsequently assigned to minesweeping, salvage, and rescue duties or used only for coastwise trips south from Brest when the weather permitted. French forces became responsible for escorting ships between ports.

On June 30, 1917, there were 16,220 A.E.F. personnel in Europe and by December 31, 1917, 183,896 had arrived. These troops landed at Liverpool then went by train to English Channel ports then across by boat or ship to France. A total of 1,127,160 troops were transported in ships controlled by the British government. Under an agreement signed on February 11, 1919, the United States agreed to pay for this service. A soldier transported in first class cost $176.30; second class $128.65; and third class, or steerage, $76.24.

The *Navajo*

The Standard Oil of New Jersey tanker was registered to their foreign subsidiary, the West India Oil Co., established in 1902 to facilitate importing and refining operations in Latin America.

On July 3, 1917, Capt. John F. McDougall was off the Irish Coast when a sub was sighted. A few shots were exchanged and the sub disappeared. The next day, gunfire was heard and wireless operator Clarence Cisin picked up a distress call. Awhile later they saw a large sailing vessel being chased by a submarine. The sub sank the sailing vessel and then headed for the tanker. Chief Mate Harry L. Ham, of Chicago, reported:

> On July 4, at 9:20 a.m., heavy gunfire was heard to starboard and shortly afterward the *Navajo* ran out of the mist and sighted a submarine firing on a British topsail schooner about two miles away. The *Navajo* changed her course, the fog shut down again and the sub was lost sight of. This was about 55 miles west of Cape La Hague, France.
>
> About 2:55 p.m. the same day the fog lifted and two shots were heard from a point 1,500 yards distant. Upon observation a submarine was seen firing with both guns at the *Navajo.* The first shot dropped 50 yards short on the starboard beam. The second one went over the ship. The *Navajo* swung, bringing the submarine about three points on the starboard quarter and opened fire with her after gun.
>
> The submarine fired about 40 shots during the engagement, which lasted 40 minutes, one of which hit the *Navajo* underneath the port counter. This shell exploded before hitting the ship and displaced some of the plates, causing the *Navajo* to leak.
>
> The *Navajo* in return fired 27 shots, the last two of which were hits. The twenty-seventh shot struck the submarine just forward of the conning tower where the ammunition hoist was located, causing an explosion on board the submarine which was plainly heard on the *Navajo.* The men who were on the deck at the guns and had not jumped overboard ran aft. The submarine then carted forward at almost 40 degrees and the propeller could be seen lashing the air. Nobody was seen coming up through the conning tower and jumping into the sea, nor were any survivors seen. It is my opinion that the submarine was sunk.

The sub had two 6" guns, the tanker had one 4" gun on the stern and an Armed Guard crew of 13. Edward Schafer was 3rd mate in the crew of 29. A British seaplane came out and forced the sub to dive but that afternoon when the fog had lifted and the seaplane returned to its base, the sub surfaced again and commenced firing. A British destroyer responded to the tanker's S.O.S. and the captain confirmed the sinking after observing a large amount of oil and debris. The sub was reported to be very large, some said 400'. The tanker lost a lifeboat and suffered damage to the stern hull plating. They went to France for repairs and on August 1, one day off the coast of Ireland en route home, an explosion occurred in a fuel oil pipe and the ship caught fire and sank. There were no casualties and all were landed at Limerick. They left Liverpool on August 15 and arrived at New York on the liner *New York* on August 24.

President Wilson's Requisition Order

On July 11, 1917, President Wilson issued Executive Order 2664, which essentially nationalized the merchant marine:

> By virtue of authority vested in me in the section entitled "Emergency Shipping Fund" of an act of Congress entitled "An act making appropriations to supply urgent deficiencies in appropriations for the Military and Naval Establishments on account of war expenses for the fiscal year ending June 30, 1917, and for other purposes," approved June 15, 1917, I hereby direct that the United States Shipping Board Emergency Fleet Corporation shall have and exercise all power and authority vested in me in said section of said act, in so far as applicable to and in furtherance of the construction of vessels, the purchase or requisitioning of vessels in process of construction, whether on the ways or already launched, or of contracts for the construction of such vessels, and the completion thereof, and all power and authority applicable to and in furtherance of the production, purchase, and requisitioning of materials for ship construction.
>
> And I do further direct that the United States Shipping Board shall have and exercise all power and authority vested in me in said section of said act, in so far as applicable to and in furtherance of the taking over of title or possession, by purchase or requisition, of constructed vessels, or parts thereof, or charters therein; and the operation, management, and disposition of such vessels, and of all other vessels heretofore or hereafter acquired by the United States. The powers herein delegated to the United States Shipping Board may, in the discretion of said board, be exercised directly by the said board or by it through the United States Shipping Board Emergency Fleet Corporation, or through any other corporation organized by it for such purpose.

On August 3, 1917, the Shipping Board formally requisitioned all vessels over 2,500 deadweight tons then under construction, "with the materials, machinery, equipment, or commitments for materials, machinery or equipment necessary for their construction or completion." This amounted to 437 ships. Of the 437, 13 were subsequently released, 10 were cancelled, and 12 contracts were transferred to the Shipping Board. Fifth Amendment "just compensation" settlements were completed with American owners, leaving some dissatisfied. On or shortly after August 4 most shipbuilders received a telegram:

> By virtue of an Act approved June 15th, 1917, and authority delegated to the Emergency Fleet Corporation by Executive Order of July 11th, 1917, all power driven cargo-carrying and passenger vessels above 2,500 tons, deadweight capacity, under construction in your yards, and materials, machinery, equipment and outfit thereto pertaining, are hereby requisitioned by the United States and will be completed with all practical dispatch. Letter follows.

It was decided by the Corporation to retain control of foreign-flag vessels being built in the United States and operate them under the American flag for the duration of the war plus six months. The owners could then reclaim their title by paying the Board for all expenses incurred in completing the ship. This decision created significant legal issues and possible treaty violations. French and British owners objected very strongly. The Board's reasoning was stated:

> The members of the Shipping Board feel that the primary obligation of the American government is to its own troops in France, or to the troops which soon will be there; that the government must leave nothing

> undone to insure maintenance of the line of communications across the Atlantic Ocean and that all ships now being built must be kept at the disposal of the United States. Where the needs of the Allies prove greater than those of the United States, exceptions may be made, but each case will be considered on its merits, and the policy of the American government will be to finish all ships in American yards with the idea that they shall fly the American flag.

It was not contemplated in the Act creating the Shipping Board that the government would operate ships or by any corporation, "in which the United States was a stockholder unless it should prove impossible to procure private enterprise to purchase or charter them under proper terms and conditions."

On August 8, 1914, Congress repealed the law forbidding foreign-built ships from being admitted to American registry and by June 30, 1915, 147 had been admitted. Of these, 96 were British, 30 German, 5 Cuban, 6 Belgian, 5 Mexican, 1 each from Rumania, Uruguay, and Chile, and 2 from Norway. These ships were operated by American companies.

While many steamship companies operated one or more "tramp" steamers that roamed from port to port as cargo became available, most companies maintained a specific trade route for passenger service, a specific cargo, or general merchandise. The company had established relationships with producers, agents, commission houses, freight brokers and forwarders, and customs officials the shipmaster and agent routinely dealt with in the course of the company's business.

The war changed the established patterns of trans–Atlantic shipping and created different priorities. In its first annual report, the Shipping Board stated in regard to private operators:

> To supply these needs, it often has been necessary to take vessels out of trades which in normal times would have been regarded as of great importance. The Board in this connection is pleased to testify to the patriotic spirit in which shipping interests have met the situation, even where it has entailed serious individual loss.
>
> The Board has continuously borne in mind, however, that too drastic a curtailment of commercial transportation would defeat its own ends. Raw materials essential to the production of military supplies and nitrates vitally needed to sustain the output of farms could not be neglected. With the acute congestion of rail traffic, coal must be transported by water, so that the war industries of the Atlantic coast shall not be deprived of sustenance.

Government contract or requisitioned ships were turned over to a Manager or a Manager/Operator. Some steamship companies only managed ships, others managed and operated them. A Manager took care of the business end of things for an Operator whereas the Operator got the ship from port to port. The larger steamship companies usually performed both functions.

Any period of a speculative nature, and especially wartime, always brings out "fly-by-night" operators and the Great War was no different. The so-called "mushroom" steamship lines sprang up and many failed to deliver cargoes as agreed. Forwarding agents spent a lot of time investigating companies for their clients and the Shipping Board did the same.

Only those firms with sufficient foreign voyage experience were selected as Operators for offshore routes. A company that owned and operated ships suitable for ocean service but only used them for coastwise or inland trade would not be considered a suitable Operator for a foreign voyage.

Because Great Lakes steamers were urgently needed, 21 were requisitioned for ocean service. Of these, 16 were too long to pass through the Welland Canal so they were cut in two and re-attached on the St. Lawrence River, 12 of them while still afloat, "an achievement never before accomplished." Newly constructed ships were also brought down.

In normal times, it was against the law for a foreign-built ship to engage in American coastwise trade but on October 6, 1917, Congress suspended the law to free up American ships for foreign military service. The suspension was good for up to 120 days after the war ended.

The Norwegians

In early 1917, Wilhelm Jebsen, of Norway, signed a contract with Union Iron Works in San Francisco to build a 10,500 deadweight ton tanker for $78.57 per deadweight ton. The total cost would be approximately $1,035,000. The keel was laid on April 11 and delivery was to be in June or July 1917. Work was slower than expected and on August 3, 1917, the ship was requisitioned by the Shipping Board at 93 percent complete. Jebsen had then paid $776,250 toward the ship. It was delivered to the Shipping Board on October 18.

The seizing of foreign-owned vessels caused concerns over possible treaty violations. The Christiana Group hired Charles S. Haight, a New York attorney, to represent Norwegian shipowners in claims against the United States when it was alleged the Norwegians had defrauded the government out of $23 million. When Haight testified before Congress, his position was the that the ships were seized illegally pursuant to the treaty of 1827, as he stated to a Congressional committee:

> In my judgment it involved a clear violation of our treaty with Norway if we seized, as we eventually did, the vessels building under Norwegian contracts in American yards.
>
> When the war started in 1914 we had no American shipyards at all for the construction of trans-Atlantic tonnage. Our yards were confined to Cramps, Newport News, and one or two other yards which did practically nothing but build naval vessels and coastwise vessels, which under our law, had to be built in American yards. We had no standardization in ship construction, and nobody in the commercial world ever thought of coming to an American yard to build an ordinary ocean steamer.
>
> When the war started, however, American shipyards became a necessity if individuals or corporations wanted ships built at all and the Norwegians were the first, and for many months, practically the only shipping people who had the courage to place contracts with American yards at the prices that those yards demanded if they were going to turn out a ship on a profitable basis.... Up to January 1, 1917, the Standard Shipbuilding Corporation had made 7 contracts, 6 of which were with Norwegian owners and 1 with a Swedish owner. The Chester Shipyard had made 14 contracts, 12 of which were with Norwegian owners and 2 with Americans. I think it is safe to say that Norway placed about 90 per cent of all the contracts that were placed in American yards and, in truth, their money built the most of our new yards.
>
> But you will find as to the Cunard Co., the French government, and others, when they came to America and began buying contracts, it was the Norwegian contracts that were in existence and which they bought.
>
> When we requisitioned the Norwegian tonnage, you gentlemen will remember that we were in the midst of a period of utmost confusion in Shipping Board affairs. Between the time that the Congress authorized the President to requisition contracts and ships and yards and everything else and the date upon which the requisition order was finally promulgated the controversy had come between Mr. Denman, the chairman of the board, and General Goethals, who was the general manager of the Fleet Corporation. Just before the requisition order of August 3, 1917, Mr. Denman had resigned and General Goethals had resigned, and with General Goethals, overnight you might say, the entire legal staff of the Fleet Corporation departed. Mr. Brent, vice chairman of the board, had resigned, and Captain White had resigned about the same time.

Haight believed that as the result of all the internal turmoil no one then knew of the existence of a treaty with Norway nor of any of its amendments or provisions. With regard to the *Wilhelm Jebsen* he stated:

Tank boats were almost unprocurable and yet, as one of the glaring illustrations of fraud perpetrated on this government that Wilhelm Jebsen received $2,045,000 for a boat of 10,500 tons, when that contract could have sold for not less than 3½ million dollars to 40 buyers on the 2nd day of August. There was a large payment still due upon the completion of the contract.

The government requisitioned 182,500 tons of shipping. Haight stated:

To go back to my statement for just a moment. Even taking $168, which was the average price paid per ton on the contract that had been assigned and converting the unassigned contracts to that same low value, you have, with two years' interest added at 6 per cent, $34,560,000. Discarding all of the tremendous values that we know were to be had for all the ships that were ready or nearing completion.

Another way of analyzing these figures is to take the tonnage, 182,500 tons, divide $34,500,000 by that tonnage, and you will find the United States government paid, by and large, for all these contracts, an average of $188.95 with no interest at all.

I know, and I urged over and over again that the taking of those Norwegian boats would be a willful waste of money. That the United States government could never get the boats completed as we could do at our plain, flat, contract price. And it turned out that they could not. I never had a single boat building in an American yard for a private owner which was finished before August 3, 1917, upon which the builder even requested an increase in the price of his ship. Not one. And when the contracts had started at $90, $95, $100, and the boats, upon completion, were worth $200, we got our ship and we got her at the contract price. But what happened when the United States government requisitioned everything? Just what always happens. The price was boosted for the benefit of the government. Instantly the yards said, "Materials are costing more and labor is costing more. You have got to do it on a cost-plus basis." And the boats that we would have had finished at our contract price the government was obliged to pay hundreds of dollars more for, sometimes literally. Our contract price of $165 or $175 probably cost the Shipping Board $270 or more per ton. I did my best and said, "Let us have the ships under our own contracts. We will get them at a contract price. We will then turn them over to you for your wartime operation." They wouldn't have it.

That is the reason why the cost-plus method of funding shipbuilding during World War II was outlawed.

The Enemy's Piers for Troopships

On October 6, 1917, Congress passed the Trading with the Enemy Act (40 Stat. L. 411) and on October 12 the War Trade Board was established as an independent agency under the Act. The War Trade Board was granted the authority to issue export licenses for everything but coin, bullion, or currency. Secretary of Commerce William C. Redfield had the power to "review the refusal of any Collector of Customs under the provisions of sections 13 and 14 of the Trading with the Enemy Act to clear any vessel, domestic or foreign, for which clearance is required by law."

Section 6 created the Office of the Alien Property Custodian and on October 22 President Wilson appointed lawyer and ex–federal judge Alexander ("A. Mitchell") Palmer, 45, as Custodian to hold in trust all property seized.

The amendment of March 28, 1918, to the Trading Act gave Palmer "the general power of sale" and a Bureau of Sales was created with Joseph F. Guffey as director. The intention was to turn over the proceeds of any sale after the war to the owners, but the property itself would be forever lost to the owner. After the war, Palmer stated:

When it is recalled that Germany's great hope for the future lay in the industrial conquest of this continent—a conquest which she believed in 1914 she was in a fair way to accomplish—it will readily be seen

> with what dismay her financial and commercial interests must have viewed the sudden and direct offensive which was designed to forever loosen the grip which she had obtained upon American industry and commerce.

The general idea was to transfer enemy property to America's war machine. As Palmer later stated:

> When the armistice was signed the Alien Property Custodian was supplying the Government with magnetos for aeroplanes and automobile motors, with cloth to make uniforms for the soldiers and the dyes with which the cloth was dyed, with medicines, surgical instruments and dressings, with musical instruments, with ball bearings, telescopes, optical instruments and engineering instruments, with coconut charcoal for the making of gas masks, with glycerin for the making of high explosives, and a large number of other and varied products.

Over 1.5 million pounds of nickel and several hundred thousand pounds of ferrovanadium was given to the War Department and 98,000 bags of coffee and 10,000 bags of rice were turned over to the Food Administration. Chronometers, binoculars, and astronomical instruments were turned over to the Navy Department.

A policy of this nature caused lawyers to salivate profusely over 5th Amendment settlements but during the war, when patriotism and nationalism was at a fever pitch, lawyers were generally reluctant to take these cases and only one or two reached the courts before the Armistice. One that did was filed by the Hamburg-American Line.

In 1917, piers in and around New York were in great demand. When the Army went shopping for a marine terminal from which to embark troops, they couldn't find any suitable property available to them at rates the government felt were reasonable and properties that had reasonable rates were unsuitable.

Hamburg-American had a terminal at the foot of 33rd Street in Brooklyn, which the company leased from New York City for $120,000 a year. Hamburg-American and North German Lloyd both had property at Hoboken, New Jersey, and at St. Thomas in the Virgin Islands.

Hamburg-American's Hoboken piers were operated by a wholly-owned American subsidiary, the Hamburg-American Line Terminals & Navigation Co., organized under the laws of New Jersey. William G. Sickel was president, Carl R. Schurz secretary, and Emil Lederer treasurer.

The company had three piers within the Port of New York. Pier No. 1 was built in 1900, of fireproof construction, 951' × 90', and double-decked for about 900'. Pier No. 2 was a wood pier built in 1882, 904' × by 81', single-decked and covered for about 930'. Pier No. 3 was another wooden pier built in 1882, 767' × 50' and covered for about 717'. At the shore end of each pier were large sheds, a 100' × 100' bonded stores warehouse, and a residence for the superintendent.

The slips were kept dredged to accommodate deep-draft steamers but after the war started no ships berthed there. Two steam tugs, one steam launch, one floating coal hoist, one covered barge named *Columbia,* and three open ash scows were stored there. When America entered the war everything was in good repair.

Adjoining property belonged to the North German Lloyd Dock Co., a subsidiary the North German Lloyd Steamship Co. formed to operate their docks. North German Lloyd Steamship held a 999-year lease on the property from the Dock Co. and paid a nominal cost to cover carrying charges, taxes, insurance, and interest on an outstanding mortgage. Charles von Helmolt was the manager at 5 Broadway. When the Army had this property assessed, it was valued at $4,831,705. On April 17, 1917, Col. John M. Carson confiscated the property and paid $4,831,705 to Palmer's office. From there the money was deposited in the Treasury to be invested in Liberty Bonds.

When Army surveyors looked at the Hamburg-American property they described as "the largest and best equipped docks for ocean liners not only in the port of New York but in the United States." It was just what the War Department needed as a port of embarkation.

The same day he seized North German Lloyd's property, Col. Carson notified Hamburg-American officers William Sickel, Karl Bunz, and Julius P. Meyer that he was going to physically occupy all of their property at Hoboken for War Department purposes and he also seized the floating property there.

On March 28, 1918, President Wilson, acting under the authority of an act of Congress, "took over title and possession of these docks for the United States" and on April 16, A. Mitchell Palmer seized the assets of the Hamburg-American Line Terminals & Navigation Co. over the company's fierce protest. Physical possession of the property took place on June 28 and the same day, by proclamation, the value of Hamburg-American's property was declared to be $2,314,877.

The Hamburg-American Line Terminal & Navigation Co. was a New Jersey corporation that owned not only the title but the real value of the piers themselves. The company remained as trustees of the corporation and a just compensation payment of $2,314,877 was paid to the company on January 5, 1919, but the company felt they were owed $2,400 a day plus interest on rent for their property from April 17, 1917, when the Army occupied it until June 28, 1918, when it was formally seized by Palmer.

The attorney general did not dispute the validity of the company's claim, only the amount due. The government thought $918.83 a day was a fair price so the case went to trial. The company claimed their property was taken on April 6, 1917, the date the United States entered the war, but the court held it was taken when Col. Carson gave them official notice and determined the fair rental value was $966.86 a day. The company was awarded $422,517.82 on May 2, 1932.

The fair value of all the floating property in April 1917 was assessed at $199,682.98. The original purchase price was $120,659. No money was paid to the company for this property so the company was awarded $199,682.98.

When their company was seized, Karl Bunz was jailed at Fort Oglethorpe and the British seized 14 of the company's ships.

The Hamburg-American Line Terminals & Navigation Company ceased to exist in 1926 but the principal officers remained as trustees to settle and collect debts.

As of February 15, 1919, about 35,400 reports of enemy property had been received and property valued at $502,945,724.75 was held by 32,296 separate trusts. About 9,000 of the reported properties had not yet been assessed but was expected to bring the total to around $700 million. Other companies seized were:

Atlas Shipping Co., 11 Broadway, Rm 507, New York. C. Gordon Rose and William M. Rose, managers. 100 percent enemy owned.

Brynhilda Shipping Corp., 17 Beaver St., Rm 16, New York. Sidney E. Wolfe, president, Charles W. Lowack, treasurer, Estelle E. Frank, secretary. 93 percent enemy owned.

Cargo Transportation Corp., 60 Wall St. 100 percent enemy owned.

Foreign Transport & Mercantile Corp., 17 Battery Place, New York. 96.1 percent enemy owned. (a subsidiary of Richard Wagner's American Transatlantic Co.).

Gans Steamship Line, 12 Broadway, Rm 701, New York. Richard Meyer, president, Herman Strangfeld, secretary, Max J. Lehmitz, treasurer. 15 percent enemy owned.

K. & E. Neumond, 15 Broad St., Room 1418, New York. 100 percent enemy owned.

Lutz Shipping Co., Millville, Florida, incorporated February 13, 1917, 503 Thiesen Building, Pensacola. Edward Lutz, president in Bremen, Germany. Ignacio J. Diaz, the Mexican vice-consul, was involved. Lutz was apparently the shipping agent for the German-American Lumber Co. in Millville. 100 percent enemy owned.

Seguranca Steamship Corp., 42 Broadway, Rm 1045, New York. Max H. Schultze, president, Hans Roehl, treasurer. 100 percent enemy owned.

Vogemann Shipping Co., 1119 Whitney Building, New Orleans. W. Dickman, manager. 100 percent enemy owned.

The Ship Operational Order

On October 6, 1917, pursuant to President Wilson's Executive Order 2664, the Shipping Board sent a telegram to all shipowners with instructions to acknowledge receipt:

> The United States Shipping Board gives notice to all owners of ships registered and enrolled under the laws of the United States that the requisition of all American steamers described below and of which previous announcement has been made will become operative and effective on October 15, 1917, at noon.
>
> 1. The ships affected by said requisition and included therein are (a) all cargo ships (and tankers) able to carry not less than 2,500 tons total dead-weight, including bunkers, water, and stores; (b) all passenger steamers of not less than 2,500 tons gross register.
>
> 2. (a) As to all steamers in or bound to American ports on October 15th, 1917, requisition becomes effective after discharge of inward cargo and ship is put in ordinary good condition.
>
> (b) As to steamers which have started to load their outward cargo, requisition becomes effective at noon on October 15, 1917, and accounts as to hire and expenses will be adjusted from time steamer began to load.
>
> 3. Steamers trading to and from American ports that have sailed on their voyage prior to October 15, 1917, at noon, are to complete that voyage as promptly as possible and report for requisitioning.
>
> 4. Steamers that are occupied in trades between foreign ports shall be requisitioned as of October 15, 1917, at noon, and accounts adjusted accordingly.
>
> 5. (a) Owners whose steamers are operating in their regular trades are to continue the operation of their steamers for account of the Government as they have been doing for themselves until they receive further instructions.
>
> 6. Owners whose steamers are chartered to others will apply to the Shipping Board for instructions regarding the future employment of said steamers. Acknowledge.

Not all freighters were physically taken and the tankers were operated by their owners on routes designated by the Shipping Board. The Shipping Board determined the route and use for every vessel. A time-charter could be for one specific voyage or a specific period of time for operation on the chartering company's account.

At this time the rates for commandeering vessels under the Act of June 15, 1917, "Emergency Shipping Fund," was tentative. The Shipping Board's goal was to set rates that applied to classes of vessels over 2,500 deadweight tons rather than to each individual vessel based on deadweight tonnage and speed.

On September 27 the Shipping Board decided that $175 per deadweight ton was the average value of requisitioned vessels. A profit margin of 25 percent per year and .50¢ per ton was added for overhead expenses and .01¢ per ton for convenience in calculations resulted in a deadweight tons rate of $4.15. Vessels making over 11 knots got .50¢ per ton per month for each knot or part of a knot over 11.

Time charters rates for cargo ships and tankers were set: over 10,000 deadweight tons, $5.75; 8,000–10,000, $6.00; 6,001–8,000, $6.25; 4,001–6,000, $6.50; 3,001–4,000, $6.75; and 2,500–3,000, $7.00.

Passenger vessels were rated on their gross tons and were either Class A, 150 passengers or greater, or Class B, 75–150 passengers. Normal charter agreements were time charters but vessels for the Army and Navy were taken on a bareboat basis. The tentative rates announced by Bainbridge Colby of the Shipping Board were to become effective on October 15, 1917.

A Naval Base and Haven for Merchant Seamen

In the fall of 1917 the converted yacht *Carola IV* was found to be unseaworthy and was moored at Brest under the ruins of a fortification built by Julius Caesar and renovated by Napoleon. It was located "above the narrow neck that leads to the harbor" and was known to locals as "The Chateau." The yacht was used as a floating hotel and makeshift receiving station but this soon proved to be inadequate.

SC-1, the first of the new 110' wooden subchasers, was commissioned on October 1, 1917, with many more to follow. Each had a crew of two officers and 25 enlisted men and the converted yachts and fishing boats along with the new destroyers being built added to the need for a receiving barracks that would serve as a transfer site for naval personnel as well as a facility to house and care for survivors from torpedoed merchant ships.

Adm. Wilson acquired the old fort and detailed Lt. John D. Pennington to transform it into a modern naval station, patrol headquarters, and training center. There was no galley or food storage, no electricity, water, sewer, or dormitory. A range was borrowed from the local hospital. Fifty sailors from the ships at Brest began cleaning and bringing in equipment while engineers got to work. Help was provided by the Red Cross and the Y.M.C.A. Medical, Dental, and Supply departments were staffed and on February 8, 1918, Carola Barracks was commissioned. The facility could bunk 2,500 and feed 5,000. During July 1918, over 800 survivors from torpedoed merchant ships were cared for.

At first, almost all the ships coming from the United States directly to France put in at Saint-Nazaire but when the big seized German liners *Mount Vernon*, *Agamemnon*, *America*, and *Von Steuben* became available they had no choice but to land at Brest. While Army engineers were working at Brest and the big transports arrived, troops had to be ferried ashore by the converted yachts and fishing boats that were serving as mine sweepers.

Likewise, when large numbers of cargo ships began coming in to Brest it became necessary to discharge cargo onto lighters and use tugs to bring them in. Three boats were acquired:

The 139' tug *Luckenbach No. 1* was chartered and commissioned as *Barnegat* on October 12 under Lt. (jg) Patrick S. Farley. Departed Philadelphia with a convoy on December 20 and arrived at Brest on April 13, 1918.

The 140' tug *Concord* was purchased from the Staples Transportation Co., of New York, and commissioned on November 20 as *Concord* under Lt. (jg) Enoch M. Gracie.

The 135' tug *Gypsum Queen* was chartered by the Navy from the J.B. King Transportation Co. at New York, commissioned on December 4 as *Gypsum Queen*, and stationed in the River Gironde. The tug reported to the Rochefort District on July 11, 1918.

The *Concord, Gypsum Queen*, and *Galatea* left Philadelphia on December 15 and headed for Bermuda to tow three French subchasers to the Azores. The *Galatea* sustained damage and remained in the Azores while the other two proceeded to Brest.

Between January 31, and February 19, 1918, the *Rambler, Corsair, Noma, May, Nokomis, Aphrodite*, and *Hubbard* were sent to Rochefort.

These boats were manned mostly by members of state naval militias, reserves, and volunteers. According to Adm. Wilson, "It was these yachts, together with the French vessels, that bore the burden on the west coast of France during the first months after our entry into the war."

About 49 percent of A.E.F. personnel landed in England, the rest in France.

A Few Escapees

The *J.L. Luckenbach*

In October 1917, the freighter was en route to Le Havre, France. Early on the morning of the 19th lookouts spotted what they thought was another steamer. The craft came to within half a mile before they realized it was a sub. At 7:40 a.m., at 48°08'N/09°31'W, both ships commenced firing. Chief Boatswain's Mate Joseph B. Trautner was in charge of the Armed Guard. The ship was hit numerous times topside causing heavy damage and disabling the engine. Destroyer *Nicholson*, 82 miles away, picked up a distress call and responded at flank speed.

Messages were sent and returned:

> S.O.S.—*J.L. Luckenbach* being gunned by submarine.
> We are coming.
> Our steam is cut off. How soon can you get here?
> Stick to it. Will be with you in three hours.
> Shell burst in engine room. Engineer crippled. Fire in our forehold. They are shooting at our antennae. How far away are you? Code books thrown overboard. How soon will you arrive?
> In two hours.
> Too late. Look out for boats. They are shelling us.
> Do not surrender.
> Never.

Chief Boatswain's Mate Henry Boljes acted as spotter from the top of the wheelhouse while Trautner supervised the guns. Trautner reported, in part:

> After firing about 35 rounds from the after gun, the breech-block jammed. The plug could not be opened. Shortly after that a shot landed near the stern and exploded, throwing pieces of shrapnel all around the after gun, one piece striking the pointer's sight and knocking it off the gun, putting the after gun out of commission entirely. The same shot carried away the after telephone. As soon as the after gun was put out of commission the smoke boxes were used to make a smoke screen. Both the boxes and machine were used,

> but as the wind was blowing in the wrong direction they did not do much good. The ship was slow answering her helm to try and maneuver to keep in the track of the smoke.
>
> The submarine had the range down to perfection and maneuvered in such a manner as to keep out of range. The *Luckenbach* was firing at extreme elevation but all the shots were falling about 1,500 yards short. It appeared as though the submarine was also firing at extreme elevation from the manner in which the shells fell. At least 225 rounds were fired by the submarine, out of which there were only nine clean hits. The shells were mostly shrapnel, as pieces were falling all around the deck. Two shots landed on the port side forward, striking the oiler's room and putting a large hole in the side. One landed forward on the port side at the water line, hitting the fresh water tank and destroying most of the fresh water supply. Another landed in the petty officers' mess room and exploded putting the ash hoist out of commission and bursting the steam pipe, also wounding two men, one mess-boy and one fireman. One shot passed through the weather screen on the bridge and landed on the deck in the cargo, exploding but not starting a fire. Pieces of shrapnel hit Houther, one of the armed guard, in three different places. One of the ship's crew who was carrying ammunition forward was hit. Another shell exploded in the engine room, wounding the first and third engineers and two oilers and putting the engine out of commission. Bell, one of the gun crew, was going up the forward ladder carrying ammunition when the shell that landed in the quartermaster's room exploded. The fumes from the shell blinded him for about two hours.

Around 11 a.m. lookouts on the greyhound spotted smoke and then saw the U-boat. *Nicholson* opened fire while the U-boat fired again at the freighter. When a third shot from the destroyer came very close, the sub submerged at 11:40 a.m. and got away. The freighter fired 202 rounds, 167 of those from the forward gun after the stern gun was wrecked.

A doctor was sent over to tend the wounded. Two Armed Guard sailors were hurt. Engineers got the ship going and they proceeded until running aground at 3 a.m. the next morning. The tide floated them off and they got in port at 11.

It was the longest recorded battle between a merchant ship and a U-boat in which the ship escaped. The sub was later identified as the *U-62*.

Boljes was on the *Luckenbach* from August 4, 1917, to December 16, 1918. He was born in Germany on November 6, 1867, came to America on April 10, 1884, and was naturalized on October 14, 1897. Two of Chief Boatswain's Mate Joseph B. Trautner's Armed Guard crew were wounded. The wounded merchant crew were:

Fermin Arnet, seaman, French, 71 Woodrow St., Marseilles, France.
Vasile Radulescu, seaman, Roumanian, 51 Warren St., Brooklyn.
Drew B. Saunders, 1st engineer, American, 18 Grove St., New York.
John Carpenter, 3rd engineer, American, father A.P. Carpenter, Bessemer City, North Carolina.
John Charisses, oiler, Greek, 57 Van Brunt St., Brooklyn.
Nick Zaranis, oiler, English, 201 E. 30th St., New York.
Juan Ramos, messboy, Porto Rican, 393 Gold St., Brooklyn.

The *Nyanza*

At 9:30 on Sunday morning, January 13, 1918, about 30 miles off Penmarch, France, the Kerr freighter *Nyanza* sighted a periscope about 1,000 yards away at the same time a torpedo track was seen heading for them. A sharp turn to port caused the fish to miss by ten feet. Chief Gunner's Mate Benjamin H. Groves' crew opened fire. The sub surfaced and commenced a zigzag course astern of the freighter. Groves reported:

At first her shots fell short, but eventually he got our range and hit us five times. One shot passed through the after gun platform, through the wood shelter house, through the iron deck, breaking a deck beam, exploding in the hold, and passing out through the side of the ship. One shot exploded in the armed guard's mess room, wrecking the place completely. Two shots exploded in a steam locomotive on deck, doing some damage. One shot hit the stern of the ship, but did not go through.

About 11:15 the submarine had our range good again. The ship zigzagged a little, which caused his shots to fall a little to the right or left of our ship. At the same time, I had his range and fired four shells quick at 7,800 yards, causing him to come broadside to and keel over, then suddenly disappeared just as he had our own range good. This leads me to think he did not quit from choice, but from necessity.

The engagement lasted two hours and 30 minutes. I fired 92 rounds, and the submarine fired approximately 200.

On May 19, 1918, at 12:14 p.m. the *Nyanza* sent an S.O.S. about 300 miles off Maryland saying they were being fired on by a submarine heading west. They managed to escape. This was the first indication that a U-boat was in American waters.

The *Kermanshah*

In July 1918, Capt. Robert H. Smith's Kerr freighter *Kermanshah* left Le Havre, France, in ballast for New York via Plymouth. The ship had a 21-man Armed Guard force under Chief Gunner's Mate Matthew Coffey.

At 11:45 a.m., on the 18th at around 38°24'N/68°41'W, Capt. Smith was in the chart room on the port side when he noticed what appeared to be a torpedo wake out the porthole. He ordered the ship turned 4 points to starboard and an "A.L.L.O." and S.O.S. sent out. He ran out on deck and watched the torpedo pass astern by about 10 to 15 feet. Just then another torpedo was seen approaching from starboard and Capt. Smith put the helm hard over to parallel the missile's course and it passed by about 5 feet away. Chief Coffey's men were already on the 4-inch stern gun and they fired a shot "at a spot the wake started from" and heard an explosion "like the bursting of the shell against a hard object." Capt. Smith proceeded north at full speed, 9.5 knots, and about 15 or 20 minutes later the sub was sighted on the horizon about 4 miles away. Coffey's crew immediately commenced firing but after a few rounds it appeared the sub was out of range and they stopped. The sub made no further attacks and Capt. Smith concluded they had been damaged by the first shot.

Later that day, one of the Armed Guard sailors reported seeing a small, two-masted vessel. Capt. Smith expressed the opinion that it was a disguised sub on the surface but nothing came of it. At about 7 p.m. that evening, when they were about 60 miles north of their noon position, Capt. Smith was just about to turn west when lookouts sighted a periscope and the gun crew again opened fire. The periscope disappeared and they made port. It was later determined the sub was the *U-140*.

The *Westward Ho*

The brand-new turbine freighter was managed by Luckenbach. In August 1918 the ship was in a New York-to-Brest convoy escorted by the French light cruiser *Dupetit Thouars*. The cruiser would normally have turned back at 22°W to pick up another convoy but the French Ministry changed the plans and kept the cruiser on station.

At 10 p.m. on August 7, 1918, the *Dupetit Thouars* was torpedoed and soon sank and at

6:40 the next morning the *Westward Ho* was torpedoed by *U-62* in the Bay of Biscay about 350 miles off the coast. A group of destroyers were en route to meet an incoming troopship convoy and responded to the distress calls. They picked up all the crew from the freighter and continued west-bound leaving the freighter in a sinking condition. Meanwhile, Lt. Cdr. Charles C. Windsor's *May*, Lt. Cdr. Howard H.J. Benson's *Noma*, and Capt. Douguet's French corvette *Cassiopee* had left a west-bound convoy and were headed back and also responded. Lt. (jg) Thomas Blau, USNRF, led a crew aboard the sinking ship and they managed to get steam up and pumps going despite having no experience with oil-burning machinery. The ship was down by the head and they decided that attempting to steam ahead would not only be too slow but would likely put more water into the ship so they steamed 315 miles stern-first back to Brest with the assist of tugs. The ship was repaired by crews from the Navy tug *Favorite* and the Repair Unit.

This salvage effort was considered one of the finest feats of seamanship during the war. The ship was subsequently requisitioned by the Navy at Brest.

The *Montoso*

On the evening of August 27, 1918, Capt. Arthur O. Forsyth was heading home from France in a convoy when a lookout reported a "suspicious looking wake" on their port beam. Capt. Forsyth was eating and by the time he reached the bridge, the wake was gone. He reported the sighting to Lt. Cdr. Paul C. Grening, the convoy commander in the Navy transport *Rondo*. Lt. Cdr. James Jonas Madison was the captain of the nearby transport Navy *Ticonderoga*. Capt. Forsyth asked Grening if it wouldn't be wise to "make a radical change in our course after dark in order to prevent our being followed successfully." Grening then ordered a 20° turn to starboard.

At 9 p.m. Capt. Forsyth observed "the outline of a long, low dark object on our port beam about 2,000 yards away, showing about 4 feet." Capt. Forsyth reported:

> I called the attention of the chief boatswain mate, commanding the Armed Guard, to it. He immediately went to the after gun and opened fire. In the meantime I ordered course to starboard to bring the object on our port quarter, leaving it a little open so that forward gun would bear on it. We fired six shots from aft and four from forward, and the *Ticonderoga* fired two. The *Rondo* was unarmed. Submarine making about 10 knots, we about the same.
>
> Just after we opened fire, two shots fell a few hundred yards astern of us. Evidently the submarine had lost sight of us and was firing at the flashes of our after gun. If so, his range was good, but direction poor.
>
> I did not see him after the first shot was fired from our forward gun as the flash of the gun blinded me, but when I first observed him he was directly under the star "Antares" and I could see that our shots were good as far as direction was concerned. If we did not hit him, we prevented his surprise attack and scared him badly.
>
> Lost sight of *Ticonderoga* and the *Rondo* for the balance of the night but picked them up the next morning. Signaled the *Ticonderoga* and compared notes. He said he was of the opinion that there were two submarines. I saw only one, however. He also said submarine had fired three or four shots. I only observed two under our stern and they were so close I had to see them. I can not give too much credit to the crew of this ship for the manner in which they conducted themselves, both the Armed Guard and the civilians, every man being in his place in the shortest possible time and absolute discipline prevailed.

Lt. Cdr. Madison's *Ticonderoga* was the ex–German freighter *Kamilla Rickmers*, commissioned on January 5, 1918. On September 30 they met Kapitänteutnant Adolf Franz's

U-152 and after a two-hour gun battle the outgunned transport sank. Only 31 of the 236 aboard survived. Capt. Madison was badly wounded. Lt. (jg) Frank Louis Muller, the executive officer, and Junius Harris Fulcher, 1st assistant engineer, were put aboard the sub and taken to Germany. The other survivors were picked up by the British steamer *Moorish Prince*.

Capt. Madison never recovered from his wounds and died at the Brooklyn Naval Hospital on Christmas Day in 1922. He was awarded the Medal of Honor and became the namesake for the destroyer *Madison* (DD-425). Fulcher was not in the Navy. It appears he was a 42-year-old marine engineer with the Lighthouse Service at Norfolk, Virginia. How he ended up aboard the *Ticonderoga* is unknown. It's possible he was hired by Capt. Madison to fill a vacancy. Muller, 27, and Fulcher left Liverpool on December 7, 1918, on the Cunard liner *Caronia* and arrived at New York on the 17th. Muller was traveling under naval orders and Fulcher under government orders.

Junius lived with his wife Grace at 319 W. 17th St. in Norfolk. After the war he became the Superintendent of Lighthouses at San Juan, Puerto Rico. In 1940, at age 63, he was aboard the lighthouse tender *Orchid* at Portsmouth, Virginia. Frank was from Nebraska and lived at 1684 83rd Ave. in Oakland, California.

Part Five

The Shipbuilders and Operators

11. The Steamship Companies

This section lists the steamship companies that played a part during the war. Most owned ships or managed and operated ships for the Shipping Board. The companies that both built and operated ships are in the Ship Construction chapter.

A.D. Carver
11 Water St., New York

Amos D. and Clifford N. Carver were partners in Baker, Carver & Morrell, a marine industry and railroad supplier at 39–41 Water St. in New York.

Solon E. Turner and Albert M. and Scott Blanchard were in the stevedoring business as Turner & Blanchard at 414 Whitehall St. in New York.

In early 1917, Solon Turner and Albert and Scott Blanchard partnered with Capt. Phineas B. Blanchard and Capt. James B. Parse, of Searsport, Maine, to enter into a $160,000 contract with the Seaborn Shipyards Co. at Tacoma, Washington, to build the 243', 5-masted lumber schooner *Betsey Ross*. Capt. Parse would be the master. Construction started on March 19.

The demand for shipping was so great that on August 30, while the ship was still on the ways, the owners signed a charter agreement with Jefferson M. Moore, of San Francisco, to carry a load of lumber from Chemainus, British Columbia, to Melbourne, Australia, at the rate of $39 per 1,000 board feet. The charter agreement stated, "This charter is subject to charter for return cargo being signed in New York, promptly after advice that this charter has been signed here."

All departures and overseas charter voyages of vessels required the approval of the Shipping Board's Chartering Committee. No vessel was to return in ballast or with a commercial cargo if a cargo could be loaded for the benefit of the government.

In accordance with the law, another charter for the return voyage was made with the Mutual Chemical Company of America at New York to transport a load of chrome ore from New Caledonia, 800 miles east of Australia, for transport to either New York or Baltimore at the rate of $60 per long ton.

The ship was delivered on October 8, 1917, at a total cost of $176,000 and registered 100-A-1 by British Lloyds. Capt. Charles Wirstrom reported aboard at Tacoma on October 31 along with his first officer, John Lundquist. On November 3, August Boche, the cabin boy, and two sailors, Richard P. O'Connell and Ray Lambert signed on at Tacoma.

On November 27, about the time the ship was ready to sail, the owners' agents at New York, D.B. Dearborn & Co., applied to the Shipping Board for the approval of the ore charter and that was granted the same day.

The ship reached Melbourne on March 9, 1918, the lumber was discharged, and on the 27th the ship went into drydock for cleaning, painting, and alterations to carry ore. In order to carry a bulk cargo, the owners had to have "trunks" built in the holds. This work was completed on April 3 at a cost of $5,889.35 and the crew began the final preparations for loading ore and securing for sea. Capt. Wirstrom wanted to leave no later than April 10.

On or about April 5, the Shipping Board, the U.S. War Industries Board, and the British and Australian governments met to discuss the idea of requiring or requesting the owners of American-flag vessels to carry grain on their return voyages from Australia and customs officials in Australia were told to hold all American ships in port. When Capt. Wirstrom applied to customs at Melbourne for clearance to New Caledonia his request was denied.

At this point, the legal concept of demurrage began, that is the amount of money a common carrier could have made while being detained lawfully or unlawfully by official authority. It cost the owners of the *Betsey Ross* about $500 a day while Capt. Wirstrom sat in port twiddling his thumbs. Most of the crew paid off and signed on other vessels.

On the 12th, the Commonwealth sent a telegram to Secretary of State Robert Lansing requesting instructions regarding the *Betsey Ross.*

On the 15th, the owners' agents at Melbourne, Wm. Haughton & Co., sent a bill to Australia's Commonwealth Shipping Board at 155 William St. requesting payment for demurrage. The bill was returned the same day along with a note saying they had received a cable from Secretary Lansing:

> With reference to your telegram of 12th April, American Shipping Board consider *Betsey Ross* suitable for wheat and requests she loads wheat not chrome ore. Doubtless the master will receive instructions in regard to this vessel.

On May 9, Capt. Wirstrom was ordered to load wheat and to await instructions. On or about May 15 officials with the U.S. Food Administration Grain Corp. met with Amos Carver in New York and presented him with an Australian Government Form Wheat Charter agreement, which Amos signed. Amos's decision to sign was the result of negotiations with the Shipping Board as to his options. His other option was to have his ship seized and loaded with grain.

The charter rate for the wheat was 125 shillings per ton on 2,144 tons loaded at Williamstown, near Melbourne. The wood used for the ore trunks was used as dunnage for the wheat.

Between April 3 and June 20 seven new sailors signed on at Melbourne and the ship departed Sydney on June 28 and arrived at New York on November 5, 1918. Capt. Wirstrom had to pay a head tax on his cook, Henry Pain, 39, who was a British subject. His father Charles lived at 165 Merton Rd., Wimbledon, London.

Amos Carver was paid $63,784 for freight on the wheat. No payment for demurrage was received and the owners lost $177,000 on the ore charter.

The owners sent the Shipping Board a bill for $113,216 on a claim of $177,000 minus the $63,784, with interest at 6 percent from May 9 as just compensation for the loss of the chrome ore contract. On February 17, 1920, the Shipping Board denied the claim so Amos Carver sued in federal court. He claimed the $5,889.35 spent on building the trunks for the ore; interest of $15,000; and the loss of the chrome contract.

On January 17, 1927, the owners of the *Betsey Ross* were awarded $113,216. The court disallowed the expenditures for the trunks since they were used effectively in the transport of the grain and the interest on 30 days idle time was disallowed on the grounds that, while they did lose the chrome contract, the expenses they would have incurred for that voyage were also eliminated.

The company owned the *Betsey Ross* and the *Nancy Hanks*.

A.H. Bull & Co., Inc.
17 Battery Place, Rm 2333, New York

Archibald Hilton Bull (1847–1920) was founder and president. Ernest M. Bull was vice-president, Abram L. Burbank secretary, William H.R. Killeen treasurer. Before the war Bull owned the steamers *Carolyn, Dorothy, Eva* (British-registered), *Millinocket, Ruth, Wilhelmina,* and nine schooners.

Bull added the freighters: *Beatrice, Clare, Edith, Evelyn, Hilton, Jean,* and *Margaret*. The company's agents were the Arrow Line and the Bull-Insular Line, Inc. Bull was a major World War II operator.

A.L. Crowley
2 Brattle St., Boston

Capt. Arthur L. Crowley was vice-president of the Coastwise Transportation Co. and apparently operated the *Lake Washburn* on his own account as A.L. Crowley.

Acme Steamship Corp.
11 Broadway, New York

Incorporated in 1918 with a capital of $100,000. Robert E. Miller was president and Bartholomew L. Stafford vice-president. The company's freighter *H.M. Whitney* was requisitioned by the Shipping Board and bareboat chartered to the Oriental Navigation Co. for operation on the French government's account.

Alaska Steamship Co.
1107 Colman Building, Seattle

Incorporated on November 12, 1907, and as of December, 31, 1915, was controlled by the Kennecott Copper Corp. Stephen Birch was president, William E. Bennett and Robert W. Baxter vice-presidents, and Robert G. Crouch cashier and treasurer. The company owned 14 steamers, including the 230' motorship *Oregon*.

A vessel's deadweight tonnage, the actual cargo-carrying capacity, is used to calculate charter rates, construction costs, and sale amounts and for that reason is recorded very accurately during the admeasurement process. The *Oregon*'s deadweight tonnage was registered at 2,614.42.

At 2 p.m. on December 2, 1917, the Shipping Board requisitioned the *Oregon* at Seattle on a time-form charter basis of $5.75 per deadweight ton per month.

The ship was turned over to the Matson Navigation Co. for operation and in February 1918 left San Francisco for Honolulu. During the trip the engine broke down and the ship was towed to Honolulu. At that time the government ceased charter payments but the agreement was still in effect. Adequate repair facilities were not available in the Hawaiian Islands so the *Oregon* was towed back to San Francisco and in April was returned to Alaska Steamship while still under charter.

On May 15, Alaska Steamship's attorney, Cornelius H. Hanford, met with Shipping Board officials in San Francisco to request the release of the *Oregon* from the charter agreement. The Board agreed but no action was taken and Alaska could not legally use the ship until it was officially released from the charter party.

During the mechanical survey, certain "serious questions" were raised as to the origin of the mechanical trouble. The engines were repaired at a cost of $102,506.64 and between May 1918 and June 1919 additional design improvements and mechanical changes totaling $57,712.24 were also made at Alaska's expense. During this time, on July 25, 1918, San Francisco Shipping Board surveyor Edward S. Hough, designer of the Hough-type wood freighter, listed the deadweight tons as 2,500 on his report.

On October 21, Carey W. Cook, a Shipping Board official in Washington, telegraphed Alaska Steamship to say the *Oregon* would be released from the charter as of May 15, 1918, as requested, provided the Shipping Board's responsibility and liability ceased as of that date and Alaska would take no further action. Alaska instructed Hanford to "take the money" and on the 22nd Hanford accepted a check for the charter period from February May 15. But the money received was based on Hough's figure of 2,500.

Matson had also incurred salvage expenses after the ship broke down and these had not been paid. Attorney Hanford was sent to Washington and on October 31 met with Carey Cook. Hanford was told Alaska was owed nothing after February 1918 but offered a settlement for the period prior to that.

On November 26, 1918, Alaska applied to the Shipping Board to sell the *Oregon* to the Alabama Steamship Co. for $500,000 cash and this was authorized on December 16.

Alaska filed suit in federal court to recover money on the charter from May 15, 1918, but their petition was dismissed.

On August 18, 1919, Alaska contacted the Shipping Board stating $3,310.82 was owed based on Hough's error. Hough filed an amended report on the 31st to correct the error and which Alaska sent to the Board. The Board paid the additional money on May 28, 1923.

An important seaman's case involving Alaska Steamship began in November 1916 when Oscar Forstrom signed articles on the company's steamer *Valdez* for a voyage from Seattle to Southeastern Alaska and back. Forstrom became intoxicated and fought with a mate and on November 16 he was discharged and put ashore at Juneau. Forstrom lost his appeal to the Shipping Commissioner in Seattle so he sued in district court for breach of contract, lost wages, and expenses. Judge Neterer reversed the Commissioner's decision on the grounds that Forstrom returned to work after the altercation and could have been, and should have been, fined for disobeying orders but the offense was not serious enough to warrant discharge prior to the end of the voyage.

Alfred H. Post & Co.
11 Broadway, Rm 601 New York

Alfred H. Post and Sidney Hoey set up shop in 1895 as foreign traffic specialists and ship brokers. In 1918, Post owned the 3-masted lumber schooner *Leora M. Thurlow*, built at Bath, Maine, in 1891. The schooner was homeported at Key West, Florida, and used to ship wood products from the Gulf of Mexico to the West Indies.

On February 8, 1918, an all-British crew signed on at Tampa under Capt. Malachi R. Yates. His mate was Norman Macarthy Yates. Dolman Parsons was bosun, Henry Ebanks cook, and the seamen were John A. Henning, Lee Haylock, and Timothy, Dallis, Reginald, and Robert Ebanks.

On April 26, 1918, the schooner was at Havana when Capt. Yates was ordered by the Shipping Board Joint Committee on West Indian Transportation and the U.S. Food Administration to take a cargo of sugar to Savannah, Georgia. The schooner was fitted only for carrying lumber and was neither designed nor classed to carry perishable cargo. Capt. Yates protested but the imperious, uncomprehending mind of the bureaucrat prevailed and he was directed to enter into a charter agreement with Galbán, Lobo y Cia. to transport 1,900 bags of sugar consigned to the Savannah Sugar Refining Corp. at Port Wentworth, Georgia, at the prevailing rate of carriage.

Around 1912, sugar production in the United States began to decline and importers started buying from Cuban growers. In 1916, Benjamin A. Oxnard was president of the Adeline Sugar Factory Co., Ltd., at 527 Canal St. in New Orleans. William S. Pardonner was secretary and manager. Oxnard lived with his wife Bobbie at 1730 Palmer Ave.

In 1916 Oxnard and Pardonner established Savannah Sugar with Oxnard as president, Pardonner secretary-treasurer, and Benjamin O. Sprague superintendent. Company offices were in room 701 in the Savannah Savings Bank & Trust Building. The Oxnards moved to 718 Drayton St. in Savannah and the company began melting sugar from Cuba on July 7, 1917.

The *Leora M. Thurlow* was loaded with 1,623 bags of sugar and departed Havana on April 30, 1918. En route, they ran into heavy weather and Capt. Yates, believing his ship was in danger, put into Key West, Florida. The sugar was soaked with salt water and was unloaded there. Damage to the sugar was estimated at $10,000.

In early June, when the *Thurlow* was at New York, a libel was preferred against the vessel by Savannah Sugar and the vessel was arrested by the U.S. marshal. Bail was provided by the National Surety Co. in the amount of $4,600, effective June 5th. While action was pending in the District Court for the Southern District of New York, National Surety conducted an investigation of the circumstances leading up to the libel and as the result of this, the bond was increased to $5,500 on the 26th.

The schooner arrived back at Tampa on September 11, 1918, from Grand Cayman, British West Indies, and sometime later, National Surety sued the government for the cost of the bond. The attorney general's office argued the schooner was not taken over or requisitioned by the Shipping Board, National Surety's claim was not brought within the 6-year statute of limitation, there was no contract, and the master acted voluntarily in contracting with the sugar company.

The court found the schooner was never in the government's possession nor were the owners ever deprived of its use. National Surety argued the schooner would not have been cleared from Havana if the master had not complied with the government's "request" pursuant to President Wilson's requisition order. Too bad, the court said: "It is a well-settled rule of law that losses resulting from such acts of sovereignty are consequential losses and are not recoverable."

In August 1921, the *Thurlow* was homeported at Philadelphia when it ran onto a reef near Cozumel Island off Yucatan with a load of fruit and sank. The crew of seven was saved.

Alpha Steamship Corp.
11 Broadway, Rm 824, New York

Charles W. Harrison was president, Franklin D. Mooney vice president, John E. Craig secretary, and Wilson C. Peck treasurer. Owned one steamer, the *Seward*, sunk in 1917. The company then dissolved.

American & Cuban Steamship Line
27 William St., Rm 704, New York

Arthur R. Lewis was president and general manager, Harold J. Charlwood treasurer. Owned one ship, the freighter *Pinar Del Rio*, sunk in June 1918.

American-Hawaiian Steamship Co.
10 Bridge St., Rm 713, New York

A company founded in 1899 by William D. Burnham. In 1917, George S. Dearborn was president, Lewis H. Lapham vice president, Henry E.D. Jackson secretary, and Edward D. Douglas treasurer.

Thirteen of the company's 17 freighters were requisitioned and bareboat chartered to the Navy for operation by the Army: *Alaskan, Arizonan, Dakotan, Floridan, Hawaiian, Iowan, Kentuckian, Mexican, Minnesotan, Ohioan, Panaman, Scranton, Texan*, and *Virginian*.

American International Corp.
120 Broadway, New York

American International was incorporated in New York on November 22, 1915, by a group of investors as a securities trading firm, a holding company, and an international export vehicle for American products "to further the development of American activities in foreign countries." The articles of incorporation stated they "may engage practically in any kind of business except banking and operation of public utilities."

Charles A. Stone was president, Philip W. Henry vice-president, and Richard P. Tinsley secretary. Company offices were on the 37th floor at 120 Broadway.

As the company grew, Stone remained president, George J. Baldwin, Thomas W. Streeter, and Matthew C. Brush joined Philip Henry as vice-presidents. Gordon C. Carson became secretary, and Gordon H. Balch treasurer. Frank A. Vanderlip was chairman of the board. The directors besides the officers were Newcomb Carlton, Charles A. Coffin, William E. Corey, Robert Dollar, Pierre S. Dupont, Robert F. Herrick, Otto H. Kahn, Robert S. Lovett, Philip A.S. Franklin, Henry S. Prichett, Percy A. Rockefeller, John D. Ryan, William L. Saunders, James A. Stillman, Guy E. Tripp, Edwin F. Webster, Albert H. Wiggin, Beekman Winthrop, William Woodward, and Daniel G. Wing. Ryan was president of the Andes Copper Mining Co.

In 1916 the group acquired the International Mercantile Marine Co. and United Fruit. The United States Rubber Co. was formed to invest in Far East rubber plantations, "which would lead to shipping through International Mercantile Marine."

In November 1916 the group took control of the New York Shipbuilding Corp. through stock ownership. In December 1916 the company went in with W.R. Grace & Co., to purchase a controlling interest in the Pacific Mail Steamship Co. Other major investments were made in petroleum, sugar, machinery, balsa wood, steel, exporting firms, and a large number of companies engaged in all manner of business.

On April 30, 1917, the Boston News Bureau stated:

> American International stock is not an investment. It is a speculation and more or less a rich man's speculation. The company is paying dividends which are being several times earned. In the year to Dec. 31 the company earned $5 per share, or 10% on the $50 paid in on the 400,000 shares of common stock. There is $1,000,000 preferred outstanding, but it is an inside issue, in which the public has no participation.
>
> It is an ambitious proposition, the exact scope of whose operations and whose "ne plus ultra" is not clearly defined. The company started out to colonize American dollars in distant lands. It has done some of this work, but the major portion of its energy and capital has been devoted to investments in steamship properties like Pacific Mail, United Fruit, and Mercantile Marine.

The May–August 1917 issue of the Standard Corporation Service stated:

> When the American International and Mercantile Marine made their investment in the New York Shipbuilding Co. and revamped its capitalization, that concern had on the books some $25,000,000 of steamship construction orders. At the end of the half year this total of unfilled orders had enormously expanded to $70,000,000. And if government work piles up, as seems likely, the company may have $150,000,000 of business on its hands by fall.

The company eventually controlled the American Line, the Atlantic Transport Co., and formed the American International Shipbuilding Corp. at Hog Island near Philadelphia.

American Italian Commercial Corp.
6 Wall St., Rm 36–37, New York

The Banca Commerciale in Italy wanted to purchase coal, steel, and ships in the United States and incorporated the American Italian Commercial Corp. in New York on August 23, 1917, with a capital of $2,000,000. Frank J. Egan was president, Roland S. Klein vice-president, and Vincent F. Handley secretary-treasurer. Capt. Angelo Ruspini was the general manager and marine superintendent in New York. The company also acted as agent for the Navigazione Generale Italiana steamship company in foreign ports.

Lester H. Livingston and Aaron H. Hecht, president of importers Hecht & Hecht, Inc., at 200 5th Ave. in New York, apparently had an agreement to act as brokers to purchase ship steel for Navigazione Generale and ships for American Italian. Livingston introduced Capt.

Ruspini to Hecht and Hecht sold Ruspini 5,000 tons of ship steel. Hecht then paid the captain a "bonus" of $923.60 for the order and another $5,272.82 "rebate."

Hecht, apparently on his own again, obtained a line of credit from the Guaranty Trust Co. and purchased an option to buy the Sinclair Oil tankers *Albert Watts, Charles Braley, Joseph Cudahy,* and *William Isom.* He then sold the option to Frank Egan and Egan bought the ships for $1,533,000 each. The *Isom* was still on the builder's ways. Hecht realized a profit of just under $400,000 on the sale of the ships. Livingston subsequently filed suit against Hecht for his share of the profits on the steel and the tankers. Guaranty Trust received a commission and Banca Italiana got $142,640, a 2 percent commission, for financing the sale to American Italian.

Ricardo Gualino, of Turin, Italy, obtained contracts to furnish a large amount of coal to the Italian government and apparently signed a contract with American Italian. In August 1917 American Italian purchased the collier *Deepwater* from the Darrow-Mann Coal Co., at the time the largest collier ever built, with a capacity of 12,785 tons. Darrow-Mann's contract was for $750,000 and they sold the ship for $1,250,000. The company also owned the *Plymouth* and the *Tidewater*. On March 17, 1918, the *Tidewater* fought a gun battle with a U-boat and possibly sank it. The incident was reported in May by the Navy Department.

American Line
9 Broadway, New York

Controlled by the International Mercantile Marine Co., which in turn was controlled by the American International Corp. The company operated from Pier 62, North River. Before the war the company owned the big liners *New York, Philadelphia, St. Louis,* and *St. Paul.*

Five of the company's cargoliners were requisitioned and bareboat chartered to the Navy for operation by the Army: *Finland, Harrisburg* (ex–*Philadelphia*), *Kroonland, Louisville* (ex–*St. Louis*), and *Plattsburg* (ex–*New York*). The 10,230-ton *St. Paul* was operated by the company. The company operated the ex–German ships *Andalusia* on the Pacific to United Kingdom route; the *Appeles* (ex–*Elsass*) to the United Kingdom; and the *Black Arrow* (ex–*Rhaetia*) to Greece.

American Metal Transport Co., Inc.
61 Broadway, New York

William E. Brady was president, Harold K. Hochschild vice-president, Julian S. Beaty secretary, and William H. Brady treasurer. The company managed and operated the seized German freighter *Yazoo* (ex–*Maruda*) for service to the west coast of South America.

American Shipping Corp.
662 E. Bay St., Jacksonville, Florida

Incorporated on August 1, 1918, with a capital of $25,000 to operate ships. James A. Campbell was president and treasurer, Ralph B. Tupper vice-president and secretary. Campbell was treasurer of the Logan Coal & Supply Co. and Tupper was president of the Tupper Coal Co. (coal, crushed rock, cement, sand, and dynamite). Operated several Shipping Board steamers.

American Star Line, Inc.
25 Beaver St., Rm 713, New York

Incorporated with a capital of $100,000 shortly before the war. Moses Ginsberg was president, Abraham Frankel vice-president, Charles Kanner secretary, and Howard Abrahams treasurer. Directors besides the officers were Sarah Ginsberg, Isadore Sirota, and Joseph A. Rubin. Owned the barkentine *Herdis* and in December 1916 the ex–Canadian laker *Algonquin*. Operated Shipping Board steamers.

American Steamship Corp.
18 Broadway, New York

Incorporated on May 27, 1918, with a capital of $10,000,000 by Leonard E. Wales, Edward W. Cooch, and Andrew A. Gray. Purchased the wood freighters *Mazama* and *Coyote*.

The American Transatlantic Co.
17 Battery Place, New York

From Mülheim to Milwaukee

The long, convoluted story of American Transatlantic contains all the elements of German influence, intrigue, disloyalty, and fraud that could possibly be found in any wartime tale.

The story starts on November 9, 1853, when Julius G. Wagenknecht, a 20-year-old German blacksmith, arrived in New York from London on the American packet *Devonshire*. He settled in Milwaukee, Wisconsin, where he apparently had relatives in the metal trades. He became a safe and ornamental fence maker and in 1859 married a girl named Marie. In 1860 they had a daughter, Laura, and on March 30, 1862, a son, Richard George arrived, followed by Oscar, Matilda, and Edward.

On December 21, 1867, he became a naturalized citizen and in 1869 went into business with John Hornbach as Hornbach & Wagner at 518 Market St. where they made iron railings and vault doors. Julius lived at 521 Market.

Richard changed his name to Wagner and was educated in engineering at the Rensselaer Polytechnic Institute at Troy, New York, married a girl named Mary, and had three sons, Julius P., born on December 6, 1890, in Milwaukee, Richard, and Milton. He worked for his father as a machinist and civil engineer and lived at 549 8th St. Oscar was a mechanical engineer in the company.

By 1886 Julius was on his own as the J.G. Wagner Co., Architectural Iron Works, 514–520 Market St. and Julius and Mary lived at 530 Galena. At some point, Richard's wife died and in 1888 he and his two sons and Oscar lived at 530 Galena.

By 1900, the company was at 428 Grand Ave. and included an iron foundry. Richard was vice president and Wendelin J. Hahn secretary-treasurer. At this time the Wagners were very successful in the bridge building business and had three servants at their home on Galena St.

Around the turn of the century, Richard saw an agricultural report that touted Wisconsin as an ideal growing region for sugar beets so in 1901 Richard, Jacob B. Whittaker, and John S. Lawson incorporated the Wisconsin Sugar Co. at 428 Grand St. Richard was president, Whittaker vice president, and Lawson secretary-treasurer. The trio went on to found the Chippewa Sugar Co. and the U.S. Sugar Co.

Business was good and in 1910 Richard was 44, Julius 19, Richard, 18, and Milton, 16. They lived at 2827 Grand Ave with their 36-year-old servant, Kate Fisher.

Soon after, the sugar business began to falter and by 1913 Richard was living at 538 Milwaukee Ave. He had sold most of his stock and resigned as president of Wisconsin Sugar but retained a small amount of stock in the Chippewa Sugar Co., which was all but inactive. His finances continued to decline and in early 1914 he assigned his life insurance policy to cover promissory notes he had taken out to cover debts but by the fall of 1914 he found himself in debt once again.

That August war had broken out in Europe and Richard's life would soon be turned upside down in a way he never anticipated. He had an uncle on his father's side, Edmund Karl Wagenknecht, who had settled in Montevideo, Uruguay, where he was known as Eduardo. Eduardo married a girl named Caroline and in 1872 they had a daughter, Clara, Richard's first cousin. On June 15, 1895, Clara married Hugo Dieter Stinnes in Wiesbaden, Germany.

Hugo was born in Mulheim, Germany, on February 22, 1870, the son of a wealthy shipping and mine operator. In 1890 Hugo inherited his father's estate and went into business for himself with a vengeance. He soon held large interests in coal mines, coke ovens, steel, patented fuels, and the Ruhr Valley's electric, gas, and water systems. He was known as the "King of Coke" and also owned over 60 newspapers. He was a director of many companies and had businesses in England, Italy, Germany, Russia, and Scandinavia but his headquarters always remained in Mülheim.

In 1907 he founded Hugo Stinnes Schiffahrt Gesellschaft, a steamship company that eventually operated 15 ships on trans–Atlantic service.

Being an international businessman, Stinnes always had his finger on the world's pulse and by October 1912 he was becoming nervous over increasing tensions throughout Europe and especially in the Balkans. He knew that if war broke out and Germany became a belligerent, which he adamantly opposed, his ships could be seized and lost to him. As the situation grew dimmer, he began formulating a plan to do three things: prevent the loss of his ships, make as much money as possible in shipping from a war, and provide all the aid and support to Germany if it came to that. He knew if Germany's enemies prevailed, all could be lost, including his ships.

His plan was simple: sell his ships and replace them with new ships actually owned by him but registered under the American flag or the flag of neutrals. He knew this would require considerable deception, fraud, and money-laundering.

His business in England was Hugo Stinnes, Ltd. His first step was to request his manager in London to retain a solicitor and find out how quickly his ships could be sold to British interests in the event of any "international complications" and how long it would take the Crown to recognize the new registry. He gave power of attorney to the manager to sell all 15 ships at prices suggested by Stinnes and they were sold.

The Copenhagen Coal & Coke Co. was established in Denmark, in 1909 and by 1914 Hugo owned 95 percent of the company's stock. The company was also the agent for his fleet when one of his ships arrived at Copenhagen. The manager of the company was Albert Jensen, a Dutch citizen, who lived in Copenhagen.

To replace his ships, Stinnes instructed Jensen to purchase ships and form a separate Danish company for each one, a not unusual practice in shipping, and Jensen's name would appear on the title. He would then obtain a temporary registration. Jensen placed one of the titles in the name of a Danish company of which Stinnes was a stockholder and also the managing director. Shortly before the war started in August 1914, Stinnes directed Jensen to transfer the title to Stinnes himself while the ship was in a German port.

While the Netherlands remained neutral during the war, it became well known in shipping circles that Stinnes' money was behind the purchase of these ships and that he was profiting from them. Stinnes acquired a number of neutral-flag ships from various countries and their titles were placed in Norwegian and Danish stock companies, all the stock of which was owned by Albert Jensen. Stinnes deposited the money for operating these ships with the Revisionsbanken in Copenhagen and either directed the bank to pay Jensen directly or he established a line of credit in Jensen's name.

In the spring of 1914 Stinnes set up his own freight forwarding business but after the war started, the manager of his freight forwarding business was drafted into the German army. He then requested a business acquaintance, Theodor Lahr of Rotterdam, Holland, to manage the new company. Lahr was a German by birth but a citizen of Holland. He worked for Lenders & Co., a freight forwarding and shipping house in Rotterdam and had previous dealings with Stinnes. The war soon reduced the new company's business so much that Stinnes closed it and Lahr's old company took over the work. Theodor wouldn't be unemployed for long.

Dr. Heinrich F. Albert was the agent for the Zentral Einkaufs Gesellschaft (Central Purchasing Agency) in Berlin and came to New York to purchase American goods. The Zentral Einkaufs Gesellschaft was managed in Germany by a board of directors made up of prominent industrialists, businessmen, and German officials including Hugo Stinnes, who was a director.

Soon after the war began, Stinnes and Albert Jensen created a secret telegraph and cablegram code by which Stinnes could order contraband goods for shipment to Germany. The code had words and phrases common to the coal and shipping industries but the words referred to commodities vital to Germany's war effort which were barred from entry by the British. The goods were concealed in non-contraband shipments, particularly of cotton.

Officials in England knew Stinnes' was purchasing neutral-flag ships and approached the Danish government about requiring Danish registry on all the ships Albert Jensen had purchased. The Dutch cooperated and Jensen then had difficulty getting permanent Danish registry on the *Djursland*, *Finland*, *Gotland*, *Gronland*, *Hogland*, and *Lapland*, which sat in limbo at Copenhagen. without papers, Stinnes' ships weren't making any money and he wasn't happy.

Meanwhile, Dr. Albert was buying all the supplies he could for the German government, but no American business would ship their own goods directly to Germany or to neutral neighbors because of the British blockade. In order to get merchandise to Germany, Albert would consign them to a neutral with a delivery guarantee before they left the United States. Stinnes' ships were used for these shipments but not cargoes all got through.

Stinnes then decided it was time to contact his American relatives. On December 4,

1914, Richard Wagner received a cablegram from Copenhagen charged to Albert Jensen, but was in fact from Hugo Stinnes. The cable requested Richard's immediate assistance. Richard was decidedly pro–German and his own finances weren't too good so he wasted no time.

Richard applied for a passport on which he stated the purpose of his trip was to purchase a supply of beet sugar seed. He left New York and arrived at Rotterdam on January 10, 1915, visited family in Berlin, then went to Mülheim to meet with Hugo. Hugo outlined a scheme to register his ships in an American company Richard owned. It would be bankrolled by Hugo, of course, but his name would never be seen. Richard agreed and was sent to Copenhagen to meet with Albert Jensen and iron out the details. Before Richard left Germany he attempted to purchase a tanker then building and prevailed on the American consul to cable Washington "to ascertain the attitude of the Commissioner of Navigation concerning American registry." This proposition apparently failed.

On February 4, 1915, workers began loading barrels of scrap copper onto a small vessel at Copenhagen. Danish authorities visited the ship on the 6th and noticed the cargo manifest did not list the contents as scrap copper. Further investigation revealed that Albert Jensen and two accomplices had purchased the copper and arranged for its shipment to Germany. It was also learned that Jensen worked for Hugo Stinnes. The copper was seized and Jensen was charged with violating Danish export laws. Jensen's trial began on February 13 and ended on the 24th. Jensen was convicted on March 2 and sentenced to 60 days in jail. He served his sentence from March 18 to May 19.

Although the arrest and trial of Albert Jensen was big news in England, Germany, and Denmark and brought much unwanted publicity to Hugo Stinnes and the Copenhagen Coal & Coke Co., the incident did not dishonor or hurt Albert Jensen's business associations or jeopardize his citizenship in any way.

While Jensen was in the slammer, urgent business needed attending to. Jensen instructed Richard to wire Theodor Lahr in Rotterdam and ask him to meet him in Germany. Lahr did not know Richard Wagner and wired back that he was unable to travel to Germany at that time. Wagner then went to Dusseldorf and on February 17, 1915, executed a power of attorney document that allowed Theodor Lahr to purchase ships in Wagner's name in Holland, Belgium, Great Britain, and Ireland. He mailed the papers to Lahr but a problem arose when Lahr noticed the document did not show Wagner as an American citizen and it had been executed by a German official. Lahr, as a citizen of a neutral country, was forbidden to sell vessels to citizens of belligerent nations. All of this, Lahr said, would be noticed quite readily if he tried to do business in England. Lahr's power of attorney was then extended to include Norway and Denmark. During this time, Lahr and Wagner never actually met in person.

On February 25, 1915, Wagner boarded the steamer *Duca degli Abruzzi* at Genoa, Italy, for home and arrived at New York on March 8. On March 20 he organized the American Transatlantic Co. in Delaware with a capitalization of $200,000 and 2,000 shares of stock with a par value of $100. Richard Wagner was president and remained so until August 1918. His son, Julius P. was secretary and the stock in American Transatlantic was reported to be held by five people: Richard Wagner, 24,992 shares; Julius P. Wagner, Richard's son, 3 shares; Walter J. Berger, Richard's nephew, 3; Matilda W. Berger, Richard's sister, 1; and John S. Lawson, Richard's former sugar business partner, 1 share.

Walter Berger was a partner in William Berger & Co., a furniture manufacturer and bedding store at 392 Florida St. in Milwaukee. He lived at 883 Lake Drive.

Lahr proceeded to purchase five ships for either Richard or American Transatlantic but Wagner never furnished any money of his own and eventually about $2,500,000 of Hugo Stinnes' money was spent on 11 ships. Each vessel was owned by a subsidiary company whose stock was held by American Transatlantic. The ships, bought from Dutch, Greek, or Norwegian owners were the *Haugerland, Laura, Rygja, Solveig,* and *Spyros Vallianos.* Lahr turned over ten percent of the money used to buy the Dutch ships to Jensen who deposited it in the Revisionsbanken in Copenhagen to Stinnes' account.

Wagner then had to register all the ships under the American flag. By law, this required a certain amount of operating capital and on March 31, 1915, Richard Wagner cabled Albert Jensen to let him know the company existed on paper and to send him $40,000 for deposit with the Bankers Trust Co. in New York City. When no money arrived, Wagner sent three or four cablegrams requesting the money. Wagner then apparently contacted Count von Bernstorff for assistance then waited until April 11 to cable Jensen again at Reichskanzler, Platz 4, Berlin, to send the money.

In 1915, Percival H. and Robert J.S. Graham opened a freight forwarding office at 17 Battery Place in New York. Richard's son, Julius, of Milwaukee, was secretary-treasurer. Before Richard returned to Milwaukee he asked the Grahams to check with the bank on a daily basis to see if the money had arrived. The money was finally sent to the Hanover National Bank in New York and on April 15 Hanover received instructions from the Revisionsbanken in Copenhagen to transfer the $40,000 to the Bankers Trust Co. to the account of Richard Wagner. Percival Graham advised Wagner on April 16 that the money had arrived. Wagner withdrew the money and paid the American Transatlantic Co. for 400 shares of stock. He then rented office space in New York at 17 Battery Place and had Graham, as a salaried employee, equip and furnish it.

In 1915, England declared certain items, including foodstuffs, as contraband. They had also protested to President Wilson that Americans were purchasing German ships in their ports and sending goods to Holland, a fellow neutral, and which made their way to Germany. The British compiled a list of American firms with German connections and said any goods seized at sea would be tried in court.

In May 1915 Denmark passed laws which made it highly unlikely the Stinnes's ships awaiting full Dutch registry would ever receive it. In response to this, the titles were transferred to the American Transatlantic Co. and given American names: *Djursland* renamed *Maumee*; *Finland* renamed *Genesee*; *Gotland* renamed *Muskegon*; *Gronland* renamed *Hocking*; *Hogland* renamed *Winneconne*, and *Lapland* renamed *Kankakee*. The title to the *Kankakee* was held by Jensen's Norwegian company. The five Dutch company ships cost 3,650,000 kroner. Another vessel originally intended for American Transatlantic, the *Vinland*, had been replaced by the *Hogland* when the *Vinland* was granted permanent Dutch registry. At the time the vessels were transferred the Norwegian government had refused to insure the *Lapland*.

With titles in hand, Wagner then sought to register all the ships in the United States but he could not convince any consular officers that the transactions were in good faith and that the ships were entitled to the flag.

On May 16, 1915, Wagner called at the Department of Commerce in Washington to see Eugene Tyler Chamberlain, the Commissioner of the Bureau of Navigation, and enquire about the procedure for transferring the titles. The Commissioner had not as yet received any notice of the impending transfer of the registrations Wagner was asking about. Wagner explained their plan was to pay for the ships partly with stock to be issued by Albert Jensen and partly in cash from himself or with money realized from the sale of his American Transatlantic stock. A list of the vessels and all the particulars was furnished but Chamberlain thought the whole thing stunk and was "suspicious of the whole transaction."

Wagner then tried to "bulldoze and deceive him" but Chamberlain already knew that Jensen, as supposed purchasing agent for Wagner, was in reality Hugo Stinnes' agent and that Jensen had been convicted of smuggling. Wagner then brought in an attorney who told Chamberlain he had no legal grounds to deny American registration since all the company's officers were U.S. citizens. Chamberlain then demanded to know where the $2,500,000 came from and Wagner was forced to admit it came from Albert Jensen and that he was entitled to only 50 percent of the company's stock.

Chamberlain asked where Albert Jensen got that kind of money and on the 22nd he informed Wagner that he didn't have enough information about the deal to grant American registration to the American Transatlantic Co. ships. The fly in the ointment was Albert Jensen's little stint in jail. News of these things travels fast in the shipping world, particularly during wartime, and it was an open secret that Albert Jensen worked for Hugo Stinnes. Besides that, the April 30 edition of the Lloyd's List contained the names of all the ships purchased by Jensen with Hugo Stinnes's money and the names of some of the Stinnes's ships were on the list submitted by Wagner.

When Theodor Lahr had arranged to purchase a ship in Norway, Denmark, or Holland he visited the American consul in each particular city to secure temporary registration but the consuls would not act without instructions from the Commissioner of Navigation in Washington.

Wagner's next task was to somehow eliminate Albert Jensen's interest in any of the ships and on June 5, 1915, he informed Commissioner Chamberlain that Jensen was being removed from the company, would not own any stock in it, and no money from the sale of Jensen's stock would be used to pay for the ships. The deal would be all-cash.

At this same time, Secretary of State Robert Lansing had directed the American consuls in Holland and Denmark to personally investigate the source of the money used to buy the ships for Richard Wagner.

On June 7 Wagner gave Jensen a promissory note for the purchase of the 11 ships, payable in installments over five years and presented that "evidence" to Chamberlain but he was still not convinced the controlling interest in American Transatlantic would be American.

In order to convince the American consuls and Chamberlain that the money for the ships was really American, Hugo Stinnes concocted a money-laundering scheme whereby funds would be sent to Wagner with no paper trail from Jensen or Stinnes attached to it.

Hugo Stinnes' first tangible trouble came in June 1915 when the French government seized American Transatlantic's freighter *Saginaw* at Marseilles as a prize of war.

On July 6 Jensen asked a local bank in Malmo, Sweden, to write to the Bankers Trust Company in New York regarding the transfer of funds and enclosed were four checks totaling $725,000. Two of the checks were on National City Bank accounts and two were on National Bank of Commerce accounts. The letter stated the money was from Mr. Eduardo Wagenknecht, of Uruguay. The next day the Swedish bank sent another letter and a National City Bank check for $54,628.23 to Bankers Trust for the account of Richard Wagner, at the request of Eduardo Wagenknecht.

On July 9 Wagner cabled Jensen to say that he was being eliminated from the company and sent by letter his unsecured promissory notes amounting to $2,654,000. Wagner then went back to Chamberlain and said he was the "beneficial owner of the stock" but Chamberlain still wasn't satisfied and insisted on seeing more proof. Wagner returned shortly afterward and told Chamberlain he had sent Jensen $1,300,000 "in real American money" and eliminated him. Chamberlain wanted to know where he got that money and Wagner said he sold $900,000 of company stock to American investors. He repeated this story to federal officials, a British prize court, and anyone else who would listen.

On July 19, 1915, the Swedish bank sent two Hanover National Bank checks totaling $556,818.18 to Bankers Trust to the credit of Richard Wagner. It became known that all the money came from Albert Jensen when Wagner's son, Julius P., foolishly wrote to Jensen on American Transatlantic letterhead asking how much the money from Malmo cost.

On July 28 Wagner wired the Department of Commerce to report the seizure of the *Saginaw* and said the French took the action in response to Secretary of Commerce William C. Redfield's report that the sale of the ship to American Transatlantic was bogus. Wagner sent the original bill of sale for the ship to Secretary Redfield to support his claim. Redfield told Wagner to take the matter up with the French government. Wagner insisted Redfield take action on his behalf and Redfield reiterated it was Wagner's problem but Wagner continued to seek assistance in getting the ship back and to acquire American registration.

In late July, American Transatlantic's lawyers filed briefs with Secretary Redfield and Secretary of State Lansing concerning the registry of the ships and on the 29th Lansing told Redfield there was no diplomatic reason to deny American registry, but stated:

> Of course the Department cannot say the transfer of flag in these cases will not be challenged by belligerents or that the ships will not be arrested on the high seas and taken before a prize court for judicial examination.

Between July 1 and December 31, 1915, Secretary Redfield granted American registry to all of the company's ships. Provisional registration was granted by the American consuls in foreign ports with permanent registration to be granted after the ship's arrival in an American port.

A newspaper wire service article published on August 24, 1915, mentioned the *Laura*:

> WASHINGTON, Aug. 23.—Acting Secretary Sweet of the Department of Commerce authorized the issuance of American registry today to the Dutch steamer *Laura* which is now at Rotterdam. The *Laura* is the ninth to be admitted of twelve ships proposed for American registry by the American Transatlantic Company, headed by Richard G. Wagner.
>
> According to information at the Department of Commerce, the corporation owning the vessels was financed by Hugo Stinnes, a German capitalist.

All of the ships were registered between July 1 and December 31, 1915, under new names. The following list shows the gross tons, official number, radio call sign, year built, and former names:

Winnebago—4,362 tons, 213528, LFMW. Built in 1900 at Stockton, England, as the British freighter *Heathering*. Name changed to *Hampton* then became the Norwegian *Haugerland*.

Allaguash—3,826 tons, 213606, LFND. Built in 1905 at Sunderland, England, as the Norwegian freighter *Rygia*.

Manitowoc—4,471 tons, 213621, LFNG. Built in 1902 at Port Glasgow, Scotland, as the Greek freighter *Spyros Vallianos*.

Kankakee—3,718 tons, 213628, LFNC. Built in 1899 at Sunderland, England, as the Norwegian freighter *Hero* then name changed to *Lapland*.

Muskegon—3,323 tons, 213636, LFPG. Built in 1897 at South Shields, England, as the British freighter *Afonwen*. Became the Greek steamer *Leonidas Cambanis* then the Danish *Gotland*.

Genesee—2,892 tons, 213659, LFNJ. Built in 1900 at Sunderland, England, as the British freighter *Avristan*. Became the Greek steamer *Condylis* then the Danish *Finland*.

Maumee—2,556 tons, 213660, LFNK. Built in 1897 in West Hartlepool, England, as the British freighter *Mayland*. Became the Dutch steamer *Verhaven* then the Danish *Djursland*.

Winneconne—1,869 tons, 213661, LFNM. Built in 1907 at Sunderland, England, as the Swedish freighter *Groveness* then became the Danish *Hogland*.

Ausable—3,153 tons, 213697, LFNR. Built in 1901 at Alblasserdam, Netherlands, as the Norwegian freighter *Laura*.

Hocking—2,553 tons, 213705, LFPQ. Built in 1905 at West Hartlepool, England, as the British freighter *Cassandra* then became the Danish *Gronland*.

Between August 17 and October 25, 1915, Wagner withdrew a total of $1,285,420 from Banker's Trust and sent $40,000 to Rotterdam and the balance to the American Transatlantic office in Copenhagen. Of the total transmitted to Europe, $1,179,657.78 was listed on American Transatlantic's New York office books as having been credited on Wagner's account, not to Jensen, including $24,807.78 in interest, but the money was actually deposited in Hugo Stinnes's account in the Revisionsbanken in Copenhagen. Wagner was telling the Commissioner of Navigation the money was being advanced by Jensen from his personal funds and the notes would be paid by Wagner with his own money and from the sale of stock in American Transatlantic.

After all 11 ships had been acquired, Wagner increased the capitalization of the American Transatlantic Company to $2,500,000 and stock of the par value of $2,459,000 was issued on the ships, based on their book value, but at an agreed upon valuation of $2,654,850. The difference between the book value and the agreed upon value was deposited into Wagner's personal bank account. The $2,654,850 was $105,200 over the purchase price of the ships, due to the substitution of the *Hogland*, purchased at Kroner 700,000 as opposed to the *Vinland*'s sale price of Kroner 1,100,000. But the money Wagner sent to Jensen in June 1915 included the dollar equivalent of the *Vinland* when the lesser-priced *Hogland* was actually transferred.

The *Saginaw* had received American papers while under arrest in France but the ship was turned over to a prize court and condemned as lawful prize on October 8. The court held that at the time of seizure the vessel had no right to fly the American flag and had no papers on board confirming American registry. The subsequent granting of American registry was irrelevant since the ship's status on the date of seizure was all that mattered. Furthermore, the court had at hand documents from the American Consul General at Marseilles obtained by him prior to the seizure which showed the State Department believed the ship "belonged to German interests."

The *Kankakee* was granted permanent American registration on September 29, 1915. W.R. Grace chartered the ship and it left Baltimore on October 3 with a cargo of coal for Buenos Aires. On the 31st it was seized by a British warship off Lopos Island, South America, and brought to an anchorage in the River Plate. On November 26 a prize crew got the ship underway for Gibraltar, via St. Vincent, then to Marseilles, France. The cargo was discharged and then went to Algeria for a cargo of iron ore. The ship arrived in England in February 1916 and was finally delivered to the Procurator General of Great Britain who issued a writ against the ship on the 21st. American Transatlantic's attorneys appeared on March 11 and on June 10 filed a formal claim for the release of the ship citing American registry and claiming its seizure was illegal.

The *Genesee* was granted permanent American registration on October 9, 1915. The ship was under charter to W.R. Grace when it left Norfolk with a load of coal for Montevideo, Uruguay. On November 6 the ship was seized by a British warship off the coast of South America, escorted to Port Castries, St. Lucia, and delivered to the British authorities. Proceedings against the ship were started then transferred to the High Court of Justice, Probate, Divorce and Admiralty Division in London. American Transatlantic attorneys appeared before the court on April 1, 1916, and on June 10 filed a formal claim for the release of the ship citing American registry and claiming the seizure was illegal.

The *Hocking* was granted permanent American registration on October 27, 1915, after arriving in New York. The ship was under charter to W.R. Grace & Co. when it left there the next day for Norfolk to load coal. The ship was soon seized by the British and taken to Halifax. A writ was issued against the ship, served on November 2, 1915, and the prize case was transferred to the High Court in London. American Transatlantic's attorneys appeared before the court on May 2, 1916, and on June 10 they filed a formal claim for the release of the ship citing American registry of the ship and claiming its seizure was illegal.

In the meantime, an elaborate scheme was concocted to get food to Germany. Dr. Albert purchased a cargo of canned meat and papers were drawn up showing Richard Wagner had chartered the ship to a New York export company and the meat was consigned to companies in Gothenburg, Sweden, and Copenhagen, Denmark. Dr. Albert paid the freight and the *Maumee* left New York on October 16, 1915. The master was directed to sail north of Iceland to avoid the Royal Navy and stop at Bergen, Norway, where he would notify the American Transatlantic office at Copenhagen of his arrival and they would call Stinnes. The master would then tell the port authorities he had orders to wait for instructions. Dr. Albert and Wagner had already made arrangements for agents from the Zentral Einkaufs Gesellschaft to bring a new set of papers to the ship showing the cargo was consigned to the Russian government. The ship would then clear Bergen and be "captured" on the high seas by a German warship and seized on the grounds that it was carrying goods for an enemy.

But before the *Maumee* got to Bergen, the *Genesee, Kankakee,* and *Hocking* had been seized and this presented a complication for Dr. Albert, being in the United States, and on November 8 he cabled the Zentral Einkaufs Gesellschaft in Berlin that the capture of Stinnes's ships—and he referred to them as Stinnes's ships—was a problem and recommended that a cable be sent through ambassador Gerard in Berlin to the German Embassy immediately after the *Maumee* was "captured." He sent along his draft of the cable:

> The S.S. *Maumee* has been captured by a German man-of-war on suspicion of ultimate destination to Russia, for the reason that the steamer has changed its original destination to Gothenburg and for other reasons.

In order to avoid complications and to show the usual good will towards the American Government, the ship will be released at once. As to the goods, establish ownership and if neutral, make arrangements with the shippers for purchase. Cable result.

When Stinnes heard the *Maumee* had arrived at Bergen, he was reluctant to go ahead with the plan. He went to the German Foreign Office in Berlin to see Zimmermann and explain the situation. It was decided to abandon the venture since the cargo was not worth the risk to the rest of Stinnes's fleet. Stinnes sent representatives to tell the Zentral Einkaufs Gesellschaft office in Copenhagen the plan had changed.

A new plan was then drafted to send the *Maumee* to Copenhagen, discharge the canned meat there, and ship it to Germany later. A new set of phony papers was made up for the *Maumee* indicating the canned meat was destined for Switzerland. The cargo was unloaded onto boats owned by Stinnes at Copenhagen and the meat eventually got to Germany.

The scheme of being "captured" was also planned for the *Winnoconne*. That ship was loaded with a general cargo belonging to the Zentral Einkaufs Gesellschaft, including rubber, at New York. Stinnes became nervous about the rubber and had it removed but the same scheme was used to get the rest of the cargo to Germany through Copenhagen.

After that, the New York office of the Zentral Einkaufs Gesellschaft stopped using American Transatlantic ships to get goods to Germany. The company's ships were then used mainly to aid Germany and her allies by subverting English cargo and business laws.

The State Department immediately intervened after American Transatlantic's formal claims for release were filed. When the demand for the release of the ships was ignored, Stinnes and his agents devised a plan to sell all of the ships to American companies the British approved of and then time-charter them to British companies. This was agreeable to the British and just when a price was agreed to, the Shipping Board refused to allow the ships to be sold. Two of Albert Jensen's Dutch-flag ships had previously been seized. In this case the British approved of the sale but also demanded the sale of another, larger ship at the same time. British authorities were convinced the legal requirements of the sales would demonstrate conclusively the ships were German-owned but Britain was so desperate for ships they were willing to have to pay German owners if necessary.

The *Saginaw* was formally condemned by the French prize court on December 28, 1915.

In late December 1915, it came to the attention of the Treasury Dept. that the *Ausable, Maumee*, and *Winnebago* were being used to supply coal and gold to German raiders at sea.

After the ships were seized, Wagner was desperate to prove American ownership. He called on friends and associates around Milwaukee to try and get them to buy stock in the company, even one share if that's all they could afford. He even offered to give out shares as gifts but no one took him up on that. Some paid for the stock, others did not. Whenever one or two shares had been sold, an affidavit was prepared and submitted to the prize court in such a way that gave the impression the person named owned a large number of shares in American Transatlantic. The idea was to make it look like the company was just a regular American company owned by a large number of shareholders instead of just Richard Wagner and his family.

Wagner managed to get ten shares of stock worth $1,000 into the hands of nine of his acquaintances. Each was told he would be getting other shares in his name and they would be paid for out of the company's profits.

He attempted to transfer $200,000 worth of stock to Edward B. Elcock, of the Hansell,

Elcock Co. in Chicago, who was a personal friend of his. Wagner told him a story about getting out of the sugar beet business "that has no future" and into shipping and said in a letter dated October 30, 1916:

> For certain confidential reasons, it is desirable that a part of this stock should be distributed, and I have taken the liberty of putting $200,000 in your name. There is no liability of any kind or nature connected with the ownership of this stock and the only question that may come up is whether or not you own it and according to the records, you do.

Elcock replied on November 4 to the American Transatlantic Co. at 17 Battery Place in New York in no uncertain terms that he didn't want anything to do with it. Wagner begged him to accept one share and allow him to use his name as a director but got no reply. Wagner wrote again and Elcock responded with a strong "No." Wagner again wrote another sob story and Elcock again said "No."

It was reported by the collector of customs at Baltimore that the *Ausable* cleared for Buenos Aires, Argentina, via Norfolk and Newport News on December 23, 1916, with 1,019 tons of coal and $375,000 in gold. The coal was shipped by the Matlack Coal & Iron Corp. in New York and consigned to the Campania Alemanna de Depositis de Carbon, Limitada, in Buenos Aires. The gold was shipped by the Equitable Trust Co. in New York and consigned to Ernesto Tornquist & Co., Ltd., in Buenos Aires.

The *Maumee* left on December 29 with 3,442 tons of coal and $414,000 in gold. The only difference was the coal was going to Campania Alemanna Transatlantica Electrictad. The *Winnebago* left on January 7, 1917, with the same goods.

In 1917 Richard Wagner organized the Foreign Transport & Mercantile Co. and transferred the seven remaining ships to the new company.

On January 16, 1917, the company declared a stock dividend of $1,623,349.70 to be paid to shareholders of record on March 1, 1917. This was subsequently postponed and the actual payment was made one year later. No dividends were paid by the company. All the money for the dividend payout was credited to Richard Wagner's personal account on June 17, 1918, and he paid certain stockholders with his personal checks. Some of the checks were endorsed and cashed and the payee in turn wrote a personal check back to Wagner for the same amount as the dividend check. Others simply returned the endorsed check or cashed the check and gave Wagner a like amount in Liberty bonds plus their personal check to equal the dividend payment. Over $172,400 of face value Liberty bonds were given to Wagner in payment for the extra shares Wagner said the investors would be getting. Some of the investors actually kept small amounts of the dividends paid out. One investor paid Wagner $12,500 for 100 shares with his own money and not from dividend money.

On June 14, 1917, the *Kankakee* was sunk by a German submarine while in British hands.

Besides the stock certificate book showing the shares owned by Wagner and his family, the company kept another book which showed the real distribution of shares: Richard Wagner, 14,082; Richard's son, 2,003; Richard's nephew, 203; Richard's sister Matilda, 2,001; John S. Lawson, secretary-treasurer of the Chippewa Sugar Co., 501; Franklin P. Mann, president of the Mill Bag Co., 2,001; P.K. Herzog, 1,001; Kurt R. Froedtart, 1,001; Rudolph E. Giljohann, owner of the Palace Theater in Milwaukee and a cigar and tobacco retailer, 501; C.W. Frazier, 1; sugar partner Jacob Whittaker, 101; Edmund J. Archambault, president of the American Grinder Manufacturing Co., 201; E.J. Neidecken, 201; Fred C. Pritzlaff, president of the John

Pritzlaff Hardware Co., in Milwaukee, 1,002; and Emil Dubreuil, a cigar maker in Milwaukee, 200.

Wagner had listed all these people as shareholders in the minutes of the board meeting in September 1915 before any of them had agreed to accept any shares.

After the United States entered the Great War on April 6, 1917, the seven remaining ships of American Transatlantic were requisitioned by the Emergency Fleet Corporation.

On October 6, 1917, Congress passed the Trading with the Enemy Act.

After the deal with the *Hocking*, *Kankakee*, and *Genesee* was squashed by the Shipping Board the prize trial in England went forward and on February 4, 1918, the court condemned the ships as lawful prizes, holding the vessels "belonged at the time of seizure thereof to enemies of the Crown."

On July 19, 1915, the *Maumee* received a certificate from the Bureau Veritas at Copenhagen stating the ship was "in good, efficient working order, classed in first division."

On November 18, 1917, Capt. Jorgen C. Jorgensen left Rotterdam, Netherlands, under charter to Norton Lilly and arrived at New York via Halifax, Nova Scotia, on December 21. One passenger, Robert Woodward, was detained by immigration authorities for questioning.

In January 1918 the *Maumee* was one of seven American steamers assigned to the Belgian Relief Commission and in February carried a cargo of wheat from New York to Rotterdam. On the voyage the ship sustained some damage due to heavy weather and went into drydock on February 27 at Rotterdam. The ship was surveyed by Meyer & Young of the Bureau Veritas and on March 8 a certificate stating the ship was sound was issued and they returned to New York uneventfully.

On April 2, 1918, the Shipping Board's Central Committee ordered the ship to Hampton Roads to load coal for the Panama Canal and to contact W.R. Grace & Co. to transport a cargo of nitrate of soda for the next voyage.

On April 4, 1918, American Transatlantic, as agents for the Shipping Board, chartered the ship to W.R. Grace & Co. to carry "not more than 4,200 and not less than 3,780 tons—English gross weight" of nitrate of soda from Caleta Buena, Chile, to Wilmington, North Carolina. On April 8 another certificate was issued by surveyors at New York and the ship was warranted in the charter party agreement to be "tight, staunch and strong, well and sufficiently manned and victualed, and in every respect fit to perform the voyage."

The coal was loaded, the passage was very rough, and the ship reached Colon on April 18. They proceeded to Caleta Buena and commenced loading on May 2.

On May 9 they departed and arrived at Balboa on May 19, transited the Panama Canal northbound, and left Colon on May 20 for Wilmington. Nitrate of soda is a very heavy material that was packed in bags and piled in a pyramid form in the holds to keep pressure off the sides of the ship. The cargo and holds were inspected every day.

The ship loaded 40,118 bags and departed on May 9. On May 26, en route from Colon, Panama, Chief Engineer Carl R. Hensen reported there was water in the shaft alley, apparently coming from No. 4 hold. Hensen and the chief mate quickly made an inspection and concluded, "a leak suddenly occurred which resulted in some sea water entering the after hold of the vessel and doing some damage to cargo stored therein." It was determined that frames between No. 3 and 4 holds had broken and rivets popped. Damage control efforts were made and on arrival at Wilmington on the 28th it was found that 328,035 pounds of nitrate of soda was destroyed. The value was set at $13,449.43. Harbor Master Capt. Edgar Williams went aboard along with other surveyors who examined the ship. They all found extensive rust, broken ribs, and the hull in a generally poor, unmaintained condition. W.R. Grace sued and recovered the cost of the lost cargo.

The *Maumee*, along with hundreds of other vessels, was inspected by naval constructors at New York, found suitable for naval service, and given the inspection number SP 1339.

In June 1918, Richard G. Wagner bought a seat on the New York Stock Exchange after his company, American Transatlantic, was seized by the Alien Property Custodian. The governor of the Exchange provided a statement to the Justice Department:

> Mr. Wagner was vouched for by two members of the exchange, and was subjected to rigid questioning by the committee on admissions. His answers were satisfactory to the committee at that time. If further developments warrant it, the exchange will make further investigation and take such steps as are found advisable.

On August 1 and 2, 1918, Richard Wagner was interviewed in the presence of his attorney. On the 1st he stated that all the money he used to set up the American Transatlantic Co. was derived from the sale of his own securities or from the sale of stock in the company and none of the money had come from Europe. He agreed to furnish a list of all the stockholders who, he said, had paid over $650,000 for stock prior to July 29, 1915, but on the 2nd he admitted there was no list of stockholders and he hadn't sold any stock at all. He admitted the original $40,000 sent to Rotterdam in April 1915 and approximately $1,300,000 paid out in July and August had been furnished by his South American cousin, a Wagenknecht, who had then been in Europe and was sent to buy out Albert Jensen's interest in American Transatlantic. He stated that Jensen had sent the $40,000 in response to his wires.

In August 1918, A. Mitchell Palmer, the Alien Property Custodian, seized the stock of the American Transatlantic Co. and the stock of the Foreign Transport & Mercantile Corporation. At one time he also had set up individual corporations for each ship, which was not uncommon in the shipping business, but Wagner's obviously had other motives. Out of 25,000 American Transatlantic shares of issued and outstanding preferred stock and 100 shares of common stock and 25,000 shares of preferred stock and 240 shares of common stock of Foreign Transport, all but 403 shares were seized by Miller as being stock belonging to German interests.

New directors were appointed: Henry L. Doherty, 60 Wall St.; Henry Thompson, 141 Broadway; Charles H. Sabin, 140 Broadway; Frederick B. Lynch, 8 Bridge St.; John Quinn, 91 Nassau St.; Joseph S. Qualey, 25 Pine St.; and Francis P. Garavan, 110 W. 42nd St.

In response to this, Richard Wagner and his son Julius P., an officer with American Transatlantic, sued Thomas W. Miller, Mitchell's successor, in the United States District Court for the Eastern District of New York. No evidence was taken in the case but it went before a grand jury in 1920. At the same time, John Lawson, J.B. Whittaker, C.W. Frazier, E.J. Niedecken, Edmund Archambault, and M.W. and W.J. Berger filed suit against Miller in U.S. District Court for the Eastern District of Wisconsin to get their stock returned. The court ordered evidence taken and in 1920 and 1921 depositions were taken in Copenhagen and London.

In 1920 Richard was a 57-year-old widower. Julius was 29 and also widowed. Both worked in shipping and lived at 85 N. Beach in Rockaway Beach, Queens.

On July 28, 1920, Albert Jensen arrived in New York from Kristiania, Norway, on the steamer *Stavangerfjord* to meet with Thomas Miller.

Jensen and his attorneys claimed that Wagner owed him $1,500,000 plus interest on the promissory notes from 1915 and he wanted Wagner to get his stock and property back so he could collect his debt. Miller released approximately $360,000 in cash and stock in American Transatlantic and the Foreign Transport and Mercantile Corp. This included $85,000 in outstanding freight charges owed him by Wagner. Jensen then purchased the interests of all the shareholders, with the exception of the Wagners, for .20¢ on the dollar and "secured assignments of all their right, title, and interest to the Alien Property Custodian and the Treasury of the United States." Some of the shareholders immediately gave the money received from Jensen to Richard Wagner. Jensen then took out a surety bond in the name of the shareholders "against any future claims for taxes or other claims arising against them out of the transaction" and deposited $200,000 with a surety company.

Jensen obtained an assignment from Richard, his son Julius P. and another son to whom stock had been issued on condition the stock be held by Jensen as collateral on the debt Wagner owed Jensen. After, this Jensen arranged for the two lawsuits against Miller dismissed and he filed a claim with Miller for the release of all other assets.

An appeal from the British prize court was filed with the Privy Council but was dismissed on July 23, 1920, and approved by His Majesty the King, George V, on August 13, 1920.

On June 16, 1923, Acting Attorney General Augustus T. Seymour advised President Coolidge that the property held by Miller's office should be released, paid, and delivered to Albert Jensen. Three days later that was done. Seymour's report contained remarks about the legality of the seizure of the property by Miller:

> The sole objection to the return of these properties by the Custodian to Jensen, as assignee, arose from the fact that these ships which were transferred to the American Transatlantic Company by Jensen in 1915 were purchased with moneys furnished by Hugo Stinnes and the firm of Robert M. Sloman, Jr., enemy aliens. It was believed that these two German subjects and citizens retained all along an interest in the profits of these two companies, and it was for that reason that the properties were seized by the Alien Property Custodian in 1918 as the property of Hugo Stinnes, Hugo Stinnes, Jr., Robert M. Sloman, Jr., Eduard Wagenknecht, Albert Jensen, and Zentraleinkaufs Gesellschaft.
>
> We have now before the Department the report of Special Assistant to the Attorney General Adna R. Johnson, Jr., who went to Europe in March 1923, for the purpose of ascertaining, if possible, whether or not Hugo Stinnes and Robert M. Sloman, Jr., and the other alien enemies above named had any interest in or to the property of these two shipping companies at the time of seizure. It now appears from Mr. Johnson's report that the loan which Hugo Stinnes made to Albert Jensen in 1914 and 1915, amounting to substantially six million kronen, was entirely repaid by Jensen, including interest and commissions, on or before July 6, 1917. It is, therefore, established without question that at the time of the seizure of the stocks of these two American shipping companies neither Hugo Stinnes nor the firm of Robert M. Sloman, Jr., had any interest, financial or otherwise, in the ships and properties of these shipping companies. In addition to the evidence establishing the repayment of these loans as above stated, Hugo Stinnes, Hugo Stinnes, Jr., Robert M. Sloman, Jr., Zentraleinkaufs Gesellschaft, and Eduard Wagenknecht have all filed sworn statements disclaiming any interest in or to the properties of these two shipping companies.
>
> Being satisfied that there was no enemy interest of any character whatsoever in or to the properties of these companies which were seized by the Alien Property Custodian in 1918, I recommend that the properties now in the hands of the Alien Property Custodian to the credit of this trust be returned to Albert Jensen, the claimant herein.

The officers of American Transatlantic Company thereupon petitioned Congress for relief for damages caused by the seizure of the *Hocking, Kankakee*, and *Genesee* by Great Britain.

Senate Bill 3396, introduced on January 6, 1930, made a grant of $15,000,000 to American Transatlantic and referred the matter to the federal Court of Claims for fact finding and recommendations. The bill stated:

> *Be it enacted by the Senate and House of Representatives of the United States of America in Congress assembled,* That there is hereby granted to the American Transatlantic Company, a corporation organized under the laws of the State of Delaware by American citizens, $15,000,000, for the injury sustained by said company to its business and property on account of an "Arrangement effected by exchange of notes between the United States and Great Britain for the disposal of certain pecuniary claims arising out of the recent war," signed May 19, 1927 whereby the rights of said corporation to prosecute, through the intervention of the United States, its diplomatic claims against Great Britain, or to request international arbitration thereof, for reimbursement of the losses or damages sustained by said corporation growing out of the illegal seizure upon the high seas of its steamships *Kankakee, Hocking,* and *Genesee,* during the months of October and November 1915, by British men-of-war, while said vessels were engaged in neutral trade and lawfully flying the American flag, said illegal seizure being subsequently followed by the unlawful and wrongful condemnation of said vessels as prizes of war by a British prize court, were surrendered by the United States; the United States, in surrendering the rights of the said corporation to prosecute its claims against the British Government, receiving in exchange a waiver from Great Britain of its right to receive from the United States a net cash payment on account of certain claims recognized by the United States as just and proper and also the right to prosecute other claims for liability for which it had not yet been formally admitted by the United States but which involved considerable amounts in exchange for which the Government of the United States agreed that said waiver was intended for the satisfaction of those claims of American nationals which the United States regarded as meritorious and in which the claimants had exhausted their remedies in British courts, in which no legal remedy was open to them, or in respect of which for other reasons the equitable construction of the agreement called for a settlement; and that the Secretary of the Treasury be, and he is hereby, authorized and directed to pay, out of any money in the Treasury not otherwise appropriated, to the said American Transatlantic Company the sum of $15,000,000.

The Court decided the company had no legal or equitable claim against the United States, that the three ships were owned by Hugo Stinnes, and if not by him then by Albert Jensen, a Dutch national, and the claims agreement between the United States and Great Britain as set forth in Senate Bill 3396 did not waive any claim against Great Britain by American Transatlantic because the seizure of the three ships did not violate any international law.

The government filed a counterclaim for $1,215,530.89 in unpaid income and excess profits taxes for 1918, 1919, and 1920, which the company had not paid. This was dismissed as barred by the statute of limitations set forth in the Internal Revenue Code, Section 276.

The value of the *Kankakee* when seized was $434,000; the *Genesee,* $339,500; and the *Hocking,* $287,350. From the date of seizure to November 1, 1917—the average date on which most other American vessels were requisitioned by the Shipping Board—the value of the *Kankakee* was $372,000; the *Genesee,* $291,000; and the *Hocking,* $246,300. Owners of vessels were paid on a bareboat basis of $4.15 per deadweight ton per month. The deadweight tons of the *Kankakee* was 6,200; the *Genesee,* 4,850; and the *Hocking,* 4,105.

On June 14, 1917, the *Kankakee* was sunk by a German submarine while in British hands. A Reparations Committee was established at the end of the war to set values for Allied ships. The rate settled on was $365 per deadweight ton. This placed the value of the *Kankakee* at $1,357,070. American Transatlantic as unable to prove the dollar amount of the consumable stores, equipment, and supplies on board as well as the cost of repatriating the crew after the ship's seizure by Britain.

The court stated, in part:

> The only part of the evidence in this case which might be regarded as complete and satisfactory is the evidence concerning the conduct of Wagner in this country. That evidence shows a succession of misrepresentations and subterfuges resorted to for the purpose of deceiving the British authorities and our own, for the purpose of getting the fleet of ships into American registry, with its advantages, and, later, of getting the ships released from British seizure. We pass no moral judgement on Wagner's conduct in this regard. He was obviously strongly pro–German, as many other people, including the American shipping officials with whom he dealt, were strongly pro–Ally, though the United States was neutral. We stress Wagner's misrepresentations and subterfuges, not for the purpose of denying that "all is fair in love and war," but only to show why we place no reliance whatever upon the form and appearance into which Wagner's transactions were cast. We have the same impression of Jensen, the Stinnes agent, willing, at Stinnes' behest, to risk violating the laws of his own country and going to jail for it. We are obliged, therefore, to look through the appearance of things and try to determine what the relations were.
>
> We are asked to believe that Stinnes, an enormously rich and successful German businessman, loaned some two and one-half million dollars to Jensen, without security, to enable Jensen to engage in the highly speculative business of buying, selling, and operating ships. Jensen says that, at the time, Stinnes was dubious, though Jensen was optimistic, about the soundness of investment in ships. Jensen's books seem to show that he had very little money, and it is incredible that Stinnes would have risked such a huge sum in a venture where it might well have been lost, and where his only gain, if the venture succeeded, would have been ordinary interest.
>
> Our conclusion as to the asserted Stinnes-Jensen loan is that it was not a loan, but an advance by Stinnes to Jensen as his agent, to buy and operate ships in his name, or in the name of some other neutral person or corporation, for the benefit of Stinnes.
>
> We have concluded that the three ships seized by Great Britain as prizes of war belonged, at the time of seizure and condemnation, to Hugo Stinnes, a German national. If they did not belong to Stinnes, they belonged to Albert Jensen, a Danish national.
>
> In our opinion the American Transatlantic Company does not, on the basis of the evidence presented to us, have any legal or equitable claim against the United States.

The decision was handed down on May 2, 1949.

Hugo Stinnes died on April 10, 1924, in Berlin. He was a cofounder of the German National People's Party (D.N.V.P.) in 1919, one of many parties formed after the Great War, and was elected to the Reichstag in 1920, holding one of the 66 seats his party won that year. He despised the Versailles Treaty and the very harsh reparations imposed against Germany. He was anti-union and joined forces with other noted German industrialists, including Gustav Krupp, to lobby against the 8-hour day.

Two American Transatlantic ships were allocated to the Naval Overseas Transportation Service, the *Ausable* and the *Winnebago*. The 320-foot *Ausable* was commissioned at Cardiff, Wales, on October 1, 1918, with Lt. Cdr. Charles L. Arey in command. The ship carried coal for the Army, brought back ammunition after the war, and was decommissioned at Brooklyn on April 19, 1919, then returned to American Transatlantic.

The *Winnebago* was commissioned on March 6, 1918, at New York, after Belgian Relief Commission duty. Lt. Cdr. Austin Russell Cushing was in command and apparently made only one voyage. The ship was returned to American Transatlantic on June 11, 1918, and in the early 20s was sold to Albert Jensen Aktieselshab (Albert Jensen Joint Stock Company) of Copenhagen and renamed *Fie Jensen*. In 1926 the ship was renamed *Ontario*.

Richard and Mary retired to Florida and where died in 1940. When the war started, their son Julius was and officer with American Transatlantic and lived with his wife Edina at 560 W. 165th St. in New York. When he was drafted into the Army on September 4, 1918, he was working for the Pan American Terminal & Dry Dock Corp. in Brooklyn. He served as a private with the 157th Depot Brigade until discharge on March 10, 1919. Julius died in 1967.

Andrew F. Mahony
Fife Building, 1 Drumm St., San Francisco

Andrew "Andy" Mahony was a shipping man and treasurer of the Olson-Mahony Lumber Co. He lived with his wife Rose at 871 Clayton St. and owned four ships including the steamer *Girlie Mahony*.

Apex Navigation Corp.
63 Wall St., New York

Incorporated with a capital of $50,000. Willard U. Taylor was president, William S. Scott treasurer, and John Potts secretary. Arthur F. Walter was a director. The company owned and operated schooners.

Archibald L. Stark
12 Agassiz, Judith Point, Boston

In 1918 Archibald Stark was treasurer of the Stark Supply Co., restaurant supplies at 1895 Washington and a restaurant at 1797 Washington. The next year he was vice-president of the Fidelity Trust Co. and in November 1919 treasurer of the newly-formed Inter-Coast Steamship Co. in room 516 at 148 State St. Capt. Arthur L. Crowley, former vice-president of the Coastwise Transportation Co. was president and general manager but Shipping Board records cite Stark as the contracting operator.

Argonne Steamship Co., Inc.
17 Battery Place, Rm 1524, New York

A subsidiary of the States Marine & Commercial Co. incorporated to manage and operate the Japanese-built steamer *Argonne*. James Donald was president, Henry C. Sheridan vice-president, Adolph J. Michel secretary, and Thomas C. Dawson treasurer.

Armour Fertilizer Works
Union Stock Yards, Chicago

Philip Danforth Armour revolutionized the meat packing industry by creating a centralized operation that turned out massed-produced products. Stockmen always had a buyer and consumers had a reliable selection at lower prices than the Main St. butcher could offer. Armour dramatically improved sanitation in the industry and devised ways to utilize the large amount of animal by-products, which included the manufacture of fertilizer.

Charles H. MacDowell started out at Armour as a stenographer and became president of Armour Fertilizer. The company owned the schooners *Charles H. MacDowell*, *De Witt Brown*, and the 357-ton *City of Baltimore*.

Arthur A. Miller
37 Wall St., New York

Miller was president and treasurer of the Miller-Cummings Co. with Cecil H. Cummings, of Boston, and Clifford C. Grow, a produce commission house at 330 Washington. Miller lived at 136 Christopher St. in Montclair, New Jersey, with his wife Julia, son Robert H., and two servants. He owned the schooners *Charles G. Endicott, Eva B. Douglas,* and *J. Manchester Haynes* and the steamers *James S. Whitney, Minneapolis,* and *Vigo.*

The company created the "Crowned Lyon" cantaloupe and expanded the brand name to Iceberg lettuce, asparagus, peas, and other products grown in the West.

Arthur Sewall & Co.
Bath, Maine

In 1589 Henry Sewall became the mayor of Coventry, England. In 1634 he decided to emigrate to America and he settled in Newbury, Massachusetts. Henry's son was the Salem witchcraft trials judge. Dummer Sewall and Lydia Marsh settled on the Kennebec in 1762.

William Dunning Sewall was a pioneer Bath shipbuilder in business with Freeman Clark as Clark & Sewall at Kennebec. William married Rachel Allyn Trufant and in 1835 they had a son, Arthur.

Arthur started out as a lumber buyer on Prince Edward Island and in 1854 went into business with his older brother Edward as E. & A. Sewall. Edward died in 1876 and Arthur took on his son, William and his nephew Samuel Sewall and became Arthur Sewall & Co.

Shipbuilding came to a halt during the first Cleveland administration but Sewall built four large wooden ships anyway and in 1894 built the first steel sailing vessel in America, the Scotch-designed 4-masted bark *Dirigo.*

Sewall was part-owner of Bath Iron Works, the New England Shipyard, president of the Main Central Railroad, the Bath National Bank, and a director on several boards. He was always a strong advocate for the merchant marine and died of a stroke at his summer home at Small Point, about 12 miles outside Bath, on September 5, 1900.

Sewall built the *William P. Frye* in 1901 and operated it on his own account.

Associated Oil Co.
55 New Montgomery St., San Francisco

A large company that refined, marketed, and transported California petroleum. Prior to the war they owned the steamers *W.S. Porter, Navigator,* and *Rosecrans;* the ship-rigged sailing vessels *Marion Chilcott* and *Falls of Clyde;* schooners *Monterey, Roderick Dhu,* and *Santiago,* and tugboats *Milton* and *Rover.*

In 1918, William Sproule was president, P. Grant Williams secretary, and Walter A. Sloan treasurer. The company owned ten ships including the *Frank H. Buck.*

Astoria Veneer Mills & Dock Co.
Lewiston, Maine

William H. Williams established a lumber business in 1876 at Astoria, Long Island, to manufacture fine veneers. The firm became the Astoria Veneer Mills and on January 3, 1911, was incorporated as the Astoria Veneer Mills & Dock Co. with 2,800 feet of waterfront adjacent to other lumber mills including the Steinway Piano Co.'s plant.

In 1917, H. Frank Packard was president and treasurer, C.F. Packard secretary, general manager, and purchasing agent. William Williams was a director at Lewiston.

The Shipping Board chartered the French sailing vessel *Cornil Bart* on February 27, 1918, and delivered the ship to the company on August 9 for operation to Africa.

Astral Shipping Co.
24 State St., New York

Oscar R. Houston was secretary and a director of the North American Chemical & Engineering Co. He incorporated Astral on May 28, 1917, at 64 Wall St., Room 803, New York, with a capital of $10,000 to own and operate the gas freighter *Asta,* a refrigerant and chemical transporter. Houston was a partner with Harrington, Bigham & Englar.

Atlantic & Pacific Steamship Co.
7 Hanover Square, New York

Incorporated in Maine in 1912 with a capital of $4,500,000 to operate from both coasts under the management of W.R. Grace & Co. Joseph P. Grace was president, Maurice Bouvier vice-president, and J. Louis Schaefer secretary-treasurer. Lawrence H. Shearman, John H. Rosseter, George E. Gordon, and Wellington Gregg were directors. They owned the four *Santas—Clara, Cruz, Catalina,* and *Olivia.*

In July 1918, Joseph Grace stepped down and was succeeded by Maurice Bouvier, senior vice-president of W.R. Grace.

Atlantic Chartering Co., Inc.
25 Beaver St., Rm 224, New York

Incorporated as steamship agents. Douglas Donald was president, Erik Marcussen vice-president, and George Schalz secretary-treasurer. The company operated nine Shipping Board steamers during the war.

The Atlantic Coast Co., Inc.
131 State St., Rm 1023, Boston

Incorporated in July 1916 in Maine by officers of the Crowell & Thurlow Steamship Co. to acquire the Townsend Marine Railway & Construction Co. at Boothbay Harbor. Walter D. Noyes was president, William A. McKenney vice-president, Ivan H. Joslin secretary, and Lewis K. Thurlow treasurer.

Also owned a shipyard at Thomaston and the big schooners *Anna Laura McKenney, Ella Pierce Thurlow,* and *Jessie G. Noyes.*

Atlantic Fruit Co.
61 Broadway, Rm 1218, New York

Norman A. MacLeod was president, Henry B. Mendenhall secretary, and Thomas O. Muller treasurer. Operated from Pier 36. Owned four ships and managed eight Shipping Board steamers during the war.

Atlantic Gulf & West Indies Steamship Lines
11 Broadway, Rm 1256, New York

Galen L. Stone was president, Alexander R. Nichol secretary-treasurer. Incorporated with a capital of $40,000,000. Edward J. Berwind, William A. Bradford, Henry Hornblower, Henry R. Mallory, Hubert E. Rogers, Frederick Fletcher, Harry H. Raymond, Alfred G. Smith, and Franklin D. Mooney were directors. Prior to the war the company owned the steamers *Altamaha, Brunswick, Ocmulgee, Ogeechee, Ossabaw,* and *Satilla.*

Atlantic Refining Co.
3144 W. Passyunk Ave., Philadelphia

Atlantic Refining was a subsidiary of Standard Oil. John Wesley Van Dyke worked his way up from roughneck to become manager in 1903. In 1911 the Supreme Court busted up Standard Oil and Atlantic Refining became a separate company to refine, market, and transport petroleum products. Van Dyke became president, Walter P. Cutler and William M. Irish were vice-presidents, Henry S. Mustin was treasurer, and William D. Anderson secretary.

The company owned seven ships when the war started and operated eight Shipping Board ships.

Atlantic Transport Co.
9 Broadway, Rm 268, New York

Charles Joseph Baker, of Baltimore, was president of the Franklin Bank, Baker Bros. & Co., and the Chemical Company of Canton. Baker Bros. were dealers in Lykens Valley, Sunbury and hard coal, "By the ton or cargo," and they manufactured Baltimore Window Glass, Flint & Green Glassware, chemicals, paints, oils, and glue and imported French window glass. Baker Bros. & Co. and the Chemical Company of Canton were at 32–34 S. Charles St.

Bernard Nadal Baker was secretary of the Chemical Company of Canton and in 1888 was in business with James S. Whiteley and Ashby L. Baker as the Baker, Whiteley Coal Co. at 409 Second St. and Bernard was president of the Baltimore Storage & Lighterage Co., steamship agents and ship brokers at 409 Second St. Charles G. Helm was secretary.

At that time, several steamship companies provided service from Baltimore to Europe and the British Isles. Bernard wanted to expand his business and in 1890 started the Atlantic Transportation Line in the office of the Baltimore Storage & Lighterage Co. Baker was president, James Whiteley vice-president, Charles G. Helm secretary, and James C. Gorman manager.

Because the cost to build and operate American ships was about 30 percent higher than in England, Baker chose to build and operate under the British flag, but English law prohibited direct foreign ownership of English vessels so the company hired an agent, Alfred S. Williams, and formed the Atlantic Transport Co., Ltd., in London to own the ships and lease them to the Atlantic Transport Co. in New York. All the stock in the London company was owned in New York. The first British ship was the *Surrey* and eventually the company operated 16 foreign-built ships. About 80 percent of the stock ownership was American with Baker owning just over 50 percent of that. Next was Harlan & Wolf, Alexander Brown & Sons, and investors in New York, Chicago, and Baltimore.

In 1891 the Atlantic Transport Line of Steamers offered weekly sailings between Baltimore and London on the *Michigan, Mississippi, Minnesota, British Crown, Maryland, Maine, Memphis,* and *Montana.* The company acted as agents for the Hamburg American Packet Line and the Lord Line. Freight was handled by Baltimore Storage and the company opened offices in Chicago and St. Louis.

In 1897 Atlantic Transport was at 403–409 Water St. and had offices in New York, Chicago, Philadelphia, and Minneapolis. The *Maine, Massachusetts, Manitoba, Mohawk, Mobile, Minnesota, Mississippi, Maryland, Missouri,* and *Montana* ran between Baltimore, Philadelphia, New York, London, and Swansea. The company added the Bristol Channel Line and the Empire Line to its agency roster and had offices in Chicago, New York, St. Louis, and Philadelphia.

In 1898 the U.S. government purchased Atlantic Transport's ships for the Spanish-American War. To replace his ships, Baker found investors and incorporated the Atlantic Transport Company of West Virginia with a capital of $3,000,000 and moved to offices on the 5th floor of the Continental Trust Building. Charles Helm was treasurer and Wade Newcomer was secretary.

Baker replaced the confiscated ships with ships from the Wilson & Furness-Leyland Line, the famous "Minnies": *Minneapolis, Minnehaha, Minnetonka,* and *Minnewaska* and acquired other ships.

In 1901 Baker sold his stock to the International Mercantile Marine Corp., but not the ships. The voting trustees were John "J.P." Morgan, J. Bruce Ismay, Clement A. Griscom, and Peter A.B. Widener. Four new ships were ordered from American yards for Atlantic Transport with IMM money: the *China, Korea,* and two big liners *Manchuria* and *Mongolia.*

Financial difficulties ensued and in 1904 the IMM bloc sold the *Manchuria* and *Mongolia* to the Pacific Mail Steamship Co. Atlantic Transport was then reorganized. Baker moved to an office at 504 Union Trust Building. James Whiteley became president of the Baker-Whiteley Coal Co. and the L.H. Miller Safe & Iron Works. James Gorman was the resident manager in Baltimore at 201–207 Chamber of Commerce Building and Arthur W. Robson was the passenger agent at 127 E. Baltimore St.

In August 1915, Pacific Mail announced that the La Follette Seamen's Act of 1915 made it impossible for them to compete against Japanese operators in the Pacific so they were selling their liners *Korea, Siberia, China.* The Atlantic Transport Company of West Virginia would buy back the *Manchuria,* and *Magnolia* for $5,250,000. The last trip to the Orient would be made by the *Mongolia* on August 25. Philip A.S. Franklin said the liners would be put on trans–Atlantic service "under the American, British, or any other flag depending on which gives the lowest cost."

Pacific Mail said they would put their eight remaining ships on coastwise routes between San Francisco and Panama and sell them before the end of the year.

Bernard Baker was born in Baltimore on May 11, 1854, to Charles Joseph and Elizabeth Baker. He was president of the Atlantic & Pacific Transport Co., the Baltimore Trust & Guarantee Co. and with Albert Ballin was the agent for the Hamburg-American Line. He died at Santa Barbara, California, on December 20, 1918, where he had moved on advice of his doctors.

In 1918 Philip A.S. Franklin was president, John J. McGlone secretary, and Horace B. Phillips treasurer. The company's three ships, *Manchuria, Mongolia,* and *Minnesota* were requisitioned by the Shipping Board and bareboat chartered to the Navy for operation by the Army.

Axim Transportation Co., Inc.
Anchorage, Kentucky

Charles C. Mengel established the Mengel Box Co. in 1877 at Louisville, Kentucky, to manufacture wooden boxes. In 1899 the lumber division became a separate entity at 311 W. Main St. as C.C. Mengel & Bro. Co. with Clarence R. Mengel as president. The company also distributed Santa Rosa Coffee and imported chicle and cocoa.

The company had a large logging operation at Axim on the west coast of Africa and in 1915 they incorporated Axim Transportation to handle increasing shipments by water. The company owned and chartered vessels.

Balestrand Navigation Corp.
27 William St., Rm 814, New York

A Norwegian firm that owned one ship, the *Balestrand,* which they sold in February 1918.

Baltimore & Carolina Steamship Co.
607 Union Trust Building, Baltimore

Owned three ships, managed two, and operated five Shipping Board steamers. Provided service from Baltimore to Wilmington, Charleston, and Georgetown in the Carolinas. Mason L. Weems Williams was president and general manager. The company owned the *Georgeanna Weems, Matilda Weems,* and *Sarah Weems* after the war.

Barber & Co., Inc.
17 Battery Place, Rm 1026, New York

Incorporated on July 13, 1917, in New York with a capital of $1,000,000 'to engage in a transportation and navigation business." James Barber was president, Edward J. Barber, Francis M. Wilson, Frank B. Mackay, and Aaron Houtman were vice-presidents. Later incorporated with a capital of $500,000 to act as steamship agents. Philip P. Barber and Oakley Wood were directors.

The company owned five freighters: *Dochra, Macona, Onega, Satsuma,* and *Suruga*. The *Macona* and *Onega* were incorporated under their own names. The *Dochra* was incorporated as the La Plata Steamship Co., and the *Satsuma* and *Suruga* were incorporated as the New York & Oriental Steamship Co.

The company's freighter *Suruga* was requisitioned by the Shipping Board on a time charter basis for operation by the company on the Army's account.

The company owned four ships in 1914, managed 6 Shipping Board steamers, operated seven, and managed and operated another 24 ships. The company operated the Dutch freighter *Tjikembang* under Shipping Board management. On January 9, 1918, the Shipping Board chartered the Chinese steamer *Hwah Yih* and delivered the ship to Barber on February 1 for operation to the Orient.

The Shipping Board chartered the Norwegian steamer *Bjornefjord* on September 28, 1918, and assigned the ship to the company on January 23, 1919, for operation to the east coast of South America for 12 months.

The Shipping Board bareboat chartered the Uruguayan steamer *Artigas* (ex–*Wiegand*) and *Paysandu* (ex–*Bahia*) on May 24, 1918, and assigned the ship to the company on November 7, 1918, for service from the United States to Montevideo.

Barnes-Ames Co.
201 Board of Trade Building, Duluth, Minnesota

Julius H. Barnes and Ward Ames, Jr., were school chums who were both born on the same day, Julius in Arkansas and Ward in New York. They went into the grain business and in 1914 incorporated the Barnes-Ames Co. in Minnesota to engage in grain exporting. Barnes was president, Ames vice-president, and Percy H. Ginder secretary.

In 1918, Julius, 43, lived at 12 S. 26th Ave. E. with his wife Harriet, 42, daughter Gertrude, 18, and son Robert L., 16, both in school. Ward lived at 1618 Vermillion Rd. with his wife Helen, two daughters, and three servants.

The company owned one ship, the *R.L. Barnes*. In 1920, Julius was the federal grain administrator in Duluth.

Battery Steamship Corp.
17 Battery Place, Rm 136, New York

A subsidiary of the States Marine & Commercial Co. James Donald was president, T.T. Hunter vice-president, Joseph Scott treasurer.

The company's steamer *Borinquen* left Baltimore in December 1916 and departed Cowes, England, for Boston but in February 1917 the ship was reported 15 days overdue and was feared torpedoed.

On January 10, 1917, the *Borinquen* was in the Bay of Biscay when the *U-84* stopped the nearby Norwegian cargoliner *Bergenhus*. Fred Fredbold, of Boston was chief mate on the *Bergenhus* and reported the sub's captain came aboard and stayed for several hours until a troop transport was sighted about five miles away. They couldn't tell the nationality but said the sub took off and shots were exchanged between the sub and the transport. Several shots hit the transport, which caught fire then broke in two and sank. The sub then returned and ordered the *Bergenhus* crew onto the *Borinquen* then sank the *Bergenhus*. The *Borinquen* took them to the Azores where they boarded the Fabre Line passenger ship *Roma* and on February 9, 1917, arrived at Providence, Rhode Island. The *Roma* also brought four survivors from the Greek steamer *Dimitrios Goulandris,* torpedoed 30 miles off Spain on January 4. One of the survivors was 3rd engineer John Rereis, of New York.

On the night of October 30, 1917, Capt. Austin Walton's *Borinquen* was in the Bay of Biscay when they encountered a sub sitting on the surface. Chief Gunner's Mate Thomas J. Beerman thought they were "receiving news from her headquarters" and reported:

> The submarine was getting under way, headed for the ship, in the effort to get under the bearing of the after gun. The captain of the *Borinquen* swung the ship around, placing the submarine directly astern of the ship and in the trough of the sea. The second shot carried away the conning tower of the submarine, which I saw myself. The third shot also hit the submarine and I saw the explosion which was very distinct. The submarine appeared to stop dead in the trough of the sea after this shot. The last seen of the submarine she was going down on swell, listed to port, with her bow sticking in air and her stern down. she was going down in an upright position.

The ship had an Armed Guard crew of 20 and arrived at New York from La Pallice on July 28, 1918.

Belgian Relief Commission 165 Broadway, New York

Around 150,000,000 bushels of wheat had piled up in Australia owing to a lack of shipping. The situation in Belgium gradually deteriorated under German occupation and rumors of starvation were heard. Pleas for aid reached future president Herbert Hoover in London and in 1914 he helped organize a relief effort.

In April 1918 the U.S. Food Administration and the Shipping Board's Ship Control Committee decided to send 3,300,000 bushels of wheat to Belgium and northern France. Fifteen American ships and Norwegian sailing vessels were chartered to carry wheat, flour, oats, meal, and general provisions. Any ship in an Australia port was required to bring wheat back.

Julius H. Barnes, head of the Federal Grain Corporation, consigned all shipments to Brand Whitlock, the American ambassador to Belgium, even though Germany agreed to allow safe conduct for any vessel employed in relief work.

Boston Molasses Co. 175 Milk St., Boston

Felix Taussig was born on June 9, 1864, into a molasses importing family, N.W. Taussig & Co., in New York City. In 1883 he joined the company and in 1897 became president of his own company, Boston Molasses. The company owned the tanker *Amolco* and the 296-ton *Pedrito.*

Bradley Transportation Co. Rogers City, Michigan

Carl D. Bradley was manager of the Michigan Limestone & Chemical Co.'s Calcite Plant.

Broughton & Wiggins Navigation Co.
910 Yeon Building, Portland, Oregon

George Broughton and William S. Wiggins sold and transported cedar piling and poles.

Bulk Oil Transports, Inc.
32 Broadway, New York

Incorporated on June 6, 1916, by Christopher Hannevig with Ralph James Bullowa as president. Hannevig controlled through stock ownership Bulk Oil, the Pennsylvania Shipbuilding Co., and the New Jersey Shipbuilding Co. Bulk Oil owned the steamers *Flirt* and *Flush.*

Bull-Insular Steamship Co., Inc.
17 Battery Place, Rm 2333, New York

A subsidiary of A.H. Bull. Archibald H. Bull was president, Ernest M. Bull vice-president, Abram L Burbank secretary, William H.R. Killeen treasurer.

C.H. Pattengill Corp.
78 Wall St., Rm 1013, New York

Charles H. Pattengill was president and Edith Brooks was secretary-treasurer of their importing firm. The Shipping Board chartered the Norwegian sailing vessel *Maella* in November 1918 and assigned the ship to the company for service to the east coast of South America. Charles lived in Glen Ridge, New Jersey, and Edith lived at 385 Ft. Washington Ave. in New York.

C.B. Richard & Co.
29 Broadway, Rm 101, New York

Established in 1847 as general passenger agents. In 1918 Oscar L. Richard, Leonard W. Simmons, Albert F. Egelhoff, and George N. Richard were steamship agents and bankers. The company managed and operated two Shipping Board freighters.

C.H. Sprague & Son
70 Kilby St., Rm 39, Boston

Charles Hill Sprague was born in 1827, graduated from Harvard in 1846, served in the Navy, and then went to work in the coal mining business as a shipping agent for the Chesapeake & Ohio Railroad at Boston. Around 1870 he began shipping coal from West Virginia on the C. & O. on his own account and then began exporting it to Europe, South America, and Japan.

In 1882, Charles' son Phineas joined his father and the firm became C.H. Sprague & Son. They expanded by purchasing their own mines in the New River district of West Virginia. and operating their own ships. Charles died in 1904 and the next year Phineas established the Penobscot Coal & Wharf Co. in Searsport, Maine.

In 1918 Phineas W. Sprague was president. The company owned three ships during the war, managed four Shipping Board freighters and managed and operated 11 ships.

Sprague continues today as a Northeast liquid fuel and natural gas distributor.

Callaghan, Atkinson & Sears, Inc.
11 Broadway, Rm 568, New York

Cornelius H. Callaghan was president, Harry G. Atkinson vice-president, and Judah H. Sears secretary-treasurer. The ship brokers managed five Shipping Board steamers, operated three, and managed and operated seven.

The Carolina Company
20 Exchange Building, Charleston, South Carolina
"Carolina Line"

Incorporated at Charleston on June 18, 1913, with a capital of $100,000 to engage in the real estate business. Alfred Huger was president, Joseph M. Whittsitt secretary. By 1918 they were ship brokers and Whittsitt was vice-president and treasurer. Operated one Shipping Board freighter and managed and operated four.

The Carpenter-O'Brien Co.
8 W. 40th St., Rm 1508, New York

In 1913, Sam J. Carpenter was a 43-year-old lawyer who lived by himself in a rooming house at 165 Manhattan Ave. in New York and Mark L. Fleishel, 36, was a wholesale lumberman who lived with his wife Marie, 35, daughter Demetria, 9, and son Marc Jr., 4, in Fullerton, Louisiana, when the two bought a tract of land in Florida and on May 19, 1913, organized the Carpenter-O'Brien Lumber Co. on St. Andrews Bay.

Dwight F. Brooks, of St. Paul, Minnesota, and Michael Joseph "M. Joseph" Scanlon were involved with several lumber businesses and were headquartered in Minneapolis. Scanlon lived at 25 Groveland Terrace in Minneapolis.

Brooks was president of the Brooks-Scanlon Lumber Co., Brooks-Scanlon-O'Brien Co., Ltd., and the Scanlon-Gipson Lumber Company. Scanlon was president of the Brooks-Robertson Lumber Co., and the Brooks Timber Company. The partners had holdings all over the country.

On December 31, 1913, the corporate name of Carpenter-O'Brien was changed to the Brooks-Scanlon Corp. but Carpenter-O'Brien continued to do business under its own name.

In 1916, Carpenter-O'Brien transported long leaf yellow pine logs 145 miles from Carbur, in the middle of Florida, to their mill at Eastport near Jacksonville. When asked why the mill wasn't at Carbur, general manager Marc L. Fleishel explained that it cost the company $2.25 a thousand to ship raw logs on their rail line but $3.15 a thousand for finished lumber. That saved the company between $500 and $600 a day. Carbur was a company town with all the amenities. A second mill was in the works to meet the demand for lumber. Their Edward S. Hough designed lumber steamer *William O'Brien*, built by the New York Shipbuilding Corp. in 1915 brought 3 million board feet to New York in one trip.

Castner, Curran & Bullitt, Inc.
1 Broadway, Rm 32, New York

Samuel Castner, Sr., Henry Curran, and Charles A. Hardy established Castner & Co., Ltd., at 203 Walnut in Philadelphia to distribute Pocahontas coal.

In 1900, Samuel Castner, Jr., Henry Curran, and William C. Bullitt were in business as Castner, Curran & Bullitt at 328 Chestnut St., a large coal buyer and distributor that had developed the C.C.B. Pocahontas and C.C.B. New River brands of high-quality smokeless steaming coal used by the Navy and steamship companies. Just prior to the World War they ordered their own steamships and owned eight vessels when the war began. Large stocks of coal were kept at Hampton Roads and Newport News and shipped all over, coastwise and foreign.

In 1918 the president was Lemuel Burrows, Phillip A. Castner secretary. The company managed one Shipping Board freighter and managed and operated one.

Charles Kurz & Co.
1095 Drexel Building, Philadelphia

In 1909 Charles Kurz went into business as a steamship agent. He then began chartering ships to transport his own goods and acted as a forwarding agent for other shippers. Kurz managed two Shipping Board freighters and managed and operated one ship. The company evolved into Keystone Shipping.

Charles Nelson Co.
230 California St., San Francisco

Charles Nelson was born in Denmark in 1830. Around 1843 he went to sea and returned home around 1849. While there, he got gold fever and headed for California and arrived in mid–1850. He found mining unprofitable and started an inland water transport and wood fuel business. He soon started operating sailing vessels to carry lumber along the California coast and established the Charles Nelson Co. in 1901.

Before the war he owned the steamers *Falcon, Lakme, Charles Nelson, North Fork,* and *Riverside,* five schooners, three barkentines, one bark, and one brig.

Charles T. Magee & Co.
155 S. 2nd St., Philadelphia

Prior to the war owned the schooner *David Baird* and two barges. The company managed eight Shipping Board freighters and managed and operated six ships.

Chile Steamship Co., Inc.
120 Broadway, Rm 3523, New York

Incorporated in New York with a capital of $10,000 after 1912. John K. McGowan was president, Charles K. Lipman vice-president, William E. Bennett secretary, and Leopold Frederick treasurer. Harry F. and Edmond A. Guggenheim, of Guggenheim Bros., were directors.

The company's freighter *Republic* was requisitioned by the Shipping Board on a time charter basis for operation by the company for the west coast of South America trade.

Cities Service Company
60 Wall St., New York

Henry L. Doherty was president, E. Mackay Edgar was 1st vice-president in London, Frank Frucauff, Ernest H. Johnston, and Thomas I. Carter were vice-presidents, Paul R. Jones was secretary, and Louis F. Musil treasurer. The company's gross earnings for 1917 were $19,252,493. Their very extensive holdings were held by several subsidiaries that earned $14,509,841 for the year. In 1917 the company incorporated the Empire Transportation & Oil Co., which owned their tankers *Holden Evans* and *John M. Connelly*. Each ship was registered as its own corporation.

The Clyde Steamship Co.
Pier 36, North River, New York

In 1899, William P. Clyde & Co., 5 Bowling Green, New York, ran freight between Jacksonville and Boston and operated a general steamship agency in Philadelphia. The Clyde Steamship Co. carried passengers as the "New York, Charleston, and Florida Lines" with tri-weekly sailings between Jacksonville and New York.

Incorporated in Maine before the war with a capital of $7,000,000. Harry H. Raymond was president and Alexander R. Nichol treasurer. Most of the officers and directors were the same as the Atlantic Gulf & West Indies Steamship Lines. Owned 19 freighters. The company managed one Shipping Board freighter, operated one, and managed and operated two ships.

Coastwise Steamship & Barge Co.
605 Burke Building, Seattle

Capt. James Griffiths was president. Before the war he owned the ship-rigged sailing vessels *America* and *St. David*, the schooner *Bangor*, three barges, and a scow. He added the steamer *Anyox* and barkentine *Louisiana* during the war. James Griffiths & Sons were ship operators during World War II.

Coastwise Transportation Co.
40 Central St., Rm 800, Boston

Incorporated on March 9, 1903, in New Jersey. John G. Crowley was president, Capt. Arthur L. Crowley vice-president and general manager, and George S. Priesing secretary-treasurer. Prior to the war the company owned the schooners *William L. Elkins* and *Isaiah K. Stetson*. By 1917 the company owned ten ships and managed four Shipping Board freighters, operated two, and managed and operated 16 ships during the war.

In September 1917 Coastwise hired shipbroker Arthur P. Friend, of Walter, Friend & Co., 15 India St., to charter their five-year-old, 369' freighter *Middlesex* when and where he could at a commission rate of 5 percent of the gross charter amount.

On September 17 he arranged a 12-month charter of the vessel with E.J. Lavino & Co., of Philadelphia, to commence on or about the 15th of November at $80,000 a month, payable monthly in advance. Besides the normal provisions of charter agreements, an additional clause was inserted: "In case of requisition or commandeering by the United States Government or under its authority, it is mutually agreed that this charter is cancelled."

On November 1, the Shipping Board chartered the ship for Army service. The ship was armed and a Navy Armed Guard unit placed aboard and headed for France. On January 2, 1918, the ship was transferred to the Navy at Norfolk and commissioned *Middlesex*. Coastwise was paid for the use of their vessel and it was returned to them in good condition on February 7, 1919.

Arthur Friend was upset over losing his commission and blamed the government. He sued to recover the amount he would have made for the one-year charter agreement but his claim was dismissed.

Coastwise owned ten ships. The company's freighter *Bristol* was requisitioned by the Shipping Board on a time charter basis for operation by the company on the New England Coal route and the *Norfolk* was requisitioned for operation by the company on the Army's account.

The company managed and operated the seized German passenger liner *Wyandotte* (ex–*Willehad*) to France. On March 3, 1920, American-Hawaiian acquired a controlling interest in the company and in June 1921 a merger was proposed.

Crowley was also president and general manager of the Consolidation Coastwise Co., Inc., at the same address. Royal C. Gillespie was vice-president. The company owned one ship and managed and operated three Shipping Board freighters.

Coos Bay Lumber Co.
Marshfield, Oregon

Charles Axel Smith was born in Sweden in 1852 and came to America in 1867. He got into the lumber business in Minnesota and, like many others, saw the riches of the Western forests. He came west, bought timberlands, and met Albert Powers. In 1907 they opened a mill in Marshfield and the next year founded the Smith Lumber Co. Troubles ensued and in 1916 the company was reorganized as the Coos Bay Lumber Co. The company owned the schooners *C.A. Smith* and *Johanna Smith*.

Cosmopolitan Shipping Co., Inc.
115 Broadway, Rm 415, New York

Augustus F. Mack was president, Harry H. Raymond 1st vice-president, George M. Pynchon 2nd vice-president, James R. Munoz 3rd vice-president, Charles H. Crocker 4th vice-president, Ross R. Sigsbee secretary, and Wilfred B. Bremner treasurer. Operators, agents, and managers who owned the sailing ship *Brynhilda* and schooner *Rostellan*. The company managed one Shipping Board freighter, operated two, and managed and operated 21 ships.

The *Rostellan* arrived at Le Havre, France, on May 31, 1917. Apparently the master was unable to continue the voyage and the American consul, John Ball Osborne, appointed Capt. Albert M. Miller to take the schooner home. Through a miscommunication, the schooner was reported sunk by a U-boat on July 10 with all aboard.

Capt. Miller arrived at Philadelphia on September 3, 1917, and cabled the consul that he had arrived. He wrote a letter on the 3rd:

Dear Sir:–

As cabled you sat. the 1st inst. arrived here all well, 56 days passage. We witnessed the sinking of the *Hildegard* on July 10th off Start Point, English Channel. The crew were all saved. Capt. Bragg is now master of the Am. schr. *Laura Holt*, bound to France. Can't give you the port. We were reported being sunk with loss of all hands on July 12th, and when we appeared coming up the Delaware on 1st inst. both New York and Phila. Got quite a surprise. The owners of the *Rostellan* had given her up for lost with all hands. I have been asked by Maritime Transportation Co. to retain the command and am of the opinion the Dept. of C. & L. will grant me the privilege of again taking her to France & return. We had a very anxious time getting away from the submarines. I have met one of the crew from the *Hildegard* who states a Dutch schr. that was moored alongside *Hildegard* & had wireless gave our sailing date to the German U boat, for he followed us from Havre till she got the *Hildegard*. I fear it is a truth & we shall have to be careful when in Havre again & guard against our sailing date being known.

You will please convey to the Naval Authorities my sincere appreciation for the Convoy #316, services rendered us, but on acct. of winds and weather we were unable to carry out their orders as to entering Cherbourg. Must again thank you for the appointment to command the *Rostellan*. I go to New York on the 5th to confer with owners & will write you again as to my definite movements.

With very best wishes for your future welfare & health. Believe me

Sincerely yours,
Albert M. Miller.

The *Hildegard* was becalmed, a sitting duck for the U-boat. The crew of the *Rostellan* got off into their boats in the belief they were next. Just then, British patrol boats arrived, attracted by the sound of the sub's gunfire, and forced the U-boat to submerge. The patrol boats were too small to tow the schooner out of danger and sent a message for larger boats. The *Rostellan* was then "towed to safety" before proceeding on.

Capt. Miller, 41, was born in Canada of Scottish parents. His wife, Nettie, 41, and their children James, 11, Albert, 9, Annette, 5, and Rebecca, 2, were mourning the loss of their father at their home on Porter St. in Pascagoula, Mississippi, and Nettie was then pregnant with another daughter, Frances. Capt. Miller's widowed aunt, Rebecca Mosley, 72, also lived with them. He sent a telegram to Nettie the instant he got back.

Consul Osborne wrote to the State Department regarding Capt. Miller's statement about the Dutch espionage incident. The consul had some doubts about the truth of the matter but reported it anyway.

The company managed and operated the seized German freighters *Neuse* (ex–*Madgeburg*) to France; the *Oconee* (ex–*Maia*) to the United Kingdom; and the *Wabash* (ex–*Tubingen*) to France.

The Shipping Board chartered the French sailing vessel *La Fontaine* on February 27, 1918, and assigned the ship to the company on June 21, 1918, for operation to Africa.

The Shipping Board chartered the Norwegian steamer *Regulus* on October 10, 1918, and assigned the ship to the company on the 15th for operation to the Netherlands for 12 months.

Crowell & Thurlow Steamship Co.
131 State St., Boston

Peter H. Crowell and Lewis K. Thurlow set up shop as ship brokers at 33 Broad St., Rm 84. In April 1912 they incorporated the Crowell & Thurlow Steamship Co. prior to the war they owned 18 schooners and the bark *Onaway*.

In 1917, Edward Pierce was president, Walter D. Noyes vice-president, and Lewis Thurlow treasurer. The company owned nine American-registered steamers and controlled the Atlantic Coast Co. through stock ownership. Their ships were the *Edward Peirce, Felix Taussig, Lewis K. Thurlow, Peter H. Crowell, Stephen R. Jones, Tampico, Walter D. Noyes,* and *William A. McKenney*. The company managed nine Shipping Board freighters and managed and operated seven ships.

Cuba Distilling Co.
40 Exchange Place, New York

Alcoholic beverage distillers used Cuban cane blackstrap molasses in their manufacturing process and around the turn of the century were importing about 500,000 tons annually. This business was dominated by the Distilling Company of America through two subsidiaries, the United States Industrial Alcohol Co., and its subsidiary, the Cuba Distilling Co. Cuba Distilling was the ship operating division.

On September 18, 1902, a group of investors established the Distillers Securities Corp. to acquire the stock of the Distilling Company of America. Shortly after the turn of the century, the use of cane blackstrap molasses to sweeten and enrich livestock feed started to become popular and the importation of molasses eventually increased to over 1,100,000 tons.

In 1915, the stock of the United States Industrial Alcohol Co. was sold to the Rockefeller interests. At that time, the price of blackstrap was around 6 cents a gallon.

In 1916, a proposal was made to dissolve Distilling Company of America and merge the two subsidiaries into the Distillers Securities Corp., making the Distilling Company of America an operating company. Opponents obtained an injunction against the dissolution and merger.

Between 1915 and 1918, due to the increased demand for alcohol and animal feeds during the war, the price of blackstrap went up to 7 or 8 cents a gallon. By early 1918 the Distillers Securities Corp., had acquired 99 percent of common stock and 97 percent of preferred stock in the Distilling Company of America and owned over 100 plants including distilleries, distribution houses, bonded and free warehouses, and ships. On January 9, 1918, the United States Industrial Alcohol Co. went from a holding company to an operating company "with direct control over production" and the company focused on "chemicals urgently needed in munition manufacture."

Gilbert E. Rubens was president, Adrian H. Larkin vice-president, Joseph Malone secretary, Samuel H. Atchison treasurer. Horatio S. Rubens was chairman. The company owned the *Cubadist, Mielro, Nelson,* and *Sucross*.

Diria Ship Corp.
672 Bullitt Building, Philadelphia

The American Manganese Manufacturing Co. was incorporated on December 31, 1913, with Edward E. Marshall as president, William Selfridge and Radcliffe Romeyn vice-presidents, William H. Locker vice-president and treasurer, and William E. Arnold secretary.

The company incorporated Diria Ship on July 16, 1917, with a capital of $300,000. In

November 1917 the 5-masted schooner *Diria* loaded lumber on the West Coast for San Salvador. The schooner then proceeded to Costa Rica to load ore for Philadelphia where two, twin 300-hp Craig oil engines were installed. The auxiliary schooner then carried Army supplies from New York to London and France. The vessel foundered on October 15, 1920, off Guantanamo Bay, Cuba.

Dollar Steamship Line
230 California St., San Francisco

Robert Dollar was a pioneer lumberman in California, ship operator, and steamship agent. In 1912 he owned the steamers *George Dollar, Melville Dollar,* and *Stanley Dollar.* The company operated three Shipping Board freighters.

Donald Steamship Co., Inc.
17 Battery Place, New York

Philip De Ronde was president of the Donald Agency, Inc., ship brokers at 39 Broadway. Frank H. Hall was secretary and Ernest Martin assistant treasurer. De Ronde was president of Donald Steamship, Alexander R. Nichol vice-president, Frank Hall secretary, and Ernest Martin assistant treasurer.

Prior to the war the company owned the steamers *Alice, Amelia, Annetta, Bella, Lillie, Norhilda,* and *Querida.*

Earn-Line Steamship Co.
139 S. 4th St., Philadelphia

The Pennsylvania Steel Co. was incorporated on March 23, 1866. In 1883, Luther S. Bent was president at the 208 S. 4th St. office in Philadelphia.

On May 1, 1883, The Juragua Iron Co., Ltd., was chartered for a term of ten years in the office of Pennsylvania Steel. Luther S. Bent was chairman and a manager. Josiah Monroe was secretary-treasurer. Other managers were Alfred Earnshaw, Robert H. Sayre, Eben F. Barker, and Robert P. Linderman. The company purchased 4,000 acres in Cuba to extract iron ore, built a plant at the mine, about 16 miles southeast of Santiago de Cuba, and commenced building a railroad to Santiago. In August 1884 they began shipping ore to the United States.

On March 18, 1889, the Earnford Steamship Co., Ltd., was chartered in England to "run steamships in general trade." The corporate office was at Newcastle-on-Tyne River, England, and the general office was at 208 S. 4th St. in Philadelphia. Alfred Earnshaw was the managing owner, his brother George E. Earnshaw was secretary-treasurer, and the directors were Joseph R. Foard, Robert P. Linderman, Luther S. Bent, Eben F. Barker, and Robert H. Sayre. The company ran the steamer *Earnford.* On October 6, 1889, the steamer *Earndale* was acquired and on February 28, 1887, the *Earnwell* went into service under their own corporate names.

On December 30, 1891, the Earn-Line Steamship Co. was incorporated in Delaware with a capital of $1,000,000. The corporate office was at Camden, New Jersey, and the general office at 204 S. 4th St., Philadelphia. Alfred Earnshaw, 401 Chestnut St., was president and George Earnshaw secretary-treasurer. Luther S. Bent, Robert Linderman, Harry A. Berwind, and Joseph Foard, of Baltimore, were directors. By then Juragua Iron had shipped about 1.4 million tons of ore.

When Alfred died in July 1896, George became president. Harry S. Sharp was vice-president and treasurer, Ambrose B. Umstead secretary, and William C. Barratt assistant treasurer.

The company managed two Shipping Board freighters and managed and operated nine ships.

East Coast Transportation Corp. 17 Battery Place, Rm 2029, New York

Incorporated on June 27, 1914, by the Brooks-Scanlon Corp. R. Lawrence Smith was president, Scott McLainahan vice-president, and Richard D. Sutton secretary. John Jannotta and Dunham Wheeler directors. The company owned nine ships during the war and became embroiled in a complicated just compensation claim between the New York Shipbuilding Corp. and the Emergency Fleet Corp. over the seizure of the *M.J. Scanlon.*

Eastern Steamship Co. India Wharf, Boston

Owned 23 steamers prior to the war. Calvin Austin as president, John W. Brophy secretary, and Josiah W. Hayden treasurer. The company's 16-knot cargoliners *Governor Dingley* and *Governor Cobb* were requisitioned by the Shipping Board on a bareboat charter basis for recruit training. The *Dingley* had a crew of 90 and accommodations for 670 passengers, the *Cobb* 92 crew and 545 passengers. The company owned nine ships in 1917 and managed and operated one Shipping Board freighter.

Eastern Transportation Co. 1331 Munsey Building, Baltimore

In 1910, Harry F. Hooper was the City of Baltimore Controller and lived at 2205 N. Charles. Timothy J. Hooper was manager of the Southern Transportation & Towing Co. at 516 E. Lombard and lived at 3608 N. Charles.

In 1911 Timothy J. Hooper was manager of the Southern Transportation Co. and lived at 2635 N. Charles and in 1912 opened the T.J. Hooper Realty Co.

In 1912 Harry Hooper was secretary-treasurer of the Eastern Transportation Co. at 811 American Building and secretary-treasurer of the T.J. Hooper Realty Co. John T. Donohue was president of Hooper Realty.

In 1913 John T. Donohue was president of Eastern Transportation at 1808 Thomas and in 1914 the company moved to 53 S. Gay.

In 1915 Timothy Hooker was president of Eastern Transportation, Frank R. Diggs secretary-treasurer, and Harry was a ship broker. In 1917 the company owned seven schooners, the small steamer *Eastern,* and managed and operated two Shipping Board freighters.

Edgar F. Luckenbach
8 Bridge St., New York

Edgar F. Luckenbach was president, John L. Luckenbach vice-president, Paul Kuhne treasurer, Alexander C. Thackara secretary. Incorporated in New York with a capital of $25,000,000. Prior to the war owned the steamers *Edgar F. Luckenbach, F.J. Luckenbach, Harry Luckenbach, Jacob Luckenbach, Julia Luckenbach, L.L. Luckenbach, Lewis Luckenbach, S.V. Luckenbach, San Mateo* and 17 schooners.

The Lewis Luckenbach company was at the same address and the steamers *Edward Luckenbach, M.E. Luckenbach, City of Atlanta,* and *William H. Conner.*

The Luckenbach Steamship Co., Inc., was incorporated at 44 Whitehall St., Room 1200, in New York with a capital of $800,000 with the same officers at the same address.

Seven of the company's freighters were requisitioned by the Shipping Board on a bareboat charter basis to the Navy for operation by the Army: *Edgar F. Luckenbach, Edward Luckenbach, Julia Luckenbach, F.J. Luckenbach, Katrina Luckenbach,* and *Walter A. Luckenbach.*

The company managed four Shipping Board freighter, operated seven, and managed and operated 12 ships. Luckenbach Steamship owned the steamers *Pleiades* and *F.J., Frederick, K.I., Katrina,* and *Walter A. Luckenbach.*

Edwin K. Wood Lumber Co.
112 Market St., San Francisco

Manufacturers and dealers in fir and spruce products. Edwin K. Wood president, Walter T. Wood vice-president, Clarence A. Thayer secretary. The yard was at Frederick & King Sts. in Oakland.

Export Steamship Co.
25 Beaver St., New York

Charles C. Torpy was president, Arthur W. Knowles secretary-treasurer. Managed and operated seven Shipping Board freighters. In July 1928 the company was awarded the first 10-year mail contract under the Merchant Marine Act of 1920. The company provided freight and passenger service between New York, the Mediterranean, and Black Sea ports.

F.D. Dimmick & Co.
2129 Land Title Building, Philadelphia

Steamship agent Frederick Dimmick was president, Harry G. Antone vice-president, Albert L. Detwiler secretary, and Bertram M. Ainesworth treasurer. The company operated ten Shipping Board freighters.

Federal Line, Inc.
78 Broad St., Rm 216, New York

The Boultons were building contractors, merchants, bankers, and brokers in Brooklyn. In 1913 Frank owned a stevedoring company at 78 Broad St., Room 211, in Manhattan and lived at 7604 14th Ave. in Brooklyn. When war came Frank decided to go into the shipping business as a steamship agent and opened the Federal Forwarding Co. in his Broad St. office. He was also a general contractor as Frank F. Boulton & Co. at the same address. Albert W. Boulton was treasurer of Federal Forwarding and M.E. Boulton was secretary.

When freight rates started going through the roof, Frank decided it was time to actually own and operate ships and in 1917 the Boultons incorporated the Federal Line with a capital of $100,000. Frank was president, Alexander H. Erickson was vice-president, and Albert Boulton secretary-treasurer. The company ran several ships on a regular schedule from New York to Genoa. The firm did very well during the war years carrying caustic soda in metal drums for the manufacture of soap and explosives. Freight rates could change dramatically overnight but in 1917 varied between $5.00 and $7.00 a hundred pounds. The soda sold for 5 to 8 cents a pound.

The company's freighter *Neptune* was requisitioned by the Shipping Board on a bareboat charter basis for operation by the French government.

During this time Federal chartered the British steamer *Gaspesian* from the Canadian Maritime Corporation of New York City. The deal was approved by the Shipping Board on the 29th and Federal paid Canadian Maritime $25,000 in cash up front with a balance of $100,000 due before the ship sailed from New York. Federal obtained freight contracts on the open market but before the ship could leave New York Harbor, Federal had to secure a "bunker license" from the War Trade Board. On July 1 the Board informed Federal that permission to issue the license had to come from the Shipping Board. Federal applied to the Shipping Board and was told they would have to pay the Shipping Board $47,688.58 to get the license. Federal paid under protest the next day. The check was certified by the New York Trust Co. and sent to the Treasury Department. Treasury presented the check to the New York Clearing House and the money was deposited into the Treasury on the 31st.

Federal objected vehemently to the necessity of obtaining a bunker license and protested the requirement of having to pay money to get it. Federal had no choice but to comply or be in breach of contract with the shippers of the cargo they were obligated to carry. Federal demanded their money back from the Shipping Board and the Board said "No."

When the war ended, freight rates fell off and by 1921 Federal was under receivership. For some reason Frank waited until July 21, 1924, to file suit. The court dismissed the petition for exceeding the statute of limitations but noted the Shipping Board was clearly wrong in its actions.

Fife Shipping Co.
351 California St., San Francisco

Joseph Fife was a lumberman in San Francisco and in 1909 built the Fife Building at 1 Drumm St. In 1917, Robert C. Reid was president, Robert F. MacLeod secretary, and James C. Fogarty, marine superintendent. The company engaged in the general lumber trade and owned the 240' schooner *Thistle* and 260' barkentine *Monterey*. Operating manager and agent was Balfour, Guthrie & Co., which eventually controlled Fife. The company owned eight ships at wars' end.

Florida East Coast Car Ferry Co.
26 Broadway, New York

Henry Morrison Flagler was a co-founder of Standard Oil of Ohio. Flagler saw opportunity on the east coast of Florida and engaged in hotel building and organized the Florida East Coast Railway. Henry then set his sights on Cuba and extended the line to Key West. Henry died on May 20, 1913, but his project continued with the Florida East Coast Car Ferry Co. Railroad cars were brought from Key West to Havana, Cuba, on the steamers *Henry M. Flagler, Joseph R. Parrot,* and *Estrada Palma.*

Foreign & Domestic Transportation Co.
11 Broadway, Rm 1159, New York

Frederick R. Eaton was a lumber dealer in New York and lived 110 Lorraine Ave. in Montclair, New Jersey. In 1917 he was president of Foreign & Domestic. Jesse Myers was secretary. The company engaged in the Southern lumber business. In September 1918 the company sold their 182', 4-masted schooner *J. Edward Drake* to William Knox & Co. for $125,000.

France & Canada Steamship Corp.
120 Broadway, Rm 2623, New York

Francis R. Mayer was president, Stuart H. Macintosh secretary. Charles Mayer was a director. Capitalized with $4,000,000. The company owned nine schooners and managed four Shipping Board freighters and managed and operated 14 ships.

The company managed and operated the seized German passenger ship *Ascutney* (ex–*Pisa*) to France and operated the freighter *Pawnee* (ex–*Harburg*) to France under Shipping Board management.

Freeport Sulphur Transportation Co.
61 Broadway, Rm 1801, New York

Incorporated in Delaware with a capital of $25,000. Eric P. Swenson was president, Florence M. Altz secretary, Swen A. Swenson treasurer. Owned the *Freeport Sulphur No. 1* and the *Freeport Sulphur No. 11*.

Gailey, Davis & Co.
316 Walnut St., Philadelphia

Robert J. Gailey and Charles E. Davis, Jr., were steamship agents and brokers. Davis was also the manager of the Red Star Tug Co. Operated only one ship, the *Yakima*, to the Netherlands.

Galena-Signal Oil Co.
Franklin, Pennsylvania

Incorporated on November 7, 1901, with the merging of the Galena Oil Co. and the Signal Oil Company. Manufactured lubricating, signal, and illuminating oils and greases for steam and electric railways. The refinery was in Houston, Texas, with plants at Franklin; Parkersburg, West Virginia; Elizabeth, New Jersey; Boston; Whiting, Indiana; Toronto; and Rouen, France.

Lauren J. Drake, president, James E. Linahen, vice-president, William P. Wescott vice-president and treasurer, and J. French Miller secretary in Franklin; Larkin F. Jordan and William A. Trubee vice-presidents in New York; William J. Walsh vice-president in Chicago; and George L. Morton vice-president in Atlanta.

Garland Steamship Corp.
511 Fifth Ave., New York

George G. Allen was president, William M. Campion vice-president, James P. Watson treasurer. Capitalized with $1,900,000. Wesley W. Cheek was a director. The company owned the steamers *Carolinian, Grayson, Javary, Justin,* and *Norlina.*

Gaston, Williams & Wigmore Steamship Corp.
Guaranty Trust Building, 39 Broadway, New York

Electrical engineers George A. Gaston, William H. Williams, and James A. Wigmore incorporated in 1914 with $5,000 to export war supplies to the Allied governments as Gaston, Williams & Wigmore, Inc. They established offices in London, Paris, Petrograd, Madrid, Rome, and Havana and set up Gaston, Williams & Wigmore, Ltd., in Canada to operate English and Canadian ships. They developed a heavy trade to the Far East and Russia.

On April 19, 1916, they incorporated their own steamship company with $5 million from the Guaranty Trust Co. to handle their trade from San Francisco to Japan, China, and Vladivostok, Russia, and from New York to Archangel, Russia, when that port was open. They acquired the steamers *Virginia, Lord Dufferin, Carolyn, Eskasoni, Maryland,* and *O.H. Brown.* They also bought the contracts for two ships on the ways, chartered six more, and ordered four to be built.

George Gaston was president, Richard Henry Lee Martin secretary, and A.W. Frank treasurer. Charles H. Sabin, president of Guaranty Trust, was a director and primary financier of the operation. All the stock was held by officers and directors in the company. In June 1916, the company declared a dividend of $1 a share out of earnings of over $400,000, "a remarkable illustration of the great profits there are in the shipping business at the present time."

The company managed one Shipping Board freighter and managed and operated seven ships.

General Navigation Co.
120 Broadway, Rm 456, New York

Henry C. Ulen was born in Indiana in 1871. He married Mary Dutch in 1890 and in 1894 began practicing law. In 1899 they moved to Indianapolis where he founded the American Light & Water Co., a construction firm, and in 1908 they moved to Chicago.

In 1914, when the war began, Henry and Mary lived at 4147 Sheridan Rd. and he was president of the Ulen Contracting Co., Ulen & Co., and Henry C. Ulen & Co., bankers at 38

S. Dearborn. Earl C. Ulen was manager of Ulen & Co. In 1916 Ulen began building water systems in South America and bought the schooner *Alice M. Colburn*.

Henry C. Ulen was president of General Navigation, Earl G. Ulen vice-president. The company operated three Shipping Board freighters and managed and operated three ships.

George E. Billings Co.
312 California St., San Francisco

Billings was a wholesale lumber dealer, agent for the Hall Bros. Marine Railway, and marine agent for the Fireman's Fund Insurance Co. The Billings Co. was an insurance brokerage and marine loss average adjusting firm. Billings was president, Roy C. Ward vice-president, and James W. Dean secretary. Billings lived in Mill Valley.

Prior to the war the company owned 19 schooners including the *Winslow*, the barkentine *John Smith*, and the small steamer *Shna-Yak*.

George Hall Coal & Transportation Co.
52 Water St., Ogdenburg, New York

In 1880, George Hall, William L. Proctor, and James S. Bean raised $21,000 to purchase the tugboat *William Gardner*, three barges, a store in Parish, and a supply of coal at auction from the bankrupt Seymour & Co. The partners went into business as George Hall & Co. at Ogdensburg to serve the Great Lakes and Western Rivers. Bean was president of the Ogdensburg Bank and the new firm was able to borrow all the working capital they needed. The firm thrived and when the war started the company owned five steel steamers and the Rochester & Pittsburgh Coal & Iron Co. Hall was president, John C. Howard vice-president and treasurer.

In December 1915, Howard was asked about the new La Follette seaman's act and stated:

> We are only prepared to discuss the matter from the viewpoint of the canal-size lake steamer and we are free to say that the enforcement of present regulations will be disastrous to American bottoms.
>
> Before the seamen's law became effective, the owner of American ships was handicapped in that his crew cost him $2,500 per boat more than his Canadian neighbor. The actual figures were as follows:
>
> Wages of American crew, season of navigation, $9,620.
>
> Wages of Canadian crew, season of navigation, $7,125.
>
> Balance in favor of Canadian vessel owner, $2,495.
>
> Now the Department of Commerce and Labor orders us to equip our boats forthwith for five extra men, dividing our engineers' watches up into units of four hours instead of six. In addition to the cost of equipment, the extra crew required, including the meals, will cost us $2,500 per steamship, or a total expense of $5,000 more for each ship than the same capacity boats in the same trade under Canadian registry are obliged to pay.
>
> This means that the owner of American bottoms cannot compete with his Canadian cousin. But Canada allows American bottoms to go in under its registry except for coasting, that is, trading from one Canadian port to another, without the exaction of duty, and as the principal commodities are coal east-bound from Lake Erie and Lake Ontario to Montreal and Quebec, with return loads of pulp wood from lower Canadian ports to American ports, American ships thus Canadianized can operate with Canadian crews on a parity with their own ships and thus save the American shipper from complete annihilation.
>
> The law, so far as the Great Lakes is concerned, is iniquitous and should be repealed. Neither captains nor engineers want the extra men and in our opinion it is the "full crew bill" operating to kill American commerce on the inland lakes.

The company's new steamer *Adrian Iselin* made 93 trips during the 1915 season bringing 300,000 tons of coal from Charlotte to Ogdensburg.

Girard Trust Co.
Broad & Chestnut Sts., Philadelphia

An old financial and real estate firm going back to the famous Philadelphia shipping man Stephen Girard. Effingham B. Morris was president, William Newbold Ely, Albert Atlee Jackson, and Edward Sydenham Page were vice-presidents; Samuel W. Morris was secretary, and George H. Stuart, III, was treasurer. Owned the steamers *Robert M. Thompson* and *Ruby*.

Grace Steamship Co.
7 Hanover Square, New York

Grace was incorporated in Delaware on February 14, 1916, to manage ships in the South American trade. Maurice Bouvier was president, Lawrence H. Shearman was vice-president, and J. Louis Schaefer was treasurer. Joseph P. Grace and John H. Rosseter were directors.

Besides nine *Santa* ships: *Alicia, Ana, Barbara, Elena, Ines, Luisa, Paula, Rita,* and *Rosa*, the company owned the *Cacique* and *Colusa*. These ships were under charter to W.R. Grace & Co. from April 1916 to April 1928 at $240,000 a year. Grace Steamship also operated the Nafra company's steamer *Chincha*.

Gravenhorst & Co.
90 Wall St., Rm 704, New York

George A.N. and Frederick Gravenhorst were importers of general merchandise from around the world. Both lived in Brooklyn.

Gulf & Southern Steamship Co.
Foot of Liberty St., Jacksonville, Florida

William M. Tupper president, J. Horatio Nelson secretary-treasurer. The company owned two ships, managed two Shipping Board freighters, operated one, and managed and operated four ships. Tupper was vice-president of the U.S. Trust & Savings Bank and the Florida Talking Machine Co. He lived with his wife Alvaretta at 131 E. 7th.

Gulf Export & Transportation Co., Inc.
Ft. of Milam St., Beaumont, Texas
"The Bowie Line"

Regular freight service on Route 1 from Houston to Tampico and Progreso, Mexico, calling at Veracruz and other ports when sufficient freight was carried. William A. Bowie was president and general manager, Clarence E. Barry vice-president, Wallace E. Sprouse secretary. The company owned the *Col. Bowie* and the *Kecchi*.

Gulf European Steamship Co.
60 Michael St., Mobile

Apparently a wartime operating subsidiary of the Gulf States Shipping Co., a lumber exporter. The company operated one Shipping Board freighter. William K. Syson was president, Joseph M. Walsh secretary-treasurer, James L. Coleman manager. Syson was also president of the Holt Lumber Co. Walsh was president of Walsh Stevedoring & Contracting Co. and the Mobile Towing & Wrecking Co., vice-president of the Alabama Iron Works and Bay City Export Co., and treasurer of the Mobile Ship Chandlery Co.

Gulf Refining Co.
West end of 7th St., Port Arthur, Texas

John W. Tryon was manager at Port Arthur. The company operated 26 different types of vessels. Their ocean tankers were the *Gulfland, Gulflight, Gulfmaid, Gulfoil, Gulfstream,* and the big 7,807-ton *Gulf of Mexico*. Tryon was also president of the Gulf Commissary Co., the Gulf Gun Club, vice-president of the Merchants State Bank, and mayor of Port Arthur. John and his wife Emily were natives of Pennsylvania and lived at 1725 Lake Shore Dr. with their son Joseph.

H.F. Ostrander
712 Leary Building, Seattle

Harry Fargo Ostrander was born in Wisconsin in 1876 and became a very successful importer and exporter of general merchandise in Seattle. He owned the steamer *Frances L. Skinner* and the schooner *Levi W. Ostrander*. He lived at 200 40th Ave. N. with his wife Lenore, of Illinois, daughters Catherine, who was born in Japan, Olive, Alice, and Rosemary, son Wills, his father Levi, three servants, and a chauffeur. The company operated one Shipping Board freighter.

Hammond Lumber Co.
260 California St.

Andrew B. Hammond was a pioneer California redwood lumberman along with Robert Dollar, Charles McCormick, and Charles Nelson.

Around the turn of the century, demand for Oregon pine (Douglas fir) from Oregon and Washington became quite strong and operations expanded further north. Prior to the war the company had an office at 60 Wall St. in New York and owned the steamers *George W. Fenwick, Francis H. Leggett, Nehalem,* and *Ravalli*. They added the small steamers *Flavel, Halco,* and *Santiam* and built ships at Samoa, California.

Harby Steamship Co.
15 William St., Rm 1410, New York

Richard T., Robert M., and William L. Harriss joined Robert F. Irby and Alden H. Vose as cotton merchants Harriss, Irby & Vose. In 1915 they incorporated the Harby Steamship

Co. in Room 1415 at the same address with a capital of $10,000. William Harriss was president, George Hoyer vice-president, Robert Harriss, secretary, and Charles E. Warren, Jr., treasurer.

Harriss, Magill & Co., Inc.
35 S. William and 50 Broad St., New York

Steamship agent James P. Magill was in business with the Harriss brothers. The company operated one Shipping Board freighter and managed and operated 15 ships.

Hasler Bros.
Produce Exchange Building, New York

Frederick E. and Thomas B. Hasler were steamship agents in the Produce Exchange Building in New York City as Hasler Bros., Inc. Frederick was president, Thomas vice-president, and James McFarland secretary. Frederick lived in Little Silver, New Jersey, and Thomas at 301 W. 106th St. The Haslers managed one Shipping Board freighter, operated one, and managed and operated 16 ships.

The Hudson Navigation Co.
Pier 32, North River

In 1903, Hudson Navigation was run by Charles W. Morse, who felt it was necessary to have a big steamer named after him so he contracted with Harlan & Hollingsworth of Wilmington, Delaware, build a big 427', 16-knot passenger liner with his name on it. Morse spent $913,000 to build and furnish the ship. The *C.W. Morse* transported passengers and freight between New York City and Albany. The ship could accommodate 3,000 passengers total and sleep 900.

In 1917, Harry F. Morse was president, George W. Loft vice-president, Benjamin W. Morse secretary, Nehemiah H. Campbell treasurer. William Guggenheim was a director. During the 1917 season from March 30 to October 3 the ship's net earnings were $227,368.33.

On December 12, 1917, the ship was requisitioned by the Navy for use as a receiving ship at the Brooklyn Navy Yard. During this time there was a severe housing shortage in and around New York City and the *Morse* served as a floating hotel.

When the *Morse* was taken, Morse put the steamers *Trojan* and *Rensellar* on the New York to Albany service.

On December 13, 1917, the Board of Appraisal of Merchant and Private Vessels determined that a charter rate of $8,000 a month was equitable for the *C.W. Morse*. The rate set for ocean-going and coastwise steamers with accommodations of 150 or more passengers and with a speed up to 11 knots was $5.75 per gross ton per month with an additional .50¢ for each knot, or part of a knot, over 11 knots.

Morse said $8,000 a month was too low so the Navy reduced the charter rate by 75 percent during the 14-month charter period. The company accepted the amount, made no further protest, and reserved the right to sue later in federal court.

On April 30, 1918, the Bureau of Navigation wrote to Hudson Navigation to inform the company that as a matter of vital military need the Navy would not be returning the steamer to the company for the duration of the war owing to the extreme need for ships of all kinds. The war ended on November 11, 1918.

The ship was returned to Hudson Navigation on February 11, 1919, and the company spent two months putting the ship back into her prewar service condition. The Just Compensation Board determined Hudson should receive a further $8,000 a month during the two-month reconditioning period but the Navy paid Hudson only $12,000, or 75 percent of the charter rate. Hudson maintained the *Morse* was worth $12,000 a month. The company was paid a total of $95,806.46. Based on a rate of $12,000 a month for 16 months they would be owed $192,000.

Hudson filed suit to recover a balance due of $96,193.54 on the 16-month period at $12,000 per month but before that action commenced, Hudson Navigation went into bankruptcy on April 16, 1921, ostensibly due to Morse's financial dealings. The company's receivers, Middleton S. Borland and James A. Emerson, continued the claim. On June 12, 1922, the court stated:

> In considering questions of taking and just compensation therefor, the owner must be compensated for what is taken from him, and not the value of the property to the Government for its particular use, and the compensation must be a full and perfect equivalent for the property taken, and the owner must be reimbursed for the full value of the services, and the compensation is to be estimated by references to the uses for which the property is suitable.
>
> The determination of what shall be the measure of compensation is a judicial question, and the executive has not the power to fix the compensation which will be given when private property is taken for public use. And, therefore, this court in determining what is just compensation can not be governed in its determination by the action of a board appointed by the President; while giving such weight to the findings of the board as the evidence in the record will justify yet the court must determine what is just compensation from the acts proven; and the court in determining these cases arising out of war conditions must also bear in mind that the war power of the United States, like its other powers, is subject to applicable constitutional limitations, and where private property is taken for public purposes during war the power exercised must be in subordination to the fifth amendment; and in arriving at what is just compensation, under the constitutional provision, that provision should be literally construed.
>
> The Government contends that there was no market value for ships or for the use of ships, because it says this market was under the influence of war, and was restricted by the fact that the United States had requisitioned all ships, and that in consequence of that action there was no market from whose prices a market value could be determined. In other words, it is contended that the market value of the property of the citizen can be destroyed, and the provisions of the fifth amendment rendered nugatory by an act of the legislature. The mere statement of such a contention refutes it. It is in the teeth of every principle of law established by a long line of the decisions of our highest court, and is obnoxious to every tenet of Anglo-Saxon law and justice.
>
> But the contention that there was no market value for ships or for their use at the time this ship was taken is not tenable. The demand for ships was universal; the demand for their use was equally so. It is true that the market was affected by the war; war always affects market conditions, either enhancing prices or lowering them, but the fact that war has such affects does not affect the principles which must govern the courts in ascertaining values. The facts in our opinion show that just compensation for the use of the *C.W. Morse* is the sum of $12,000 per month. The amount to which the plaintiffs are entitled is $96,193.54.

The *C.W. Morse* name was changed to *Fort Orange* and Hudson River Navigation continued to operate from Pier 52, North River.

On February 27, 1922, Charles W. Morse and his sons, Ervin, Benjamin, and Harry; George M. Burditt, a Morse Co. attorney; Colin H. Livinston, president of the Boy Scouts of America and former president of the Virginia Shipbuilding Corp.; Leonard B. Christie, treasurer of Virginia Ship; Nehemiah M. Much, assistant treasurer of Virginia Ship; William

W. Scott, of the District of Columbia and a Virginia Ship attorney; Philip Reinhardt, an Emergency Fleet Corp. auditor, and Robert O. White assistant treasurer of Groton Iron Works and president of the United States Transportation Co. were indicted by a Washington, D.C. grand jury for conspiracy to defraud the United States and the Emergency Fleet Corp. in connection with wartime contracts. Morse said the indictments were "secured because of personal feelings against me on the part of Attorney General Harry M. Daugherty and certain present officials of the United States Shipping Board."

Huron Navigation Corp. 20 Exchange Place, New York

In early 1917, the Brooks-Scanlon Corp. decided to sell their 2-year-old, 5,211-ton, 9½-knot, Hough-designed oil-burning lumber freighter *William O'Brien*.

To a banker like Sylvester W. Labrot, vice-president of the National Park Bank of New York, it seemed like a great time to get into shipping. Ship values were rising rapidly and freight rates to Europe were way up.

On February 9, 1917, he and his partners, attorney James H. Carpenter, John J. Reagan, and stock broker Percy R. Pyne, incorporated the Delaware Navigation Co. 55 Wall St., Rm 200, New York, with a nominal capitalization of $25,000 to purchase the *William O'Brien* for $1,600,000. Labrot was president, Carpenter vice-president, Reagan secretary, and Pyne, 2nd treasurer. The partners planned to pay the balance with promissory notes and money borrowed from banks.

On April 13, 1917, right after the United States entered the war, Delaware Navigation entered into a general agency agreement with the Cosmopolitan Shipping Co. to manage and operate their ship on "Route 20" from New York to Genoa, Italy, under the direction of Delaware's board of directors "as opportunity offers."

On April 16, in order to increase the company's capitalization, the shareholders of Delaware Navigation created the Huron Navigation Corp. with Sylvester Labrot as president. All 10,000 shares of Delaware stock were distributed proportionately among the shareholders based on the assessed value of the ship.

While the ship was on Voyage No. 3 under Capt. John Vaughn, Cosmopolitan made an agreement with the French government to carry miscellaneous goods from Le Havre to New York on the ship's homeward voyage.

On October 12, 1917, while the ship was at sea, the Shipping Board sent the Requisition Order telegram to Huron Navigation in care of Cosmopolitan, with instructions to acknowledge receipt. This was followed up the next day with a letter outlining the proposed charter rates.

The Shipping Board proposed two forms of charters: a "time-form" and a "bare-boat form." Under the time-form rate, owners of vessels over 6,001 deadweight tons and under 8,000 deadweight tons, the class the *O'Brien* was in, would be paid $6.25 per deadweight ton per month with the owner bearing the costs of operation. The bare-boat rate depended on the speed of the ship. The base rate was $4.15 per deadweight ton per month up to and including 11 knots and for each additional knot, or fractions of a knot over 11 knots, an additional .50¢ per deadweight ton per month was allowed.

Around October 17, Huron and Cosmopolitan wired the Shipping Board wanting to know that if they were to enter into a charter agreement with the government, what form would it be, how long would it last, and who would the charterer be? They said they had a contract to purchase fuel oil from the Standard Oil Co. and asked if they should continue the contract or would the charterers provide the fuel? If stevedoring services were paid for by the government, were they authorized to work on Sundays and holidays and pay overtime? Would the government provide war risk insurance for the full value of the ship and since the Government War Risk Bureau required requisitioned vessels on which they provided war risk insurance to be painted in a camouflage scheme, carry smoke boxes, and provide accommodations for gun crews would they incur those expenses?

The ship was due in New York from Italy within 48 hours and Huron was anxious for answers. Cosmopolitan wrote a note on behalf of the owners stating their desire to cooperate with the government.

The Shipping Board wired back authorizing the ship to load for France and proceed on Voyage No. 4, "on account of the government," in other words in government service. Capt. Vaughan arrived in New York on the afternoon of October 19 and began discharging cargo then began reloading government cargo for France.

On or about November 22, 1917, 90 percent of the stock of Huron Navigation was purchased by the France & Canada Steamship Corp.

The France & Canada Steamship Co., Ltd. of Montreal, Canada, was organized in 1915 to operate ships of various flags, but not American ships. The company had a subsidiary in New York, the France & America Steamship Co., Ltd., at 120 Broadway, Rm 2623. Charles Mayer was president, Francis R. Mayer vice-president, and Carlos Mayer secretary.

In 1917, the company set up the France & Canada Steamship Corp. specifically to operate American registered ships. Francis Mayer was president and Stuart N. McIntosh was secretary-treasurer. France & America Steamship and France & Canada Steamship were in the same office but kept separate books and payrolls.

When the *O'Brien* was ready to leave, Cosmopolitan reported to the Shipping Board and Capt. Vaughn departed on the 24th for Le Havre on Voyage No. 4. Sylvester Labrot, writing on behalf of Cosmopolitan, had some concerns for Edward F. Carry, the Shipping Board's director of operations, in a letter written on the 24th:

> Dear Sir: Complying with your instructions as follows:
>
> We have ordered the steamship *William O'Brien* to proceed to sea as you have directed.
>
> In taking this action we do so under protest regarding the item of war risk insurance. We think it is most unfair that the value of the ship is not specifically stipulated. This vessel cost us only a few months ago $1,650,000 in cash.
>
> The Government has repeatedly insured this vessel for $1,600,000 and $1,300,000, respectively. At the time of insuring the vessel for the smaller sum we went into the market and bought additional insurance to a total of $1,600,000.
>
> We request that the Government specify the value they place upon the vessel, and should this be less than $1,600,000, it is our intention to go into the market and buy additional insurance, and the cost of said insurance shall be an operating expense of the vessel.
>
> Any other course would be most unfair to the owners of this property, for unless the Government protects them in their insurance value, they should be permitted to protect themselves. Kindly advise in regard to this matter.

Carry wrote back to say the official policy of the Shipping Board was to determine the value of a vessel only at the time of loss—that's what the role of the Just Compensation Board was for—and the company's objection was duly noted.

Cosmopolitan then elected not to take out any supplemental war risk insurance on Voyage No. 4. The ship arrived at Le Havre on November 19 and discharged the cargo from New York.

The same day the ship arrived at Le Havre, Lt. Col. Joseph L. Gilbreth, of the Army Quartermaster Corps, wrote to Captain Vaughn:

> From: Commanding Officer.
> To: The master of the American S.S. *William O'Brien.*
> Subject: Requisition of steamer.
> 1. By authority of the commander in chief American Expeditionary Forces in France, I hereby requisition the United States steamer *William O'Brien* for duty as an Army transport. On and after the date of delivery of these papers to the master of the *William O'Brien* the ship, its crew, supplies, and stores pass into the service of the United States Government.
> 2. Sailing instructions will be communicated to you later.

Technically, Cosmopolitan's position as agent for the *William O'Brien*, and their income from that role, would be terminated when the ship was taken over by the Army.

The American consul, John Ball Osborne, came aboard and took the ship's papers to Col. Gilbreth and then reported the action to the State Department. Capt. Vaughan was ordered to proceed to Cardiff, Wales, for a survey and any necessary repairs made. This was done by Clover, Clayton & Co., Ltd. for £2,281 7s 5d, or $8,669.20.

All expenses for crew, fuel, food, lubricating oil, etc., were paid by the company. Huron Navigation realized gross profits of $682,684.20 less $98,788.98 for expenses and $6,470.22 for a standard marine risk insurance policy making their net earnings $583,895.22 for Voyage No. 4. Of that, they used $350,000 for debts and to pay off the last promissory note for the purchase of the ship but they did not make this final payment until permission was granted by the Shipping Board. When the final payment for ship from Delaware Navigation was made, Delaware Navigation was dissolved.

On November 27, 1917, the *O'Brien* officially came under Army control. Capt. Vaughan and his crew were now officially War Department employees. The Army technically operated the ship on a bareboat charter basis and they it turned over to the France & Canada Steamship Corp. At that time there was 144,000 gallons of fuel on board and a large quantity of stores and provisions which cost Huron $13,774.92, and some of these stores were consumed by the Army.

Some time before December 8, 1917, the Shipping Board sent Cosmopolitan a requisition agreement form to be signed by Huron's officers. On the 11th, Cosmopolitan replied that Huron Navigation had a new president, James Carpenter, who requested the forms be sent directly to him since Cosmopolitan was no longer the agent for the *William O'Brien*. Some delay occurred in signing and returning the charter to the Shipping Board and on the 14th, the Board telegraphed Huron to find out what the problem was. On the 19th, Stuart McIntosh wrote to Carry at the Shipping Board to say they never received a charter form.

In the meantime, the Army planned to load coal at Cardiff for Le Havre but found the ship unsuitable for that, being a lumber carrier.

On January 21, 1918, the Army entered into an agreement with the British Admiralty whereby they would trade the services of the *William O'Brien* in exchange for two smaller vessels suitable as colliers. The *William O'Brien* was then placed on a time-charter basis with the Admiralty but the crew remained War Department employees.

On February 14, 1918, Director Carry telegraphed Huron wanting to know how long it would be before the agreement was signed. McIntosh replied the same day that a board meeting had been called for the 19th and that no action had been taken up to that time due to the absence of a quorum of directors. On the 19th, two copies of the charter agreement were signed and returned to Carry noting November 19, 1917, as the date the charter agreement commenced and requesting he sign one and return it to Huron. Carry wrote back:

> We are advised by the War Department that this steamer was taken delivery of by General Pershing on bare-boat basis at Havre November 27, 1917, at 1.10 p.m. We are advising you of this in the event you have not already received notice to this effect, of which please take due notice.

McIntosh said OK and requested a revised form. On February 25, 1918, Carry wrote to Huron to say the charter actually began on the 25th of October and to send any outstanding bills prior to that date, if any and to "render any accounts of the voyage made before the vessel was taken by the Army." Huron made no objection to the date change and forwarded the documents. Huron subsequently received a "public voucher" in the amount of $127,022.01 to cover the charter period from 1:10 p.m. November 27, 1917, to midnight March 31, 1918. On April 2, 1918, Huron sent the voucher back requesting a payment of $144,810.71 saying they had recomputed the charter time period on a bare-boat basis instead of the time-charter basis the Shipping Board had used. Huron had signed off on the original voucher as "correct and just." After April 2, 1918, Huron calculated all payments due on a bare-boat basis.

The ship was returned to Huron Navigation on February 21, 1919, but no settlement for Voyage No. 4 was made and that voyage was clearly a time-charter. From November 19, 1917, to February 21, 1919, the charter was a bare-boat charter.

On a bareboat basis from October 19 to November 19, 1917, the rate would total $30,867.70 and from November 19, 1917, to February 21, 1919, the amount would be $465,073.34. On a time-charter basis the amount due from October 19 to November 19, 1917, would be $46,487.50.

When the ship was returned, Huron was owed $17,000, in lieu of overhauling and reconditioning by the Army—minus the value of the fuel and stores left on board—which left a balance due of $13,219.25, but this was never paid to Huron. In addition to this, Huron was owed $1,794.87 for the required installation of gun mounts.

Clover, Clayton & Co. did not send their bill to the Army until February 1919 with a note demanding the payment be made by Huron Navigation, which Huron paid in December 1919. Huron also paid $213.73 to their agent in Le Havre, the France & Canada Steamship Corp., for a total of $8,882.93.

In December 1919 Huron Navigation paid France & Canada $37,914.49 reimbursement along with $924.74 for commission fees at 2½ percent on the bills. The rate of exchange during this time was about $4.75 per pound sterling.

On April 15, 1920, the *O'Brien* left New York for Rotterdam with a cargo of coal. On Sunday, April 18, the ship was in heavy weather when they became disabled and radioed for assistance. The steamer *Baltic* picked up the S.O.S. and was told the hatches had opened and they were sinking at 39°50'N/65°50'W. On Thursday, April 20, the Coast Guard cutter *Achushnet* picked up an empty lifeboat, cushions, part of a deck house, and other debris from the *O'Brien*. On the 22nd a large oil slick and the ship's nameboard was discovered about 500 miles out. Neither the ship or any of the 40 crewmen aboard were ever seen or heard from again.

In March 1920, France & Canada Steamship Corp. owed the Shipping Board about $1,000,000 for debts unrelated to the *William O'Brien*. Francis Mayer held 8,996 shares of Huron Navigation's 10,000 shares of capital stock, which was purchased for $700,000 on November 22, 1917, to buy the *William O'Brien*. The balance of the 9,000 shares France & Canada originally bought were distributed to other investors. Mayer gave Huron Navigation $475,000 as credit toward the Shipping Board's debt for the charter of the *O'Brien* and the balance was paid directly to the Board.

In August 1920, Huron Navigation unsuccessfully sought to recover its costs from the Shipping Board. By December 1920, France & Canada Corp. had not paid France & Canada, Ltd. the loans made to purchase the stock in Huron Navigation so on December 31, France & Canada, Ltd. presented a demand promissory note to the Corporation for $1,011,304.17 at 6 percent interest. The collateral on the promissory note was 10,000 shares of Huron Navigation stock.

On December 6, 1920, the France & Canada Steamship Corp. filed for bankruptcy and they were declared bankrupt on April 7, 1922.

Huron sued the Shipping Board and the court found they were entitled to recover $46,487.50 based on $6.25 per ton per day on a time-charter basis from October 19 to November 19, 1917, and $465,073.34 on a bare-boat charter basis of $4.15 per ton per day after November 19, 1917. They were also owed $37,914.49 for collision expenses; $8,882.93 for damage repair at Cardiff; $13,774.92 for fuel used; and $13,219.25 for provisions on board. The total was $385,352.43. The government was credited the $583,895.22 net proceeds from Voyage No. 4, New York to Le Havre, and $1,794.87 for an advance for the installation of gun mounts. Total credits, $585,690.09, leaving Huron Navigation with $337.66 on May 12, 1925.

Ichabod T. Williams & Sons
Staten Island, New York

Thomas Williams went into the lumber business, married Sarah Waldron, and in 1826 they had a son, Ichabod Thomas Williams, in New York and in 1838 Thomas started his own lumber business on Broad St. in New York.

Ichabod joined his father's business and married Elizabeth Skelding, of Stamford, Connecticut, and they had four sons: Thomas on August 10, 1854, in Stamford; Henry K.S., Waldron in 1862, and Lloyd.

Around 1870 Ichabod took over the business, which by then specialized in imported mahogany and fine cabinet woods and manufactured veneers.

In 1880, Ichabod's son Thomas became partner and on October 21 of that year Thomas married Emma Wells Stott, of Stottsville, New York, and they had a son, Thomas Resolved Williams. The wealthy family lived at 18 W. 36th St.

In 1882 Henry K.S. joined the firm and at some point Waldron and Lloyd went in. Ichabod died on March 1, 1899, and by that time the company was one of the largest hardwood importing, milling, and exporting firms in the world with its own fleet of Great Lakes schooners. Many moves and expansions occurred over the years.

In 1900, Thomas Resolved Williams, eldest son of Thomas, joined his father and in 1905 became a partner. Henry K.S. retired in 1910 and Waldron died on July 19, 1917. Henry and Waldron were succeeded in 1918 by Frederick C. Leary and Thomas Blagden.

The Shipping Board chartered the French sailing vessels *Dieppe Dalle* and *La Bruyere* on February 27, 1918, and delivered the *Dalle* to the company on August 22 and the *Bruyere* on November 4 for operation to Africa. The company had a large works and docks on Staten Island.

The Independent Steamship Corp.
6 Hanover Square, New York

Incorporated on May 8, 1918, in Delaware, as managing agent for American-Hawaiian, American Ship & Commerce Navigation Corp., and the Shipping Board. The company name was changed on December 3, 1919, to Livermore, Dearborn & Co., Inc.

William A. Harriman was chairman. Richard Hallett Meredith Robinson was president, J.D. Tomlinson, E.C. Tobey, and W.B. Ryan were vice-presidents, Charles M. Fedderman treasurer, F.W. Anderson secretary. Henry Dearborn was a director.

The company managed one Shipping Board freighter, operated one, and managed and operated 11 ships. Managed and operated the seized German freighter *Suevia* (renamed *Wachusett*) to the Dutch East Indies.

The company name was changed to United American Lines, Inc., on August 16, 1920.

International Freighting Corp.
170 Broadway, Rm 1105, New York

Harry J. Lesser was president, James H. Graves vice-president, Irving L. Ernest secretary-treasurer. The company operated one Shipping Board freighter and managed and operated 12 ships. Graves was also secretary of the Fuel Export Corp. of America at the same address. He lived in Orange, New Jersey.

International Mercantile Marine Co.
9 Broadway, Rm 268, New York

Philip A.S. Franklin was president, Edward C. Grenfell, Frederick Toppin, and John H. Thomas vice-presidents, Emerson E. Parvin secretary, Horace G. Phillips treasurer. Incorporated with a capital of $120,000,000. Otto T. Bannard, Harry Bronner, Donald G. Gedden, J.P. Morgan, Lord Pirrie, John W. Platten, Charles W. Sabin, Frederic W. Scott, Charles Steele, John M. Perry, Charles A. Stone, and Frank A. Vanderlip directors.

Owned six big passenger liners: *Finland, Kroonland, New York, Philadelphia, St. Louis,* and *St. Paul.* President Cleveland and his entire cabinet attended the launching of the *St. Louis,* which was regarded as a rejuvenation of the merchant marine.

The company owned nine ships in 1917. Managed eight Shipping Board freighters, operated two, and managed and operated 67 ships.

J.A. Elliot & Co.
116 Broad St., Rm 61, New York

John Elliot was a ship broker who lived in Orange, New Jersey. The company managed two Shipping Board freighters and managed and operated four ships.

J.B. King Transportation Co.
17 State St., Rm 1002, New York

George B. King was president, Jerome A. King vice-president, Joseph C. Seguine secretary, and Edward D. King treasurer. Horace F. Hutchinson and George T. Hay were directors. Capitalized with $200,000 prior to the war. Owned the schooners *Newburgh* and *Lewis H. St. John,* three schooner barges, and the tugboat *Gypsum King.*

J.E. Chilberg

Nelson and Charlotte Chilberg were born in Sweden and settled in Iowa where son John Edward and daughter Mary were born. The family came out to Seattle where dad was a grocer and mom operated a knitting house. Nelson's business became N. Chilberg & Son, wholesale and retail groceries, crockery, and household items at 713 Front. John was manager.

John got into shipping and by 1898 he was secretary of the Puget Sound & Central American Steamship Co. and in 1906 he was president of the Seattle Shipyards Co. In 1914 he was vice-president of the Scandinavian-American Bank, president of the Seattle Trust Co., and vice-president of The Chilberg Agency, steamship and insurance agents run by Andrew Chilberg.

J.F. Whitney & Co.
8 Bridge St., Rm 317, New York

George T. Hay, Frederick C. Lockhart, and Theodore Dougherty were steamship agents who owned the schooners *Coral Leaf, Emperor Gypsum, Empress Gypsum, Invictus,* the bark *Calburga,* and barkentines *Ensenada* and *Malwa.*

The company managed two Shipping Board freighters, operated four, and managed and operated four ships. The Shipping Board bareboat chartered the Swedish steamer *Osterland* on July 10, 1918, and assigned the ship to the company for service to the east coast of South America.

J.H. Baxter & Co.
Merchants Exchange Building, Rm 536, San Francisco

John and Alfred Baxter dealt in lumber, poles, piles, creosoted materials, and shipping.

J.H.W. Steele Co., Inc.
Fox Building, 220 21st St., Galveston, Texas

J. Harvey W. Steele was president and manager in Galveston. Neal M. Leach in New Orleans, Semmes Steele in New York, and Herbert Y. Cartwright in Savannah were vice-presidents. George L.H. Koehler was secretary-treasurer. Steamship and forwarding agents.

Forwarding agents attended to all the details of getting a manufacturer's goods to market. They negotiated freight rates, booked space on outgoing steamers, handled bills of lading, consular invoices and documents in any language, other necessary paperwork and numerous details, and arranged for the transfer of cargo to railroad cars at its destination.

Determining the legal, net, and gross weights of goods was a headache as many standards were used.

Pilfering of goods on the dock was especially common in South America, Australia, and Russia and less so, but still a problem, in England and European ports.

The company operated seven Shipping Board freighters and managed and operated 18 ships. The Shipping Board bareboat chartered the Cuban steamer *Bavaria* and delivered the ship to Steele on August 22, 1917, for operation to the United Kingdom.

J. Harvey was also president of the Steele Towing & Wrecking Co., vice-president of the Texas Bank & Trust Co., and general agent for the Southern Steamship Co.

J.H. Winchester & Co., Inc.
358 Produce Exchange Building, New York

Winchester Noyes was president, Jacob B. Smull vice-president, Anthony J. Mouris secretary, and Everett Hollenbeck treasurer. The steamship agents managed two Shipping Board freighters, operated two, and managed and operated seven ships. Managed and operated the seized German freighter *Isomnia* (ex–*Nassovia*) for service to the east coast of South America.

J.R. Hanify Co.
21 Market St., San Francisco

John R. Hanify was president of the Hanify Lumber Co. in Raymond, Washington, and the J.R. Hanify Co. in San Francisco. Hanify owned the steam schooners *W.J. Patterson, Salem, Santa Barbara, Santa Monica,* and *Zampa, Ryder Hanify, Lucinda Hanify,* and *Francis Hanify.*

James Rolph, Jr.
230 California St., San Francisco

James "Sunny Jim" Rolph, Jr., was the mayor of San Francisco, president of the Rolph Steamship Co., the Rolph Navigation & Coal Co., and the Rolph Shipbuilding Co. John D. Stelling was secretary and David F. Ewart was manager.

James W. Elwell & Co.
17 State St., Rm C, New York

James, David H. Elwell, and Howard E. Jones were steamship agents who owned the big barkentine *Aryan*. The company later owned five ships and managed five Shipping Board freighters, operated one, and managed and operated 14 ships. The company's freighter *Evangeline* was requisitioned by the Shipping Board for operation by the French government. The *Aryan* burned in the western Atlantic on December 24, 1918, with the loss of 8 out of 22 on board.

Jewett M. Scott
1445 Dauphin St., Mobile, Alabama

Jewett McLellan Scott was born on Grand Cayman Island, British West Indies, on May 21, 1873, became a sea captain, and married a widow, Amie Ida Irene Eden, who had four children.

On May 1, 1906, the family left their home on the schooner *Bartholdi* and arrived at Mobile on the 18th. The family lived at 105 Dexter Ave. and Jewett became a citizen on January 12, 1911. He eventually quit the sea and became a successful lumber exporter and shipowner and by mid–1918 owned eight vessels. In 1920 the family was living in rented quarters at 44 W. 73rd St. in Manhattan.

John S. Emery & Co., Inc.
114 State St., Rm 70, Boston

Steamship agents and ship brokers who owned seven schooners, two barks, and five barkentines including the *John S. Emery* and the *Rachel Emery*. Managed two Shipping Board freighters and managed and operated 13 ships.

Jonathan May & Sons
500 S. Delaware Ave., Philadelphia

Jonathan May was born in Delaware in 1820, became a sea captain, and lived at 218 German St. In 1874 he quit the sea and opened a ship chandlery at 500 S. Delaware Ave. with his sons Jonathan, Thomas, and George. Jonathan senior died on March 10, 1898. In 1918 Jonas Steelman and Clarence D. and William C. May ran the chandlery and owned the schooner *A.B. Sherman*.

Kerr Navigation Corp.
17 Battery Place, Rm 1215, New York

Edward F. Geer was in the steamship business in New York and lived at 241 Washington St., Brooklyn with his wife Blance and daughter Rosalie. When the war started he saw an opportunity to acquire German ships that were laid up in American ports at very reasonable rates. In March 1915, he incorporated the Housatonic Steamship Co. with a capital stock of $125,000 to purchase the Hambug-American freighter *Georgia*. Geer was president, George F. Allison vice-president, Frank N. Southard secretary, Henry M. Hewitt treasurer, and Edward Sandford, a Hamburg-American attorney. The office was at 21 State St., Room 714.

The Hamburg-American's book value for the ship was $83,000 and the purchase price was $85,000. The purchase was made on April 16, 1915, and was registered under the American flag as the *Housatonic*.

At that time, the British government refused to recognize the transfer of German ships to American registry in the belief that the new owners would try and use them to benefit Germany and as such they were liable to seizure on the high seas.

To avoid this unpleasant possibility, Geer elected to charter the ship to a British steamship company for operation and on February 23, 1916, executed a charter-party agreement with Brown, Jenkinson & Co. in London "for the term of the present war" at a rate of 7 shillings per month per deadweight ton. The going rate was then about 32 shillings per deadweight ton. In exchange for the low charter rate, the charterers were obligated to "secure immunity against attack or seizure on the part of the British and Allied Governments." While the *Sacramento* in San Francisco was blacklisted, the *Housatonic*, by virtue of this charter agreement, was not. The owners paid for provisions, crew wages, equipment, and insurance.

Geer and company incorporated the Jason Navigation Co. to own and operate the *Argonaut*.

On July 12, 1917, Geer incorporated the Kerr Navigation Co. at 17 Battery Place with a capital of $800,000. Geer was president, Edward B. Vanderveer vice-president, Frank Southard secretary. James A. Smith and Edward M. Strothers were directors. The company owned the *Keresan, Keresaspa, Kerkenna, Kerlew, Kermanshah, Kermoor, Kerowlee,* and *Kerwood.*

The company's freighter *Kerowlee* was requisitioned by the Shipping Board on a bareboat charter basis to the Navy for operation by the Army. The company managed one Shipping Board freighter, operated seven, and managed and operated 18 ships.

On August 14, 1917, Geer incorporated the Shoshone Navigation Co. at the 24 State St. office with a capital of $603,000 to operate the company's freighter *Shoshone*. Geer was president, Frank Southard vice-president, Edward M. Strothers secretary. Incorporated with a capital of $603,000. Noel McVicker and Lacy Lee Bright directors.

The *Shoshone* was requisitioned by the Shipping Board on a bareboat charter basis to the Navy for operation by the Army.

The Shipping Board bareboat chartered the Uruguayan steamers *Salto* (ex–*Harzburg*) on May 24, 1918, and assigned the ship to the company on January 9, 1918, for service from the United States to the River Plate and the *Riviera* (ex–*Silvia*) on February 27, 1919, for service to Montevideo.

Kerr Steamship Co., Inc.
17 Battery Place, Rm 1937, New York

Harry F. Kerr president, Alfred E. Clegg and Edward F. Geer vice presidents, Charles H. Melsome secretary. Incorporated with a capital of $1,500,000 to manage and operate ships.

The company managed and operated the seized German freighter *Nyanza* (ex–*Esslingen*) to Italy and the passenger ship *Orion* (ex–*Prinz Oskar*) to Belgium.

On May 21, 1918, Davis Z. Donnahoe, an American seaman, was accidentally shot and killed by William B. Jones while he was asleep in his bunk on the *Orion*.

Lewis K. Thurlow
55 Broad St., Boston

Lewis Thurlow and Peter H. Crowell were steamship agents. Thurlow owned a fleet of schooners and ships on his own account. He was born in Maine in 1868. He lived with his wife Eva, son Paul, daughter Florence, and three servants at 9 Stratford Road.

Libby, McNeill & Libby
420 Merchants Exchange Building, San Francisco

Canners and provision packers headquartered at the Union Stockyards in Chicago. Founded in 1869 by Archibald McNeill and Arthur and Charles Libby. In 1918, William F. Burrows was president and Harry Williams secretary. The company operated four sailing vessels and two motor freighters.

Lind Navigation Co.
17 Battery Place, Rm 2029, New York

Peter and Katrina Jonasson lived in Kånna, Sweden, and on March 25, 1854, had a son John. Peter and Katrina emigrated to Minnesota in 1868 and Peter changed his surname to Lind.

John became a lawyer, got into politics, and in 1877 married Alice Shepard from Wisconsin. Norman G. was born in 1880, Jenny in 1884, and in 1887 Lind was elected to Congress. Daughter Winifred was born in 1890 and 1893 he left Congress. In 1898 he was elected governor of Minnesota and the family lived at 651 Fairmount Ave. in St. Paul.

Norman went into the lumber and shipping business and in 1918 he was president of Lind Navigation. Richard D. Sutton was vice-president and John J. Jannotte secretary.

The company managed and operated four Shipping Board freighters. In 1920 Lind was 39, single, and living in the Gramatan Court apartments in Eastchester, New York.

Lowrance & Co.
109 State St., Boston

George L. Crooker, president and treasurer of the Providence, Rhode Island, Chamber of Commerce, wanted "to secure a fair and more general distribution of port calls" for Providence. He and Providence attorney Frank L. Hinckley contacted an attorney in Boston, John W. Lowrance, apparently well known in shipping, and incorporated Lowrance & Co. to manage and operate ships. The company managed one Shipping Board freighter and managed and operated five ships.

Lykes Bros.
425 Security Building, Galveston, Texas

Steamship brokers, agents, importers, and exporters. James M. Lykes was president. The company managed one Shipping Board freighter, operated one, and managed and operated 17 ships. James lived with his wife Genevieve at 1416 Ave. H.

M.E. Kinsley & Co.
17 Battery Place, Rm 237, New York

Merton Elroy Kinsley was born in Stoneham, Massachusetts, on February 5, 1874, and became clerk in Boston then a railroad passenger agent in Manhattan. He married Sara Frances Gilmore from Sandersonville, Georgia, in 1895 and by 1900 he was the district agent. He eventually became a ship broker and operated six Shipping Board freighters. He and Sara lived at 226 W. 70th St. They apparently never had any children.

M.H. Tracy & Co., Inc.
17 State St., Rm 110, New York

Capt. Marcus Hunter Tracy and David T. Lamond were steamship and freight brokers, agents, and freight forwarders. They managed six Shipping Board freighters, operated one, and managed and operated nine ships.

Capt. Tracy was Chairman of the Board of Governors of the New York State Nautical School, served on the Board of Pilot Commissioners, and was treasurer of the State Board of Commissioners for Licensing Sailors' Hotels & Boarding Houses. He lived at 23 Clinton St. in Brooklyn.

McIntyre Lumber & Export Co.
705 Antwerp Building, Mobile, Alabama

Incorporated in Mobile on January 3, 1908, by James McIntyre, Whitney W. Cleveland, and Edgar D. Flynn with a capital of $25,000 to engage in the lumber business.

In 1918, Cleveland was president, Flynn secretary-treasurer. Owned the schooner *Magnus Manson,* the bark *Anna Maria D. Abundo* and the steamer *Itasca.*

Magnolia Petroleum Co.
1202 Ave. A, Galveston

In the 1890s, Joseph S. Cullinan built an oil refinery at Corsicana, Texas, and on Christmas Day, 1898, began production. The next year he formed the Corsicana Petroleum Co. to extract and deliver crude oil. Cullinan's operation subsequently became the Navarro Refining Co.

In 1901 George A. Burts began taking oil from the Spindletop field and formed the Security Oil Co. That same year, John Henry Hutchins, Sr., 79, and George Sealy, 66, were bankers at 2328 Strand in Galveston, Texas, as Hutchings, Sealy & Co.

John lived with his wife, Minnie, 64, sons, John Henry, Jr., 33, and George B., 28, daughter-in-law Lelia C., and four servants at 2816 Ave. O. Another son, Sealy Hutchings, 32, lived with his wife Mary, and five children at 2805 Ave. O.

George lived with his wife Magnolia, 46, and seven children including George, 20, at 2424 Ave. J. An older son John, 30, worked in the bank and lived at 522 Vermont St. On December 14, 1901, George died.

1905, Sealy Hutchings, John and George Sealy, and Herman O. Stein were unincorporated bankers at 2328 Strand as Hutchings, Sealy & Co. In 1909 they purchased Security Oil and Navarro and on April 24, 1911, formed Magnolia Petroleum with John Sealy as president. John Sealy was a very busy fellow. In his maritime related operations, he was president of the Galveston Wharf Co. and treasurer of the Galveston Maritime Association.

Mallory Steamship Co.
Pier 36, North River, New York

Harry H. Raymond was president, Clifford D. Mallory vice-president, Nicholas W. Schaack secretary, and Alexander R. Nichol treasurer. Incorporated with a capital of $7,000,000. Henry R. Mallory, Robert Mallory, Alfred G. Smith, and Galen L. Stone were directors. Mallory was succeeded by J.B. Dennison.

Owned 13 steamers before the war. The company managed five Shipping Board freighters, operated one, and managed and operated 14 ships.

Maritime Transportation Co.
25 Beaver St., Rm 517, New York

George W.F. Green was president and treasurer, William H. Hax vice-president, and William R. Virtue secretary. The company operated one Shipping Board freighter and managed and operated one.

Maru Navigation Co., Inc.
5 State St., New York

Peter McDonnell was president, Dominic A. Truda vice-president, and Alexander J. McDonnell treasurer. The company purchased the 1,367-ton freighter *St. Charles* from the Mexican Constitutional government and registered it on June 4, 1917. The ship was requisitioned by the Shipping Board on a bareboat charter basis to the French government for operation on the French account. The same men were officers of the American Mediterranean Steamship Co.

Massey Steamship Co., Inc.
1713 Broadway, Superior, Wisconsin

In 1915, Chester A. Massey was a lake captain who lived at 1927 Hughitt Ave. in Superior, Wisconsin. Victor M. Jarrett was a marine engineer in Superior. In 1917 the two incorporated Massey Steamship with Chester as president and treasurer and Victor as secretary. The company managed and operated two Shipping Board freighters and had an office at 17 Battery Place in New York.

Matlack Coal & Iron Corp.
52 Vanderbilt, Rm 1710, New York

Roy A. Rainey was president, Howard C. Matlack vice-president, and Cipriapo Andradeji treasurer. The Shipping Board chartered the Norwegian sailing vessel *Siam* in November 1918 and assigned the ship to the company for service to the east coast of South America.

Matson Navigation Co.
120 Market St., San Francisco

Edward D. Tenney was president in Honolulu, William P. Roth was vice-president and general manager, John A. Buck and Wallace M. Alexander were vice-presidents, Alfred M. McCarty treasurer, and Frazer A. Bailey secretary. Roth lived with his wife Lurline at 2251 Jackson.

Owned the steamers *Enterprise, Hilonian, Hyades, Lurline,* and *Wilhelmina* prior to the war along with the five sailing vessels. Three of the company's cargoliners were requisitioned by the Shipping Board on a bareboat charter basis to the Navy for operation by the Army: *Matsonia, Maui,* and *Wilhelmina.*

The company managed six Shipping Board freighters, operated 19, and managed and operated 11 ships. Also managed and operated the seized German freighter *Rajah* and the passenger ship *Sachem* (ex–*Coblenz*) to Hawaii.

Merchants Navigation Co.
540 S. Spring St., Rm 603, Los Angeles
"Merchants Line"

Apparently established in 1894 as the Northern Pacific Division of W.R. Grace & Co. for passenger and freight service from Peru to the West Coast and as far south as Manzanillo, Mexico. They managed and operated 11 ships from Municipal Pier A and operated three Shipping Board freighters.

Minnehaha Motor-Schooner Co., Inc.
32 Broadway, New York

Incorporated in Manhattan in April 1918 with a capital of $100,000 by R.J. Bellows, Edward Thomas Wrack, and attorney Alexander Begg.

Wrack was an Englishman born in Norway who worked for the Anglo-Norwegian Shipping Agency, Inc., managing shipowners, ship brokers, and forwarding agents at 17 Battery Place in New York. He lived in Asbury Park, New Jersey.

Mississippi Shipping Co., Inc.
812 Gravier St., Rm 501, New Orleans
"The Delta Line"

Matthew J. Sanders was manager of Frederick Leyland & Co., Ltd., (Leyland Steamship Line) and the federal manager of the U.S. Railroad Administration's Mississippi-Warrior Water Way. Thomas J. Cunningham operated the Commercial Warehouse at 537 S. Peters, and Rudolph S. Hecht was president of the Hibernia Bank & Trust Co. in New Orleans.

The three incorporated on March 24, 1919, with Sanders as president, Cunningham vice-president, and Hecht secretary-treasurer. The company operated one Shipping Board freighter and managed and operated five ships.

Moore & McCormack Co., Inc.
29 Broadway, Rm 900, New York

Steamship agents with Albert V. Moore was president, Henry P. Molloy secretary, and Emmet J. McCormack treasurer. The company managed one Shipping Board freighter, operated five, and managed and operated 15 ships including the seized German freighter *Honolulu* (ex–*Setos*) to the east coast of South America.

Munson Steamship Lines
82 Beaver St., Rm 1501, New York

Franck C. Munson was president, Alfred H. Bromell vice-president, John W. Reynolds secretary, and Charles M. Dimm treasurer. Incorporated with a capital of $250,000.

Owned the *Munalbro, Munamar, Mundale, Mundelta, Munisla, Munplace, Munrio, Munsomo, Munwood,* and *Muskegon*. Managed two Shipping Board freighters, operated 11, and managed and operated 29 ships. Managed and operated the seized German passenger ship *Coosa* (ex–*Loongmon*) to the West Indies.

Nafra Co., Inc.
120 Broadway, Rm 2334, New York

Joseph Mercadante was president, George Hurd vice-president, H.G. Robinson treasurer. Incorporated with a capital of $1,000,000 to engage in importing.

Joseph was born in Palermo, Italy, on November 27, 1884, arrived in New York on November 26, 1905, and started an exporting business. He was naturalized on September 5, 1916, and had a wife, Mary, and daughters Dorothy and Victoria. The family lived at 375 Park Ave.

Nafra owned the *Chincha, Eurana,* and *Santa Cecelia.* Two of the company's freighters were requisitioned by the Shipping Board on a bareboat and time charter basis respectively to the Navy for operation by the Army: *Eurana* and *Santa Cecilia.*

The company managed two Shipping Board freighters, operated five, and managed and operated five ships. Managed and operated the seized German freighter *Tunica* (ex–*Sambia*) to France. The Shipping Board chartered the Danish steamer *Hans Maersk* on October 14, 1918, and assigned the ship to Nafra on the 26th for a 6-month operation on a Swiss account.

National Shipping Corp.
25 Broad St., Rm 622, New York

Ship broker George A. Carden was president, F. Lee Williams secretary, stock broker Harvey C. Sickler was treasurer. Incorporated in 1917 with a capital of $1,000,000. The company operated two Shipping Board freighters and managed and operated eight ships.

New England Fuel & Transportation Co.
111 Devonshire St., Boston

On June 30, 1917, the officers of the Massachusetts Gas Co., the New England Gas & Coke Co., the Boston Tow Boat Co., the Federal Coal & Coke Co., and the New England Coal & Coke Co. formed an association whereby all the assets and liabilities of each firm were vested in the New England Fuel & Transportation Co.

New England Coal & Coke owned the steamers *Everett, Malden,* and *Melrose,* three tugboats, and eight barges before the war.

James L. Richards, of Newtonville, was chairman of the board. Robert Grant, of Brookline, was president; William E. McKay, of Boston, and Edward Page, of Newtonville, were vice-presidents; Henry S. Lyons, of Brookline, was secretary, and Edwin M. Richards was treasurer.

The new company acquired other assets and eventually owned 24 schooners and steamers. The company's big freighter *Everett* was requisitioned by the Shipping Board on a time charter basis for operation by the company on the European Food Relief mission.

The company managed five Shipping Board freighters, operated one, and managed and operated 11 ships.

New Orleans & South American Steamship Co.
802 Queen & Crescent Building, New Orleans

Incorporated on July 31, 1918, in Louisiana with a capital of $500,000 to operate the former Grace Line steamer *American* and motor vessels *Santa Cristina* and *Santa Flavia* from New Orleans to Cuba, the French West Indies, Columbia, and Venezuela.

In 1912 Arthur S. Chesebrough was president of the California-Atlantic Steamship Co. and Albert G. Bates was treasurer. Bates became the chartering manager at W.R. Grace and treasurer of the Pacific Mail Steamship Co.

Bates was elected president of New Orleans & South American. Albert R. Williams was vice-president and general manager, Frederick W. Beyer secretary, and Emil C. Drewes treasurer. The company was one of the group formed by the American International Corp., the National City Bank, and W.R. Grace.

The company managed two Shipping Board freighters, operated one, and managed and operated seven ships. Drewes was the agent for W.R. Grace. The company ran two steamers to the Caribbean after the war.

New York & Baltimore Transportation Line
112 Liglet St., Baltimore

John Monks, Jr., was president, Bertram R. Roome secretary. In 1912 the company owned the freighters *Chesapeake* and *Manna-Hata*, both built at Wilmington, Delaware, in 1900, and the *Porto Rico*.

At some point they acquired the *Baltimorean*. The *Chesapeake*, *Manna-Hata*, and *Baltimorean* were purchased from the Compañía Marítima, Manila, Philippines, for $250 per deadweight ton for each vessel. The money was deposited with the Philippine National Bank in New York pending authorization of the sale by the Shipping Board. This was not forthcoming and the money sat in the bank. In the meantime, the vessels were placed in service between Baltimore and New York City, apparently under some form of contract. The company leased a wharf on Pier 10 in the East River and had a dock at Baltimore. They also had a number of tugs and lighters.

The company had an office at 61 Broadway in New York and on August 26, 1918, they were informed that the Navy Department was appropriating the *Chesapeake* and *Manna-Hata* by purchase. Five days later, Acting Secretary of the Navy Franklin Roosevelt wrote to inform the company that the President had determined a just compensation for the two vessels was $175,000 and a further sum for whatever consumable stores were on board at the time of delivery.

The company wrote back on November 8 to request a reappraisal of the vessels, which was denied, and the company then requested a payment of 75 percent of the offered price so they could prosecute a claim in court. Capt. James H. Glennon, commandant of the Third Naval District, requested a deed of transfer for each vessel before payment could be made. This done, the company's attorney, Joseph Diehl Fachenthal, received two copies of proposed bills of sale on December 7, 1918, for $131,250 which required signatures and the return of one copy to Glennon.

On the 10th the company held a special meeting of the board of directors to consider the recommendation of the company's president and secretary to accept the 75 percent reserving the right to sue. This was not approved and the president was directed to execute a $175,000 bill of sale for the *Chesapeake* and a $175,000 bill of sale for the *Manna-Hata*. This was done and delivered to Assistant Paymaster Lt. Albert E. Ackerman in the commandant's office on or about December 11.

The company then received payments totaling $352,481.39 on or about January 13, 1919. Of this, $2,481.39 was for consumable stores on board.

After the two larger vessels were gone, the company could not conduct business with the *Baltimorean* and was forced to sell it for $71,000 or $85 per deadweight ton as there was no market for a vessel like it. During the time it operated with the other two, it was worth about $175 a deadweight ton. The company lost its lease at Pier 10 and on its terminals in Baltimore.

On April 8, 1920, the company's attorney, William A. Wimsbish, wrote to Secretary of the Navy Daniels:

> Please permit me to hand you herewith a brief memorandum of facts relating to the commandeering by the Navy Department of the steamships *Chesapeake* and *Manna-Hata*, then owned by the New York & Baltimore Transportation Line, a Maryland corporation.
>
> May I invite your special consideration of the fact that these ships had been sold for more than $300,000 each to a Manila, P.I. Corporation; that the approval of the sale was withheld by the Shipping Board only at the insistence of the Director General of Railroads because of the then severe traffic congestion; that this objection was subsequently withdrawn, and the sale about to be approved, when, in August, 1918, the ships were taken over by the Navy under an award of only $175,000 for each vessel; and that, after fourteen months' use of the ships, they were sold by the Government at auction in England, in November, 1919, for 160,000 pounds, which, at the then current rate of exchange was the equivalent of $667,200, or more than twice the amount that had been awarded and paid the former owner.
>
> We feel assured that the Government does not wish to profit at the expense of its citizens, and we therefore present this matter to you in the hope that its moral justice is such as will commend itself to you, and dispose you to aid us in securing just compensation for the loss of our property to the extent that you may become satisfied that such compensation has not been accorded.

Apparently no reply was made to this and the matter was left to rest but on July 28, 1923, the company decided to sue. The government filed a counterclaim against the company to recover two outstanding debts owed for loss of goods shipped on the company's rail line during the time the railroad was under government control.

The court found the sale of the *Baltimorean* to be voluntary on the company's part and related to contractual obligations. A judgement of $274.02 and $275.25 was awarded to the government on May 6, 1929.

The Navy spent $328,430.65 to convert the *Chesapeake* into a salvage ship and $330,538.18 on the *Manna-Hata* for the same purpose.

The *Chesapeake* was commissioned on March 22, 1919, under Lt. Melvin C. Kent. *Manna- Hata* was commissioned the same day under Lt. Harry Huxford. The *Manna-Hata* sailed for Brest, France on April 2, 1919, and the Chesapeake on May 12. Both reported for work with the First Salvage Division. They spent the summer and part of fall bringing equipment from Brest and Liverpool to Kirkwall in the Orkney Islands, the base for the extensive clearing of the North Sea Mine Barrage.

They were decommissioned at Brest on October 25, 1919, and sold at public auction on November 3, the *Chesapeake* for $430,540 and the *Manna-Hata* for $238,260. The *Manna-*

Hata was sold to Maritime Salvors, Ltd., of London and went into commercial service as the *Reliant*.

New York & Cuba Mail Steamship Co.
Pier 14, East River, New York

Alfred G. Smith was president, William D. Macy vice-president, Henry E. Caland secretary, and Alexander R. Nichol treasurer. Capitalized with $10,000,000. Henry R. Mallory, Robert Mallory, Harry H. Raymond, Galen L. Stone, and Franklin D. Mooney were directors. The company owned 21 steamers in 1912, including the ill-fated *Morro Castle*.

The company operated two Shipping Board freighters and managed and operated 37 ships and managed and operated the seized German passenger liner *Tacony* (ex–*Staatssekretar Kraetke*) to Cuba and the *Wacouta* (ex–*Prinz Waldemar*) to the West Indies.

New York & Oriental Steamship Co., Ltd.
17 Battery Place, Rm 1026, New York

Incorporated by Barber & Co. with a capital of $750,000 to run the *Satsuma* and *Suruga*. Barber & Co. director Oakley Wood was elected president, Francis M. Wilson vice-president, Edward J. Barber secretary, and Frank A.E. Cott treasurer.

In 1904 Wood was president of the Produce Exchange in Manhattan, became secretary of a stevedoring firm.

New York & Porto Rico Steamship Co.
11 Broadway, Rm 750, New York

Franklin D. Mooney was president, Wilson C. Peek secretary, Alexander R. Nichol treasurer. Capitalized with $5,000,000. Henry R. Mallory, Robert Mallory, Galen L. Stone, John E. Berwind, Harry H. Raymond, Franklin D. Mooney, and Hubert E. Rogers were directors.

The company owned 14 steamers before the war and managed one Shipping Board freighter, operated one, and managed and operated 13 ships.

In 1895 the Plant Investment Co. signed a $497,000 contract with the Newport News Shipbuilding & Dry Dock Co. to build the big 5,093-ton cargoliner liner *La Grande Duchesse*. The ship was launched on January 30, 1896, and delivered in November. Boiler and propeller problems immediately surfaced and in November 1897 the ship went back to the yard for new boilers. Work was completed on April 9, 1899, and the ship was chartered by the War Department as a transport. In November 1901 Plant sold the ship to the Ocean Steamship Co., of Savannah, Georgia, and it was renamed *City of Savannah*.

Ocean Steamship said the ship was unprofitable for the company due to faulty construction and they sold the ship to the New York & Porto Rico Steamship Co. in 1905 for $257,801.28. New York & Porto Rico changed the name to *Carolina* for operation between New York and Puerto Rico. On November 21, 1907, the ship was in dry dock at Shooters Island, New York, when it caught fire.

In 1913 the company took the ship back to Newport News for major reconstruction. During the time New York & Porto Rico operated the ship they spent $248,995.77 on repairs.

The cargoliner *La Grande Duchesse* in 1896 (author's collection).

After spending $418,207.54 for reconditioning, the ship went back into service and became profitable. In May 1914 the ship collided with the steamer *Cleveland* in New York Harbor.

On October 12, 1917, the Shipping Board requisitioned the ship under a "Requisition charter—Time form." The United States assumed the war risk as contained in the agreement:

> In case of loss or damage due to the operation of a risk assumed by the United States, just compensation for such loss or damage, with interest at 6 per cent per annum, commencing 30 days after proof of such loss or damage, shall be paid by the United States in the District of Columbia, interest not to be paid, however, if the owner resorts to any court for the purpose of establishing the amount of the just compensation.

The *Carolina* operated uneventfully until Sunday, June 2, 1918, when the ship was shelled and sunk by *U-151* about 62 miles off Tucker Beach Light, Atlantic City, New Jersey.

On July 3, the Shipping Board wrote to New York & Porto Rico to inform them a hearing before the Just Compensation Board would be held on the 9th. The ship obviously was a total loss but the surviving lifeboats were valued at about $1,500. The board took testimony from five witnesses and experts and assessed the value of the *Carolina* at $1,892,500.00 as of June 2.

On July 18, 1919, the Shipping Board notified New York & Porto Rico their compensation would be $1,250,000.00. The company protested and accepted 75 percent of that amount, $937,000, on April 21, 1920, reserving the right to sue in federal court.

The company waited until January 24, 1921, to file its petition, didn't furnish any evidence or depositions until June 6, 1921, and finally filed its brief on June 21, 1923, which, the court noted, would cost the government huge amounts of money in interest if honored. The government then amended the amount of interest that accrued due to the delay in filing.

The court said the government's clause in the agreement prohibiting court action to recover was illegal (40 Stat. L. 182) and New York & Porto Rico recovered $1,351,381.81 including 6 percent interest from April 21, 1920, the date the 75 percent was received, to October 29, 1923, the date of judgement.

The company managed and operated the seized German passenger ships *Watauga* (ex–*Governeur Jaeschke*) and the *Yuma* (ex–*Tsintau*) to the West Indies.

New York & South American Line
11 Broadway, Rm 1201, New York

Organized around 1904 by the U.S. Steel Products Co., 30 Church St., for monthly sailings from Brooklyn to Chile and Peru on the "Hall" ships: *Craster, Charlton, Fallodon, Foxton,* and *Newton,* etc. Thomas P. Alder was secretary, John W. Ryan agent.

The company operated one Shipping Board freighter and managed and operated three ships. The company managed and operated the seized German ship *Casco* (ex–*Elmshorn*) to the west coast of South America. The company was the only line that refused to operate Shipping Board steamers after the war due to the low rates offered.

Norden Steamship Co.
(Dampskibsselskabet Norden)

On April 22, 1918, the Shipping Board signed a charter agreement with Dampskibsselskabet Norden, a Danish company, to charter their ship, the *Norden,* for one voyage between Atlantic Coast ports and the east coast of South America. The charter rate was 35 shillings, British sterling, per deadweight ton per month on the ship's summer marks.

This voyage commenced and on July 23 the Board again chartered the ship for about six weeks after the voyage was completed "for direct continuation with no interruption of time between the two charters." This charter would formally begin at 6 p.m. on August 3rd and would end at 8 p.m. on September 6, 1918, at New York City.

The payment for the charter was made to the company but $1,271.99 was withheld. The Shipping Board claimed that for 20¾ hours between 7 p.m. on August 3 and 3:45 p.m. on August 4, the ship, although ready to sail on schedule for Norfolk, Virginia to load cargo, sat at Boston waiting for two crewmembers to come back aboard. At 7 p.m. the *Norden* was towed out to an anchorage in the harbor where they waited for the two crewmen.

After the previous voyage the sailors left the ship, which was customary, and by sailing time, 6 p.m. on August 3, all the crew, with the exception of one fireman and one donkeyman, were back aboard. They showed up at 3:45 p.m. the next day and the *Norden* proceeded on the voyage.

At the end of the voyage the company was paid in full except for $1,271.99 for the 20¾ hours between 7 p.m. on August 3, and 3:45 p.m. on August 4. The company accepted this without protest.

On September 25 the steamship company's agents sent the Shipping Board a bill for $5,766.95 for coal used on the voyages. The Board paid $4,062.05, less $1,704.90 for the 20¾ hours. The company then claimed the $1,271.99 for the hire of the vessel during the waiting period. The board disallowed the claim so the company sued in federal court pursuant to treaties between the United States and the Kingdom of Denmark.

The government claimed the ship could have gone to Norfolk with the two missing crewmen and the two could easily have joined the ship there while loading cargo. The court found that, while the government controlled the destinations of the vessel, the ship itself remained under the control of the ship's officers and the petition was dismissed.

Northern Shipping Corp.
17 Battery Place, Rm 203, New York

John Dowd was president, George H. White secretary-treasurer. The company managed and operated two Shipping Board freighters. The Shipping Board bareboat chartered the Peruvian steamer *Eten* (ex–*Rhakotis*) on August 4, 1918, and assigned the ship to the company for service to Scandinavia.

Northern Transportation Co.
610–611 Munsey Building, Baltimore

John T. Donohue was president, Thomas J. Donohue treasurer. Coastwise transportation service. The company owned five vessels in mid–1918.

Northwestern Shipping Co.
Colman Building, Seattle

Capt. Harry W. Crosby was president of the Washington Tug & Barge Co. and lived at 1516 10th Ave. with his wife Bertha. James D. Esary was president of the Inland Navigation Co. and lived with his wife May at 523 16th Ave. Howard B. Lovejoy was the general manager of Inland Transportation and lived with his wife Lillie at 427 20th Ave.

Northwestern Shipping operated the *H.B. Lovejoy*. Crosby was president, Essary vice-president, and Lovejoy superintendent and general manager.

Norton Lilly & Co.
Produce Exchange Building, Rm C-10, New York

In 1841, John Norton shipped a load of kerosene to South America and established Norton & Co. In 1907, Joseph T. Lilly, 35, was a steamship broker and agent in Brooklyn when he became a partner and the firm became Norton Lilly & Co.

During the war Joseph Lilly, Skeffington S. Norton, John B. O'Reilly, and John J. Farrell were steamship agents. The company managed and operated two Shipping Board freighters. The company survives today as Norton Lilly International.

Ocean Steamship Co.
37 Bull St., Savannah, Georgia

In 1904 John F. Hanson, of Macon, was president, William H. Pleasants vice-president. By 1917, Pleasants was president, Walter C. Askew was treasurer, and Charles S. Groves secretary. The company operated from Pier 35, North River, New York. The Eastern passenger agent, George R. Angevine, was at 140 W. 42nd. The company owned eight coastwise steamers and two tugboats before the war. Managed and operated eight Shipping Board freighters.

Oceanic Steamship Co.
601 Market St., San Francisco

Incorporated in 1881 by John D. and Adolph B. Spreckels, sons of sugar magnate Claus Spreckels. Owned the passenger ships *Sierra*, *Sonoma*, and *Ventura*. The *Sierra* was requisitioned by the Shipping Board on a bareboat charter basis to the Navy for operation by the Army. The company managed and operated one Shipping Board freighter.

Old Dominion Steamship Co.
Pier 25, North River, New York

William L. Woodrow was president, Charles J. Ahlstrom treasurer. Capitalized with $1,500,000. David R. Daly, George A. Elliott, Frederick H. von Slade, Henry Walters, Robert H. Richards, Frank Trumbull, and S. Davis Warfield were directors. Incorporated with a capital of $1,500,000. Prior to the war the company owned 19 steamers and four tugboats.

Oneida Navigation Corp.
115 Broadway, Rm 2822, New York

George C. Sprague was president, John W. Crandall secretary, Edward A. Lyon treasurer. Incorporated with a capital of $20,000. The company owned the *Percy R. Pyne II* and the *Perry Stetzer*.

Ore Steamship Co.
111 Broadway, Rm 601, New York

John R. Gross president, Willard A. Mitchell secretary, E. S Hill treasurer. Harry W. Warley and H.S. Snyder directors. In mid–1918 the company owned the *Cubore*, *Feltore*, *Mangore*, and *Santore*.

Oriental Navigation Corp.
17 Battery Place, Rm 434, New York

Philip De Ronde president, John F. Sampson and Alexander R. Nichol vice-presidents, Frank H. Hall secretary. Incorporated with a capitalization of $2,000,000. In mid–1918 the company owned the *Oriole* and the *Susanne*. Oriental was known in shipping circles as Sampson & DeRonde. The company managed one Shipping Board freighter, operated seven, and managed and operated 11 ships.

On January 9, 1918, the Shipping Board chartered the Chinese steamer *Hwah Jah* and delivered the ship to the company on the 24th for operation to the Orient.

P. Dougherty Co.
15th Floor, Maryland Casualty Building, Baltimore

Thomas F. McHugh was president and treasurer, Francis P. Dougherty secretary. The company ran 31 tugs and barges before the war. In mid–1918 they owned 17 schooner barges.

Pacific Alaska Navigation Co.
506 Perkins Building, Tacoma, Washington

The Pacific Coast Steamship Co. was incorporated in New Jersey on November 27, 1897, and in 1912 owned the steamers *City of Pueblo, City of Seattle, City of Topeka, Coos Bay, Cottage City, Curacao, Dehli, Edward May, Eureka, Governor, Meteor, Montara, President, Queen, Ramona, Santa Rosa, Senator, Spokane, State of California, Tampico,* and *Umatilla.*

The Alaska Pacific Steamship Co. was incorporated in Maine on September 27, 1905, and in 1912 owned and operated the *Buchman, Admiral Sampson,* and *Admiral Watson* between Puget Sound and Alaska. Later acquired the *Admiral Evans.*

The Alaska Coast Co. was incorporated on May 1, 1907, in Washington with a capital of $84,500 to operate ships.

The Pacific Alaska Navigation Co. was incorporated on October 7, 1911, in Maine and owned the "Admiral" steamers: *Dewey, Farragut, Goodrich, Schley,* and *Sebree.*

Built in 1882 as the passenger vessel *Queen of the Pacific,* the ship was known as *Queen* by World War I. It operated along the West Coast from San Francisco (author's collection).

The Pacific Steamship Co. was incorporated in Maine in October 1916 and owned the schooner *Admiral Nicholson* and the steamers *Admiral Wainwright*, *Yale*, and *Harvard*. The company was jointly controlled by the Pacific Coast Co. and the Pacific Alaska Navigation Co. Pacific Steamship operated its own ships and chartered the *Yale* and the *Harvard*. On March 13, 1918, the *Yale* and *Harvard* were purchased by the Navy.

On November 1, 1916, Alaska Coast ceased operating the ships and chartered the entire fleet to the Pacific Steamship Co., which, on the same date, chartered them to their subsidiary, the Pacific Coast Co.

On October 1, 1918, Pacific Steamship purchased the *Admiral Evans* and *Admiral Watson* and executed a 6 percent mortgage for $5,000,000 to purchase eight ships from the Pacific Alaska Navigation Co. and acquired ten ships from the Pacific Coast Co. for operation.

Pacific Alaska Navigation then became a holding company owning the entire capital stock of the Pacific Steamship Co., the Alaska Pacific Steamship Co., and the Alaska Coast Co.

Chester Thorne was chairman. Hubbard F. Alexander, of Tacoma, was president, F.A. Stuart, of Seattle, was 1st vice-president, Robert S. Dollar, of San Francisco, 2nd vice-president, Alexander M. Dollar, of Vancouver, B.C., treasurer, and Frank Woolsey, of Tacoma, was secretary.

The Pacific Coast Co.
10 Wall St., Rm 306, New York

The Oregon Improvement Co. owned coal mines, real property, and wharves and docks in Nome, Juneua, and Skagway, Alaska, and controlled the Juneau Steamship Co., operator of the steamers *Georgia* and *Rustler*.

In 1897 the company went into receivership and on November 6, 1897, all the property, rights, and franchises of the company were sold. Pacific Coast was incorporated in New Jersey on November 27, 1897, with a capital of $12,525,000 and on December 1, 1897, acquired all of the Oregon Improvement property.

In 1918, attorney William M. Barnum (Yale '77) was president, Walter Barnum treasurer, and John H. Kelly secretary. Washington B. Thomas, John I. Waterbury, John W. Simpson, Albert H. Wiggin, Henry W. Cannon, Robert W. Parker, Otto Bannard, and Seward Prosser were directors.

Pacific Freighters Co.
310 California St., San Francisco

Richard James Ringwood was president. The company owned ten sailing vessels in mid–1918. In 1920 the company purchased five Ferris hulls and converted one into the 5-masted barkentine *Alicia Haviside*. On February 12, 1920, the ship was towed out of San Francisco in ballast for Vancouver, B.C. with a crew of 18 to load lumber for Durban, South Africa. They picked up a tug off Cape Flattery on March 6 and made an average passage despite unfavorable winds. The voyage was watched very closely by Shipping Board officials and shipping men but it did not revive the age of sail or result in a demand for wood freighter hulls.

Pacific Mail Steamship Co.
384 Flood Building, San Francisco

Formed in New York in April 1848 to operate 12 steamers from San Francisco to Panama, Central America, Mexico, Hong Kong, and Yokohama. In November 1900 controlling interest was acquired by the Southern Pacific Co. Service from Pacific ports was discontinued by shareholders on October 13, 1915, in response to the Seamen's Act. In August 1915 the company sold five ships to the International Mercantile Marine. In December 1915, Southern Pacific's shares were purchased by W.R. Grace & Co. in association with the American International Corp. and in February 1916 the company purchased three steamers from the Dutch West Indies Steamship Co.

The company owned 21 steamers in 1912 including the big liners *Korea, Manchuria, Mongolia,* and *Siberia*. The company's ships were Dutch-registered. In October 1917 Congress suspended the laws prohibiting foreign-flag ships from engaging in the American coastwise trade and the company was the first to receive permission from the Treasury Department to transport passengers and freight between San Francisco and Honolulu. The *Ecuador, Venezuela,* and *Columbia* were put on that route. The company had previously operated on that route before withdrawing, leaving the trade to Matson.

Inn 1917 George J. Baldwin was president in New York, Gordon G. Carson was vice-president and John H. Rosseter was vice-president and general manager at 508 California St. in San Francisco. Cecil Page was secretary and Albert G. Bates was treasurer in New York. The company managed one Shipping Board freighter, operated one, and managed and operated 12 ships.

Rosseter was president of the Sperry Flour Co. at 332 Pine St. in San Francisco and was involved with several shipping companies. He lived at the Fairmont Hotel.

For the year ending June 30, 1917, it was reported that Pacific Mail earned $3 a share on 230,000 shares for net earnings of $759,217, up $268,751 for the previous fiscal year. The International Seamen's Union took notice of this and stated:

> Only a short while ago were told editorially by practically every daily paper in the country that "the Seamen's bill had driven the Pacific Mail Company out of business." And now the corpse is very much alive and earning more money than ever in history. Besides the Mail Company is now employing more white seamen and fewer Chinese than ever before. Oh how the prophets have come to grief!

Pacific Steamship Co.
112 Market St., San Francisco
"The Admiral Line"

Richard James Ringwood was vice-president. The company managed eight Shipping Board freighters, operated two, and managed and operated 54 ships. Ringwood resigned on January 1, 1918, after 20 years and started his own company, Pacific Freighters.

Page & Jones
20 Bank of Mobile Building, Mobile, Alabama

Arthur H. Page, of New Orleans, and Charles H. Jones, of Mobile, were steamship agents, ship brokers, and forwarding agents. The company managed two Shipping Board freighters and managed and operated five ships.

Pan American Petroleum & Transport Co., Inc.
510 S. Spring St., Rm 1015, Los Angeles

Incorporated on February 2, 1916, as the ship owning division of the Pan American Petroleum Company. Edward L. Doheny was president, Herbert G. Wylie vice-president, Oscar D. Bennett secretary, and Norman Bridge treasurer.

Owned 13 tankers in mid–1918: *George G. Henry, George W. Barnes, Harold Walker, J.M. Danziger, Mexoil, Norman Bridge, Paul H. Harwood, S.M. Spalding,* and *William Green*. The company owned 17 ships and managed one Shipping Board tanker. The same officers ran the Pan American Petroleum Investment Corp. and apparently the Petroleum Transport Co. They had an office in New York at 120 Broadway.

Panama Railroad Steamship Line
24 State St., Rm 314, New York

Chester Harding was president, Thomas H. Rossbottom secretary, and Sylvester Deming treasurer. Capitalized with $7,000,000. The company owned ten steamers in mid–1918, the *Advance, Alliance, Ancon, Colon, Cristobal, General G.W. Goethals, General H.F. Hodges, General O.H. Ernst, General W.C. Gorgas,* and *Panama*.

The company operated five Shipping Board freighters and managed and operated one ship.

Paragon Shipping Corp.
Pier 19, North River, New York

Incorporated in May 1918 in New York to operate ships with a capital of $5,000 by marine insurance broker Edgerton Parsons, 925 Park Ave., R.W. Wallis, and A.C. Girard. The company managed two Shipping Board freighters, operated two, and managed and operated two ships.

Pearson Engineering Corp.
115 Broadway, 17th Floor, New York

Frederick Stark Pearson was born on July 3, 1861, became an electrical engineer, and eventually got involved in numerous electric power, railroad, and mining ventures in South America, Canada, and Mexico. In 1912 the Pearson Syndicate began drilling irrigation wells in Texas. He founded the town of Natalia, named after his daughter Natalie, and is the namesake for Pearson, Texas. Pearson and his wife Mabel were lost on the *Lusitania* in May 1915 en route to see Natalie in England.

In 1918, Ward E. Pearson was president and treasurer and Louis J. Hirt secretary. The Shipping Board chartered the Norwegian sailing vessel *Svalen* in November 1918 and assigned the ship to the company for service to Brazil.

Pendleton Brothers, Inc.
130 Pearl St., Rm 707, New York

Fields S. Pendleton was president and Edwin S. Pendleton treasurer. Incorporated as ship brokers with a capital of $200,000. The company owned 20 schooners and the bark

Holliswood in 1912. The company managed four Shipping Board freighters and managed and operated three ships.

Pensacola Shipping Co.
505 American National Bank Building, Pensacola, Florida

George W. Howe was president, Frank D. Tracy, secretary-treasurer. Agents for the Memphis Line. The company operated one Shipping Board freighter and managed and operated five ships.

Phelps Bros. & Co.
17 Battery Place, New York

Charles H., Jr., and Howard Phelps were agents for the Austro-American Steamship Co., Ltd., of Trieste. The company managed three Shipping Board freighters, operated one, and managed and operated nine ships. The Shipping Board bareboat chartered the Peruvian steamer *Callao* (ex–*Sierra Cordova*) on September 6, 1918, and assigned the ship to the company for service to Greece.

Philip Shore
101½ Franklin St., Tampa, Florida

Philip Shore was the steamship agent for the Mexican Petroleum Corp., secretary of the Tampa Bay Pilots Association, and manager of the Gulf Towing & Transportation Co. His company managed one Shipping Board freighter and managed and operated three ships. Philip lived with his wife Eva at 801 S. Edison Ave. and was later president of the Tampa Inter-Ocean Steamship Co.

Pierce Oil Corp.
25 Broad St., New York

H. Clay Pierce president, Clay A. Pierce, Eben Richards, and J.N. Galbraith vice-presidents, Samuel L. Kamps assistant secretary, W.L. Wagner treasurer. Owned the tankers *Mexicano*, *Pennant*, and *Solarina* in mid–1918.

Pocahontas Fuel Co., Inc.
1 Broadway, Rm 250, New York

On April 1, 1910, the Pocahontas Consolidated Collieries Co., Inc., terminated their contracts with their selling agents and incorporated Pocahontas Fuel to sell about 2,500,000 tons of Black Land Coal from the Olcott Coal & Iron Co. mines as "Original Pocahontas" and to deal directly with the wholesale coal trade. The coal was previously sold by S.C. Schenck, of Chicago and Toledo and J.C. Wheeler of Cincinnati. The company opened branch offices in Boston, Norfolk, Cincinnati, Chicago, and Bluefield, West Virginia.

Isaac T. Mann, of Bramwell, West Virginia, was president, Charles S. Thorne vice-president, George W. Woodruff secretary-treasurer, and Thomas H. Farrell general manager.

Potter Transportation Co., Inc.
11 Broadway, Rm 662, New York

Charles H. Potter was president, Robert E. Whelan vice-president, Joseph Morcraft, Jr., secretary, William G. Edinburg treasurer. Potter owned five ships. The company managed four Shipping Board freighters, operated one, and managed and operated 12 ships.

Producers Steamship Co.
1300 Leader-News Building, Cleveland

Apparently a subsidiary of M.A. Hanna & Co., sales agents for iron ore, pig iron, coal, and coke. Leonard C. Hanna was president.

R.I. Cameron
59 Pearl St., Rm 1506, New York

In 1917, Reuben I. Cameron and Albert E. Skae were ship brokers as Cameron & Skae and apparently dissolved their partnership in 1918 when Albert moved to New Jersey. The Shipping Board chartered the Norwegian sailing vessels *Dione, Fiskjo, Fjong, Olivebank, Sirdal, Skansen I, Skarv, Skomvaer, Steinsund, Svdnaes, Vigor, Vik,* and *Vision* in November 1918 and assigned the ships to Cameron for service to the east coast of South Africa, some for only one trip. Operated 15 sailing vessels under Shipping Board control and operated and managed one steamer.

Railroad Administration
Pier 49, North River, New York

President Wilson nationalized the nation's railroads for the prosecution of the war and Congress passed the Railway Act on March 21, 1918. Wilson appointed his son-in-law and Treasury Secretary William G. McAdoo as director general. The Administration operated one Shipping Board freighter and managed and operated five ships.

Raporel Steamship Line, Inc.
17 Battery Place, Rm 2341, New York

Edward M. Raporel was president, Henry Desvernine vice-president, Charles H.C. Pearsall secretary, and Henry J. Raphael treasurer. The company owned one ship, operated two Shipping Board freighters, and managed and operated six ships.

The Reading Co.
Reading Terminal Building, Philadelphia

Incorporated as the Excelsior Enterprise Co. on May 24, 1871, in Pennsylvania. Became the National Co. on January 19, 1872, and on December 6, 1896, the Reading Co.

Reading acquired almost all the stock of the Philadelphia & Reading Railway Co., incorporated on November 17, 1896, the Philadelphia & Reading Coal & Iron Co., and later the

Reading Iron Co. and the Central Railroad of New Jersey. The company owned a large fleet of schooners.

Agnew T. Dice was president, Albert B. Bierck vice-president, Harry E. Paisley vice-president and treasurer, and Jay V. Hare, secretary.

Reed, Welt & Co.
Waldoboro, Maine

Isaac G. Reed and George L. Welt were merchants, shipbuilders and operators.

Robert Hasler & Co., Inc.
111 W. Main, Norfolk, Virginia

Robert T. Hasler was president of Robert Hasler & Co., steamship agents and bunker coal supplies. Cecil B. Austin was treasurer and Sallie L. Darden secretary. Robert was also vice-president and general manager of the Virginia-Carolina Navigation Co.

Robert Ramsay Co.
703 Keyser Building, Baltimore

Steamship agents and brokers. Robert was president, Charles E. Scarlett vice-president and secretary, and I. Joseph Bolton treasurer. Ramsay was also managing director of the Johnston Line Foreign Agency, Ltd. The company managed and operated two Shipping Board freighters.

Rogers & Webb
110 State St., Rm 203, Boston

George D. Rogers and Charles F. Webb were ship brokers who owned the schooner *John W. Dana*. The company operated and managed two Shipping Board freighters.

S. Sgitcovich & Co.
Cotton Exchange Building, 4th Fl, Galveston

Stephen F. Sgitcovich was a steamship agent for the Gulf-Havre Line and the Gulf-Barcelona Line. The company operated three Shipping Board freighters and managed and operated three ships. Stephen lived with his wife Winnie at 1610 Ave. G.

Seaboard & Gulf Steamship Co.
Pier 32, East River, New York

Edward S. Hughes was president, Elisha G. Warfield vice-president, and Felix Jackson, secretary. The company owned two ships and managed and operated three Shipping Board freighters.

Seaboard Oil & Guano Co.
Reedville, Virginia

William Blundon was president, John A. Haynie 1st vice-president, John Hinton 2nd vice-president, George Hinton secretary, and Thomas H. Jett treasurer. The company owned two small steamers.

Shawmut Steamship Co.
60 Federal St., 10th Fl, Boston

Incorporated in Massachusetts on March 17, 1914, to take over the business of the Harper Transportation Co., of Maine, and subsequently sold its entire fleet. Harris Livermore, of New York, was president, E.C. Tobey vice president, Walter Camp, Jr., secretary, and Allen W. Lishawa treasurer.

The company owned two ships and managed and operated two Shipping Board freighters, operated, and managed and operated ships. New York to Hamburg.

Shell Company of California
343 Sansome St., San Francisco

William Meischke-Smith was president, Robert A. Lewin secretary. Meischke-Smith was the representative for the Shell interests in California.

Milan J. Trumble invented a process for refining crude oil. His work was financed by the General Petroleum Co., which bought a half interest for $75,000 on condition they complete the process. Trumble used the money to secure patents for his work in 72 countries. The Trumble Refining Co. became a $5,000,000 firm controlled by General Petroleum but General Petroleum was unable to fulfill their original commitments and in October 1915 the Trumble process was purchased the Shell Company of California for $1,000,000 "for the benefit of the Royal Dutch and Shell group." Shell paid $900,000 up front but withheld the balance pending the resolution of certain patent applications. General Petroleum made $450,000 in the deal on an investment of $75,000 while still retaining the rights to use the Trumble process in California.

Soon after, the Simplex Refining Co. offered $1,000,000 in stock to Meischke-Smith "in exchange for the rights to the patents covering the Trumble process for refining petroleum."

Shell owned the tankers *Gold Shell*, *Pearl Shell*, and *Silver Shell*.

Sinclair Navigation Co.
120 Broadway, Rm 3124, New York

In 1917, Harry F. Sinclair was president of the Sinclair Central American Oil Corp. at the 120 Broadway address. Albert E. Watts and Thomas W. White were vice-presidents, Anthony G. Steinmetz was secretary, and J. Fletcher Farrell was treasurer.

In 1918 Sinclair Navigation was incorporated in Delaware with a capital of $1,000,000. Albert Watts was president, Farrell treasurer, and Steinmetz secretary. The company owned the four tankers *Joseph Cudahy*, *Albert Watts*, *Charles Braley*, and *William Isom*. They were subsequently sold to the American Italian Commercial Corp.

Sinclair Navigation owned the *Gene Crawley, Harry Farnum, Roy Hooker,* and *Walter Hardcastle.* A messboy, Archibald Raiford, was murdered on the *Harry Farnum* on November 11, 1918, by Gotthar Wiss while discharging coal at Puerto Padre, Cuba.

Smith & Terry, Inc.
11 Broadway, Rm 531, New York

In 1912 Edwin Smith was a ship broker at 116 Broadway, Room 63, in New York and Howard A. Terry was a notary in the same office. In 1913 they incorporated as Smith & Terry. Smith was president and Terry treasurer. In 1916 they incorporated with a capital of $10,000 as the Smith & Terry Transportation Co. at 11 Broadway, Room 331, to operate ships. Smith was president, Terry vice-president, William J. Kenny was secretary, and Howard A. Terry treasurer.

They owned the schooners *Grace Van Dusen, Lucie Wheatley, Lydia Middleton, Smith & Terry,* and *Thomas F. Pollard.* The company managed and operated five Shipping Board freighters.

Smith Navigation Corp.
17 Battery Place, Rm 2029, New York

R. Lawrence Smith was president, Scott McLanahan vice-president, Richard D. Sutton secretary. Capitalized with $100,000. Dunham Wheeler and John Jannotta directors. The company owned two ships.

South Atlantic Maritime Corp.
920 Munsey Building, Washington, D.C.

Matthew Hale was president. The company managed two Shipping Board freighters and managed and operated nine ships. Hale was also president of the South Atlantic Export Co. and the South Atlantic Forwarding Co. in Room 907 and president of the South Atlantic States Association. Hale lived at 1762 Corcoran NW.

The Southern Pacific Co.
165 Broadway, Rm 2506, New York
"Morgan Line"

Charles Morgan was born on April 21, 1795, in Clinton, Connecticut. At age 14 he went to work as a clerk in his uncle's grocery store where he developed an aptitude for business and began investing in shipping. On December 20, 1817, he married Emile Reeves. Their children were Emile, Frances, Charles W., Henry, and Maria Louise.

Charles invested in the New York-to-Charleston coastwise steamer *David Brown* and made enough money to order the steamers *William Gibbons, Columbia,* and *New York* on his own account. In 1835 the *Columbia* became the first steamer to run from New York to New Orleans and that year took a cargo of gunpowder to Galveston, a community consisting of one house. Thus began the Morgan Line. In the 1850s Morgan began operating between New Orleans and Brasher City, which was subsequently renamed Morgan City. Morgan acquired railroads to expand his business, incorporated the Louisiana and Texas Railroad and Steamship Co. and in 1870 put the steamers *Josephine* and *Gussie* on routes to Cuba and Mexico.

The Southern Pacific Railroad began in 1869 as an extension of the Central Pacific Railroad and by 1880 had expanded eastward to El Paso, Texas. Charles Morgan died on May 6, 1878, and around that time SP started thinking about extending its freight and passenger service to New York. Colis P. Huntington then began negotiations in New York to purchase the Louisiana and Texas Railroad and Steamship Co.

On February 1, 1883, the Southern Pacific Steamship Co. was incorporated in Kentucky as a subsidiary with a capital of $272,623,405 and in 1885 took over the Morgan operation.

In 1918 Julius Kruttschnitt was president, Henry W. de Forest, A.O. McDonald, Paul Shoup, W.A. Worthington, William F. Herrinand, and E.O. McCormick were vice-presidents, Hugh Neill was secretary, and Andrew K. Van Deventer was treasurer. William Simmons was the general freight agent in New York. The company owned 25 ships and provided service from New York, New Orleans, and Galveston to the east coast of Central America, the American West Coast, Hawaii, Philippines, China, and Japan.

The company sailed from Pier 48 on the North River. Six of the company's freighters were requisitioned by the Shipping Board on a bareboat charter basis to the Navy for operation by the Army: *El Oriente, El Sol, Cononicus, Roanoke,* and *Housatonic.* The *Canandaigua* was time chartered from SP for and turned back over to them for operation on the Atlantic coast.

Standard Commercial Steamship Corp. 120 Broadway, Rm 2351, New York

Ery Kehaya was president, Jacob Valensi, Michael Litsas, and Rafael S. Amado vice-presidents, M.S. Amado treasurer. Thalla Kehaya was a director.

The company's 4,033-ton freighter *Seguranca* was requisitioned by the Shipping Board on a time charter basis for operation by the company.

The Standard Oil Company

John D. Rockefeller and Maurice B. Clark were in the produce business in the early 1860s when they became interested in the fledgling petroleum industry. In 1863 they formed a partnership with a chemist, Samuel Andrews, and built an oil refinery in Cleveland, Ohio, to manufacture kerosene for lighting. In February 1865 Rockefeller bought out Clark and in 1867 Henry Morrison Flagler, a salt and grain merchant, became a partner.

In January 1870, Rockefeller, Andrews, and Flagler incorporated the Standard Oil Company of Ohio as a joint stock company. By then they were worth over $1 million, owned the world's largest oil refinery, and had over 1,000 people on their payroll.

Standard Oil paid the railroads a lot of money to get its products to market. In seeking total control of the oil industry, Rockefeller used his heavy railroad patronage to conspire to raise rates on competitors.

Charles Pratt and Henry H. Rogers, of New York, exposed the scheme and opposition to this "conspiracy in restraint of trade" gained steam in Pennsylvania, the center of oil production. On April 10, 1879, a grand jury indicted Rockefeller and his henchmen for efforts "to secure a monopoly of the oil industry, to oppose other refiners, to injure the carrying trade, to extort unreasonable railroad rates, and to fraudulently control prices." Rockefeller said his business was too big to completely control and managed to evade prosecution. In time, he even managed to absorb his competitors and harshest critics into his fold.

By 1880 the Rockefeller interests had bought up and controlled 41 refining, piping, and marketing competitors in numerous states. But many state legislatures prohibited a business incorporated in their state from operating in another state so in 1882, company attorney Samuel C.T. Dodd came up with the idea of having all 37 stockholders of the various firms convey their shares to a board of nine Trustees. The board consisted of John Rockefeller and his brother William, Oliver H. Payne, Charles Pratt, Henry Flagler, John D. Archbold, William G. Warden, Jabez Bostwick, and Benjamin Brewster. The Trustees issued $100 denomination Trust Certificates to the shareholders but the Trustees exercised control over two-thirds of all the shares and had total control over all the companies. The Trustees could dissolve any company and the shareholders could not contest any decision.

All of this was being watched closely and in 1890 Congress passed the Sherman Antitrust Act. In 1909, New Jersey allowed the Standard Oil Trust to re-incorporate as a holding company and the Standard Oil Company of New Jersey was organized.

This end run was challenged and in 1911 the Supreme Court found Standard Oil of New Jersey in violation of the Sherman Act. The company then controlled about 15 percent of the crude oil supply and 70 percent of refined oil. The Court ordered Standard Oil of New Jersey to be reorganized as 34 separate companies. The company operated an extensive fleet of schooners, barges, and ocean tankers for transporting their products.

The only vessel owned by Standard Oil of New Jersey was the small steamer *S.O. Co. No. 12,* homeported at Baltimore.

Standard Oil Company of California
461 Market St., San Francisco

The Pacific Coast Oil Co. was incorporated at San Francisco on September 10, 1879. The name was changed to Standard Oil Company (California) on July 23, 1906. The company owned crude oil properties, pipelines, refineries at Richmond, El Segundo, and Bakersfield, tankers and barges. The company owned 20 schooners and steamers, including the tanker *J.A. Moffett,* and managed two Shipping Board ships. The company became Chevron.

William S. Rheem was president, Kenneth R. Kingsbury, William S. Miller vice-presidents, Fred H. Hillman vice-president and director of productions, Richard J. Hanna treasurer and director of manufacturing, and Harry M. Storey secretary and director of pipe lines.

Standard Oil Company of New Jersey
26 Broadway, Rm 1500, New York

Walter Clark Teagle was president, Charles T. White secretary, Seth Bliss Hunt treasurer, Incorporated with a capital of $110,000,000. Alfred C. Bedford chairman, Frederick H. Bedford, Frederick W. Weller, Orville T. Waring, Walter Jennings, Frederick D. Asche, and George H. Jones directors. The company's 4,095-ton tanker *Caloria* was requisitioned by the Shipping Board on a time charter basis for operation by the company on British trade.

Owned 51 ships in 1917 including the big 10,000-ton tankers *A.C. Bedford, F.Q. Barstow, Fred W. Weller, H.H. Rogers, James McGee, O.B. Jennings, Standard Arrow, W.C. Teagle, Wm. G. Warden,* and many other smaller tankers and vessels.

The company managed one Shipping Board steamer and managed and operated four ships.

Standard Oil Company of New York 26 Broadway, Rm 1300, New York

Henry C. Folger was president, William R. King and Herbert L. Pratt vice-presidents, Richard C. Velt secretary, and Herman H. Stein treasurer. Capitalized with $15,000,000. Martin Carey, Charles M. Higgins, Charles F. Meyer, Howard E. Cole, and Lucien I. Thomas were directors.

The *Socony* and *Topila*. In July 1917 the United States Shipping Board made arrangements with the French Ministry to charge a basic rate of $40 per ton for delivering petroleum products from North Atlantic ports in the United States to French Atlantic or Channel ports with extra charges from $1.00 to $2.00 per ton to go to Bordeaux or Rouen or to more than one port. These rates covered all operating and managing expenses and marine and war risk insurance.

The Standard Oil of New York's 5,073-deadweight ton coastwise tanker *Socony* was requisitioned on October 12, 1917. The deadweight capacity equaled 37,000 barrels at 42 gallons per barrel and the ship also towed a barge of 30,000 barrels' capacity.

The *Socony* was requisitioned on the 12th but not physically delivered until 4 p.m. on February 23, 1918. The ship was returned to the company on July 31, 1918. The total time under government control was 5 months, 17 hours, and 20 minutes. The formal charter agreement of August 22 was not signed and was returned to the Shipping Board and wasn't signed until September 23, 1918, two months after the government returned the ship. No money was paid to Standard Oil during this time.

While Standard Oil operated the ship after requisitioning, the ship continued to supply petroleum products to the U.S. government and to the French government at the direction of the Shipping Board at the rates set.

On August 28, 1918, the Shipping Board requisitioned the company's 8,100 deadweight-ton tanker *Topila* to load oil at Bayonne, New Jersey, and Beaumont, Texas, for shipment overseas. The agreement signed by George S. Ali, of Standard Oil, stated:

> In consideration of the United States Shipping Board giving the use of the steamer *Topila* for a voyage, under charter party dated August 28th, 1918, we, Standard Oil Company of New York, agree to accept as just compensation for the use of the steamer *Socony* while in trans-Atlantic trade under requisition by the United States, the rate of compensation fixed by the United States Shipping Board ($4.15 per dead-weight ton per month) for such period of time as the steamer *Topila* may be engaged in the service of the Standard Oil Company of New York under said charter party dated August 28th, 1918, multiplied by the ratio of the dead-weight tonnage of the *Topila* (8,100) to the dead-weight tonnage of the *Socony* (5,075).

The government had the *Topila* for 31 days, 15 hours, and 31½ minutes.

After requisitioning, the *Socony* made a voyage to France for which the French High Commission paid the company $151,098.86 for a cargo of gasoline at $44.50 per ton. A second voyage for the Commission grossed $158,858.18 at $45.75 per ton. The total paid to

Standard Oil was $309,957.04, the rates being equal to the prevailing commercial rates of the day. The U.S. government calculated these rates according to the July 1917 agreement but the rates were subsequently reduced and France was reimbursed $1,123,501.

A third voyage was made for the U.S. government and if the charges were calculated at the same rate paid by the French High Commission, the company would have received $115,628.69. In complying with government regulations for this voyage, Standard Oil paid to have the *Socony* painted in a camouflage scheme, costing $718.71 for a total $116,347.40, with no allowance for operating or managing expenses. All vessels operated under Shipping Board control were required to be painted in the "dazzle" camouflage scheme devised by inventor Thomas Edison.

Standard Oil spent $115,628.69 from October 12 to February 23, 1918, while the *Socony* was under government control with no allowance for operating, managing, or other expenses. They would have paid $14,416.10 for marine insurance and the War Risk Bureau would have spent $71,050 for war risk insurance. Total operating expenses would have been $201,094.79 or $4.65 per deadweight tons.

After the outbreak of war, almost the entire world's inventory of tankers was commandeered, seized, or chartered by Allied or Triple Entente nations as were all shipyards. Standard Oil, like all shipping companies, had no opportunity to replace seized vessels, build new vessels, or buy replacement ships on the open market since there was no market.

Under the charter agreement of August 28, the hire rate for the *Socony*, while the government had the *Topila*, a period of 1.054 months, was $22,198.55. The amount due for the entire requisition period for both vessels was $160,876.85. Reasonable expenses for acting as operating agents were $3,415; additional operating expenses and repairs came to $115, 628.69 for a grand total of $280,630.25. Standard Oil reimbursed the government $29,326.79 for overpayment on the voyages to France. They refused to accept the charter rates ultimately set by the Shipping Board and sued but the court said they had received just compensation.

Standard Transportation Co.
26 Broadway, Rm 200, New York

In 1912, Richard C. Velt, Herbert L. Pratt, and Henry C. Folger, Jr., incorporated the Saddle River Oil Co. at Garfield, New Jersey, with a capital of $500,000 "to deal in crude petroleum."

Standard Transportation was incorporated in Delaware with a capital of $15,000,000. Richard C. Velt president, Howard E. Cole and Lucien I. Thomas vice-presidents, and Herman H. Stein secretary.

The company owned seven ships. Their freighter *Tiger* was requisitioned by the Shipping Board on a bareboat charter basis to the Navy for operation by the Army.

Staples Transportation Co.
404 Davol St., Fall River, Massachusetts

In 1851, shipmasters Sylvanus Nelson Staples and William H. Phillips started transporting coal from New York to Somerset and Taunton as the Staples & Philips Coal Co. In 1848, Staples had a son, Herbert Mason Staples, who joined the company in 1868. In 1869 Herbert married Alice Moore Presbrey and they had a son, Arthur C. Staples.

Around 1881 William Phillips moved on and opened the Taunton Iron Works. In 1888 the Staples Coal Co. was incorporated with Sylvanus as president, John J. Hannah treasurer, and Herbert Staples as selling agent. Sylvanus died in 1893 and in 1894 Joseph Stickney, of New York, was elected president.

In 1909 J.W. Hammond was superintendent of the Staples Coal Co. transportation department and C.G. Hannah was an officer. Staples owned a large fleet of sailing vessels and steamers. The company managed and operated two Shipping Board freighters.

States Marine & Commercial Co., Inc.
17 Battery Place, Rm 143, New York

Freight forwarders, freight contractors, and steamship managers. Thomas C. Dawson was president, Henry R. Carse treasurer, Norman B. Chambers secretary, and James Donald general manager. Owned two ships. The company managed one Shipping Board freighter, operated two, and managed and operated five ships.

Strachan Shipping Co.
1501 Savannah Bank & Trust Building, Savannah, Georgia

William A. Strachan was a ship broker. George P. Walker was president, Frank D.M. Strachan, of Brunswick, and George F. Armstrong were vice-presidents, and Robert W. Groves was secretary-treasurer. Harry Strachan was a director. The company managed one Shipping Board freighter, operated nine, and managed and operated 15 ships.

The company managed and operated the seized German freighters *Monticello* (ex–*O.J.D. Ahlers*) to France and the *Pequot* (ex–*Ockenfels*) to Greece. The same officers ran the Southland Steamship Co. and all were bankers.

Struthers & Dixon
256 California St., San Francisco

Harry E. Struthers and Andrew Dixon managed three Shipping Board steamers and one sailing vessel, operated four steamers, and managed and operated 20 steamers and five sailing vessels. The company managed and operated the ex–German sailing vessels *Araphoe* (ex–*Steinbeck*), a ship, on the Pacific to United Kingdom route; the bark *Monongahela* (ex–*Dalbek*) on the Pacific to Atlantic route; the ship *Chillocothe* (ex–*Arnoldus Vinnen*) and barks *Muscoota* (ex–*Ottawa*) and *Moshulu* (ex–*Kurt*) to the Philippines; and managed the bark *Montauk* (ex–*Matador*) for operation by the Shipping Board to Australia.

Sudden & Christenson, Inc.
110 Market St., San Francisco

Edwin A. Christenson and Charles Sudden went into the lumber, shipping, and commission business in 1899. In 1912 the company owned five steamers, five schooners, and a barge. Charles died in 1913 and in 1917 Edwin was president and D. Walter Rasor was vice-president. managed ten Shipping Board freighters and managed and operated four ships.

The Sun Company
Finance Building, 9th Fl, Philadelphia

An oil company. J. Howard Pew was president, Francis S. McIlhenny, Joseph M. Pew, and Robert C. Pew vice-presidents, and Frank Cross secretary-treasurer.

Susquehanna Steamship Co.
26 Broad St., New York

Frank Auditore president, Joseph Auditore secretary. The company's freighter *Susquehanna* was requisitioned by the Shipping Board for operation by the company in the Italian trade.

Terminal Shipping Co.
506 Keyser Building, Baltimore

Forwarding and steamship agents, brokers, towing, bags, and stevedoring. Eugene A. Slack was president, G. Stewart Wise vice-president and secretary, and Leroy W. Sperry treasurer. Managed one Shipping Board freighter, operated three, and managed and operated eight ships.

The Texas Co.
7 W. 10th St., Wilmington, Delaware

In March 1901 the Texas Fuel Co. was formed in Beaumont to purchase oil from the Spindletop field. In April 1902, Joseph S. Cullinan and Arnold Schlaet purchased Texas Fuel and organized the Texas Co. and by November were operating vessels.

William Andrew Thompson, Jr., came from the shipbuilding industry to become general manager of the Texas Fuel Oil Company of Louisiana in 1904. In July 1905 he became a Texas Co. director in Louisiana and in 1906 went to New York organize a marine department. In April 1912 The Texas Steamship Co. was incorporated as a subsidiary at 17 Battery Place, Rm 711, New York, to build and operate tankers, barges, and other marine equipment. Thompson was president, Guy Stevens treasurer, and William W. Bruce secretary. The company owned three small steamers and the 6,000-ton *Pennsylvania* and *Virginia*.

Texas Transport & Terminal Co., Inc.
11 Broadway, Rm 808, New York

Steamship agents and stevedores incorporated in New Jersey in 1895 with a capital of $250,000. Lewis H. Spence was president, Melvin P. Billups vice-president, and William F. Hayward secretary-treasurer. The company managed two Shipping Board freighters, operated nine, and managed and operated 13 ships

Tiltro Ltd, Inc.
540 44th St., Brooklyn

Incorporated in February 1916 by O. Abrahamson at 318 60th St. in New York with a capital of $150,000. Owned the freighter *St. Helens*, sunk on October 15, 1917.

Trans-Oceanic Shipping Co.
25 Beaver St., Rm 612, New York

Starr Giddings was president, Andrew Megalos secretary, and Alfred Jaretzki, Jr., treasurer. Incorporated with a capital of $1,000,000. Royall Victor was a director. The company's 1,879-ton freighter *Navahoe* was requisitioned by the Shipping Board on a time charter basis for operation by the company on the French government's account.

Triangle Steamship Co., Inc.
44 Whitehall St., New York

Freight forwarder and exporter Edward J. Dalton was president. The company opened numerous offices and eventually operated 15 Shipping Board freighters from Pier 74, North River. By May 1920 the company had a new president and declared bankruptcy with liabilities of $113,648.

Tropical Fruit Co.
61 Broadway, Rm 910, New York

Salvatore Di Giorgio was president, Rosario Di Giorgio vice-president in Baltimore, and Angelo C. Bertolini secretary. Operated the *Wishkah* and owned the *Yuma* in 1920.

Union Oil Company of California
1114 Union Oil Building, 215 W. 7th, Los Angeles

On October 17, 1890, Thomas Bard, Wallace Hardison, and Lyman Stewart merged their southern California oil companies to create Union Oil.

In 1912, the company was comprised of the Union Steamship Co., the United Steamship Co., the Mission Transportation & Refining Co., the Union Transportation Co., and the Pectan Steamship Co., Ltd.

In 1918 William L. Stewart was president, John McPeak secretary, and Robert J. Keown treasurer. Owned the tankers *La Brea, Los Angeles, Lyman Stewart,* and *Oleum.*

Union Sulphur Co.
17 Battery Place, Rm 2832, New York

Henry Whiton was president, Erskine Hewitt vice-president, and Clarence A. Snider secretary. Incorporated with a capital of $200,000. Owned the *Freida, Herman Frasch,* and *Hewitt.*

United Fruit Co.
131 State St., Rm 900, Boston

Incorporated in New Jersey on March 30, 1899, to import bananas, sugar, and coconuts with the acquisition of the Boston, American, Quaker City, Buckman, Banes, Domnican, and Sama Fruit Companies, the Snyder Banana Co., the Columbian Land Co., the Tropical Trading

& Transport Co., Ltd., the New Orleans Belize Royal Mail & Central American Steamship Co., Ltd., and controlled through stock ownership the Nipe Bay Co., the Northern Railway of Costa Rica, and the Tropical Fruit Steamship Co.

Andrew W. Preston was president, Minor C. Keith, William Newsome, Victor M. Cutter, Crawford H. Ellis, and attorney Eugene W. Ong were vice-presidents, Bradley W. Palmer was secretary, and John W. Damon treasurer. The company's selling agent in the United States was the Fruit Dispatch Co.

The company had "vast possessions" in the West Indies and Central America and in 1913 controlled over 85 percent of fruit imported into the country. Because of this, the company came under scrutiny over possible violations of the Sherman Anti-Trust laws. The federal grand jury investigation was precipitated by Attorney General Wickersham and James di Giorgio, president of the Atlantic Fruit Co., United's main competitor.

In 1917 the company owned 22 vessels. Three of the company's cargoliners were requisitioned by the Shipping Board on a bareboat charter basis to the Navy for operation by the Army: *Calamares*, *Pastores*, and *Sixaola*. The company operated four Shipping Board freighters and managed and operated six ships.

United States & Australasia Steamship Co.
8 Bridge St., Rm 211, New York

Frederick W. Lincoln was president, Welding Ring secretary, William A. Street, treasurer, and Robert G. Moran manager. The company managed one Shipping Board freighter, operated three steamers and six sailing vessels, managed and operated seven steamers, and operated the seized German bark *Tonawana* (ex–*Indra*) to Italy under management of the Shipping Board. The Shipping Board bareboat chartered the Cuban steamer *Olivant* and delivered the ship to the company on August 22, 1917, for operation to Africa. The Shipping Board chartered the Norwegian sailing vessel *Bragdo* on November 11, 1918, and assigned the ship to the company for one trip to Australia.

United States & Brazil Steamship Co.
27 William St., Rm 704, New York

Arthur N. Lewis was the New York agent. The company managed and operated seven Shipping Board freighters. The company managed and operated the seized German freighter *Yucca* (ex–*Portonia*) and operated the *Osage* (ex–*Serapis*) under management by the Shipping Board to the east coast of South America.

United States Steamship Co., Inc.
50 Broad St., New York

Organized in Maine in December 1913 by the Charles W. Morse combination with a capital stock of $10,000,000 to acquire several Eastern steamship companies and operated them as the Steamship Operating Co. Morse was president, attorney Brodie G. Higley was vice-president and general counsel, Benjamin W. Morse vice-president, and Nehemiah H. Campbell secretary-treasurer.

The name United States Steamship Co. was adopted on December 27, 1915, and the company was incorporated in New York on January 10, 1916. Morse had acquired about 65,000 tons of shipping by the time the country got into the war in 1917. After some of his ships were requisitioned he felt shipbuilding was a safer bet than owning ships so he sold his fleet and purchased all of the capital stock of the 38-acre Robert Palmer & Sons Shipbuilding & Marine Railway Co. at Noank, Connecticut, to build Emergency Fleet Corp. ships. He changed the name to Groton Iron Works and also controlled the Virginia Shipbuilding Corp. through stock ownership.

United Steamship Co.
2215 Ave. C, Galveston, Texas

Heinrich Mosle was president, traffic manager, commission merchant, and steamship agent as H. Mosle & Co. He was also the consul for Costa Rica and Ecuador. His company ran four ships from Pier 27 and 28 to Cuba, Jamaica, and West Indies ports and managed and operated the four Shipping Board freighters *Lake Fairlie, Lake Flournoy, Lake Gera,* and *Lake Hector*. He lived with his wife Margaret, son John, and daughter Anna at 1628 Ave, J. The Mosle Building was at 2213 Ave. C.

V.Q. Peterson

Andrea Sanguineti fu Giuseppe (son of Giuseppe), a citizen of the Kingdom of Italy, owned the sailing vessel *Jolanda*. During all of 1918 the vessel was chartered to the Italian government to bring coal from the United States to an Italian coaling station at Dakar, Republique du Senegal, West Africa. On the westbound leg of the voyage back to the United States, the ship was permitted to carry private cargo with the Italian government getting 20 percent of the gross earnings.

On May 19 Sanguineti received permission from the Italian government to charter the vessel to V.Q. Peterson, a citizen of Denmark, to carry a cargo of cocoa beans from Accra, Gold Coast, West Africa to New York. The eastbound leg from New York would be to Dakar for coal for the Italian government. At this time, neither Andrea Sanguineti, Peterson, or the Italian government had applied to the United States War Trade Board or to the United States Shipping Board for an import license for cocoa beans or for approval of the charter. The Italian Ministry of Shipping in New York, acting on behalf of the owner, did contact the Shipping Board's chartering committee to request approval for the charter. The ministry was informed on May 28 that the committee would not approve a charter rate higher than $22 per ton and the committee would seek approval from higher authority in Washington since the vessel was foreign-owned. They also advised the ministry that nothing further could be done until the vessel arrived in New York and they would "endeavor to straighten out the matter, together with the Food Administration and the War Trade Board."

Peterson's charter rate was £15 per English ton (2,240 pounds). He paid £5,000 up front, £8,000 when the vessel finished loading, and was to pay the balance at the Italian consulate at the port of unloading. All expenses, licenses, and fees were Peterson's responsibility. Andrea Sanguineti held a lien on the cargo to insure the final payment for freight.

On June 30 Peterson sub-chartered the vessel to F.W. and W. Swanzy of London, England, at £20 sterling per ton to carry the cargo of cocoa beans, which was consigned to Alexander Roberts & Company of New York. Roberts was to have a pier ready in New York for unloading.

On July 8 the War Trade Board revoked all outstanding import licenses on goods shipped after the 20th. On the 17th, Roberts & Company applied for a renewal of their original license, which was granted on August 13 subject to approval of the Shipping Board's chartering committee.

The ship finished loading on or before August 13 and Peterson collected the money due at loading. The Italian government was owed 20 percent so £13,000 was paid at the Italian consulate at Dakar and the ship departed on the 13th for New York. Peterson paid Andrea Sanguineti $61,880.

The ship arrived at New York on October 13 but was not permitted to unload the cargo pending approval of the chartering committee. Peterson's agent in New York was Lykes Bros., Inc.

The $61,880 Peterson paid Sanguineti was $5,880 over the chartering committee's allowable maximum of $22 per ton so Peterson felt he overpaid and demanded a refund from the owner. The Italian Ministry of Shipping in New York held the lien on the cargo for Andrea Sanguineti and demanded $87,917.20 due on the freight and would not permit the vessel to be unloaded until it was paid. The *Jolanda* was due to load a cargo of coal for Africa and the Italian Ministry and the Shipping Board wanted the matter settled quickly.

An oral agreement between the Italian Ministry, Lykes Bros., and the chartering committee was reached whereby Lykes Bros. would pay the freight charge to the Ministry, who would then turn the money over to the Chartering Committee pending the resolution of the whole matter. The $87,917.20 was an estimate; the actual total would be known when the cargo was discharged and the bean counters could arrive at a final amount.

On November 8 the Italian Ministry delivered the check to the chartering committee, payable to the Treasurer of the United States. The check was returned for endorsement to the Shipping Board. If the money were deposited into the Treasury, it would take an act of Congress to withdraw it.

Now the only person owed money was Andrea Sanguineti. The money that was paid in Africa was also in excess of the $22 a ton limit but the chartering committee waived its demand on the Italian Ministry to collect it from Peterson.

When the cargo was finally unloaded and an accounting made, it was found the $87,917.20 was $2,412.46 over the actual amount due for freight. The ministry refunded that amount to Lykes Bros. on January 21, 1919. When the Italian Ministry sought to recover the freight money from the Shipping Board, no action was taken to refund it.

On May 23, 1924, the commercial attaché from the Italian embassy in Washington presented a claim to the Shipping Board for $87,917.20 on behalf of the *Jolanda*'s owner. Many letters and conferences produced no results. Andrea Sanguineti died on June 30, 1929. His heirs, who were probably his children, Angela, Olga, and Guiseppe Sanguineti, sued in federal court through Andrea's executors, Luigi Mazzino and Alessandro Fava.

The Act of July 18, 1918, granted the President comprehensive power in regulating the chartering of American vessels. All vessels were sorely needed by the United States and Italy at that time. All Andrea Sanguineti fu Guiseppe had to do in this case was to sail his ship to a neutral port, unload the cargo, and sell it since he held a lien on the cargo for the payment of freight. This would have tied the vessel up for weeks or months and since the vessel's services were needed by both governments, the oral agreement was made to unload in New York pending formal approval.

The court had difficulty with the Shipping Board's concern with the charter rate and why the money was held at all. The plaintiffs could not recover interest since the money was not taken for a public use. They were awarded $87,917.20 on April 6, 1936.

Ghana is a large producer of cocoa and experienced an economic boom after the Great War due to the export of cocoa. Accra is the largest city and a major deepwater port.

Vacuum Oil Co.
61 Broadway, Rm 1701, New York

Charles M. Everest was president in Rochester, Edward Prizer, George P. Whaley, and Charles E. Bedford were vice-presidents in New York, Wendel M. Smith was secretary in Rochester, and Herbert Baker was treasurer in Rochester. Charles L. Jones was manager in New York. Owned six ships in 1917. *Bramell Point, Gargoyle, Olean,* and *Paulsboro.* The company managed and operated one Shipping Board ship.

Vacuum Oil Company Proprietary, Ltd.

Soon after joining the new War Trade Board as representative of the chairman of the Shipping Board, Frank C. Munson joined Navy Department officials to express the necessity of forbidding sailing vessels to transit war or danger zones.

Jens Alexander Zachariassen, of Nystad, Finland, owned the big 4-masted, 264' bark *Albyn.* The *Albyn* arrived at Hampton Roads, Virginia, on October 23, 1917.

Around the beginning of November 1917 collectors of customs were advised that every vessel cleared for departure from an American port had to have a license for all bunkers, fuel, ship's stores, and supplies and that no license would be issued to any vessel sailing under a neutral flag without instructions from Lowell L. Richards, director of the War Trade Board's Bureau of Transportation.

The Vacuum Oil Company Proprietary, Ltd., of New York, sought to charter the *Albyn* and obtained a revocable export license from the War Trade Board. On December 18, 1917, they signed an agreement with J.A. Zachariassen & Co. to carry petroleum products in cases from New York to one or two ports in South Africa between Capetown and Delagoa Bay. Vacuum Oil agreed to pay $1.15 a case if all the cargo was discharged at one port or $1.16 a case if the cargo was discharged at two ports.

The next day, December 19, the charter agreement was submitted as required to the Chartering Committee of the Shipping Board and was approved on the 21st. It was then submitted to the Inter-Allied Chartering Committee in London and was approved there on the 24th.

Formal notice of the new export rules was published in the *Journal of the War Trade Board* of January 1, 1918.

On February 12, 1918, the *Albyn* was towed to New York and joined the Finnish-flagged vessels *Kensington, Professor Koch, Prompt,* and *Rowena.* In time, the Finnish-flagged ships *Fahrwohl, Glenard, Grace Harwar, Pampa, Parchim, Port Patrick, Vidylia,* and *Woodburn* would be in New York.

On March 14 the *Albyn* commenced loading 85,256 cases of petroleum products. Four days later, Zachariassen's agent, J.F. Whitney & Co., applied to the War Trade Board for a bunker and stores license for the *Albyn*—which they described as Russian-registered—for a voyage from New York to Durban, South Africa. The stores on board were sufficient for ten months for the 24-man crew.

Loading was finished on the 22nd, Vacuum Oil paid freight charges of $98,897.25, and the *Albyn* was ready to sail the next day. That same day, Whitney applied for a license for the ship's stores from the collector of customs at New York. The application included the bunker license obtained from the War Trade Board but the license was refused because no War Trade Board license for the stores was presented. They didn't have it yet.

On April 3, 1918, the War Trade Board wrote to the Shipping Board's director of operations, Edward F. Carry, concerning the *Albyn, Kensington, Professor Koch, Prompt,* and *Rowena*:

> These five Russian barks are being held, and we are awaiting reply from the Shipping Board as to their wishes regarding same, it being my opinion that this is a very serious matter, and I am strongly in favor of these Russian barks being taken under control of the Shipping Board or one of our Allies before being allowed to proceed. The present personnel of the crews would undoubtedly make them a very undesirable risk no matter what guarantees were given.

Instructions were given to a Shipping Board staff member to look into the matter:

> You will note that some of them are carrying oil and it is our feeling that these vessels should be prevented from continuing their voyages with Russian crews. Mr. Polk [lawyer] of the State Department has expressed this view. Will you kindly look over this situation and advise what your recommendations would be? In the meantime the ships will be held.

This person contacted Whitney & Co. and suggested the vessels might be bareboat chartered by the Shipping Board for the safety of the ships and cargo. The agents expressed "some doubt as to the responsibility of the crews of the vessels" but no action was taken.

On April 9 the War Trade Board took up the matter of the ship's stores licenses for the five vessels and their proposed voyages to Africa and Australia. The Board referred it to the State Department. Capt. John Foster Dulles of the State Department expressed the opinion that the vessels should not be allowed to proceed.

On April 12 the Shipping Board's Chartering Committee responded to an inquiry from Capt. Thomas Fisher of the British War Mission at Washington:

> When you were last here, we talked over the question of the Russian Sailing Vessels *Albyn, Prompt,* and *Professor Koch.*
>
> These vessels were all approved by the Chartering Committee some months back and we could not understand why the War Trade Board was holding them up. They now tell us that all Russian sailing vessels are being held and the matter has been the subject of many discussions before the War Trade Board. It seems to be a question of these vessels being manned by Russian crews and the American Government do not seem to be willing to trust these vessels to go to sea fully manned by Russians.
>
> This matter you might take up with Mr. Munson and he will, no doubt, be in a position to advise you fully.

On April 17, the Shipping Board staffer filed his report:

> I took up the matter with Mr. Hay of Messrs. J.F. Whitney & Company who are handling the ships.
>
> As suggested by you to Mr. Andrews I offered them the bare-boat charter basis at the rate of $1.60. They have cabled to their Agents in London yesterday asking for authority to close the deal with us.

> Mr. Hay informed me that the greatest part of the crew and Masters had the worst reputation. Some of them are openly pro-German and, not considering political feelings, they are far from reliable in many questions, and spend most of their time drinking. Mr. Hay will be glad to see these men depart, but asked me to keep the matter secret until a definite decision is taken, as he fears that the crew will make trouble if they hear that they might be dismissed.

On April 20 the War Trade Board authorized its Bureau of Transportation to refuse licenses for bunkers to "the owners, masters, or officers whereof are suspected of enemy affiliations or activities." The Shipping Board and the War Trade Board Bureaus of Exports, Imports, Intelligence, Transportation, Administration, and Research had been suspicious of certain members of the crews.

On April 25 the Shipping Board's Edward Carry wrote to the Chief of Naval Operations:

> The Shipping Board intends to charter some Finnish sailing vessels at the port of New York, a list of which is attached herewith.
>
> The Representative of the owners are in London and J.F. Whitney and Company, 8–10 Bridge Street, New York City, to whom the vessels are consigned and who have been of great assistance to us in this matter.
>
> We have been informed that the Masters and crews of these vessels are far from being very reliable, some of them being openly pro-German.
>
> As our principal purpose is to protect the vessels and cargo from possible damage, we have proposed to take them over on a bare-boat basis.
>
> We hear now from Messrs. J.F. Whitney and Company that the Masters and crews of these vessels having heard of our intention, might make some trouble. On the other hand, the agents in London have not yet given a definite answer to our offer.
>
> In view of this I believe it is very necessary to have the vessels and the crews well observed until a definite decision is reached in their respect.
>
> Will you be so kind as to let me know of your decision?

On May 6 while the agents in London were considering the charter offer, Whitney sent a copy of a cable they sent to London to Dr. Julio N. Reuter, head of a Finnish delegation at the St. Regis Hotel in New York, who had inquired about the matter on the 4th:

> Ninetythree Reuter after interview Washington officials and conference New York with ten Finnish Masters consulted leading international Admiralty Lawyers states masters crew cannot be retained and advises owners accept shipping boards offer but to strongly insist on generous allowance salary compensation and repatriation expenses."
>
> We hardly expect though response from cable before tomorrow.
>
> We understand the Masters of the *Albyn* and *Pampa* are in receipt of further instructions from London to take legal steps to protect their owners' interests so possibly they may get in touch with your good-self before engaging counsel. When you wrote the letter under review possibly you did not go into the conditions of Charter, Bare Boat Form, which we think covers very well the matter of stores onboard, but of course makes no reference to compensation and repatriation expenses you refer to, which really in our way of thinking would be a gratuitous act on the part of the U.S. Government. Has it occurred to you that possibly some arrangement might be made with the Washington Officials to make use of the Masters and Officers of the ships above referred to in positions centering around the manning or navigation of Ocean-going tonnage, and if in your position we would certainly suggest some sort of arrangement even if it were necessary for the men to undergo a probationary period so as to, in a measure, satisfy the 'Powers that Be' that Captains are not Pro-German, or in other words, can be depended upon to work in interest of the Allies of U.S., Great Britain, France, etc.

Sailing vessels were also now prohibited from transiting the war or dangerous zones. The War Trade Board considered the war zone to include "in general all areas within the radius of submarine activities, in particular all European and Mediterranean ports, and all Atlantic Islands lying east of 59 degrees west longitude and north of 20 degrees north latitude." On April 26, the Board adopted the same definition of the war zone used by the Navy Department, as published on May 10, 1918:

> No sailing vessels shall be allowed to proceed into the area which is enclosed by a line joining the following positions: Running South along 37 degrees West longitude; thence Southeast to 10 degrees North latitude, 29 degrees 30 minutes West longitude; thence Southeast to Cape Palmas, West Coast of Africa (4 degrees 22 minutes North latitude, 7 degrees 44 minutes West longitude.

In the voluminous correspondence between the owners, agents, and government bureaucrats, the *Albyn* was referred to as Russian, Russian-Finnish, and Finnish.

The Shipping Board as a whole never formally considered chartering the vessels. The only communication about the possibility was between the War Trade Board and the Shipping Board's Edward F. Carry. On June 10, 1918, the Shipping Board told Whitney & Co. it was no longer interested in chartering the vessels. On the 17th, Whitney notified London and the agents there contacted the American War Trade Department at the American embassy in London to see about releasing the ships. The next day, the charge d'affaires at the embassy cabled Secretary of State Robert Lansing about the inquiry and wanted to know what was needed to free the ships. The War Trade Board and the State Department took up the question of granting export licenses so the vessels could depart.

On June 20 the War Trade Board wrote to the head of their Bureau of Transportation:

> Upon agitation with the State Department regarding the following Russian sailing vessels, *Albyn, Fairwohl, Professor Koch, Kensington,* and *Prompt,* they have decided that perhaps it will be all right to let them proceed to South Africa, provided we are perfectly sure that none of the crews are Bolsheviki or sympathizers with the Bolsheviki regime.
>
> Please take this up with the Navy and have the crews gone over with a fine tooth comb, preparatory to taking some action towards allowing them to go.

Shortly after this, the departure of the other eight ships was considered.

On June 21 the director of the War Trade Board's Bureau of Transportation wrote to the head of Naval Intelligence saying he had been directed to release the vessels contingent on the crews who were Bolsheviks taken off and replaced and would he see to the investigation. On the 27th, the director of Naval Intelligence reported:

> 1. Interrogations of the captains of the above mentioned barks regarding their Bolsheviki tendencies, were conducted by representatives of this Office and are appended herein.
>
> 2. It will be noticed that these vessels and their crews have been out from Finland from three to ten years, and as communications from Finland with any of the captains or crews have been very infrequent, there is no doubt that newspaper accounts are the basis upon which they form their opinions of the existing conditions in that country.
>
> 3. As a result of these interrogations and from personal observations of the captains and crews of the barks in question, it is the opinion of this Office that there are none of the captains or members of the crews of these barks that have Bolsheviki inclinations. The Bolsheviki or Red Guards have maltreated or killed relatives or friends of nearly all of these crews.
>
> 4. Interviews with Captain E.A. Snell, of the Finnish Bark *Woodburn*; Captain Frans F. Rinne, of the Finnish Bark *Port Patrick*; Captain Orvar Blom, of the Finnish Bark *Glenard*; Captain John E. Johannsen, of the Finnish Bark *Vidylia*; Captain Arthur Anderson, of the Finnish Bark *Parchim*; Captain A.J. Henrikessen, of the Finnish Bark *Pampa*; and Captain V. Erikssen, of the Finnish Bark *Rowena*; were also obtained, and their opinions coincide with the attached interrogations.

5. J.F. Whitney & Co. of 8–10 Bridge St., New York City, are the chartering brokers and American agents for these Finnish barks and it has been the custom of these captains to meet in Whitney's office each day. Mr. George T. Hay, Frederick C. Lock and Theodore Dougherty of J.F. Whitney & Co., have become well acquainted with the views of each captain and often engaged them in conversation as to the situation in Finland.

6. These twelve barks are being operated by J.F. Whitney & Co., for the London agents, Eggar, Forester & Parker, of 2–3–4 Billiter Avenue, London, E.C., who in turn chartered them from the owners' agents in London, H. Clarkson & Co., of 60 Fenchurch St., London, E.C., and Anderson, Becker & Co., of 22 Billiter Avenue, London, E.C.

7. Five of these barks are loaded, or being loaded in New York Harbor, and the other seven are in ballast awaiting cargo at the same place.

The War Trade Board received this report on July 8, 1918, and the next day approved the export license for bunkers for the *Albyn*, *Professor Koch*, and *Prompt*, which were loaded and ready to sail for South Africa, and also for the *Fahrwohl* and *Kensington*. The remaining barks were cleared on the 12th. The *Rowena* was sailing for Australia and there was sight delay in getting a guarantee of the return of the vessel.

On July 12, 1918, the War Trade Board ruled the goods and commodities were not allowed to be taken out of the country, clearance was refused, and the export license was refused.

The *Albyn* was one of 13 Finnish ships detained at the Port of New York between March 8, 1918, and November 26, 1918. The others were the *Fahrwohl*, *Glenard*, *Grace Harwar*, *Kensington*, *Pampa*, *Parchim*, *Port Patrick*, *Professor Koch*, *Prompt*, *Rowena*, *Vidylia*, and *Woodburn*.

After the war, Zachariassen claimed their ship was unlawfully detained by the United States for 130 days from March 23 to July 31, 1918.

Finland was a Grand Duchy of Russia and all the Zachariassen sailing vessels were registered in Finland but flew the Russian flag. On December 6, 1917, Finland declared its independence from Russia. All the ships arrived in the United States after Finland declared independence and upon their arrival the masters of the vessels asked their agent, J.F. Whitney & Co., what the new Finnish flag looked like and fly it. The flag was rectangular with a pale blue cross on a white background, the vertical part of the cross offset 1½ its own width toward the staff.

But the United States did not recognize the new government of Finland or the new Soviet government. The *Albyn*'s bunker license, applied for on March 18, was issued on the 12th and the licenses for the other ships were issued on various dates after that. The *Albyn* was officially released on the 24th and after reconditioning and replacing certain sailors, they left New York on the 31st for East London and Durban, South Africa, and arrived at East London on December 6, 1918, at Durban on the 12th, and departed Africa on February 13, 1919.

After the *Albyn* completed her voyage, the owners promptly sought compensation for unlawful detention of 130 days. The daily net earnings of the *Albyn* under the charter agreement with Vacuum Oil was $385.06. This was arrived at by dividing the net freight of $81,244.59 by the 211 days it took to make the voyage for a total of $50,055.91.

While the ship was held in New York, the company kept the master, two mates, and six seamen on board to guard the ship and cargo and to shift the ship as necessary. Their wages were to $1,150 a month for a total of $4,914 and their subsistence and other expenses totaled $1,300. The company spent $88.85 on cablegrams and telephone calls; $493 to reship the crew; $537.36 to store and repair sails; $50 on additional security; $95 for a survey for repairs; $2,612.00 on repairs; $25 for attorney's fees; $58 for transporting the master to and from the ship; and $910 for towing. Total for these items is $4,869.71.

Vacuum Oil paid total freight costs of $98,897.25. Out of this, Zachariassen paid $4,094.86 in brokerage commissions; $6,657.80 to load the ship at New York; $3,000 for insurance; and $3,000 to unload the cargo for a net freight of $81.244.59.

The company sued to recover $75,856.26 plus six percent interest from July 31, 1918. The ship was chartered to Vacuum Oil at .25¢ per net registered tons per day, or $498.25 per day. The amount on 130 days was $64,772.50. The other vessels were also included in the action and all were dismissed. The company claimed that mere suspicion of the crew was not a just cause for detention. The government argued it was detaining the ships because the cargo was not allowed to be exported. In a case involving Cunard, the Supreme Court stated:

> A merchant ship of one country voluntarily entering the territorial limits of another subjects herself to the jurisdiction of the latter. The jurisdiction attaches in virtue of her presence, just as with objects within those limits.

Another federal court ruled in an old Hamburg-American Steam Packet Company case:

> In the United States the power to regulate commerce with foreign nations and among the several states is vested in Congress, and that body has the right to determine the conditions under which ships, or persons, or merchandise may enter or depart from the ports which it has established, and those conditions must be conformed to. The ship before it leaves port must obtain its "clearance" from the collector of the port. Clearances have a history in the maritime law extending over hundreds of years. A clearance is an important document, even in time of peace. It is particularly so in time of war. It certifies to the fact that a vessel has complied with the law and is authorized to leave port.

The *Albyn* was lost at sea on March 20, 1921, between Newport News, Virginia, and Gothenburg, Sweden. The ship's logbook was lost with the vessel and the court had to rely on the logs of the *Prompt*, *Rowena*, and *Vidylia*, which made essentially the same voyage as the *Albyn*, to determine the duration of the passage and unloading and foreign costs.

On October 6, 1941, the court concluded the company could not recover as no laws were broken in detaining the ships. Subsequent appeals were denied.

W.A. Blake & Co.
365 Marine Bank Building, Baltimore

Warren A. Blake was a ship broker who operated one Shipping Board steamer and managed six others. The Shipping Board bareboat chartered the Cuban steamers *Adelheid* and *Constantia* and delivered them to Blake on August 22, 1917, for operation to France.

W.C. Richardson & Co.
816 Leader-News Building, Cleveland

Wesley Richardson, Clarence E. Richardson, Tracy H. Paine, and John T. Kelly were vessel owners, agents, and ship and marine insurance brokers.

W.F. Spice & Co.
316 Chamber of Commerce Building, Baltimore

Wilbur F. Spice was a steamship agent and broker. The company operated two Shipping Board freighters and managed and operated eight ships. He lived at 13 Midvale Rd. in Roland Park.

W.J. Grandfield & Co.
308 Walnut St., Philadelphia

In 1910, William J. Grandfield was a ship agent who lived at 1934 N. Broad St. He died in September 1913 and in 1918 the company was run by ship broker John J. Turney. The company owned no ships. Managed one Shipping Board freighter and managed and operated five ships. The company operated the Dutch freighter *Noord Brabant* under Shipping Board management. The Shipping Board chartered the Norwegian sailing vessel *Souverain* on May 18, 1918, and assigned the ship to the company for one trip to the east coast of South America.

W.R. Grace & Co.
7 Hanover Square, New York

Importers and exporters. Joseph P. Grace was president, J. Louis Schaefer vice-president and treasurer, and Harold J. Roig secretary. The company owned 19 ships in 1917, managed nine Shipping Board freighters, operated four, and managed and operated 22 ships.

W.R. Weist & Co.
409 Marine Bank Building, Baltimore

William Read Weist was a steamship agent and broker. The company managed one Shipping Board freighter, operated one, and managed and operated five ships. Weist lived with his wife Maude Owen Weist at 628 N. Gilman St. in Baltimore.

Walker & Daly
10 Hanover, New York

Francis A. Daly and William H.A. Walker were steamship agents. They managed and operated nine Shipping Board freighters.

Warren Transportation Co.
35 Congress St., Rm 1008, Boston

George E. Warren was a coal dealer. The company owned one ship, managed three Shipping Board freighters, and managed and operated seven ships.

Wessel, Duval & Co.
25 Broad St., Rms 1004–1013, New York

Pedro M. Wessel in Denmark, Harry L. Wessel, Guilermo E. Purcell, Teodosio F. Budge, and Edmundo Eastman in Chile, George L. Duval, Robert Jaffray, and Paul W. Alexander in New York were commission merchants and exporters and importers.

The company managed two Shipping Board freighters and managed and operated four ships. On September 6, 1918, the bareboat chartered Peruvian steamer *Paita* (ex–*Anubis*) was assigned to the company for service to Peru, the United States, England, and France.

West India Steamship Co.
26 Beaver St., 14th Fl, New York

B. Rush Stoddard was president, Richard M. Goddard vice-president, William F.S. Pell secretary, and David E. Oak treasurer. The company managed two Shipping Board freighters, operated nine, and managed and operated 15 ships. The company operated the seized German freighter *Yadkin* (ex–*Clara Mennig*) to the West Indies under Shipping Board management. The Shipping Board chartered the Norwegian steamer *Thelma* on September 28, 1918, and assigned the ship to the company on October 21 for operation to the West Indies for 12 months.

Western Mercantile Marine Corp.
110 Market St., San Francisco

A subsidiary of Sudden & Christenson. Edwin A. Christenson was president.

Westmoreland Coal Co.
224 S. 3rd St., Philadelphia

S. Pemberton Hutchinson president, George McCall vice-president, Howard R. Yearsley secretary, and Herman Roll treasurer. Owned ten ships in mid–1918.

Whitney Steamship Corp.
11 Broadway, New York

Established in December 1917 at 19 Cedar St., New York, with a capital of $109,000 by R.E. Miller, attorney Richard L. Stafford, and Louis B. Eppstein. The company's freighter *James S. Whitney* was requisitioned by the Shipping Board on a bareboat charter basis to the Luckenbach Co. for operation by the Army between New York and the West Indies.

William H. Randall & Co.
60 Federal St., Boston

The company owned three ships, managed two Shipping Board freighters, operated two, and managed and operated nine ships. The Shipping Board chartered the Norwegian steamer *Times* on September 28, 1918, and assigned the ship to the company on October 18 for operation to the West Indies for 6 months.

Williamson & Rauers
26 Bay St. E, Savannah, Georgia

William W. Williamson, John J. Rauers, and John J. Magee were forwarding agents. The company managed and operated five Shipping Board freighters.

Wilson Bros. & Co.
325 Berry St., San Francisco

Gustaf Fredrik Johasson was born in Bohus Lard Bokeoräs, Sweden, in 1842, went to sea, came to America, and became a Columbia River pilot as Fred Wilson. Fred returned to Sweden for a visit and arrived at San Francisco from Göteborg on April 13, 1870, on the steamer *Orlando* with his older brother John, 33, and sister Berta, 18. They went to Portland and Fred began transporting lumber on his schooner barge *John Francis*. The business thrived and he was soon joined by brothers Charles R. and Henry. Their interests ranged from Washington to San Francisco and by the World War they owned a large fleet of sailing vessels and steamers including the *Idaho*, *Oregon*, and *Sveo*.

12. *Ship Construction*

During the World War I years, two new types of freighters not previously seen in American yards, wood and concrete, were produced. Neither one had any future.

The vast majority of privately built and Shipping Board Emergency Fleet Corporation contract ships were delivered after the war ended.

In 1917, 40 big schooners and barkentines were delivered by yards on both coasts. Yards in the Pacific Northwest, where Douglas fir was plentiful, delivered 125 barkentines or 4- or 5-masted schooners between 1916 and 1919. Sailing ship building on the East Coast was concentrated in Maine and these vessels primarily operated coastwise or to Africa, South America, Australia, or the Far East. In 1919, 33 sailing vessels were delivered and by 1921 all the contracts had been fulfilled and the age of sail was over for good, never to be revived again.

When the Shipping Board Emergency Fleet Corporation was created and Gen. Goethals took the helm, he needed someone to oversee the technical aspects of ship design. The name of Theodore E. Ferris kept popping up. Ferris was born in Stamford, Connecticut, on August 17, 1872, went to work in a shipyard on Long Island and while there took an intense interest in wood construction. Around 1890 he apprenticed to naval architect A. Cary Smith and began his design career. In 1903 the two formed Carry Smith & Ferris and in 1910 Ferris hung out his own shingle as a naval architect and marine engineer. Over the years he earned a reputation for excellent steel merchant vessels. On August 12, 1912, he married Lois Davis and they had two sons, Nathaniel and Theodore L.

By 1917 his large and busy office was at 30 Church St. in New York. On April 27, 1917, Gen. Goethals asked him to serve as chief naval architect for the Emergency Fleet Corporation but he declined owing to his current workload of 35 ships worth around $17 million then under construction. He was earning $2,500 a month plus expenses. Goethals turned up the heat and prevailed upon his patriotic duty and Ferris finally agreed, "at great personal and financial sacrifice."

At that time each shipyard seeking a government contract submitted its own plans and Ferris became the final technical authority for approving all designs. But he realized very quickly that speedy deliveries would require a standardized design and prefabrication where units of a ship were built and transported to the ways where the ship was assembled. This would be much faster than building a ship piece by piece on the ways. This concept was initially rejected by shipbuilders as "factory methods" unsuitable for the shipbuilding industry.

The Emergency Fleet Corporation's plan was to construct three large yards for building prefabricated steel ships, one at Hog Island on the Delaware River just below Philadelphia; one at Bristol, Pennsylvania, on the Delaware near Trenton, New Jersey; and one at Newark Bay, in New Jersey.

The American International Shipbuilding Corp. operated the Hog Island yard, The Merchant Shipbuilding Corp. ran the plant at Bristol, and the Submarine Boat Co. operated the yard on Newark Bay. These large yards held contracts for about 75 percent of the total Shipping Board steel ship output and naturally required a large workforce. But in their haste to put ships in the water, planners overlooked a few minor details: where would the workers come from, where would they live, and how would they get to work? This problem manifested itself early on when workers began quitting because they couldn't get to work due to traffic congestion. Turnover was very high, about 30 percent each week. To fix the problem, $10 million was appropriated for mass transit services and $75 million for housing. A total of 9,443 houses, 60 dormitory buildings, 95 apartment buildings, 27 boarding houses, and 11 cafeterias were built.

The Wood Freighters

It became obvious very quickly that the construction of modern steel freighters would not be able to replace lost tonnage fast enough so the idea of an entirely new type of wooden freighter was proposed. No such ship had ever been built before in an American yard. Ferris designed the basic 268', 3,500 deadweight-ton freighter propelled by a triple expansion steam engine that bore his name.

Edward S. Hough, a consulting engineer at 16 California St. in San Francisco, came up with Design No. 1003. The 286' × 46' ship had two forward holds, a raised fo'c'sle, a large midship house, one after hold and an after deckhouse. The 1,400-hp triple expansion engine gave a design, loaded, sea speed of 10.5 knots. A total of 35 Hough ships were delivered.

Eventually nine versions, most designed by shipyard naval architects, and all approved by Ferris, were in production. Most were built on the West Coast of green Douglas fir by inexperienced builders, some using machines that beveled and precut the planking very quickly. Entire frames were prefabricated and lifted onto the ways and the whole assembled using pneumatic tools. A serious shortage of oak timber in East Coast yards slowed production there.

The Shipping Board contracted for 703 of these freighters but only 323 were built. The cancelled contracts left builders holding the bag for materials purchased, plants expanded, and workers hired to fill the contracts. It then became necessary for Congress to enact a bill to "adjust the equitable claims of wooden-ship builders arising out of the prosecution of the war." There were approximately $300,000,000 worth of claims filed against the Shipping Board.

No ship built of green lumber is going to last very long in the tropics as seams opened and rot set in. There was little to no demand for the wooden freighters after the war so even finishing them on the builders' account would have been unprofitable. Most were laid up in storage yards and then burned later for their scrap metal.

John A. Fahey, president of the St. Johns River Shipyard Co. at Jacksonville, Florida, outlined the whole situation for a group of representatives in Washington in January 1920:

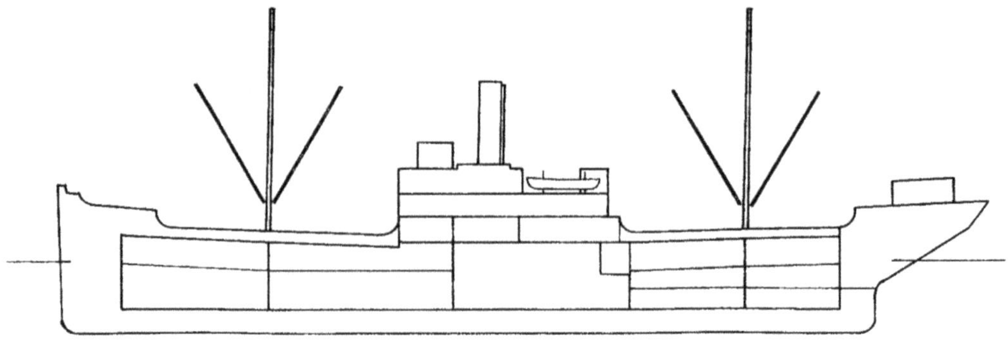

Outboard Profile of a Ferris Freighter. Overall length: 281'-6"; estimated deadweight tons: 3,500; sea speed fully loaded: 10 knots with 7,000 gallons of fresh water.

This was a straight stem, schooner-rigged, wood masted vessel with four main holds and a small hatch on poop. It was built of Douglas fir, grade select and common "reasonably free from sap, shakes and large knots, and other defects." Stern post, rudder post, and rudder stock iron bark or other approved hardwood. Keel shoe of oak. Fir keel 16" × 20" with scarfs 10' long. Frames 36" on center. Side planking 5" × 10", average length 45'. Bulwarks 3" thick fir. Treenails of locust or other suitable hard, close-grained wood. After deckhouse 6" × 12" edgebolted "log" timber to support the gun.

Floors of tile, sheet lead, and linoleum. Officers' berths 28" × 72". Petty officers, ordinary and able seamen, firemen, oilers, cooks, and messmen berth in forecastle head, 12 gunners in poop quarters.

Built to American Bureau of Shipping Class A-1 for 15 years' service. Constructed, equipped and fitted to U.S. Steamboat Inspection Service rules and to comply with Seamen's Act in effect November 4, 1915. Theodore Ferris's plan was designated Design No. 1001 (drawing by the author).

You will, however, remember that the Emergency Fleet Corporation was supposed to have been created as a special corporation, free from some of the handicaps of government departments, in order that they might operate as established business, that it might go ahead and accomplish something in what was then regarded as the most needed facility of the entire war. In other words, going into the war as we did, in the spring of 1917, with the constantly increasing submarine destruction, it was idle for us to raise an army of two or four million men, to create great munitions plants, and to turn out vast supplies and to try to get food if we could not get it to the other side after it was available.

It was found that the steel plants of the country were tied up and so there came the proposal to resort to wood, which in the early days of this country had been effective. It was under these circumstances that the Shipping Board sent out a general call to all of the shipyards on the Atlantic and Gulf coasts, and indeed all over the country, and asked them to Washington for a conference. They likewise invited various contractors and builders at the same time. A conference lasting some four days was held here in Washington to consider the situation. These builders were asked what they could do to increase their facilities and to build a ship, a proposed wooden steamship for the general purpose of serving the needs of the country as a war emergency.

No such ship was designed. Nothing of the kind was on paper at the time. They had only concluded that they were going to build a wooden steamship and a wooden steamship of this type had never been built in the United States. There were no wood yards on this coast that had ever built such a ship and, as a matter of fact, no such ship had ever been designed before. These builders told the facilities they had. The Shipping Board told them that the wood-ship yards offered the only immediate relief, that the steel yards could do nothing except on the ships that were then under construction. They went so far in that meeting as to bring in engineers and other experts who said they proposed to employ the electric drive in the operation of these wooden ships.

> These builders were appealed to go back home and to begin building yards for a proposed ship of about 275 feet in length and not to wait, to get their organization together, to expand, to get their land, to raise their capital, to go ahead, and that contracts would be given to them to cover the cost of the yards built and a profit of 10 percent or a fixed fee of $20,000 per ship.

This is the reason the fixed-fee contract was outlawed during the World War II building program.

On May 1, 1917, Capt. Albert F. Pillsbury became the Emergency Fleet Corporation's District Officer in San Francisco. His district extended from Coos Bay, Oregon, to San Diego, California. In January 1918 he had 77 steamers building, 29 requisitioned ships, and contracts for 16 steel ships pending. In addition 20 to 25 wooden Ferris and Hough type ships were under construction. As to size, one expert stated in early 1918:

> I understand they have stopped building wooden ships. The difficulty with the wooden ship is that they can not develop the full strength of the members in the end connections and they can not build a large wooden ship and can not get enough power in it to drive it fast. Speed is a very important thing in dealing with the submarine situation.

The sailors themselves had mixed views on the wooden ships. Fred J. "Fritz" Harmon, of 302 2nd Ave. in Detroit, was an engineer on lakers before the war and was first assistant on the ill-fated *Dumaru*. He later stated:

> In August 1918, at Portland, Oregon, I first saw the *Dumaru*. She was a war baby, one of the hundreds built in haste, and often built badly, to take the place of the huge tonnage the German U-boats were destroying. In her camouflaged paint she was not handsome. She was cumbersome and as a steamer not much. Like many of her sisterships the *Dumaru* was constructed of green Oregon fir at Portland. She was twin screwed and well decked with a gun platform on the poop. They were built out of timber taken right out of the forests and the tropics is a terrible place to send unseasoned ships made of green wood. The heat cuts the seams wide open. In Northern waters they would have got along better. They limped along and managed to keep afloat and not much more. They had leaky seams, their pumps going day and night to keep clear of water.
>
> Experienced seamen did not want to sail on the wooden ships and the local marine men at Portland, where the wooden ships were, preferred to go to Seattle or San Francisco and sign up on a smart looking iron ship. The United States Shipping Board had to make a wide search for officers and men to sail on the wooden wonders.

On November 21, 1918, Sen. William M. Calder wrote to John A. Donald, acting chairman of the Shipping Board for an update on the wood freighters. Donald replied that as of the 20th, 98 wooden steamers had been completed and 76 had carried cargoes or had sailed in ballast for loading ports. Most voyages were between the Pacific Coast, Hawaii, the Philippines, or to the west coast of South America to load nitrate for Pacific, Atlantic, or Gulf ports.

The first vessel delivered was the *North Bend*, built by Kruse & Banks, and ran very successfully to Hawaii with general merchandise and a return cargo of sugar. The *Biloxi* was the second ship delivered and was in service to Chile.

He stated the ships were too small for trans–Atlantic service, "which required that a steamer going into the war zone should carry sufficient bunkers and stores for the round voyage and be convoyed. This would have necessitated from 1,000 to 1,200 tons of coal and stores, and after deduction of the space required for the gun crew and weight of the military features, there would probably not to exceed 1,500 to 2,000 tons of available cargo space."

Only three ships had been lost: the *Coos Bay* and the *Blackford* in a hurricane off Mexico on September 11, 1918, and the *Dumaru*, hit by lightning on October 16, and burned.

Concrete Ships?

In January 1918, the Senate Committee on Commerce held hearings in Washington, D.C., on the Shipping Board's program. Sen. Duncan U. Fletcher, of Florida, was chairman. One early witness and contributor was Roy H. Robinson, 35, of 5227 Cornell Ave., Chicago, the son in J.C. Robinson & Son, building contractors at 79 W. Monroe Street. He had extensive experience with concrete and had made a very close study of shipping since the war began in 1914. He asked to testify before the committee regarding shipbuilding and stressed that he was not seeking any shipbuilding contracts. He was admitted as a qualified witness and testified on January 6, 1918.

At that time, there was considerable dispute among the various authorities as to the amount of tonnage actually being lost to U-boats—deadweight tonnage versus gross tonnage and so forth—but what alarmed Robinson the most was the Allied and neutral merchant ship losses versus the replacement tons being built. He was not so concerned with the exact figures and stated, "Because, from what study I have made of the situation, it appears that if the shipbuilding situation is not remedied and saved in this year of 1918, I do not believe it will be much use to be talking about ships for 1919."

The Emergency Fleet Corporation had already let contracts for wood and steel ships and the committee had already discussed naval construction, which was being concentrated on destroyers and the new anti-submarine vessels.

Senator Knute Nelson, of Minnesota, began questioning Robinson:

NELSON: How many have we building now in this country?

ROBINSON: I could not tell you how many we have building. I think we have contracted for about 500 or 600.

NELSON: And they are at work on them, are they not?

ROBINSON: Some of them. They are building a yard up at Hog Island that I understand will not get any ships built this year. They have not even started to provide the housing facilities for the men. You understand building a yard or contracting for a ship is very different from building ships.

NELSON: Do you know how many we are building on the Pacific Coast?

ROBINSON: No, I do not. I would not attempt to say offhand, but all told, I believe they have contracted for about 500 or 600 ships. Last year England, under the stress of knowing that her existence was at stake in this war and dependent upon producing enough ships, produced 1,900,000 tons. That was the most England could produce, with the greatest yards and the best facilities in the world. We commenced in 1916 to build about 531,000 tons. Last year, 1917, notwithstanding the fact that we went in for shipbuilding and attempted to do our best, we built something less than 700,000 tons.

NELSON: You understand we commandeered over 49 ships?

ROBINSON: Yes, but that did not change the tonnage. It simply meant a change in who was going to run them. Our tonnage was less than 700,000.

NELSON: Most of the ships we have are commandeered ships.

ROBINSON: Yes.

NELSON: And then we got all those interned vessels.

ROBINSON: About 600,000 tons.

NELSON: And we got most of the Austrian ships, too, by purchase.

ROBINSON: The Austrian and German ships all told amount to less than 700,000 tons. These ships that we produced last year, 700,000 tons, are one of our assets. That is new construction to replace lost construction. This year the Shipping Board expects to produce, under all these contracts which it has for last year and this year, about 5,000,000 tons, as I understand. Whether they will do so or not is still a problem. Mr. Hurley has said he might be able to get 8,000,000 tons if everything was very favorable.... Of course the government has taken a position which makes it rather difficult for private individuals to do anything in private shipbuilding because the government has occupied, perhaps, unwittingly, the position of a dog in the manger. They did not do anything themselves and at the same time, with the threat of commandeering anything that was produced, they prevented private capital from coming in, taking away the bait of the large remuneration which otherwise would accrue to them. For instance, a ship nowadays is worth its weight in gold. If a man could go out and build a ship with the knowledge that if it was a success it would not be commandeered, undoubtedly much capital would be invested.

Senator Harding then turned to the discussion of concrete ships, which had engendered considerable interest. Robinson said the technique went back as far as 1849 in France and had been used very successfully for small vessels and barges there and in Sweden, Norway, Denmark, and England. A proposal had been made by William Comyn in San Francisco, California, to build one and if it was successful to construct 49 more. His San Francisco Shipbuilding Co. down the Peninsula in Redwood City was then constructing the 5,000-ton ship *Faith* on his own account.

Robinson said that in ship construction generally, the hull represented about half the total cost of a ship and that a concrete ship could be built for about half the cost of a steel ship and was "infinitely faster" to build. While plate and structural steel was tightly controlled, the steel reinforcing bar would be easy to obtain. The problem would be machinery as no machinery would be available on the open market for a private builder. But in any case, he said, not nearly enough concrete ships would be available in 1918.

Robinson stated the only way to win the war was with quickly-built, standardized concrete ships so the flow of available steel and wood could continue unimpeded to the shipyards building those vessels. The concrete ships would amount to an emergency construction program. Sen. Randell joined the discussion:

RANDELL: You would not need the steel for the construction of the ships and therefore you could use it for the other things.

ROBINSON: Exactly. It would be diverted to the other necessities of the war. Just look at the ammunition proposition. In 1915 there were over 50,000,000 shells fired in a few days by the Allies to gain a few miles. How many shells is it going to take to break that western front? We have got to use a lot of ammunition on that line and we have got to break that line at several places at once. These local offensives have proven a failure. England put 5,000,000 men in the field after France had the Germans stopped. And with a 30-mile haul, with the greatest merchant marine in the world at their command, and with those 5,000,000 men thrown in as a new force into that front, when Germany had only half the strength she has today on that front, what did they get? This year Germany has twice as many men. Colonel Reppington served a warning the other day saying that Germany is going to have two men on that front to the Allies' one. The German front has been cut from 2,000 miles to 1,000 miles. Germany has only a 1,000-mile front this year to defend as against 2,000 miles this year. She has practically all the coal and iron of Europe, outside of England, right in her hand. That is a terrible handicap. But we have got to overcome it and we have got to do it with ships. We have got to build up a great big pyramid with ships at the bottom. And if we build up a 30,000-ton pyramid with 5,000-ton ships it is going down.

Robinson wanted the Shipping Board to immediately establish five to ten shipyards just to build concrete ships and also pointed out that when a torpedo hits a steel, riveted hull it blows the whole thing apart whereas a concrete hull would absorb the blow and with adequate transverse bulkheads every fifty feet or so the ship would stay afloat.

It was also pointed out that a wooden ship has to be painted every six months, or after only two average voyages, and steel ships rust, but a concrete ship would actually improve with age.

The Emergency Fleet Corporation contracted for 11 ships total, all delivered long after the war ended. The one 3,000 deadweight ton ship, the *Atlantus,* was built by the Liberty Shipbuilding Co. at Wilmington, North Carolina. The keel was laid on May 24, 1918, and launched on December 4.

Ten 7,500 deadweight ton ships were ordered. The *Guyamaca* and *San Pasqual* were built by the Pacific Marine & Construction Co. at San Diego and the *Santa Monica,* and *Monrovia* were built at the company's yard in San Francisco. The *Twilight* and *Palo Alto* were built in San Francisco by the San Francisco Ship Co. The *Dinsmore* and *Moffit* were built by the A. Bentley & Sons Co. at Jacksonville, Florida. The *Latham* and *Selma* were built in Mobile, Alabama, by Fred T. Ley & Co.

After the war, Admiral William Sims, talking about the allocation of scarce resources early in the war, stated:

> It is quite true that there were many naval activities outside of the eastern Atlantic, such as the Caribbean, south Atlantic, Pacific, and Asiatic waters. But, considering the rapidity with which at the time in question we were losing the war in the submarine zone, these forces were of little importance—practically none as regards ultimate success. A great deal of unnecessary effort was expended in these areas. It was repeatedly pointed out that we could afford to lose some antisubmarine craft, but could not afford to continue the loss of merchantmen at the rate then being sustained.

In July 1921 the Bureau of Navigation reported that on May 1, 1921, the Shipping Board owned 1,777 vessels of all types and by July of the same year the figure had increased to 1,789.

13. Ship Construction by Shipyard, 1914–1918

This section lists every steamer, sailing, or motor vessel over 1,000 gross tons delivered between 1914 and 1918. In a few cases, ships delivered in 1919 are included when the story of the shipyard contributes to a fuller picture of the building program.

At this time the transition from coal to oil fuels was not yet complete so some steamers were designed to burn either coal or oil. England had a lot of coal but little oil while the United States had both in abundance but could not rely on an adequate oil supply overseas.

The yards are arranged geographically by Pacific Coast, Great Lakes and Western Rivers, Atlantic Coast, and Gulf Coast. The yards are arranged alphabetically within their respective regions.

Each entry records the name, tonnage, speed, fuel, type, year delivered, official number, call sign, private owner, or Manager and Operator if requisitioned by the Shipping Board and allocated to another Manager or Operator.

The delivery date is the date a contract ship was physically delivered to the Shipping Board or a requisitioned ship officially came under Shipping Board control.

Because the steel Shipping Board steamers formed the backbone of the U.S. merchant marine in the early days of World War II, notes about their losses during the war are included.

The Pacific Coast

The Pacific Coast lumber trade produced a unique type of coastwise schooner-rigged steamer called a "steam schooner" with engines and house aft that could sail if necessary. The earliest steam schooner was a coastwise lumber schooner someone fitted with a steam engine to gain some control over the very rough waters, winds, and vagaries of the Northern California and Oregon coasts. The earliest type mentioned is the 247-ton *Newport* built by the Dickie Bros. in San Francisco in 1875. The style evolved into a freighter with engines and house aft and modest passenger accommodations. They were called "stem winders" in the lumber trade.

During the war a representative of the French government came to the Pacific Coast and contracted for the purchase of seven steam schooners at $525,000 each and one for $635,000 but the sale was stopped by the Shipping Board. Around 225 of these types of vessels were built.

Aberdeen Shipbuilding Co., Inc. Aberdeen, Washington

Capt. Andrew Peterson moved his original Raymond Shipyard from Raymond, Washington, to Aberdeen when he signed a contract with the Wilson Bros. for the *Oregon*. Raymond is up the Willapa River from Willapa Bay in southern Washington.

Oregon—1,014 tons, steam schooner, 1916, 214322, LGFP. Wilson Bros. & Co.
Columbia River—1,200 tons, steam schooner, 1916, 214504, LGKQ. Fife Shipping Co.
Idaho—1,047 tons, steam schooner, 1916, 214627, LGPB. Wilson Bors. & Co.
Phyllis—1,266 tons, steam schooner, 1917, 214757, LGRW. Wilson Bors. & Co.
Santa Christina—2,119 tons, motor freighter, 1917, 215329, LHMP. Built on the company's account. Allocated to the New Orleans & South American Steamship Co.
Susanne—1,431 tons, motor freighter, 1918, 215999, LJRC. Oriental Navigation Corp.
Santa Flavia—2,113 tons, motor freighter, 1918, 216005, LJRK. Built on the company's own account. Allocated to the Grace Steamship Co.

Alaska Pacific Navigation Co. 201 Mutual Life Building, Seattle

Charles A. Burckhardt was president of the 1-way Alaska Pacific Navigation yard, president of the Cotton-Burckhardt Co., a general insurance agency in Ketchikan, and president of the Alaska Pacific Fisheries Company.

Oregon—1,616 tons, wood motor freighter, 1917, 215742, LJDW. Alaska Pacific Navigation Co., homeported at Portland.
Alabama—2,088 tons, wood motor freighter, 1918, 216577, LMDN. Jewett M. Scott. The 270', twin-screw, 4,000 deadweight ton Douglas fir "auxiliary schooner" was designed by Edson B. Shock with McIntosh & Seymour Diesel engines. The pilot house was aft and accommodations for a crew of 20 were in the forecastle. It was launched in April 1918, made a voyage to the Far East, and was sold alien on May 5, 1919.

Albina Engine & Machine Works, Inc. Portland, Oregon

Albina Engine opened for business in 1916 at Lewis and Loring Streets on the west side of the Willamette River with four building slips. William Cornfoot was president and general manager, Arnold Reimann vice-president. The company employed 2,900 workers.

Point Loma—2,119 tons, steel, 9.5 knot coal and oil-burning freighter, 216021, LJSB. Keel laid on April 18, 1917. Requisitioned from Norwegian owners while building and delivered to the USSB on March 7, 1918. Allocated to the South Atlantic Maritime Corp. for service to the West Indies under Capt. Mark Woodbury.

Point Arena—2,117 tons, steel, 9.5 knot coal-burning freighter, 216119, LJWH. Keel laid on April 25, 1917. Requisitioned from Norwegian owners while building and delivered to the USSB on March 30, 1918. Allocated to A.H. Bull for service to the West Indies.

Point Bonita—2,675 tons, steel, 12 knot coal and oil-burning freighter, 216512, LMBH. Keel laid on May 29, 1917. Requisitioned from Danish owners while building and delivered to the USSB on June 24, 1918. Commissioned *Point Bonita* in the Naval Overseas Transportation Service on October 7, 1918.

Point Lobos—2,670 tons, steel, 12 knot coal and oil-burning freighter, 216535, LMBW. Keel laid on May 29, 1917. Requisitioned from Danish owners while building and delivered to the USSB on June 29, 1918. Magnolia Petroleum Co. Commissioned *Point Lobos* in the Naval Overseas Transportation Service on October 25, 1918.

Point Judith—2,670 tons, steel, 12 knot coal and oil-burning freighter, 216656, LMGV. Keel laid on November 3, 1917. Requisitioned from Danish owners while building and delivered to the USSB on July 25, 1918.

Point Adams—2,671 tons, steel, 12 knot coal and oil-burning freighter, 216749, LMKQ. Keel laid on November 30, 1917. Requisitioned from Danish owners while building and delivered to the USSB on August 12, 1918.

Cadaretta—2,865 tons, steel, 12 knot coal and oil-burning freighter, 217153, LNPB. USSB contract ship. Keel laid on June 24, 1918, and delivered on October 31, 1918, 129 building days. Allocated to Sudden & Christenson as manager and operated by the Robert Dollar Co. for trans-Pacific service.

Caddopeake—2,865 tons, steel, 12 knot coal and oil-burning freighter, 217293, LPBJ. USSB contract ship. Keel laid on June 24, 1918, and delivered on December 4, 1918, 185 building days. Allocated to the Pacific Mail Steamship Co. for service in the Philippine Inter-Island trade.

Callabasas—2,740 tons, steel, 12 knot coal and oil-burning freighter, 217394, LPGW. USSB contract ship. Keel laid on June 24, 1918, and delivered on December 31, 1918, 190 building days. Allocated to James W. Elwell & Co. for service to Scandinavia.

Ames Shipbuilding & Dry Dock Co.
1610–16 Koge Building, Seattle, Washington

Edgar Ames was president and George W. Albin secretary-treasurer. The yard was at 26th Ave. SW and W. Hanford. John Wilson was the general superintendent and also vice president and general manager of Wilson Shipyards, a 200–300-ton motor craft builder. The company built steel ships on private accounts and steel Shipping Board freighters homeported at Seattle.

Ames was also vice president of the State Bank of Seattle and lived with his wife Anne at 1132 Harvard Ave. N.

Westerly—5,615 tons, steel, 10.5 knot coal-burning freighter, 215983, LJQP. Keel laid on March 1, 1917. Requisitioned from British owners while building and delivered to the USSB on February 16, 1918. Sunk in a collision with the British steamer *Luciline* off the French coast on April 27, 1918. There were no casualties among the 57 on board.

Westwood—5,612 tons, steel, 10.5 knot coal-burning freighter, 216045, LJSQ. Keel laid on March 20, 1917. Requisitioned from British owners while building and delivered to the USSB on March 12, 1918. Operated by the Shipping Board for service to France. Commissioned *Westwood* in the Naval Overseas Transportation Service on July 14, 1918.

West Eagle—5,614 tons, steel, 10.5 knot coal-burning freighter, 216110, LJWF. Keel laid on April 1, 1917. Requisitioned from British owners while building and delivered to the USSB on March 29, 1918. Allocated to American-Hawaiian as manager for service to the River Plate by the Oriental Navigation Co.

Westmount—5,584 tons, steel, 10.5 knot coal-burning freighter, 216333, LKRM. Keel laid on November 26, 1917. Requisitioned from French owners while building and delivered to the USSB on May 20, 1918. Commissioned *Westmount* in the Naval Overseas Transportation Service on May 21, 1918.

Westford—5,620 tons, steel, 10.5 knot coal-burning freighter, 216563, LMCT. Keel laid on February 5, 1918. Requisitioned from British owners while building and delivered to the USSB on June 29, 1918. Commissioned *Westford* in the Naval Overseas Transportation Service on July 2, 1918.

Montrolite—6,834 tons, steel, 10.5 knot coal or oil-burning tanker, 216747, LMKN. Keel laid on November 15, 1917, for Standard Oil of New Jersey. Requisitioned while building and delivered to the USSB on August 12, 1918.

Westport—5,591 tons, steel, 10.5 knot coal-burning freighter, 216902, LMVF. Keel laid on April 22, 1918. Requisitioned from French owners while building and delivered to the USSB on September 10, 1918. Commissioned *Westport* in the Naval Overseas Transportation Service on September 10, 1918.

L.J. Drake—6,836 tons, 10.5 knot coal or oil-burning tanker, 216986, LNCQ. Keel laid on January 17, 1918, for Standard Oil of New Jersey. Requisitioned while building and delivered to the USSB on September 30, 1918. Lauren J. Drake was president of Standard Oil of Indiana. Operated during World War II by Standard Oil of New Jersey and sunk on June 5, 1942, by *U-68* in the Caribbean.

West Mead—5,618 tons, steel, 10.5 knot coal-burning freighter, 217152, LNMW. Keel laid on June 4, 1918. Requisitioned from British owners while building and delivered to the USSB on October 29, 1918. Commissioned *West Mead* in the Naval Overseas Transportation Service on October 29, 1918.

West Cape—5,691 tons, steel, 10.5 knot coal-burning freighter, 217238, LNTQ. Keel laid on June 17, 1918. Requisitioned while building and delivered to the USSB on November 20, 1918. Allocated to the American Line for the European civilian food relief mission.

Babare Bros.
Tacoma, Washington

George and Nicholas Babare were builders of wooden steam and schooner yachts, fireboats, ferryboats, canal boats, houseboats, scows, tugs, steamboats, lifeboats, lighters, sailing vessels, dredges, and sand suckers in their yard with two 350' ways at the foot of Carr Street. George was chief of the mechanical department and purchasing agent. Nick was superintendent of hulls. The company signed a contract with the Emergency Fleet Corp. on February 21, 1918, to build wood ships for a flat rate of $338,500 each. The *Bottineau* was delivered in 1919. The company employed 130 workers.

Mahaska—2,524 tons, wood, 10 knot coal-burning Ferris freighter, 216763, LMNC. USSB contract ship. Keel laid on July 26, 1917, and delivered on August 31, 1918, 401 building days. Allocated to A.D. Cummins & Co. for service to the Netherlands.

Ballard Shipbuilding Co.
Seattle, Washington

Capt. Harry W. Crosby was president, James D. Esary vice-president, and Howard B. Lovejoy superintendent and general manager. The yard was at the foot of Shilshole Ave. and 24th Ave. NW. The partners built the *H.B. Lovejoy* for themselves as the Northwestern Shipping Co.

H.B. Lovejoy—1,067 tons, steam schooner, 1918, 216542, LMDC. Northwestern Shipping Co. The *Lovejoy* was the first ship in the Pacific Northwest launched at 100 percent complete—no time spent at the outfitting dock for the installation of machinery, fittings, etc. It was ready to go to work the minute it hit the water. The ship was sponsored by James Esary's wife, May.

The only other vessel Ballard built was the 277-ton sidewheel steamer *T.C. Reed* in 1918.

Benicia Shipbuilding Corp.
Benicia, California

In the 1870s, Matthew Turner and Richard T. Rundle were shipbuilders as Turner & Rundle at the foot of 6th St. in San Francisco and agents for the Tahiti Line Packets at 122 California. In 1878 Rundle passed away and in 1882 Matthew moved the yard to Benicia near the foot of W. 12th St. The last vessel built there was the schooner *M. Turner* in 1902.

James Robertson was a shipwright at the Mare Island Navy Yard who lived at 17 Eighth St. in Vallejo, California. Robertson purchased the yard in 1913 and built five vessels there. After the *Rose Mahony* was delivered he moved his operation to Alameda and built another five ships between 1920 and 1924.

La Merced—1,696 tons, wood composite-hull motor freighter, 1917, 215578, LHTN. Standard Oil Company of California.

Oronite—1,704 tons, wood motor freighter, 1918, 216067, LJTG. Standard Oil Company of California. Ordered by Andrew F. Mahony but he sold the contract to Standard Oil for $275,000 during building.

Rose Mahony—2,051 tons, schooner, 1918, 216507, LMBF. Andrew F. Mahony.

Bethlehem Shipbuilding Corp., Ltd.
Bethlehem, Pennsylvania

Bethlehem Steel's Shipbuilding Division acquired yards at San Francisco and Alameda. The San Francisco yard was the old Union Iron Works that Bethlehem acquired around 1905. The Alameda yard was the former United Engineering Works acquired in 1916.

John A. McGregor was Bethlehem's West Coast representative and president of Bethlehem of California, Joseph J. Tynan was vice-president and general manager, and Arnold

Foster was secretary-treasurer. The company's main office was at 260 California St. in San Francisco.

Union Iron went back to brothers James and Peter Donahue who opened a blacksmith shop in 1849. The firm expanded into foundry and marine work and many folks in the shipping trades, including the Scotts and Dickies, future prominent shipbuilders, honed their skills at Union Iron. The Union Iron name continued to be used for some time after Bethlehem's takeover and some Builder's Certificates from the World War years state Union Iron as the builder.

All executive and financial matters were handled by Bethlehem in Pennsylvania.

San Francisco

Frank H. Buck—6,076 tons, tanker, 1914, 212090, LDKH. Associated Oil Co. Commissioned *Frank H. Buck* in the Naval Overseas Transportation Service on January 25, 1918.

Lyman Stewart—6,054 tons, tanker, 1914, 212860, LFBW. Union Oil Co. of California.

J.A. Moffett—6,521 tons, tanker, 1915, 212994, LFGW. Standard Oil Company of California. James A. Moffett was president of Standard Oil of Indiana. At 3:50 a.m. on September 25, 1917, gases in a forward hold exploded while the ship was moored at Point Richmond, California. Two workers, Albert R. Juillerat, 30, and Ernest Rutter, 22, were in the hold at the time. The shock of the explosion was "felt in all the bay cities."

Pacific—6,034 tons, freighter, 1915, 213769, LFQS. Union Iron Works. Sold alien in 1916.

Eurana—5,915 tons, freighter, 1916, 213861, LFRS. Nafra Co. Commissioned *Eurana* on September 13, 1918, in the Naval Overseas Transportation Service. Operated by the Calmar Steamship Co. during World War II. Bombed and sunk by German aircraft on May 27, 1942, in Convoy PQ 16.

La Brea—6,945 tons, tanker, 1916, 213909, LFSV. Union Oil Co. of California.

Los Angeles—6,876 tons, tanker, 1916, 213962, LFTS. Union Oil Co. of California. Allocated to the Navy.

Acme—6,895 tons, tanker, 1916, 214173, LGCM. Standard Transportation Co. Operated during World War II by the Socony-Vacuum Oil Co. Torpedoed by *U-124* on March 17, 1942, off North Carolina. Towed in and repaired.

Davanger—5,876 tons, freighter, 1916, 214301, LGFH. Union Iron Works.

D.G. Schofield—8,651 tons, tanker, 1916, 214353, LGHC. Standard Oil Company of California. Demetrius Gustavus Schofield was president of the new Standard Oil of California.

Paulsboro—6,945 tons, tanker, 1916, 214449, LGJM. Vacuum Oil Co.

Astral—8,653 tons, tanker, 1916, 214492, LGKJ. Standard Transportation Co. Operated in 1941 by the Socony-Vacuum Oil Co. Sunk by *U-43* on December 2 in the Atlantic.

H.C. Folger—7,185 tons, tanker, 1916, 214615, LGNV. Atlantic Refining Co. Henry Clay Folger, Jr., was president of Standard Oil of New York.

Talabot—6,378 tons, freighter, 1917, 214699. Built for charter to a Norwegian operator.

J.W. Van Dyke—7,197 tons, tanker, 1917, 214715, LGQS. Atlantic Refining Co. John Wesley Van Dyke was president of Atlantic Refining.

Maui—10,261 tons, passenger, 1917, 214861, LGTH. Matson Navigation Co.

Eagle—6,273 tons, tanker, 1917, 214921, LGVP. Standard Transportation Co. Commissioned *Peerless* in the Naval Overseas Transportation Service on December 8, 1918.

Operated during World War II by the Socony-Vacuum Oil Co. Torpedoed by *U-513* on June 25, 1943, off Brazil. Made port and repaired.

George G. Henry—6,936 tons, tanker, 1917, 215154, LHFW. Pan American Petroleum & Transport Co. Commissioned *George G. Henry* in the Naval Overseas Transportation Service on August 23, 1918.

Tiger—6,273 tons, tanker, 1917, 215248, LHKN. Standard Transportation Co. Commissioned *Tiger* in the Naval Overseas Transportation Service on December 23, 1918. Operated during World War II by the Socony-Vacuum Oil Co. Sunk on April 1, 1942, by *U-754* off Cape Henry, Virginia.

Wilhelm Jebsen—7,204 tons, steel, 10.5 oil-burning tanker, 215635, LHWD. Keel laid on April 11, 1917, for Wilhelm Jebsen, of Norway. Requisitioned while building and delivered to the USSB on October 18, 1917. Operated by the Shipping Board for service to Mexico. Operated in World War II by Standard Oil of New Jersey as the *C.J. Barkdull.* Sunk by *U-632* on January 10, 1943, in the North Atlantic.

Fred W. Weller—10,626 tons, tanker, 1917, 215847, LJKP. Standard Oil of New Jersey. Weller was president of the Standard Oil Company of Louisiana.

J.E. O'Neill—7,150 tons, steel, 10.5 knot oil-burning tanker, 215903, LJNK. Keel laid on April 11, 1917, for the Atlantic Refining Co. Requisitioned while building and delivered to the USSB on January 10, 1918. James O'Neill was president of the Prairie Oil & Gas Co.

A.C. Bedford—10,614 tons, 10.5 knot oil-burning tanker, 215979, LJQM. Keel laid on April 21, 1917, for Standard Oil of New Jersey. Requisitioned while building and delivered to the USSB on February 14, 1918. Operated by Standard Oil. Alfred C. Bedford was chairman of the Standard Oil Company of New Jersey.

S.M. Spalding—7,175 tons, steel, 11 knot oil-burning tanker, 216019, LJRW. Keel laid on July 21, 1917, for the Pan American Petroleum & Transport Co. Requisitioned while building and delivered to the USSB on March 11, 1918. Allocated to the company for operation. Silsby Morse Spalding was a director.

Paul H. Harwood—7,192 tons, steel, 11 knot oil-burning tanker, 216222, LKPJ. Keel laid on March 8, 1917, for the Pan American Petroleum & Transport Co. Requisitioned while building and delivered to the USSB on May 1, 1918. Operated during World War II by Standard Oil of New Jersey. Torpedoed on July 7, 1942, off Louisiana. There were no casualties and the ship was salvaged.

Defiance—7,898 tons, steel, 11 knot oil-burning freighter, 216875, LMTD. Keel laid on May 21, 1918. Requisitioned from British owners while building and delivered to the USSB on September 5, 1918. Commissioned *Defiance* on September 5, 1918, in the Naval Overseas Transportation Service.

Dicto—3,743 tons, steel, oil-burning freighter. Built for Norwegian owners.

Oakland Estuary, Alameda

Redondo—4,113 tons, steel, 10 knot oil-burning freighter, 215963, LJPV. Keel laid on August 2, 1917. Requisitioned from Norwegian owners while building and delivered to the USSB on February 6, 1918. Redondo Steamship Co., New York.

Herbert L. Pratt—7,145 tons, steel, 10 knot oil-burning tanker, 215969, LJQD. Keel laid on May 29, 1917, for the Atlantic Refining Co. and apparently sold to Joseph Gabriel, of Philadelphia. Requisitioned while building and delivered to the USSB on March 11, 1918. The ship hit a mine off the East Coast on June 3 and was salvaged. Commissioned *Herbert L. Pratt* on June 10, 1918, in the Naval Overseas Transportation Service.

William M. Irish—7,150 tons, steel, 11 knot oil-burning tanker, 216249, LKQB. Keel laid on August 20, 1917, for the Atlantic Refining Co. Requisitioned while building and delivered to the USSB on April 30, 1918. Irish was president of Atlantic Refining.

W.M. Burton—7,170 tons, steel, 11 knot oil-burning tanker, 216508, LMBG. Keel laid on October 2, 1917, for the Atlantic Refining Co. Requisitioned while building and delivered to the USSB on June 25, 1918. William M. Burton was an officer of Atlantic Refining.

Liberator—7,951 tons, steel, 10.5 knot oil-burning freighter, 216541, LMCK. USSB contract ship. Keel laid on December 3, 1917, and delivered on July 2, 1918, 211 building days. Commissioned *Liberator* in the Naval Overseas Transportation Service on July 2, 1918. Operated during World War II by Lykes Bros. and sunk on March 19, 1942, by *U-332* off North Carolina.

Volunteer—7,955 tons, steel, 10.5 knot oil-burning freighter, 216762, LMND. USSB contract ship. Keel laid on January 8, 1918, and delivered on August 23, 1918, 226 building days. Commissioned *Volunteer* in the Naval Overseas Transportation Service on August 23, 1918.

W.S. Rheem—7,277 tons, steel, 11 knot oil-burning tanker, 216788, LMPF. Keel laid on December 10, 1917, for Standard Oil of California. Keel laid on June 5, 1916. Requisitioned while building and delivered to the USSB on August 17, 1917. Allocated to Standard Oil for service to Mexico. William S. Rheem was president of Standard Oil of California.

Challenger—7,955 tons, steel, 10.5 knot coal-burning freighter, 217003, LNDS. USSB contract ship. Keel laid on February 21, 1918, and delivered on October 4, 1918, 228 building days. Allocated to the Naval Overseas Transportation Service. Operated during World War II by the American South African Line. Sunk on May 17, 1942, by *U-155* off Grenada.

Victorious—7,881 tons, steel, 11 knot oil-burning freighter, 217060, LNHF. Keel laid on April 30, 1918. Requisitioned from British owners while building and delivered to the USSB on October 19, 1918. Commissioned *Victorious* in the Naval Overseas Transportation Service on October 19, 1918.

Invincible—7,888 tons, steel, 11 knot oil-burning freighter, 217093, LNJQ. Keel laid on July 4, 1918. Requisitioned from British owners while building and delivered to the USSB on October 17, 1918. Commissioned *Invincible* in the Naval Overseas Transportation Service on October 17, 1918.

Independence—7,893 tons, steel, 11 knot oil-burning freighter, 217172, LNPW. Keel laid on March 26, 1918. Requisitioned from British owners while building and delivered to the USSB on October 31, 1918. Commissioned *Independence* in the Naval Overseas Transportation Service on November 16, 1918.

Courageous—7,887 tons, steel, 11 knot oil-burning freighter, 217287, LPBC. Keel laid on July 4, 1918. Requisitioned from British owners while building and delivered to the USSB on November 27, 1918. Allocated to the American Line for service to the United Kingdom. Operated during World War II by Grace Line. Sunk on June 8, 1944, at Normandy, France, as part of a breakwater.

Coast Shipbuilding Co.
914 Lewis Building, Portland, Oregon

The company's yard was at the foot of Gibbs Street. Henry E. Pennell was president, Arthur M. "Art" Sherwood, Jr., vice-president, Charles E. McCulloch secretary, and Donald M. Green treasurer.

Barabos—2,564 tons, wood, 10 knot coal-burning Ferris freighter, 216886, LMTK, USSB contract ship. Keel laid on September 1, 1917, and delivered on September 12, 1918, 376 building days. Allocated to the New York & Porto Rico Steamship Co. for service to Scandinavia.

Barrington—2,564 tons, wood, 10 knot coal-burning Ferris freighter, 217025, LNFT. USSB contract ship. Keel laid on September 1, 1917, and delivered on October 23, 1918, 417 building days. Allocated to the Kerr Steamship Co. for service to the United Kingdom.

Columbia Engineering Works
Waldemere Station, Portland, Oregon

Columbia Engineering started in 1901 on Linnton Road and by 1917 the yard was on 13 acres with two covered building ways, three outdoor slips, and boiler and machine shops. The company built ships and tanks of all kinds. Samuel M. Mears was president, Walter Mackey vice-president, Art M. Mears, treasurer, and Charles W. Steele secretary. Charles Mears also ran the Portland Cordage Co. and lived with his wife Laura at 721 Flanders. The company normally employed 300 to 400 workers.

Diria—1,481 tons, 5-masted schooner, 1917, 215776, LJGB. Diria Ship Corp., New York.

Columbia River Ship Building Corp.
Foot of Arthur St., Portland, Oregon

Columbia River was incorporated in 1916 and opened for business the next year on 18 acres at the foot of Meade St. with five 550' ways. Alf F. Smith was president and also ran Smith & Watson Iron Works at 400 Front Ave. Joseph R. Bowles, president of the Northwest Steel Co., banker John C. Ainsworth, and J. Frank Watson were vice-presidents, Walter B. Beebe was secretary, and James McKinley general manager. Ainsworth was a prominent banker and president of the U.S. National Bank. The company's 6,000 workers built steel, turbine-powered Shipping Board freighters.

Westward Ho—5,814 tons, steel, 10.5 knot coal or oil-burning freighter, 215990, LJQW. Keel laid on March 26, 1917. Requisitioned from British owners while building and delivered to the USSB on March 4, 1918. Allocated to Luckenbach for the European Food Relief mission. Commissioned *Westward Ho* in the Naval Overseas Transportation Service on October 19, 1918.

Westbrook—5,748 tons, steel, 10.5 knot coal or oil-burning freighter, 216076, LJVG. Keel laid on April 26, 1917. Requisitioned from Norwegian owners while building and delivered to the USSB on March 30, 1918. Operated by the Shipping Board for service to France.

Westgate—5,799 tons, steel, 10.5 knot coal or oil-burning freighter, 216174, LKNC. Keel laid on May 26, 1917. Requisitioned from British owners while building and delivered to the USSB on April 23, 1918. Commissioned *Westgate* in the Naval Overseas Transportation Service on May 29, 1918. Sank in a collision with the steamer *American* on October 7, 1918, 500 miles off the East Coast with the loss of six crew out of 45 aboard.

Westgrove—5,815 tons, steel, 10.5 knot coal or oil-burning freighter, 216205, LKNQ. Keel laid on January 15, 1918. Requisitioned from British owners while building and delivered to the USSB on April 27, 1918. Allocated to American Line for service to the United Kingdom.

West Indian—5,828 tons, steel, 10.5 knot coal or oil-burning freighter, 216310, LKRC. USSB contract ship. Keel laid on November 20, 1917, and delivered on May 23, 1918, 184 building days. Commissioned *West Indian* in the Naval Overseas Transportation Service on May 22, 1918.

Western City—5,828 tons, steel, 10.5 knot coal or oil-burning freighter, 216372, LKSJ. USSB contract ship. Keel laid on January 29, 1918, and delivered on June 6, 1918, 128 building days. Allocated to the American-Hawaiian Steamship Co. as manager for service to Italy by the Phelps Bros.

West Coast—5,688 tons, steel, 12 knot coal-burning freighter, 216704, LMJD. USSB contract ship. Keel laid on March 2, 1918, and delivered on July 31, 1918, 150 building days. Allocated to J.H.W. Steele for service to the United Kingdom. Commissioned *West Coast* in the Naval Overseas Transportation Service on August 9, 1918.

Western Pride—5,688 tons, steel, 11.5 knot coal-burning freighter, 216787, LMPB. USSB contract ship. Keel laid on March 30, 1918, and delivered on August 26, 1918, 149 building days. Allocated to Williamson & Rauers for service to the United Kingdom.

Western Plains—5,688 tons, steel, 11.5 knot coal-burning freighter, 216976, LNBV. USSB contract ship. Keel laid on May 3, 1918, and delivered on October 17, 1918, 179 building days. Commissioned *Western Plains* in the Naval Overseas Transportation Service on October 17, 1918.

Western Belle—5,688 tons, steel, 11.5 oil-burning freighter, 217182, LNQM. USSB contract ship. Keel laid on July 22, 1918, and delivered on November 21, 1918, 124 building days. Allocated to James W. Elwell & Co. for service with the American Relief Administration. Commissioned *Western Belle* in the Naval Overseas Transportation Service on November 22, 1918.

West Cobalt—6,179 tons, steel, 10.5 knot coal or oil-burning freighter, 217341, LPDR. USSB contract ship. Keel laid on July 24, 1918, and delivered on December 22, 1918, 149 building days. Commissioned *West Cobalt* in the Naval Overseas Transportation Service on December 29, 1918.

Coos Bay Shipbuilding Co.
Marshfield, Oregon

Established in May 1917 at Marshfield on 12½ acres with four ways. Augustus E. Adelsperger was president, Ray H. Corey vice president, Al Y. Meyers treasurer, and John D. Goss secretary. James H. Polhemus was general manager. The normal workforce was 300. Polhemus was a civil engineer and after the war the controversial general manager of the Port of Portland. He was president of the Portland General Electric Co., and then owned the Willamette Tug & Barge Co.

The last of four wood freighters was the *Burnside*, delivered in 1919.

Coos Bay—3,011 tons, 10 knot coal-burning Hough freighter, 216801, LMQB. USSB contract ship. Keel laid on October 12, 1917, and delivered on August 27, 1918, 319 building days. Foundered off Lower California on September 11, 1918. No lives were lost out of the 48 on board.

Balliett—3,012 tons, wood, 10 knot coal-burning Hough freighter, 217021, LNFP. USSB contract ship. Keel laid on October 15, 1917, and delivered on September 30, 1918, 351 building days. Allocated to the Matson Navigation Co. for service to the Hawaiian Islands.

Marshfield—2,988 tons, wood, 10 knot coal-burning Hough freighter, 217354, LPFD. USSB contract ship. Keel laid on November 30, 1917, and delivered on December 19, 1918, 384 building days. Allocated to Matson for service to Hawaii.

Craig Shipbuilding Co.
Long Beach, California

John Craig was born in New York on December 24, 1838, made one sea voyage under sail, then became a ship designer and builder in Toledo, Ohio, as the Craig Ship Building Co.

In 1905, after he delivered the 199-ton *R.E. Doville*, John started thinking seriously about moving his modest shipbuilding operation as it was becoming difficult to compete with the larger yards. John told his wife Anna he wanted to move the company to sunny California.

Alfred H. Merrell was secretary of Craig Ship, and John's son, John F., 37, was treasurer and lived at 2351 Monroe with his wife Edith, 32, daughter Ruth, 13, John F., Jr., 11, and James G., 9.

The family moved to Long Beach and John and Anna settled in at 905 Atlantic Ave. The yard was located at Water and River Sts. The company's first delivery was the steam towboat *Virgil G. Bogue* in 1909 followed the next year by the 250' freighter *General Hubbard* for the Hammond Lumber Co. The *Hubbard* was the first steel freighter built south of San Francisco.

In 1915, Frederick W.G. Phillips was treasurer and George Craig was a clerk. In 1917, James G. Craig was president and treasurer, John, Jr., was vice-president and secretary, and Thomas R. Merrell general manager. In May 1917 the Craig Shipbuilding Co. was incorporated at Long Beach with a capital of $50,000. John F. was vice president and general manager.

In 1918 the Craigs established the Long Beach Shipbuilding Co. at the same location to build steel Shipping Board freighters. James G. was president and secretary. John F., Sr., was chairman of the board, John F., Jr., was vice president, married to Martha, and lived at 525 W. 8th. Thomas R. Merrell was general manager.

The company's last vessel was the motor yacht *Samona II* in 1931. The business was revived for World War II.

Alvarado—2,053 tons, wood, freighter, 1914, 212017, LDJS. Pacific Transport Co., San Francisco.

Silverado—2,328 tons, steel, 9 knot oil-burning freighter, 216304, LKJT. Keel laid on

June 26, 1917. Requisitioned while building and delivered to the USSB on May 16, 1918. Allocated to Matson for service to Hawaii.

The Foundation Co.
New York

The company headquarters were in New York City. They had Pacific Coast yards at Portland, Oregon, and Tacoma, Washington.

Portland

Theodore Knudson was superintendent. The yard was at Front Ave. at Nicholai.

Nancy—2,117 tons, 5-masted auxiliary schooner, 1918, 224628, MFJK. Built for the French and delivered before documentation.

Tacoma

Established on September 1, 1917, on 100 acres next door to Todd Drydock with ten 280' building ways. Henry A. Rhodes was president, John S. Baker vice-president, Joseph S. Whitehouse, secretary, and Arthur G. Prichard, treasurer at their 303 Tacoma Bldg. office. About 4,000 were employed to build auxiliary schooners for the French government.

Robin Hood—1,983 tons, schooner, 1918, 223102. Built for the French.

Thann—2,102 tons, 5-masted schooner, 1918, 223311. Launched on July 12, 1918. Built for the French.

Fulton Shipbuilding Co.
Wilmington, California

Everett H. Seaver was born in Salina, Kansas, on September 2, 1886, and became prominent in the grain business. In February 1909 he married Gertrude Sharp and in 1911 they moved to California. Everett became secretary-treasurer of the California Drug & Chemical Co. in Wilmington and in 1915 he was president of the Catalina Island Company. In 1917 he was general manager of the C.H. Sharp Manufacturing Co., stove makers at 1312 E. 12th in Los Angeles.

Charles Everett Fulton was born in Port Ludlow, Washington, on February 17, 1871, a descendent of Robert Fulton of steamboat fame. He went to work in shipyards and on February 16, 1896, at Tacoma married a Norwegian girl from Chicago, Carrie Linnie Johnson. In 1899 they had a daughter Clara May then a son Robert W. In 1904 the family moved to San Pedro, California, where Charles went to work for the Wilmington Transportation Co.

In the spring of 1905 Charles teamed up with Peter L. Iversen as Fulton & Iversen and delivered the gas vessel *City of Long Beach* on Terminal Island in 1906. By 1914 Charles was working with William Woodley as Fulton & Woodley, boat builders on Mormon Island at Wilmington. During this time, Fulton's wife took off and he was granted a divorce on February 8, 1916, on the grounds of desertion. He was on his own, working as the Wilmington Shipbuilding Co. and delivered the gas vessel *Aliceil* and steam schooner *Edna Christenson*.

In July of 1917 Everett H. Seaver organized the Fulton Shipbuilding Co. with Charles as president but in January 1918 Seaver took over as president and general manager. This might have had something to do with the divorce as Carrie had filed a claim against Charles' substantial estate.

The *Lucinda Hanify, Jane Nettleton, Edna Christenson,* and *Whitney Olson* were 235' "steam lumber schooners" designed on a concept devised by Edwin A. Christenson, president of Sudden & Christenson. The 16–27–46-inch triple expansion engine's boiler was on the main deck in the midship house. There were two holds forward and one aft. These two ships and two others built by W.F. Stone in Oakland were sold to the French government on December 6, 1917.

Edna Christenson—1,497 tons, steam schooner, 1917, 215494, LHRS. Western Mercantile Marine Corp. (Sudden & Christenson), San Francisco. Sold to the French.

Lucinda Hanify—1,497 tons, steam schooner, 1917. Sold to the French.

Jane Nettleton—1,545 tons, steam schooner, 1917. Sold to the French.

Whitney Olson—1,558 tons, steam schooner, 1917. Sold to the French and named *Ghislaine*. Oliver Olson purchased the ship and it arrived at San Francisco from Dunkirk in March 1922 and operated under the original name.

Yehama—3,129 tons, wood, 10 knot coal-burning Hough freighter, 216711, LMJP. USSB contract ship. Keel laid on July 15, 1917, and delivered on August 26, 1918, 407 building days. Operated by Archibald L. Stark.

Catawha—3,129 tons, wood, 10 knot coal-burning Hough freighter, 216872, LMSW. USSB contract ship. Keel laid on August 25, 1917, and delivered on October 4, 1918, 404 building days. Allocated to the Potter Transportation Co. as manager and operated by the Triangle Steamship Co. for service to France.

G.M. Standifer Construction Corp. Vancouver, Washington

In 1915, Guy M. Standifer was president of the Flagg & Standifer Co., railroad and general contractors in Portland, Oregon. He lived with his wife Marguerite at the Mallory Hotel.

In 1916 he organized the G.M. Standifer Construction Corp. with Lee B. Menefee as vice-president, Romeo V. Jones, Jr., treasurer, and James F. Clarkson general manager. Clarkson was the manager of the Oregon Bridge & Construction Co. in Portland. At the same time, Standifer and Clarkson opened the Standifer-Clarkson Shipyard, Inc., in North Portland. Guy and Marguerite moved to 475 NE 18th Ave. and James and his wife Lucy lived at 689 NE 21st.

On January 1, 1918, Standifer Construction got a Shipping Board contract and started work on a "steel shipyard" in Vancouver, Washington. The Port of Vancouver planned a special election to vote on bonding the 50-acre site and leasing it to the company but they didn't start building steel ships until 1919.

Kineo—2,573 tons, wood, 10 knot coal-burning Ferris freighter, 216940, LMWP. USSB contract ship. Keel laid on September 8, 1917, and delivered on October 16, 1918, 403 building days. Allocated to the International Freighting Corp. for service to the Netherlands.

Belding—2,573 tons, wood, 10 knot coal-burning Standifer freighter, 217141, LNMK.

USSB contract ship. Keel laid on September 6, 1917, and delivered on December 26, 1918, 508 building days. Allocated to Matson as manager for service to Spain by Luckenbach.

Geo. F. Rodgers & Co.
Astoria, Oregon

Established on August 1, 1917, with George Rodgers as president, Ignatz Lowengart vice-president, and DeLue M. Field secretary-treasurer. Field was a dentist in Salem with Thomas C. Smith, Jr., as Smith & Field. He lived in a rooming house at 248 N. Summer Street. Lowengart was a prominent wholesale milliner in Portland. The yard was at Pier 2, Municipal Docks, one block west of the Port of Astoria docks. The company built one Shipping Board freighter homeported at Astoria.

Blue Eagle—2,559 tons, wood, 10 knot oil-burning Ferris freighter, 217181, LNQK. USSB contract ship. Keel laid on November 3, 1917, launched on July 4, 1918, 245 days later, and as of May 31, 1919, had not been delivered. Allocated to the Columbia Pacific Shipping Co. for service to the United Kingdom.

Grant Smith-Porter-Guthrie Co.
Portland, Oregon

Grant Smith, of St. Paul, Minnesota, Eric V. Hauser, of Portland, and Harry H. Hunt, of Spokane, Washington, were general contractors in business as the Grant Smith Co. headquartered in the Multnomah Hotel at 275 Pine Street. Smith was president, Hauser vice-president and general manager, and Hunt secretary-treasurer. The company built railroads and Hauser was president of the Multnomah Hotel, Inc. He lived in the hotel and is the namesake for Hauser, Oregon.

John D. Porter, of Spokane, Washington, was president of the Porter Bros Co. at 524 Railway Exchange Bldg. in Spokane. Andrew Porter, of 728 Tillamook St. in Portland, was a vice-president, Richard B. Porter was a vice-president, and Art G. Osland, of Portland, was asst. secretary-treasurer. They were railroad contractors.

Grant Smith was already in the spruce lumber business with the Porter brothers as the Smith-Porter Bros. Co. in Spokane.

In July 1917 the company acquired tidal acreage on the Willamette River at the foot of W. Baltimore St. in St. Johns, north Portland, dredged almost a half a million cubic yards of mud and silt, drove over 200,000 piles, and built a complete 8-way shipyard.

The company built the most wooden ships of any yard. Between February and November 1918 they delivered 25 ships for the Shipping Board. It was a popular workplace for draft-age men from around the state.

In June 1917 Smith and Porter purchased the Aberdeen Shipbuilding Co. in Washington, finished five ships then under construction, added four ways, and on March 17, 1918, launched the *Manada*. By November 10 more Ferris type ships were completed.

Portland

Wasco—3,044 tons, wood, 10 knot coal-burning Hough freighter, 216398, LKTH. USSB contract ship. Keel laid on August 22, 1917, and delivered on June 30, 1918, 312 building days. Allocated to Callaghan, Atkinson as manager for time charter to the Triangle Steamship Co.

Biloxi—3,021 tons, wood, 10 knot coal-burning Hough freighter, 216490, LKWT. USSB contract ship. Keel laid on September 10, 1917, and delivered on September 26, 1918, 289 building days. Allocated to J.A. Elliott as manager and operated by the Triangle Steamship Co. on a time charter.

Kasota—3,017 tons, wood, 10.5 knot coal-burning Hough freighter, 216550, LMCQ. USSB contract ship. Keel laid on September 8, 1917, and delivered on June 30, 1918, 295 building days. Allocated to the Pacific Mail Steamship Co. as managers for service to Australia by Robert Dollar.

Blandon—3,017 tons, wood, 10 knot coal-burning Hough freighter, 216564, LMCV. USSB contract ship. Keel laid on October 25, 1917, and delivered on July 17, 1918, 265 building days. Allocated to Pendleton Bros. as managers and operated by Triangle Steamship.

Boilston—3,017 tons, wood, 10 knot coal-burning Hough freighter, 216602, LMFK. USSB contract ship. Keel laid on October 31, 1917, and delivered on July 26, 1918, 268 building days. Allocated to Pendleton Bros. as managers and operated by Triangle Steamship.

Callooh—3,017 tons, wood, 10.5 knot coal-burning Hough freighter, 216632, LMGK. USSB contract ship. Keel laid on November 8, 1917, and delivered on August 7, 1918, 272 building days. Allocated to Potter Transportation as manager for service to France by Triangle Steamship.

Moritz—3,019 tons, wood, 10 knot coal-burning Hough freighter, 216703, LMJC. USSB contract ship. Keel laid on November 12, 1917, and delivered on August 12, 1918, 273 building days. Allocated to Charles T. Magee as manager for service to the West Indies by Triangle Steamship.

Dumaru—3,019 tons, wood, 10 knot coal-burning Hough freighter, 216729, LMKB. USSB contract ship. Keel laid on November 17, 1917, and delivered on August 28, 1918, 284 building days. Allocated to Sudden & Christenson for service in the Pacific. Chartered by the Navy to carry gasoline and munitions to Honolulu, Guam, and Manila. Hit by lightning off Guam on October 16, 1918, caught fire and exploded. Of the 60 on board, 13 were lost.

Wakan—3,019 tons, wood, 10 knot coal-burning Hough freighter, 216783, LMNV. USSB contract ship. Keel laid on February 19, 1917, and delivered on September 4, 1918, 197 building days. Allocated to the W.F. Spice & Co. for service to the Netherlands.

Caponka—3,019 tons, wood, 10 knot coal-burning Hough freighter, 216784, LMNW. USSB contract ship. Keel laid on February 25, 1918, and delivered on September 13, 1918, 200 building days. Allocated to Sudden & Christenson as manager for operation by Struthers & Dixon in the Philippine Inter-Island trade.

Holbrook—3,019 tons, wood, 10 knot coal-burning Hough freighter, 216830, LMRH. USSB contract ship. Keel laid on March 15, 1918, and delivered on September 30, 1918, 198 building days. Allocated to the New Orleans & South American Steamship Co. for service to the West Indies.

Kuwa—3,019 tons, wood, 10 knot coal-burning Hough freighter, 216829, LMRG. USSB contract ship. Keel laid on March 7, 1918, and delivered on September 21, 1918, 197 building days. Allocated to the Texas Transport & Terminal Co. for service to the United Kingdom.

Nashotah—2,594 tons, wood, 10 knot coal-burning Ferris freighter, 217034, LNGF. USSB contract ship. Keel laid on March 28, 1918, and delivered on October 14, 1918, 199 building days. Allocated to Page & Jones for service to the United Kingdom.

Neeolah—2,594 tons, wood, 10 knot coal-burning Ferris freighter, 217096, LNJS. USSB contract ship. Keel laid on April 4, 1918, and delivered on October 23, 1918, 202 building days. Allocated to Luckenbach for service to Scandinavia.

Nupolela—2,594 tons, wood, 10 knot coal-burning Ferris freighter, 217160, LNJV. USSB contract ship. Keel laid on April 13, 1918, and delivered on October 31, 1918, 201 building days. Allocated to Luckenbach for service to Scandinavia.

Tillamook—2,594 tons, wood, 10 knot coal-burning Ferris freighter, 217174, LNQC. USSB contract ship. Keel laid on April 19, 1918, and delivered on November 6, 1918, 201 building days. Allocated to Williams-Dimond for service to the United Kingdom.

Aberdeen

Manada—2,542 tons, wood, 10 knot coal-burning Ferris freighter, 216750, LMKR. USSB contract ship. Keel laid on August 30, 1917, and delivered on August 31, 1918, 366 building days. Allocated to the Pacific Steamship Co. as manager for service to the United Kingdom by Swayne & Hoyt.

Boxley—2,532 tons, wood, 10 knot coal-burning Ferris freighter, 216944, LMWR. USSB contract ship. Keel laid on December 6, 1917, and delivered on September 19, 1918, 283 building days. Allocated to the New Orleans & South American Steamship Co. for service to the east coast of South America.

Bancraft—2,537 tons, wood, 10 knot coal-burning Ferris freighter, 216952, LNBD. USSB contract ship. Keel laid on October 26, 1917, and delivered on September 30, 1918, 340 building days. Allocated to the American Shipping Corp. for service to Italy.

Wihaha—2,537 tons, wood, 10 knot coal-burning Ferris freighter, 217173, LNQD. USSB contract ship. Keel laid on March 23, 1918, and delivered on November 1, 1918, 222 building days. Allocated to the Paragon Shipping Co. for service to the Netherlands.

Itanca—2,537 tons, wood, 10 knot coal-burning Ferris freighter, 217187, LNQS. USSB contract ship. Keel laid on April 29, 1918, and delivered on December 13, 1918, 228 building days. Allocated to J.A. Elliott & Co. for service to the Netherlands.

Moraine—2,537 tons, wood, 10 knot coal-burning Ferris freighter, 217188, LNQT. USSB contract ship. Keel laid on February 2, 1918, and delivered on December 3, 1918, 306 building days. Allocated to the New Orleans & South American Steamship Co. as manager for service to the United Kingdom by the Gulf European Steamship Co.

Grays Harbor Shipbuilding Co.
1900 3rd Ave., Seattle

Albert Schubach was a shipping and commission merchant at 609 1st Ave. and lived with his wife Julia at 1212 21st Ave. N. Eugene R. and Montegue R. "Monty" Ward, were boat builders as Ward & Sons at 24 Colman Dock.

Grays Harbor Ship was incorporated in 1916 with a capital of $50,000 with Schubach as president, Ward, 26, vice-president, attorney Bruce C. Shorts, secretary, and W.J. Patterson, treasurer. The 8-way yard was built on 12 acres at Aberdeen, Washington.

In March 1917, the company incorporated the Grays Harbor Motor Ship Co. to build wood Shipping Board freighters. They enlarged their existing facilities and constructed a new plant at a cost of $323,303.27 and another $133,565.39 for new machinery.

On May 19 and September 12 the company signed contracts to build 25 hulls. Separate contracts covered the installation of the engine and all other auxiliary machinery in each hull at a flat rate of $65,000 per ship.

The freighters were a "Grays Harbor" design by Montegue Ward as Shipping Board Design No. 1005. The well deck freighter had two holds forward and two aft with a single midship house and accommodations for a crew of 46.

Santino—2,491 tons, wood motor freighter, 1917, 214850, LGTS. Gaston, Williams & Wigmore.

Grays Harbor—2,373 tons, wood motor freighter, 1917, 215389, LHNV. Gaston, Williams & Wigmore. Burned on August 19, 1918, off Fernando Noronha Island, Brazil. Two of the 31 crew were lost.

Mount Rainier—2,397 tons, wood motor freighter, 1917, 215798, LJGQ. Gaston, Williams & Wigmore.

Balestrand—2,403 tons, wood motor freighter, 1917, 215877, LJMH. Balestrand Navigation Corp. The company sold the ship the next year.

Hjeltenaes—2,387 tons, wood motor freighter, 1918, 216104, LJWD. Launched on November 3, 1917. Engines installed in Seattle. Hjeltenaes Motor Ship Co., Portland.

Marie De Ronde—2,415 tons, 5-masted topsail auxiliary schooner, 216188, LKNF. Built for John A. Donald's Donald Steamship Co. Launched on January 24, 1918. In July 1917, Philip de Ronde purchased 85 percent of the stock of Donald Steamship and became president. De Ronde signed a $475,000 contract to build the ship but it cost $441,331.42 to build the vessel, realizing a profit of $33,668.58.

Fanestrand—2,557 tons, wood motor freighter, 1918, 216370, LKSG. Fanestrand Navigation Corp., New York. Launched on November 3, 1917, and finished at the Grant Smith-Porter yard in Aberdeen. Loaded cargo at Seattle and sailed for the Orient.

Wishkah—2,924 tons, wood, 10 knot coal-burning Grays Harbor freighter, 216576, LMDH. USSB contract ship. Keel laid on July 18, 1917, and delivered on June 30, 1918, 347 building days. Allocated to the Tropical Fruit Co. for service to the West Indies.

Kaskaskia—2,934 tons, wood, 10 knot coal-burning Grays Harbor freighter, 216653, LMGT. USSB contract ship. Keel laid on September 7, 1917, and delivered on July 27, 1918, 323 building days. Allocated to the Export Steamship Co. for service to France.

Blackford—2,950 tons, wood, 10 knot coal-burning Grays Harbor freighter, 216745, LMKJ. USSB contract ship. Keel laid on September 10, 1917, and delivered on August 23, 1918, 347 building days. Foundered off Lower California on September 17, 1918, with 54 on board. No lives were lost.

Abrigada—2,929 tons, wood, 10 knot coal-burning Grays Harbor freighter, 216880, LMTJ. USSB contract ship. Keel laid on June 4, 1917, and delivered on August 31, 1918, 453 building days. Allocated to Callaghan, Atkinson to France. Burned on December 31, 1919, at St. Michaels, Azores.

Bromela—2,962 tons, wood, 10 knot coal-burning Grays Harbor freighter, 216943, LMWQ. USSB contract ship. Keel laid on January 11, 1918, and delivered on September 21, 1918, 256 building days. Allocated to Lykes Bros. for service to the United Kingdom.

Brompton—2,940 tons, wood, 10 knot coal-burning Grays Harbor freighter, 217027, LNFV. USSB contract ship. Keel laid on February 9, 1918, and delivered on December 31, 1918, 325 building days. Allocated to W.R. Grace for service to the United Kingdom.

Aberdeen—2,917 tons, wood, 10.5 knot coal-burning Grays Harbor freighter, 217063, LNHK. USSB contract ship. Keel laid on September 9, 1918, and delivered on October 6, 1918, 27 building days. Allocated to M.H. Tracy as manager, M.E. Kinsley as operator on a time charter.

Broncho—2,940 tons, wood, 10 knot coal-burning Grays Harbor freighter, 217201, LNRP. USSB contract ship. Keel laid on February 20, 1918, and delivered on December 7, 1918, 292 building days. Allocated to W.R. Grace as manager for service to the United Kingdom by S. Sgitcovich & Co.

Toward the end of 1918, it became apparent that some of the ships contracted for would not be needed so the contract signed on September 12, No. 468 W.H., for eight hulls and the separate contracts for the installation of machinery, was modified on November 23. Four of the hulls, 2681–2684, were cancelled and all of the machinery contracts were cancelled. The loss on the cancelled contracts, the new federal income taxes, and the usual capriciousness of government officials caused a financial quagmire and endless headaches for the company.

All of the ships the company built and delivered are listed below with the date of delivery, Gray Harbor's cost to build the ship, and the profit realized. The delivery date might differ slightly from the Shipping Board date, which would probably be the formal acceptance date.

EFC Hull No.	Name	Date Delivered	Cost	Profit
55	*Abrigada*	Aug. 3, 1918	$360,607.90	$108,428.62
56	*Wishkah*	Aug. 3, 1918	$355,700.55	$120,180.10
57	*Kaskaskia*	Jul. 26, 1918	$359,275.04	$109,802.57
58	*Blackford*	Aug. 23, 1918	$362,698.75	$100,389.67
998	*Bromela*	Sept. 21, 1918	$370,481.96	$141,447.84
999	*Brompton*	Dec. 31, 1918	$380,732.38	$131,643.55
1000	*Broncho*	Dec. 7, 1918	$363,877.44	$138,610.71
1001	*Brookdale*	Jan. 9, 1919	$383,043.14	$123,660.30
1059	*Brookfield*	Apr. 3, 1919	$413,941.15	$78,122.48
1060	*Brookhaven*	Apr. 23, 1919	$404,668.54	$78,838.65
1061	*Brookland*	Jun. 10, 1919	$425,658.21	$59,000.90
1062	*Brookside*	Jul. 3, 1919	$422,713.29	$60,335.65
1892	*Grayling*	Jul. 16, 1919	$408,596.75	$74,390.19
1893	*Gray Eagle*	Aug. 6, 1919	$392,962.85	$91,017.47
1894	*Mannahocking*	Sept. 5, 1919	$423,961.62	$61,428.40
1895	*Gray Cloud*	Sept. 26, 1919	$384,612.41	$102,822.64
1896	*Aberdeen*	Oct. 6, 1919	$421,788.37	$72,018.88
2677	*Adria*	Jun. 4, 1919	$311,589.09	$141,885.05

EFC Hull No.	Name	Date Delivered	Cost	Profit
2678	*Agron*	Jun. 19, 1919	$268,772.25	$179,266.40
2679	*Agathon*	May 19, 1919	$237,883.65	$146,135.74
2680	*Agylha*	Jul. 18, 1919	$254,969.59	$190,107.61
2681–2684			Cancellations	$199,888.23

Grays Harbor was paid $452,373.38 for Hull 2677, the *Adria*; $447,515.99 for 2678, the *Agron*; $383,995.88 for 2679, the *Agathon*; and $444,562.46 for 2680, the *Agylha*, for a total of $1,728,447.71.

The company could have built the cancelled hulls, 2681–2684, for the same or less money as the previous four and they had planned to use the expected profits for plant expansion and improvements but after the cancellation the plant and machinery were junked and sold for $38,844.90.

Materials had already been purchased for the cancelled hulls and the company was instructed by the Shipping Board to inventory all of the left over material. The Board subsequently found the inventory to be correct but gave no instructions as to its disposition. The inventory included $189,473.83 worth of miscellaneous materials; $8,752.55 for lumber; and $37,894.76 in freight charges. The total came to $236,121.14. A number of clinch rings were also ordered from the National Malleable Castings Co. in Chicago for $3,809.75. They were paid for but never delivered and the company's money was never returned.

In 1918 the company claimed net earnings of $206,372.46. Their taxes were $160,218.91 of which $103,231.06 was paid.

On May 29, 1919, the company was notified that they would be held responsible for any material belonging to the Shipping Board for use on the cancelled contracts.

A huge amount of lumber was located at three different West Coast shipyards. It was anticipated that much of it would become surplus from the wood ship building program due to the cancellation of contracts but would still be useful for shipbuilding.

William A. Magee was vice-president and general manager of the Winslow Marine Railway & Shipbuilding Co. and during the war he was the Shipping Board's assistant district manager in Seattle. On June 24, 1919, Magee signed a contract with the EFC to purchase up to 5,000,000 board feet of surplus No. 1 common or better Douglas fir, at $17.50 a thousand board feet. The shipyards would use whatever they needed to complete their contracts and Magee would buy whatever was left over. A few days later, Magee assigned and sold the contract to Grays Harbor.

The EFC delivered 2,529,097 board feet of No. 1 common or better Douglas fir to Grays Harbor at $17.50 a thousand for the construction of their ships still on the ways and demanded $44,259.21. Grays sent a check for $7,371.84 as partial payment but the check was never cashed.

On August 9, 1919, Grays demanded the balance of the lumber, which was then 1,038,846 board feet, but the EFC refused to deliver it. The market price on that date was $41.08 a thousand b/f which meant Gray's was owed $19,763.22 on the contract based on market value but they were never paid.

The company's earnings for 1919 were $406,981.34. The taxes were $317,999.29 and the company paid $213,155.34.

Disputes over the vessels ensued even though government inspectors were present at all times during construction. Each ship was taken on a 200-mile shakedown cruise, dry docked and inspected again, and then taken on another trip.

A fire damaged the *Agathon* before delivery and the company collected $43,836.17 in insurance money but spent $4,938.87 cleaning up after the fire. Fire damage to the plant plus the contract cancellations cost the company $276,905.71 and $62,135.03 for machinery installation.

After the *Gray Eagle* was delivered, the EFC sent the ship to Todd Dry Dock Co. at Seattle to have fair weathers on the after side of the strut palms installed. Grays Harbor was not bound by their contract to do this work but the EFC deducted $230.44 from Gray's payment. They also wanted non-contract watertight sliding doors installed on the *Gray Eagle* at a cost of $1,436.34, which was never paid.

The EFC sent the *Brookside* up to Todd for drydocking and additional work at a cost of $1,966.25, which was deducted from Gray's payment.

The *Grayling* was sent to the Pacific Construction Engineering Co. in Seattle to have the galley modified at a cost of $197.94 of which $138.65 was at Gray Harbor's expense.

During construction of the *Gray Cloud*, the EFC ordered certain work on the hull stopped until the machinery was installed then withheld $5,000 to insure completion of the hull work but after the hull was completed, the company was never paid. They wanted non-contract watertight sliding doors installed at a cost of $1,436.34 which was never paid. All the ships were being built according to Lloyd's classification standards. The inspection survey and certificates cost $874.25. The company was to be reimbursed by the EFC but the *Gray Cloud* was never paid for.

Sometime prior to June 8, 1918, the EFC ordered extra work done at a cost of $3,350 which was never paid. Bill plates installed on the *Adria*, *Agron*, and *Agylha* at $780 per vessel was never paid and waterproof hatch cover tarps for the *Mannahocking* that came to $260.74 was never paid.

All the machinery was ordered by the EFC and was to be delivered to Grays Harbor prepaid by the EFC was very erratic and sometimes 10 or 15 railroad cars would arrive on the same day unpaid for. The railroad refused to release the shipments until they were paid and also charged demurrage for the delay in unloading. The EFC paid the freight but not the demurrage. The company paid $571.65 on this total but was never reimbursed.

The EFC sent their steamer *Kenosha*, built by Sanderson & Porter, to Grays Harbor to have some minor work done. The company billed the $36.90 but they refused to pay because they thought it was $3.00 too much.

Grays Harbor contracted for locust wood squares for fastenings. Part of the order was paid for at the agreed upon price before the Shipping Board commandeered the supplier's plant then turned around and billed Grays Harbor $4,503.87 more than the original price.

Gray's Harbor bought and paid for a supply of ship's felt. The EFC wrote to the company and ordered them to ship part of the felt to the Ames Shipbuilding Corporation which was building Shipping Board vessels on a cost-plus basis. Gray's Harbor was promised reimbursement but was never paid the $1,075.20 for the felt, crating, and shipping to Ames.

The EFC was supposed to order and supply all fittings but when they were unable to do so, Grays was ordered to buy it locally any way they could to keep production going. Grays contracted with the Seattle Plumbing Supply Co. for a shipment of custom made valves. Just as the plumbing company was ready to ship the order, the EFC obtained a supply and delivered them to Grays Harbor. Grays cancelled the order with Seattle Plumbing but the two could not agree on an equitable cancellation fee. Seattle Plumbing sued for $2,468.60, the full amount of the order and they were awarded $1,600. Grays Harbor was never reimbursed.

The contracts provided for protection of Grays Harbor against an increase in freight rates. On June 25, 1918, a 25 percent increase went into effect costing the company $11,783.50. The increase was approved for reimbursement but was never paid.

Grays Harbor was paid a total of $10,223,975.52 on the hull and machinery contracts. The company's total expenses were $7,714.534.43 leaving a net profit of $2,509,441.09. Complicated tax issues also intervened. Another complication was the loss by fire of the plant and the *Agathon*; the discrepancy over the delivery date and tax return on the *Brompton*; and the loss of the *Blackford* in a storm before official Lloyd's classification was issued. The ship was fully inspected and accepted by the Emergency Fleet Corp.

In May 1920, the company requested permission to use portions of the materials. Andrew F. Wooley, District Adjuster of the Construction Claims Board, wrote back saying that was OK with him so long as the materials were deducted from any future claims the company made. Materials used amounted to $16,000.

Earnings in 1920 were $102,996.12 and taxes $63,761.31, of which none was paid. The company went back and forth with the Commissioner of Internal Revenue over their wartime taxes for years. The company had claimed the *Brompton* was delivered during 1919 but a tax audit found it was actually delivered in 1918. More re-determining of taxes and amended returns followed.

In early 1922, the company wanted to get rid of the material to reduce the storage charges on it and received permission to auction it off. The sale was advertised in newspapers in Seattle and Portland. On February 8, 1922, with Emergency Fleet Corp. officers present, the lot was sold for $24,100 to William Rosenbranz, a dealer in secondhand and surplus items.

The company spent $16,872.22 to store and guard the material up to February 8, 1922. The Shipping Board then audited the company's books related to handling, sorting, warehousing, oiling, painting, covering, and insuring this material and offered $38,198 as a reasonable reimbursement but that was never paid. The market at that time was flooded with surplus equipment and materials since there were hardly any ships being built on the West Coast.

On December 18, 1923, Grays Harbor claimed they overpaid $116,341.57 from 1917 to 1920 from a total of $316,341.57. No action was taken by the Commissioner of Internal Revenue. Grays Harbor sued in federal court to recover the tax money.

The ships were delivered, accepted and only $3,000 was spent by the EFC to fix minor problems, of which Grays Harbor had no knowledge, as they did not do the work. They weren't even notified or contacted until their claim for overpayment of taxes was filed.

After a complicated trial, Grays Harbor was awarded $247,732.48 on the cancelled contracts and $36,066.71 for unpaid services. The government was awarded $19,793.22 on the lumber deal and $537,440.37 for unpaid taxes in 1918 and 1919. The government ended up with a net gain of $273.404.40.

On April 28, 1924, the government filed a counterclaim against Grays Harbor for $409,807.97 alleging faulty construction over the loss of the *Blackford*.

In September 1918, the *Blackford* embarked on its maiden voyage from San Francisco to the west coast of South America in ballast when they got caught in a hurricane off the border between California and Mexico along with several other vessels. When the rudder and propeller were smashed on September 17, the crew of 46 abandoned ship. The boats were picked up on the 23rd by a passing tanker. William A. Brown, 1st mate on the *Blackford*, arrived at San Diego on the 26th on a launch from Magdalena Bay and said the *Blackford* foundered 250 miles south of Magdalena Bay. The *Blackford*, however, did not sink and on September 25 went ashore on the Mexican coast but could not be salvaged.

No report of leaks was made by the master or crew and no attempt was made to fix the leak. Water was shipped into the *Blackford* faster than it could be pumped out because, as Grays Harbor alleged, faulty pumps were supplied by the Shipping Board. The ship was inspected by a Shipping Board surveyor who would not release his report to Grays Harbor citing confidentiality, but he did tell them the hull was tight and sound and that the casualty must have been the crew's fault. A board of inquiry was conducted by the U.S. Steamboat Inspection Service in Seattle and the ship's officers were all found guilty of negligence.

In 1919 the company built three barkentines on speculation, the *Forest Pride, Forest Dream*, and *Forest Friend* at a cost of $731,199.68. The vessels were sold in late 1920 for $300,000 for a loss of $431,199.68.

Hammond Lumber Co.
260 California St., San Francisco

Andrew Benoni Hammond was born in New Brunswick, Canada, in 1848 and at 16 went to work in the woods in Maine and then Pennsylvania. In 1897 he moved to Montana and became a successful merchant and railroad builder there, in Washington, and then in California. In 1900 he purchased the John Vance redwood sawmill at Samoa along with a rail line. He used steam schooners to haul lumber coastwise and tugs to tow huge log rafts.

He incorporated Hammond Lumber in 1912 and built a shipyard at Samoa on Humboldt Bay. Hammond was president. His daughter Grace married an attorney, William S. Burnett, who became vice-president. Louis C. Stewart was secretary, and shipwright William P. McDade was superintendent.

Keota—2,512 tons, wood, 10 knot coal-burning Ferris freighter, 216835, LMRN. USSB contract ship. Keel laid on August 6, 1917, and delivered on September 12, 1918, 401 building days. Allocated to J.A. Elliott & Co. for service to France.

Bloomington—2,572 tons, wood, 10 knot coal-burning Ferris freighter, 216970, LNBQ. USSB contract ship. Keel laid on August 14, 1917, and delivered on November 5, 1918, 447 building days. Allocated to the Pensacola Shipping Co. for service to the United Kingdom.

Hanlon Dry Dock & Shipbuilding Co.
244 California St., San Francisco

In 1915, Daniel J. Hanlon was a boatbuilder in San Francisco and lived with his wife Catherine at 124 Duboce. Carl J. Foard was president of the Foard-Barstow Ship Chandlers Co. at 25–29 Drumm. In 1916 the two formed the Hanlon Dry Dock & Shipbuilding Co. at the foot of 5th Ave. in Oakland. Hanlon was president, Foard vice-president and treasurer, and Everett L. Cramer secretary. David C. Young was general manager, and marine engineer Louis Wilhelmson, was chief engineer. The company built 15 vessels through 1925.

Flagstaff—2,101 tons, schooner, 1917, 214971, LGWT. Name changed to *Falketind* in 1918.

Governor John Lind—3,431 tons, steel, 11 knot oil-burning freighter, 216871, LMSV. Keel laid on January 12, 1918. Requisitioned while building and delivered to the USSB on September 10, 1918. Allocated to the Lind Navigation Co. for service to the east coast of South America.

Major Wheeler—3,435 tons, steel, 11 knot oil-burning freighter, 217046, LNGS. Keel laid on January 14, 1918. Requisitioned from British owners while building and delivered to the USSB on September 30, 1918. Commissioned *Major Wheeler* in the Naval Overseas Transportation Service on October 8, 1918. Operated during World War II by the Baltimore Insular Line. Sunk on February 6, 1942, by *U-107* off the East Coast.

J.F. Duthie & Co.
Seattle, Washington

On October 30, 1887, the steamer *British King* arrived at Philadelphia from Liverpool with John R. Duthie, 33, a shipbuilder from Dundee, Scotland, his wife Anna, 30, son John Frederick, 10, daughter Hellen, 6, and sons William, 4, and Henry, 3. The family moved to 6 Burton St. in Brisol, Rhode Island, where John F. eventually became a boilermaker.

Around 1897 the family moved across the country to Portland, Oregon, where the men went to work at the Wolff & Zwicker Iron Works. John R. was a shipbuilder, William was a machinist, and John F. a shipfitter.

On August 19, 1897, John F. married a girl from Michigan named Electa Markham Henion, and they had a son Wallace Frederick on December 25, 1899. John F. was a foreman by then and they lived at 93 E. 12th.

In 1901 John and Electa moved to Seattle where John took a job with the Moran Bros. Co. as an inspector in the naval constructor's office. In 1911 he set up his own shop at 13 W. Connecticut on the Harbor Island shore of the East Waterway in Seattle to build ships, boilers, steam and hand winches, and provide engineering services, a foundry, machining, iron workers, tinsmiths, coppersmiths, and blacksmiths.

West Point—5,676 tons, steel, 10.5 knot oil-burning freighter, 215924, LJPB. Keel laid on November 29, 1916. Requisitioned from British owners while building and delivered to the USSB on January 21, 1918. Allocated to James W. Elwell as manager for service to the United Kingdom by M.H. Tracy. Commissioned *West Point* in the Naval Overseas Transportation Service on August 5, 1918.

Westerner—5,699 tons, steel, 10.5 knot coal or oil-burning freighter, 215951, LJPN. Keel laid on December 10, 1916. Requisitioned from Norwegian owners while building and delivered to the USSB on February 4, 1918. Allocated to S. Sgitcovich for service to the United Kingdom. Commissioned *Westerner* in the Naval Overseas Transportation Service on June 20, 1918.

Westfield—5,666 tons, steel, 10.5 knot coal or oil-burning freighter, 216008, LJRM. Keel laid on March 1, 1917. Requisitioned from British owners while building and delivered to the USSB on February 24, 1918. Allocated to the Kerr Steamship Co. for service to Italy.

Western King—5,659 tons, steel, 11.5 knot coal-burning freighter, 216073, LJVC. Keel laid on May 14, 1917. Requisitioned from British owners while building and delivered to the USSB on March 26, 1918. Managed and operated by the USSB.

Westover—5,769 tons, steel, 10.5 knot coal or oil-burning freighter, 216130, LJWP. Keel laid on September 5, 1917. Requisitioned from British owners while building and delivered to the USSB on April 12, 1918. Commissioned *Westover* in the Naval Overseas Transportation Service on May 20, 1918. Torpedoed on July 11, 1918, en route to Europe with the loss of 6 out of 44 aboard.

Westboro—5,769 tons, steel, 11.5 knot coal or oil-burning freighter, 216223, LKPM. Keel laid on November 8, 1917. Requisitioned from British owners while building and delivered to the USSB on April 27, 1918. Allocated to the American Line as manager for service to Hawaii by the Shipping Board.

West Bridge—5,818 tons, steel, 10 knot coal or oil-burning freighter, 216348, LKRQ. Keel laid on December 11, 1917. Requisitioned from British owners while building and delivered to the USSB on May 24, 1918. Commissioned *West Bridge* in the Naval Overseas Transportation Service on May 26, 1918.

Western Sea—5,813 tons, steel, 10.5 knot coal or oil-burning freighter, 216465, LKWC. Keel laid on January 5, 1918. Requisitioned from British owners while building and delivered to the USSB on June 20, 1918. Commissioned *Western Sea* in the Naval Overseas Transportation Service on June 20, 1918.

Western Cross—5,778 tons, steel, 11.5 knot coal-burning freighter, 216826, LMRC. Keel laid on March 28, 1918. Requisitioned from British owners while building and delivered to the USSB on August 31, 1918. Allocated to the United States & Australasia Steamship Co. as manager for service to the East Indies by Barber. On December 4, 1918, fireman Harold N. Currier got in a fight ashore at Malta and died of injuries.

Western Star—5,628 tons, steel, 11.5 knot coal or oil-burning freighter, 216875, LMPC. Keel laid on February 20, 1917. Requisitioned from French owners while building and delivered to the USSB on August 28, 1918. Commissioned *Western Star* in the Naval Overseas Transportation Service on August 28, 1918.

Western Hope—5,676 tons, steel, 11.5 knot coal-burning freighter, 216949, LNBC. Keel laid on April 25, 1918, for the Cunard Steamship Co. Requisitioned from British owners while building and delivered to the USSB on September 25, 1918. Commissioned *Western Hope* in the Naval Overseas Transportation Service on September 25, 1918.

West Pool—5,724 tons, steel, 12.5 knot coal-burning freighter, 217127, LNKT. Keel laid on May 27, 1918. Requisitioned from French owners while building and delivered to the USSB on October 31, 1918. Commissioned *West Pool* in the Naval Overseas Transportation Service on November 2, 1918.

Kruse & Banks Shipbuilding Co.
North Bend, Oregon

Knud Vladmir Kruse was born in Denmark on July 24, 1853, came to America in 1874, and worked as a ship carpenter in San Francisco. On January 6, 1881, he married Meta Marie Jensen. Daughter Nicoline arrived a little early in March, followed by son Henrick (who died), then daughters Dorthea, Bertha, Laura, and in 1893 son Fredrick. In 1897 the family repaired to North Bend, Oregon, where Knud went to work in Asa Simpson's shipyard, eventually becoming superintendent. Simpson's shipbuilding apparently ceased in 1903 with the delivery of the 143' schooner *Alpha*. Simpson then purchased the nearby Porter sawmill and put in new equipment.

Knute rented property in Marshfield owned by Eugene O'Connor that was referred to as the Ferndale yard. He delivered the 3-masted schooner *Hugh Hogan* in 1904 and in 1905 hired Robert Banks. Banks was born in Canada in 1872 and came to America in 1880. Younger brother Neil Banks arrived in 1895 and eventually became Knute's yard foreman. The small gas freighter *Oakland* was delivered in 1905 and the *Washcalore* in 1906. Banks became a partner and they started building small steam schooners for the lumber trade. Their first vessel as Kruse & Banks was the *Casco* in 1906 followed by the *R.D. Inman*, *F.S. Loop*, and *E.P. Ripley* in 1907.

In August 1907, just as they were finishing the *F.S. Loop* for the Loop Lumber Co. in San Francisco, O'Connor gave them notice to vacate the property so he could put up a lumber mill. The *Ripley* was still on the ways.

They moved to a portion of the Porter sawmill property in North Bend and there became the Kruse & Banks Shipbuilding Co. The new firm's first delivery was the 625-ton steamer *Bandon* in 1907.

The *North Bend* was the first wood freighter delivered to the Shipping Board. The company's last delivery was the paddlewheel steamer *Roosevelt* in 1921 but the yard was revived for World War II construction.

Stanwood—1,129 tons, steam schooner, 1916, 214500, LGKN. Wilson Bors. & Co.

Port Angeles—1,358 tons, steam schooner, 1916, 214716, LGQT. Charles Nelson Co.

Florence Olson—1,185 tons, steam schooner, 1917, 214976, LHBC. Oliver J. Olson, San Francisco. Sold to French interests. Olson also had the *Mary Olson* and *Virginia Olson* built. Both were sold to the French government and could carry 1,700,000 feet of lumber.

Johanna Smith—1,884 tons, steam schooner, 1917, 215021, LHCD. Coos Bay Lumber Co.

Horace X. Baxter—1,293 tons, steam schooner, 1917, 215342, LHMT. J.H. Baxter & Co.

C.A. Smith—1,878 tons, wood freighter, 1917, 215698, LJCP. Coos Bay Lumber Co.

Fred Baxter—1,294 tons, steam schooner, 1917, 215829, LJHR. J.H. Baxter & Co.

North Bend—3,924 tons, wood, 10 knot coal-burning Hough freighter, 216317, LKRF. USSB contract ship. Keel laid on July 28, 1917, and delivered on May 27, 1918, 303 building days. Allocated to J.F. Whitney as manager for service to Scandinavia by the Triangle Steamship Co.

Quidnic—3,019 tons, wood, 10 knot coal-burning Hough freighter, 216521, LMBP. USSB contract ship. Keel laid on September 14, 1917, and delivered on June 30, 1918, 289 building days. Allocated to the Pacific Mail Steamship Co. for the Philippine inter-island service.

Kickapoo—3,017 tons, wood, 10 knot coal-burning Hough freighter, 216701, LMHW. USSB contract ship. Keel laid on October 16, 1917, and delivered on July 29, 1918, 286 building days. Allocated to the Paragon Steamship Co. for service to Prussia.

Baladan—3,004 tons, wood, 10 knot coal-burning Hough freighter, 216861, LMSK. USSB contract ship. Keel laid on November 13, 1917, and delivered on September 1, 1918,

The 257', 13-knot steam schooner *Johanna Smith* off Coos Bay, Oregon. Johanna Smith was Charles Smith's wife. The ship took lumber to San Francisco. A steam turbine engine was installed in 1921 and in 1928 the ship was sold to Albert Howard of San Francisco, who used it in the infamous offshore gambling business that flourished just outside the three-mile limit off the coast of Southern California. On July 22, 1932, the boat caught fire and sank off Long Beach (courtesy the San Francisco Maritime National Historical Park).

292 building days. Allocated to Pendleton Bros. as managers and operated by the Triangle Steamship Co. on time charter.

Coconino—3,013 tons, wood, 10 knot coal-burning Hough freighter, 216983, LNCJ. USSB contract ship. Keel laid on December 17, 1917, and delivered on September 25, 1918, 283 building days. Allocated to Matson Navigation Co. as manager for operation by Swayne & Hoyt for service to the United Kingdom.

Long Beach Ship Building Co.
Long Beach, California

The officers of the Craig Shipbuilding Co. opened the Long Beach Shipbuilding Co. in 1918 at their same location to build steel Shipping Board freighters. James G. was president and secretary, John F., Sr., was chairman of the board, John F., Jr., was vice president, and Thomas Merrell was general manager. Ernest Hadrath was superintendent, Lloyd Swayne was the naval architect, and George LaFrance chief engineer. They built 16 ships between 1918 and 1920 all homeported at Los Angeles.

Eldorado—2,310 tons, steel, 9 knot oil-burning freighter, 216615, LMGB. Keel laid on July 27, 1917, for Swayne & Hoyt, of San Francisco. Requisitioned by the USSB while building and delivered on July 19, 1918.

Ozaukee—4,045 tons, steel, 10.5 knot oil-burning freighter, 216995, LNDC. USSB contract ship. Keel laid on November 1, 1917, and delivered on September 30, 1918, 334 building days. Commissioned *Ozaukee* in the Naval Overseas Transportation Service on September 30, 1918.

Oshkosh—4,056 tons, steel, 10.5 knot oil-burning freighter, 217205, LNRV. USSB contract ship. Keel laid on February 1, 1918, and delivered on November 30, 1918, 302 building days. Allocated to Barber for service to Italy.

Los Angeles Shipbuilding & Dry Dock Co. San Pedro, California

Frederick L. Baker was president of the Baker Iron Works—engineers, founders, and machinists at 912 N. Broadway in Los Angeles. He lived at 2701 Wilshire Blvd. Samuel L. Naphtaly was president of the Nevada County Narrow Gauge Railroad and lived at 2960 Broadway in San Francisco.

The two established Los Angeles Ship in 1917 as a 6-way yard on 70 acres to build steel Shipping Board freighters. Baker was president and treasurer, Naphtaly vice-president and general manager, attorney Erle M. Leaf was secretary, and Leroy E. Caverly, of Oakland, was the naval architect and chief engineer.

Accomac—5,898 tons, steel, 10.5 knot oil-burning freighter, 216445, LKVM. USSB contract ship. Keel laid on July, 29, 1917, and delivered on June 14, 1918, 320 building days. Allocated to M.H. Tracy & Co. for service to the United Kingdom.

Wakulla—5,898 tons, steel, 10.5 knot coal or oil-burning freighter, 216489, LKWS. USSB contract ship. Keel laid on August 1, 1917, and delivered on June 25, 1918, 328 building days. Commissioned *Wakulla* in the Naval Overseas Transportation Service on June 26, 1918.

Wampum—5,898 tons, steel, 10.5 knot oil-burning freighter, 216572, LMDJ. USSB contract ship. Keel laid on August 25, 1917, and delivered on July 16, 1918, 325 building days. Allocated to the American Line for service to the United Kingdom.

Wassaic—5,898 tons, steel, 10.5 knot oil-burning freighter, 216598, LMFH. USSB contract ship. Keel laid on August 30, 1917, and delivered on July 25, 1918, 329 building days. Allocated to the Lykes Bros. for service to the Mediterranean. Commissioned *Wassaic* in the Naval Overseas Transportation Service on October 14, 1918.

West Galoc—6,000 tons, steel, 10.5 knot coal-burning freighter, 216760, LMKW. USSB contract ship. Keel laid on December 15, 1917, and delivered on August 21, 1918, 249 building days. Commissioned *West Galoc* in the Naval Overseas Transportation Service on August 21, 1918.

West Galeta—6,000 tons, steel, 10.5 knot oil-burning freighter, 216761, LMNB. USSB contract ship. Keel laid on January 19, 1918, and delivered on August 31, 1918, 222 building days. Commissioned *West Galeta* in the Naval Overseas Transportation Service on September 4, 1918.

West Zula—6,000 tons, steel, 10.5 knot oil-burning freighter, 216863, LMSN. USSB contract ship. Keel laid on February 8, 1918, and delivered on September 26, 1918, 250 building days. Commissioned *West Zula* in the Naval Overseas Transportation Service on September 26, 1918.

West Zucker—6,000 tons, steel, 10.5 coal-burning freighter, 217124, LNKR. USSB contract ship. Keel laid on February 15, 1918, and delivered on November 20, 1918, 280 building days. Commissioned *West Zucker* in the Naval Overseas Transportation Service on November 12, 1918.

West Grama—5.967 tons, steel, 10.5 knot coal or oil-burning freighter, 217220, LNSQ. USSB contract ship. Keel laid on March 13, 1918, and delivered on December 31, 1918, 293 building days. Commissioned *West Grama* in the Naval Overseas Transportation Service on January 9, 1918. Operated during World War II by Grace Line. Sunk on June 8, 1944, at Normandy, France, for a breakwater.

McEachern Ship Co.
912 Board of Trade Building, Portland

John A. "Jack" McEachern was a general contractor in Seattle. In 1916 he opened an office in Portland managed by William W. Clark. Clark soon moved to Astoria to set up an office and McEachern came down with his superintendent, John F. Gilpin, to build a 6-way yard on ten acres at the foot of 7th St. on Youngs Bay.

Axel Olaf Andersen's Danish company might have purchased or had a substantial financial interest in the yard. A.O. Andersen & Co. were shipbrokers and agents at 50 Broad St. in New York and there were several branch offices. Arnold Reimann was Andersen's U.S. manager, Charles E. Edwards was the sales manager, and company attorney was Jesse Stears. The three incorporated the Viking Shipping Co. operate a fleet of auxiliary lumber schooners built by McEachern.

In 1918 Edgar W. Wright was vice-president and general manager in Portland. Emery Olmstead was secretary-treasurer, and Fred B. Wright assistant manager. Edgar Wright was the general manager of the Port of Portland Commission. Olmstead was vice-president of the Northwestern National Bank. The company's last ship was the *Anson S. Brooks* in 1920.

Astoria—1,611 tons, auxiliary lumber schooner, 1917, 214712, LGRK. Viking Shipping Co., Portland, Oregon. Launched on October 28, 1916.

Margaret—1,613 tons, auxiliary lumber schooner, 1917, 214908, LGVJ. 3–3–17 Anglo-Norwegian Shipping Agency, New York.

Astri—1,613 tons, 266', 4-masted auxiliary lumber schooner. A.O. Andersen & Co. Purchased on the ways by Kluver & Co. of Christiania, Denmark, for $315,000. Launched on April 30, 1917, and fitted with two 320-hp Bolinger engines. Loaded 1,750,000 feet of lumber at the St. Johns Lumber Co. for Australia. Sailed under the Norwegian flag.

Madrugada—1,613 tons, auxiliary schooner, 1917, 214909, LGVK. Gravenhorst & Co.

May—1,745 tons, auxiliary schooner, 1917, 215625, LHVR. Name changed to *Pelican*. Pelican Motorship Co., New York.

Pauline—1,750 tons, auxiliary schooner, 1917, 215886, LJND. Vestkysten Shipping Co., New York.

Carmen—1,610 tons, auxiliary schooner, 1918, 215971, LJQF. Carmen Ship Co., Portland, Oregon.

Evelyn—1,584 tons, auxiliary schooner, 1918, 216126, LJWN. Anglo-Norwegian Shipping Agency, New York.

Astoria—3,020 tons, wood, 10 knot coal-burning Hough freighter, 216985. LNCP. USSB contract ship. Keel laid on August 1, 1917, and delivered on October 10, 1918, 435 building days. Allocated to Callaghan, Atkinson & Co. as managers for operation by the Triangle Steamship Co. on time charter.

Salmon—3,020 tons, wood, 10 knot coal-burning Hough freighter, 217013, LNFC. USSB contract ship. Keel laid on August 18, 1917, and delivered on November 1, 1918, 439 building days. Allocated to Wilkins & Biehl for service to the West Indies.

Matthews Ship Building Co.
Hoquiam, Washington

Gordon Frazer Matthews was born in Eureka, California, on July 4, 1877, and became a ship carpenter. In 1899 he married an English girl, Florence Barker, who came to America in 1880, and the next year they moved to Hoquiam. In 1901 daughter Mildred was born and the next year son Gordon H. arrived.

Around 1910 he established his own 3-way yard with Florence as vice-president and bookkeeper and the next year delivered the 818-ton steamer *Daisy Gadsby* followed by the *Avalon, Daisy Putnam, O.M. Clark, Rosalie Mahony, Daisy Matthews,* and in 1916 the 946-ton *Hartwood.* The company's last ship was the 227-ton motor vessel *North Beach* in 1927.

Sierra—1,034 tons, steam schooner, 1916, 214391, LGHR. Edwin K. Wood Lumber Co.

Santa Isabel—1,969 tons, schooner, 1917, 215330, LHMQ. Grace Steamship Co.

Claremont—1,291 tons, steam schooner, 1917, 215907, LJNP. Edwin K. Wood Lumber Co.

San Diego—1,487 tons, steam schooner, 1918, 216794, LMPR. Edwin K. Wood Lumber Co.

Meacham & Babcock Shipbuilding Co.
15th Ave. W. and Emerson, Seattle

William M. Meacham was born in Abilene, Kansas, on December 16, 1871. In 1900 his family lived in Cicero, Illinois, where William was an electrician and younger brother Larned E. was a bookkeeper.

Herman G. Babcock was born in Nebraska in 1881 and in 1901 was an electrician with the Electric Construction Co., electrical contractors and supplies at 111 Seneca in Seattle.

The two got together and by 1907 they were electrical contractors and engineers in Seattle as Meacham & Babcock at 825 3rd Ave. In 1908 William married Catherine "Kate" Moore and the next year son William was born. They lived at 127 N. 11th Ave.

When war came they thought their expertise could be transferred to shipbuilding even though neither one had ever built a ship before. They hired an advisor and set up shop on Salmon Bay. Meacham was president and Babcock treasurer. Larned Meacham was then secretary-treasurer of the Railways Accessories Co. in Seattle and was brought in as secretary-treasurer.

Shipping Board contracts were generally let for the construction of hulls, the installation of machinery into the hull, or for the complete ship. The work could be done by one builder or two separate contractors. The company had five contracts as explained by William Meacham:

The first contract called for two hulls only. The second contract was for four complete ships. The third contract was for four hulls only. The fifth contract for the installation of two hulls of the first contract and the four hulls of the third contract. Rather complicated.

To further complicate matters, the head accountant they hired was not a fully-certified public accountant—he was still studying—and on top of that he was sick the entire time the contracts were in force. As a result, the books were a mess.

Unfortunately, his health worsened and he died while still employed. In September 1918 an auditor came in and, according to William Meacham, "boastingly told a young man in our plant that we were broke." Meacham vehemently denied ever being insolvent but in March 1919, the company was put under receivership on application by the Northwest Steel Co. in Portland by virtue of having "creditors numbering over two hundred and located at various points" and "is and has been for some time hopelessly insolvent."

Charles G. Heifner, a prominent politician and member of the Seattle Chamber of Commerce, was brought in as president and treasurer. The company owed $250,000 for materials and the Emergency Fleet Corp. was obliged to meet the payroll.

Boulton—2,584 tons, wood, 10 knot coal-burning Ferris freighter, 216819, LMQS. USSB contract ship. Keel laid on October 2, 1917, and delivered on September 12, 1918, 345 building days. Allocated to Lykes Bros. for service to the United Kingdom.

Daca—2,579 tons, wood, 10 knot coal-burning Ferris freighter, 217005, LNDT. USSB contract ship. Keel laid on October 12, 1917, and delivered on October 19, 1918, 372 building days. Allocated to the Pensacola Shipping for service to the United Kingdom.

Moore & Scott Iron Works
670 2nd St., San Francisco, California

In 1904, Robert S. Moore lived in Oakland and was vice-president and superintendent of Risdon Iron Works in San Francisco. The next year he was president of the Paraffine Paint Co.

John T. Scott was superintendent of the Union Iron Works in San Francisco at lived at 929 Hayes. In 1905 he and Moore decided to go into business together as the Moore & Scott Iron Works and bought the National Iron Works at Main and Howard from Thomas G. Cantrell and Leon C. Marshutz. They did forgings, castings, machining, and built boilers.

After the big fire in April 1906, Moore moved to 2901 Pacific Ave. and Scott relocated to Burlingame where he lived with his wife Catherine and their Japanese cook Fondo. Their business continued but a lot of work and contracts moved south to Los Angeles while the devastated City recovered.

William A. Boole had a marine railway and shipyard in Oakland at the foot of Adeline St. as W.A. Boole & Son. Boole was president, Horace Z. Howard vice-president, and Arthur E Wellington secretary of the 15 acre, 5-way yard. Their last big project was in 1907 when they built the steamer *Sibyl Marston*.

In 1910 Moore and Scott decided to invest in Boole's yard. Their company was reorganized, apparently with help from investors, but was still the Moore & Scott Iron Works. Henry T. Scott, of Burlingame, became president, John Scott vice-president, Robert Moore treasurer, Lewis H. Cromwell, of Berkeley, assistant secretary. Moore moved to the Grenada Hotel in San Francisco.

The first vessel the new company built in 1910 was the steam freighter *Coalinga* for the Associated Transportation Co. That company operated several oil pumping lines.

On July 19, 1912, they launched their first big job, the 230' steel ferryboat *Edward T. Jeffery* for the Western Pacific Railway Co.

The company increased its interest in shipping and by September 1915 Moore & Scott Iron Works had a second shipyard at 678 Second St. in San Francisco where they did machine work, castings, and built boilers, mining machinery, steam engines, hydrants, oil-burning plants, stamp mills, sheet iron and sheet steel pipe, tugs, powerboats, and field gun carriages.

In 1915 the Moore & Scott Ship & Dock Co. was organized in Oakland with Moore as president. At this time Robert Moore was a busy man. He was also president of The Paraffine Paint Co., California Paper & Board Mills, and the Moore Investment Co.

On October 15, 1918, Moore & Scott Iron Works and Moore & Scott Ship & Dock consolidated into the Moore Shipbuilding & Dock Co. in Oakland. George A. Armes was president. Armes had been chief engineer at Union Iron Works. Andrew Moore was 1st vice-president, Joseph A. Moore 2nd vice-president, and J.H. Small secretary.

John Scott moved to Walnut Creek and became vice-president of the Pacific Coast Shipbuilding Co. in San Francisco.

Capto—4,823 tons, freighter, 1916, 214643, LGPJ. James Rolph Jr., ordered the 376' oil-burning Curtis turbine freighter *Annette Rolph* on January 1, 1916. The keel was laid in May but the same month Rolph sold the ship to B. Stolt Nielsen, of Haugesund, Norway, and the name was changed to *Capto*. The oil-burning system was developed by Moore & Scott engineers.

Frederic R. Kellogg—7,005 tons, tanker, 1917, 215424, LHPT. Pan American Petroleum & Transport Co.

Thordis—4,371 tons, oil-burning freighter, 1917. Built for Norwegian owners.

Mount Shasta—4,865 tons, steel, 10.5 oil-burning freighter, 215836, LJHW. Keel laid on November 11, 1916. Requisitioned from Norwegian owners while building and delivered to the USSB on December 7, 1917. Allocated to Cosmopolitan Shipping for service to France. Commissioned *Mount Shasta* in the Naval Overseas Transportation Service on August 26, 1918.

Coronado—5,967 tons, steel, 10.5 knot coal or oil-burning freighter, 215918, LJNW. Keel laid on January 25, 1917. Requisitioned from British while building and delivered to the USSB on January 16, 1918. Allocated to W.F. Spice for service to the by the Netherlands.

Yosemite—6,149 tons, steel, 10.5 knot coal or oil-burning freighter, 215958, LJPT. Keel laid on May 22, 1917. Requisitioned from Norwegian owners while building and delivered to the USSB on February 4, 1918. Allocated to Lykes Bros. for service to the United Kingdom.

Yellowstone—6,171 tons, steel, 10.5 coal or oil-burning freighter, 216200, LKNM. Keel laid on July 12, 1917. Requisitioned from Norwegian owners while building and delivered to the USSB on April 22, 1918. Allocated to Strachan Shipping for service to the United Kingdom.

Pasadena—6,002 tons, steel, 11.5 knot coal-burning freighter, 216303, LKJS. Keel laid on October 28, 1917. Requisitioned from British owners while building and delivered to the USSB on May 11, 1918. Commissioned *Pasadena* in the Naval Overseas Transportation Service on May 13, 1918.

Oakland—6,002 tons, steel, 11 knot coal-burning freighter, 216387, LKSV. Keel laid on November 10, 1917. Requisitioned from British owners while building and delivered to the USSB on May 30, 1918. Commissioned *Oakland* in the Naval Overseas Transportation Service on June 3, 1918.

Fresno—6,002 tons, steel, 11 knot coal-burning freighter, 216455, LKVS. Keel laid on December 13, 1917. Requisitioned from British owners while building and delivered to the USSB on June 22, 1918. Commissioned *Fresno* on June 22, 1918, in the Naval Overseas Transportation Service.

Alloway—6,165 tons, steel, 11 knot oil-burning freighter, 216557, LMCS. USSB contract ship. Keel laid on October 15, 1917, and delivered on July 11, 1918, 269 building days. Commissioned *Alloway* on July 11, 1918, in the Naval Overseas Transportation Service.

Aniwa—6,165 tons, steel, 11 knot oil-burning freighter, 216612, LMFT. USSB contract ship. Keel laid on October 19, 1917, and delivered on July 26, 1918, 280 building days. Commissioned *Aniway* on July 26, 1918, in the Naval Overseas Transportation Service.

Zirkel—6,171 tons, steel, 11 knot oil-burning freighter, 216984, LNCM. USSB contract ship. Keel laid on May 21, 1918, and delivered on September 27, 1918, 128 building days. Allocated to the American Line for service to the Netherlands.

Oskawa—6,100 tons, steel, 11 knot oil-burning freighter, 217206, LNRW. USSB contract ship. Keel laid on January 1, 1918, and delivered on December 30, 1918, 364 building days. Commissioned *Oskawa* in the Naval Overseas Transportation Service on January 9, 1919.

Zaca—6,165 tons, steel, 11 knot oil-burning freighter, 217416, LPJG. USSB contract ship. Keel laid on May 21, 1918, and delivered on December 30, 1918, 223 building days. Allocated to Barber for service to Germany.

National Shipbuilding Co.
655 Gordon St., Seattle

In 1913 Frank Garrett Fisher was president of the Tacoma Speedway Association and lived with his wife Jennie at 1102 N. 5th in Tacoma.

In 1905 Capt. Otis D. Trieber purchased the ferryboat *Bear* from Harry Olson and ran it across the Columbia River from Underwood, Washington, to Hood River, Oregon. He and his wife lived on a houseboat at the mouth of the White Salmon River at Underwood.

Edward Meath was the Washington State treasurer and lived with his wife Edith at 1213 Main in Olympia. George A. Lee was a Seattle attorney.

In December 1916 the fearless foursome incorporated National Ship with a capital of $40,000 and acquired and modernized the Johnson & Blanchard Shipbuilding Co. in Tacoma. Fisher was president, Trieber vice-president and general manager, Lee secretary and general manager, and Meath treasurer. The 2-way yard employed around 200.

In April 1917 Capt. Trieber was president and a year later the company was reported to be in the hands of trustees James F. Lane, J.L. McLean, and Loren Grimstead. McLean was general manager, Ben Murphy, plant manager and purchasing agent. Martin G. Didrickson, naval architect and superintendent. Their first vessel was the 761-ton *Apex* for fishing and banking magnate Lee H. Wakefield.

Bright—2,176 tons, 5-masted motor schooner, 1918, 217271, LNWJ. Minnehaha Motor-Schooner Co.

Brisk—2,151 tons, 5-masted motor schooner, 1918, 217272, LNWK. Minnehaha Motor-Schooner Co.

Nilson & Kelez Shipbuilding Corp. Seattle, Washington

Andrew S. Nilson and Nicholas M. Kelez were boat builders at the foot of Massachusetts Ave. in Seattle as Nilson & Kelez. John Erikson was president of the Pacific Safety Deposit Co. and Erikson-Johanson & Co., real estate agents at 96 University in Seattle.

When war came they decided to build wood Shipping Board freighters and in July 1917 opened a 3-way yard on nine acres at 2241 Whatcom Ave. Erikson was president, Nilson vice-president and superintendent, Daniel Johanson, secretary-treasurer, and Kelez, purchasing agent. The company built four Ferris Shipping Board freighters and had around 600 employees.

Bonnafon—2,542 tons, wood, 10 knot coal-burning Ferris freighter, 216847, LMSG. USSB contract ship. Keel laid on November 16, 1917, and delivered on September 9, 1918, 296 building days. Allocated to the McDonald & Truda Co. for service to the United Kingdom.

Forster—2,552 tons, wood, 10 knot coal-burning Ferris freighter, 216919, LMWB. USSB contract ship. Keel laid on November 17, 1917, and delivered on October 18, 1918, 337 building days. Allocated to Charles Kurz & Co. for service to the Netherlands.

Octorara—2,531 tons, wood, 10 knot coal-burning Ferris freighter, 217231, LNTG. USSB contract ship. Keel laid on November 19, 1917, and delivered on December 28, 1918, 404 building days. Allocated to the Texas Transport & Terminal Co. for service to the United Kingdom.

Northwest Steel Co. Portland, Oregon

Northwest Steel was established in 1903 on the Willamette River at the foot of Sheridan St. as an iron and steel wholesaler, structural steel fabricator, shipbuilder, and nut, bolt, rivet, and boat spike manufacturer. Joseph R. Bowles was president, Walter E. Beebe vice-president, William H. Cullers chief engineer, and John Murray superintendent of ship construction.

War Viceroy—5,700 tons, steel, 10.5 knot coal-burning freighter. Keel laid on August 10, 1916, for the Cunard Steamship Co. Requisitioned from British owners while building and delivered to the USSB on October 23, 1917. Returned to Cunard before documentation.

Westland—5,832 tons, steel, 10.5 knot oil-burning freighter, 215879, LJMP. Keel laid on October 8, 1916. Requisitioned from Norwegian owners while building and delivered to the USSB on December 28, 1917. Allocated to J.H. Wincherster as manager for service to the United Kingdom by the Texas Transport & Terminal Co.

Westwind—5,814 tons, steel, 10.5 knot oil-burning freighter, 215948, LJPM. Keel laid on February 1, 1917. Requisitioned from Norwegian owners while building and delivered to the USSB on February 4, 1918. Allocated to Barber for service to the Orient.

Westchester—5,725 tons, steel, 10.5 knot coal-burning freighter, 216059, LJTB. Keel laid on April 2, 1917. Requisitioned from British owners while building and delivered to the USSB on March 17, 1918. Allocated to J.H. Winchester for service to the United Kingdom. Commissioned *Westchester* in the Naval Overseas Transportation Service on August 9, 1918.

West Shore—5,831 tons, steel, 10.5 knot coal or oil-burning freighter, 216158, LKMP. Keel laid on August 3, 1917. Requisitioned from British owners while building and delivered to the USSB on April 15, 1918. Operated by the Shipping Board for service to France. Commissioned *West Shore* in the Naval Overseas Transportation Service on August 6, 1918.

Westhampton—5,832 tons, steel, 10.5 knot coal or oil-burning freighter, 216309, LKRB. Keel laid on September 17, 1917. Requisitioned from British owners while building and delivered to the USSB on May 13, 1918. Allocated to American-Hawaiian as manager for service to Italy by Robert Hasler.

Western Wave—5,832 tons, steel, 10.5 knot coal or oil-burning freighter, 216373, LKSM. Keel laid on November 5, 1917. Requisitioned from Norwegian owners while building and delivered to the USSB on June 8, 1918. Allocated to the Carolina Co. for service to France.

Western Ocean—5,869 tons, steel, 10.5 knot coal-burning freighter, 216446, LKVN. Keel laid on December 6, 1917. Requisitioned from French owners while building and delivered to the USSB on June 17, 1918. Commissioned *Western Ocean* in the Naval Overseas Transportation Service on June 17, 1918.

Western Chief—5,869 tons, steel, 11 knot coal-burning freighter, 216525, LMBR. Keel laid on January 14, 1918. Requisitioned from French owners while building and delivered to the USSB on July 1, 1918. Commissioned *Western Chief* in the Naval Overseas Transportation Service on July 3, 1918.

Western Spirit—5,869 tons, steel, 11 knot coal-burning freighter, 216627, LMGF. Keel laid on February 9, 1918. Requisitioned from French owners while building and delivered to the USSB on July 19, 1918. Commissioned *Western Spirit* in the Naval Overseas Transportation Service on July 19, 1918.

Western Light—5,869 tons, steel, 11 knot coal-burning freighter, 216631, LMGJ. Keel laid on March 5, 1918. Requisitioned from French owners while building and delivered to the USSB on July 29, 1918. Commissioned *Western Light* in the Naval Overseas Transportation Service on July 30, 1918.

Western Maid—5,871 tons, steel, 10.5 knot coal-burning freighter, 216754, LMKT. Keel laid on March 20, 1918. Requisitioned from French owners while building and delivered to the USSB on August 17, 1918. Commissioned *Western Maid* in the Naval Overseas Transportation Service on August 17, 1918.

Western Comet—5,871 tons, steel, 10 knot coal-burning freighter, 216910, LMVP. Keel laid on April 22, 1918. Requisitioned from French owners while building and delivered to the USSB on September 22, 1918. Commissioned *Western Comet* in the Naval Overseas Transportation Service on September 22, 1918.

Western Scout—5,871 tons, steel, 10 knot coal-burning freighter, 217032, LNGC. Keel laid on May 11, 1918. Requisitioned from French owners while building and delivered to the USSB on October 15, 1918. Allocated to John S. Emery & Co. for service to France.

West View—5,871 tons, steel, 10 knot coal-burning freighter, 217223, LNSR. Keel laid on May 28, 1918. Requisitioned from French owners while building and delivered to the USSB on November 19, 1918. Commissioned *West View* in the Naval Overseas Transportation Service on November 19, 1918.

West Kyska—5,871 tons, steel, 11.5 knot oil-burning freighter, 217224, LNST. USSB contract ship. Keel laid on July 9, 1918, and delivered on November 22, 1918, 135 building days. Commissioned *West Kyska* in the Naval Overseas Transportation Service on November 22, 1918.

West Zeda—5,871 tons, steel, 10.5 knot oil-burning freighter, 217225, LNSV. USSB contract ship. Keel laid on July 24, 1918, and delivered on December 23, 1918, 132 building days. Commissioned *West Zeda* in the Naval Overseas Transportation Service on December 23, 1918. Operated during World War II by Coastwise (Pacific Far East) Line. Sunk on February 22, 1942, by *U-129* in the South Atlantic.

Olympia Shipbuilding Co.
Olympia, Washington

Richard Ward was born in England in 1857, came to America in 1880, and settled in Seattle with his wife Martha at 3214 3rd Ave. W. and built boats with his sons Eugene R., William M., and Montague R. as Ward & Sons at 24 Colman Dock Yards.

In 1915 Eugene struck out on his own and started the Olympia Shipbuilding Co. at 1st & Main in Olympia. In 1916 he was joined by Seattle Attorney Dean Burkheimer, as secretary. In May 1917 they started the Ward & Burkheimer Shipbuilding Co. at Seattle with a capitalization of $100,000 with Ward as president. Olympia Ship built 290', 5-masted auxiliary schooners. The company dissolved on July 1, 1920.

Weregland—2,457 tons, wooden motor schooner, 1917, 215896, LJNC. Weregland Motor Ship Co., Portland, Oregon.

General Pershing—2,450 tons, wooden motor schooner, 1918, 216063, LJTC. Pershing Navigation Co., Portland, Oregon.

Korsnaes—2,446 tons, wooden motor schooner, 1918, 216118, LJWG. Korsnaes Motor Ship Co., Portland, Oregon.

Pacific American Fisheries Co.
Fairhaven Harbor, So. Bellingham, Washington

Established in 1905. When the war started, Pacific American was a big salmon canning operation with a boatyard, marine railway, and plant on Eliza Island. Everett B. Deming was president and general manager, Arthur W. Deming vice-president and superintendent, Samuel C. Scotten treasurer, and Christopher M. Mitchell secretary.

The prospects of obtaining ships was uncertain so the company decided to build its own cargoliners to carry supplies, passengers, and the company's products throughout its far-flung operations. They purchased property adjoining their operations at Commercial Point on Bellingham Bay and built a 4-way shipyard at the foot of Harris Ave. at 4th St. The yard employed 160.

In 1916, George H. Hitchings was a naval architect in Seattle at 3202 Irving St. He was hired and assisted by Christopher Nottley. The keel for Hitchings' first ship, the *Redwood*, was laid on June 20, 1916, but when he became too ill to work, Nottley took over. When the company got a Shipping Board contract for seven wood Ferris freighters on February 6, 1918, Nottley's own design was approved by Theodore Ferris as Design No. 1065. The 268' ship had two forward and two after holds, a single midship house, and raised fo'c'sle and poop with an after deckhouse.

Redwood—1,793 tons, wood, cargoliner, 1917, 214849, LGTR. Company's account to accommodate 120 passengers and 60,000 cases of salmon.

Firwood—1,782 tons, wood, cargoliner, 1917, 214997, LHBM. Company's account to accommodate 120 passengers and 60,000 cases of salmon.

Rosewood—1,929 tons, wood, cargoliner, 1917, 215720, LJDK. French account.

Hollywood—1,940 tons, wood, cargoliner, 1917, 215900, LJNG. French account.

Catherine D—2,224 tons, wood, cargoliner, 1918, 216216, LKNW. Company's account, named for Everett Deming's granddaughter. Accommodations for 180 passengers and 84,000 cases of salmon. The company's flagship sailed under Capt. William D. Knight.

Oakwood—1,951 tons, composite hull wood freighter, 1918, 216590, LMFG. French account. Sold to a Belgian owner on June 6, 1919.

Cruso—2,442 tons, wood, 10 knot coal-burning Nottley freighter, 1918, 216960, LNBK. USSB contract ship. Keel laid on November 10, 1917, and delivered on December 9, 1918, 394 building days. Allocated to J.F. Whitney as manager for service to Africa by the F.D. Dimmick Co.

Peninsula Shipbuilding Co.
Portland, Oregon

Fred C. Knapp was president of the Peninsula Lumber Co. and vice-president of the Peninsula National Bank. Abbott L. Mills was president of the First National Bank and vice-president of the Security Savings & Trust Co. James Kerr practiced law with Charles H. Carey.

The four established Peninsula Ship in 1916 at the foot of McKenna Ave. on 40 acres with four 300' ways. Knapp was president, Mills vice-president, and Kerr secretary. The yard had a 1,000' dock and employed 550 people.

The company came up with its own wood freighter design approved by the Shipping Board as Design No. 1004. The 287' ship had two forward holds a single midship house, one after hold, and raised foc's'le and poop. The crew berthed forward, the Armed Guard aft.

Erris—1,582 tons, composite-hull motor freighter, 1917, 215899, LJNF. Erris Motor Ship Co., Portland, Oregon.

Pechiney—1,586 tons, composite-hull motor freighter, 1917, 216464, LKWB. Albert Alexis Naphegyi. Albert was a French citizen from Paris in New York to expedite shipments of machinery, copper and other metals, and raw materials.

Adrien Badin—1,601 tons, composite-hull motor freighter, 1917, 216712, LMJQ. The company's own account.

Anoka—2,924 tons, wood, 10 knot coal-burning Peninsula freighter, 216941, LMWN. USSB contract ship. Keel laid on May 15, 1917, and delivered on October 26, 1918, 509 building days. Allocated to Williams Dimond & Co. for unspecified trade routes.

Puget Sound Bridge & Dredging Co., Inc. 811 Central Building, Seattle, Washington

Puget Sound was established in 1898. In 1917 the office and shipyard was at 2742 16th Ave. SW. Samuel H. Hedges was president, Robert M. Dyer, vice-president and treasurer, and George E. Hardenbergh, secretary. The company's 6-way, 8-acre yard on Harbor Island employed 300–350. The company delivered three more ships in 1919.

Admiral Sims—1,929 tons, wood motor freighter, 1918, 216636, LMGN. Pacific Alaska Navigation Co.

Admiral Mayo—1,934 tons, wood motor freighter, 1918, 217075, LNHW. Pacific Alaska Navigation Co.

Bluebird—1,618 tons, auxiliary schooner, 1918, 225211, MDTS. Keel laid as the French *Ypres*.

Ralph J. Chandler Shipbuilding Co. Wilmington, California

Ralph Joseph Chandler was born in Los Angeles on February 25, 1891. In 1913 he was vice-president of the Star & Crecent Boat Co.

He decided he wanted his own yard but needed capital. Moses Haseltine Sherman was president of the M.H. Sherman Investment Co. and Robert Pierce Sherman was vice-president. Robert was Moses' stepson.

Work started on the 4-way shipyard on July 6, 1917, with Moses Sherman as president, Chandler vice-president and general manager, Robert Sherman vice-president, and William Muller superintendent.

Bellota—2,620 tons, wood, 10 knot coal-burning Ferris freighter, 216702, LMJB. USSB contract ship. Keel laid on August 1, 1917, and delivered on August 13, 1918, 378 building days. Allocated to Luckenbach for service to the Netherlands.

Blythedale—2,612 tons, wood, 10 knot coal-burning Ferris freighter, 217077, LNJB. USSB contract ship. Keel laid on August 23, 1917, and delivered on November 18, 1918, 451 building days. Allocated to W.F. Spice for service to Scandinavia.

Raymond Shipyard Raymond, Washington

Captain Andrew Peterson was born in Sweden in 1853, emigrated to the United States in the 1870s, went to sea, then worked in shipbuilding at Camden, New Jersey. In 1913 he opened his own shipyard at Raymond, Washington, where he obtained a contract to build the steam schooner *Solano* for Charles Nelson. In 1914 Nelson ordered the *Mukilteo*.

In February 1916 he got an order from the Wilson brothers for the steam schooner *Oregon* and at that point he decided to move his operation to Aberdeen and incorporate as the Aberdeen Shipbuilding Co.

Mukilteo—1,163 tons, steam schooner, 1915, 213571, LFNP. Charles Nelson Co. The ship was ready for launching on February 11, 1915, when the workers filed a lien against the vessel for back wages.

Rolph Shipbuilding Co.
Rolph, California

James Rolph, Jr., was president, David F. Ewart vice-president, and John D. Stelling secretary and purchasing agent of the old 12-way Bendixen shipyard at Fairhaven, two miles southwest of Eureka. The main office was at 230 California St. in San Francisco. The town of Fairhaven was renamed Rolph and the yard employed 300. It was the largest wood shipbuilding yard on the Pacific Coast at the time. All the wartime-built vessels were for the Rolph interests.

Joan of Arc—2,360 tons, steam schooner, 1918, 216834, LMRK. Rolph Navigation & Coal.

Hesperian—1,385 tons, barkentine, 1918, 217229, LNTD, Rolph Shipbuilding.

Annette Rolph—2,361 tons, steam schooner, 1918, 217255, LNVP. Rolph Shipbuilding.

Georgina Rolph—2,354 tons, steam schooner, 1918, 217442, LPKT. Rolph Navigation & Coal.

St. Helens Shipbuilding Co.
St. Helens, Oregon

George Broughton and William Wiggins were dealers in piles, ties, and cedar poles in Portland, Oregon. In 1911 Wiggins opened a 10-acre shipyard at St. Helens on the Columbia River and in 1912 delivered the 969-ton steamer *Multnomah*. In January 1916 the shipyard, a few local mills, and a creosoting plant were taken over by Charles R. McCormick. James H. Price managed the shipyard and 250 workers for McCormick.

City of Portland—1,833 tons, 5-masted auxiliary lumber schooner, 1916, 214008, LFWC. Charles R. McCormick. The ship made a successful round trip bringing lumber to Sydney, Australia, and coal for the return trip. On the second voyage, the stevedores in Sydney were on strike so the master went to Rabaul in the Solomon Islands and loaded copra, which turned out to be more profitable than the coal would have been.

S.I. Allard—1,915 tons, wood motor freighter, 1917, 214730, LGRJ. Charles R. McCormick. Stranded on September 21, 1918, at Corona de San Carlos, Cuba, with 21 aboard. There were no casualties.

Frank D. Stout—1,226 tons, steam schooner, 1917, 215145, LHFT. Brookings Commercial Co., Marshfield, Oregon.

City of St. Helens—2,135 tons, 5-masted auxiliary lumber schooner, 1917, 215674, LJBT. Charles R. McCormick.

Thistle—1,586 tons, schooner, 1918, 216068, LJVB. Fife Shipping Co.

John W. Wells—2,527 tons, schooner, 1918, 216901, LMVD. Charles R. McCormick.

Issaquena—2,503 tons, wood, 10 knot coal-burning Ferris freighter, 217177, LNQG. USSB contract ship. Keel laid on August 11, 1917, and delivered on December 18, 1918, 494 building days. Allocated to Harriss, Magill for service to Scandinavia.

San Francisco Shipbuilding Co. Redwood City, California

William Leslie "W. Leslie" Comyn was born in Hammersmith, Middlesex, England, on October 30, 1878, came to the United States around 1900, and in 1903 was the manager of Bowring & Co., a shipping and commission house at 222 Sansome St. in San Francisco. He also had his own lumber business. In March 1906 he married Anne Gerber in Sacramento and they apparently took a honeymoon trip abroad. The big earthquake and fire was on April 6 and on April 25 they left Liverpool on the steamer *Baltic* and arrived at New York on May 8. The next year their son Jack was born followed by William and the family moved to San Rafael.

In 1910 he and Benjamin F. Mackall, president of the Weston Basket & Barrel Co., went into the shipping and commission business as Comyn, Mackall & Co. at 310 California St.

When the war started, Comyn became intrigued with the idea of building a concrete ship and in 1918 organized the San Francisco Shipbuilding Co. at 310 California St. He asked Alan Macdonald, of Macdonald & Kahn, construction managers, and civil engineer Victor H. Poss, of Berkeley, to design a ferrocement hull. During this time the Shipping Board had signed contracts for 18 concrete ships of 3,000 and 3,500 tons.

Comyn set up a yard at Redwood City and work commenced in September 1917. It was estimated the ship would displace 10 percent more than a steel ship and carry 10–15 percent less cargo. The typical cost of a steel freighter was $150– 200 a deadweight ton. The anticipated government cost to build a concrete ship was $70 a deadweight ton.

The ship was launched sideways on March 14, 1918. Trials were held on May 5 and left San Francisco for Vancouver, B.C., on May 13 under Capt. Richard Ernest Connell. Emergency Fleet concrete ship research engineers Franklin R. McMillan and Henry S. Loeffler and Lighthouse Service superintendent Charles C. Brush were aboard as observers. The ship arrived at Seattle after a very rough passage with 80-mph winds and 35' waves. No problems were reported and the Portland Cement Association reported the ship was "as dry as a Fresno roof in July." They arrived at New Orleans in October, made another voyage, and arrived at New York from Matanzas, Cuba, on November 22.

The Shipping Board announced on May 16 that 54 more concrete ships would be built on each coast at an estimated cost of $125 a ton. San Francisco Ship was the designated West Coast Yard and the East Coast yard was to be at Wilmington, North Carolina. Comyn apparently planned three more larger ships but the *Faith* was the only one he built.

Faith—3,427 tons, concrete freighter, 1918, 216271, LKQP. San Francisco Shipbuilding Co., San Francisco. The first concrete ship $750,000, about $35 a deadweight ton, and was 320' × 45', 3 holds, depth of hold 31', 5,000 deadweight tons. The bottom was 4½" thick and the sides about 4". There was 540 tons of reinforcing steel bar and extra reinforcing in the bow. The main deck was wood. The 1,700-hp triple expansion engine was powered by Scotch boilers and gave 11 knots.

Schaw-Batcher Pipe Works
356 Market St., San Francisco, California

William Schaw and John H. Batcher had operations in Sacramento and San Francisco. Schaw was president, Batcher was vice-president, Henry A. Heilbron secretary, and Frank H. Kiefer treasurer.

The Schaw-Batcher Co. at 209 J St. in Sacramento sold wholesale and retail hardware, sporting goods, mill, mining, and blacksmith supplies, brass, sheet steel, and iron pipe, and offered machining services.

The Schaw-Batcher Pipe Works was established on August 22, 1917, on 175 acres at 15th & B Sts. to manufacture sheet steel, iron pipe, water and oil tanks, and single and double-walled well casings. Batcher lived in San Francisco at 1124 Page. The shipbuilding division's 3-way yard was in South San Francisco. Batcher was president, Charles L. Moorman vice-president, and accountant William G. Aldenhagen, of Richmond, was secretary-treasurer.

Isanti—6,177 tons, steel, 10.5 knot oil-burning freighter, 216975, LNBT. USSB contract ship. Keel laid on November 27, 1917, and delivered on September 30, 1918, 308 building days. Commissioned *Isanti* on September 30, 1918, in the Naval Overseas Transportation Service.

Seaborn Shipyards Co.
624 E. D. St., Tacoma, Washington

Phillips Morrison and Charles Daley were general contractors in Seattle, Charles N. Seaborn was the foreman at Wilson's Boat & Ship Yard in Seattle, and Willis A. Mahoney was a clerk at the Great Northern Steamship Co. dock in Seattle.

Morrison was president, Seaborn vice president and general manager, and Mahoney secretary. The firm employed 600 workers and had four 300' ways. The company was the first large-scale builder of wood ships in Tacoma. Their first ship, the 242' *Seaborn*, was launched on December 12, 1916, and christened by Catherine Ostrander, 14, Harry's oldest daughter. After hitting the water, the ship continued across the slip and damaged the Eureka Dock. The ship was reportedly undamaged and went into service in February 1917.

Seaborn—1,296 tons, auxiliary schooner, 1917, 214692, LGQK. Jewett M. Scott.

Levi W. Ostrander—1,638 tons, schooner, 1917, 215629, LHVT. Harry F. Ostrander.

Betsey Ross—1,630 tons, schooner, 1917, 215660, LJBM. Amos D. Carver.

Quinault—2,531 tons, wood, 10 knot coal-burning Ferris freighter, 216588, LMFC. USSB contract ship. Keel laid on August 11, 1917, and delivered on July 7, 1918, 330 building days. Allocated to the Terminal Shipping Co. for service to France.

Wahkiakum—2,528 tons, wood, 10 knot coal-burning Ferris freighter, 216685, LMHQ. USSB contract ship. Keel laid on August 28, 1917, and delivered on August 2, 1918, 339 building days. Allocated to the Earn Line for service to Belgium.

Mojave—2,545 tons, wood, 10 knot coal-burning Ferris freighter, 216797, LMPT. USSB contract ship. Keel laid on September 12, 1917, and delivered on August 24, 1918, 346 building days. Allocated to Lowrance & Co. for service to the West Indies.

Cheron—2,543 tons, wood, 10 knot coal-burning Ferris freighter, 216828, LMRF. USSB contract ship. Keel laid on September 18, 1917, and delivered on September 13, 1918, 360 building days. Allocated to Potter Transportation as manager for trans-Atlantic service by M.E. Kinsley.

Mazama—2,542 tons, wood, 10 knot coal-burning Ferris freighter, 216955, LNBG. USSB contract ship. Keel laid on February 12, 1918, and delivered on September 24, 1918, 225 building days. The ship was sold to the American Steamship Co.

Chimo—2,525 tons, wood, 10 knot coal-burning Ferris freighter, 217067, LNHQ. USSB contract ship. Keel laid on March 12, 1918, and delivered on October 18, 1918, 220 building days. Allocated to W.R. Weist as manager for service to the west coast of South America by the American Metal Transportation Co.

Wakanna—2,541 tons, wood, 10 knot coal-burning Ferris freighter, 217162, LNPK. USSB contract ship. Keel laid on April 21, 1917, and delivered on October 31, 1918, 193 building days. Allocated to Laing & McKay for service to France.

Dungeness—2,537 tons, wood, 10 knot coal-burning Ferris freighter, 217297, LPBR. USSB contract ship. Keel laid on May 11, 1918, and delivered on December 21, 1918, 224 building days. Allocated to the Northern Shipping Corp. for service to the Netherlands.

Seattle Construction & Dry Dock Co.
Foot of Charles St., Seattle, Washington

James V. Paterson was vice-president and general manager of The Moran Co. in Seattle, famous for building the battleship *Nebraska,* commissioned in 1907. Moran subsequently delivered 13 steamers the last being the 120-ton *Paterson* in 1911.

James Paterson then established his own 6-way yard on the Moran property. He was president, Harry W. Kent secretary-treasurer. The company also offered engineering services. The storied site was adjacent to the celebrated Skinner & Eddy yard, subsequently occupied by Todd Shipyards.

Their first vessel was the 116-ton steel fishing boat *Aberdeen* in 1912 for the American Pacific Whaling Co. and the *Masuda* was the last ship they built.

Cauto—3,820 tons, freighter, 1916, 214559, LGMN. Atlantic Gulf & West Indies. Commissioned *Cauto* on July 13, 1918, in the Naval Overseas Transportation Service.

Panuco—3,832 tons, freighter, 1917, 214668, LGPS. Atlantic Gulf & West Indies. Commissioned *Panuco* in the Naval Overseas Transportation Service on September 13, 1918.

Sacramento—4,856 tons, steel, 10.5 coal-burning freighter, 215939, LJPF. Keel laid on April 2, 1917. Requisitioned from Norwegian owners while building and delivered to the USSB on January 28, 1918. Managed and operated by the USSB.

Sutherland—4,855 tons, steel, 10.5 knot coal-burning freighter, 216043, LJSN. Keel laid on May 25, 1917. Requisitioned from French owners while building and delivered to the USSB on March 16, 1918. Allocated to Terminal Shipping for service to Italy under Capt. Robert S. Sparrow.

Bremerton—4,856 tons, steel, 10.5 knot coal-burning freighter, 216182, LKMQ. Keel laid on June 14, 1917. Requisitioned from Norwegian owners while building and delivered to the USSB on April 23, 1918. Allocated to the New York & Cuba Mail Steamship Co. for service to the east coast of South America.

Walter A. Luckenbach—8,286 tons, steel, 14 knot oil-burning freighter, 216218, LKPC. Keel laid on December 20, 1916, for Edgar F. Luckenbach. Requisitioned while building and delivered to the USSB on June 10, 1918. Commissioned *Walter A. Luckenbach* in the Naval Overseas Transportation Service on June 9, 1918.

Vittorio Emmanuele III—4,855 tons, steel, 10.5 knot coal-burning freighter, 216491, LKWV. Keel laid on November 5, 1917. Requisitioned from Norwegian owners while building and delivered to the USSB on June 27, 1918. Commissioned *Vittorio Emmanuele III* in the Naval Overseas Transportation Service on June 27, 1918.

Masuda—4,844 tons, steel, 10.5 knot coal-burning freighter, 216709, LMJK. Keel laid on November 15, 1917. Requisitioned while building and delivered to the USSB on July 31, 1918. Allocated to Barber for service to the United Kingdom.

Skinner & Eddy Corp.
Seattle, Washington

David E Skinner was president of the Port Blakely Mill Co., lumber manufacturers, and lived with his wife Jennie at 725 14th Ave. N.

John W. Eddy was a lumberman who lived with his wife Ethel G. at 1117 Boylston Ave. N.

In 1916 the two formed the Skinner & Eddy Corp. to build steel ships. Skinner was president, Eddy vice-president, and Clarence B. Lamont general manager. The company's last ship was the *Robin Gray* in 1920 for the Robin Line.

S.V. Harkness—6,899 tons, tanker, 1917, 215006, LHBQ. Standard Oil of New Jersey. Stephen Harkness was a director of the company.

Josiah Macy—6,899 tons, freighter, 1917, 215180, LHGR. Standard Oil of New Jersey. Macy was an officer in the company.

Jeanette Skinner—5,715 tons, steel, 11.5 knot coal or oil-burning freighter, 215542, LHSP. Keel laid on March 17, 1917. Requisitioned from Norwegian owners while building and delivered to the USSB on September 6, 1917. Allocated to the New York & Cuba Mail Steamship Co. for service to the River Plate. Commissioned *Jeanette Skinner* in the Naval Overseas Transportation Service on April 12, 1918.

Lieutenant De Missiessy—5,746 tons, steel, 11.5 coal or oil-burning freighter, 215619, LHVP. Keel laid on April 25, 1917. Requisitioned from French owners while building and delivered to the USSB on September 19, 1917. Later returned to France.

Western Front—5,742 tons, steel, 11.5 knot coal or oil-burning freighter, 215724, LJDN. Keel laid on May 25, 1917. Requisitioned from French owners while building and delivered to the USSB on November 1, 1917. Allocated to Strachan Shipping for service to the United Kingdom. Commissioned *Western Front* in the Naval Overseas Transportation Service on May 9, 1918.

West Haven—5,699 tons, steel, 11.5 knot coal or oil-burning freighter, 215889, LJMT. Keel laid on August 9, 1917. Requisitioned from British owners while building and delivered to the USSB on December 27, 1917. Commissioned *West Haven* in the Naval Overseas Transportation Service on June 18, 1918.

Western Hero—5,564 tons, steel, 11.5 knot coal-burning freighter, 215908, LJNQ. USSB contract ship. Keel laid on August 21, 1917, and delivered on January 5, 1918, 137 building days. Allocated to the New York & Cuba Mail Steamship Co. for service to the east coast of South America.

Trontolite—6,882 tons, steel, 11.5 knot coal or oil-burning tanker, 215947, LJPK. Keel laid on July 3, 1917, for Standard Oil of New Jersey. Requisitioned while building and delivered to the USSB on February 1, 1918. Operated by the company.

Absaroka—5,568 tons, steel, 10.5 knot coal-burning freighter, 215986, LJQR. USSB contract ship. Keel laid on September 4, 1917, and delivered on February 12, 1918, 161 building days. Allocated to Harriss, Magill for service to France. Allocated to the Naval Overseas Transportation Service. Operated during World War II by the McCormick Steamship Co. Torpedoed on December 24, 1941, by *I-19* off San Pedro, California, but towed in and repaired.

West Arrow—5,852 tons, steel, 11.5 knot coal or oil-burning freighter, 216012, LJRP. Keel laid on September 18, 1917. Requisitioned from Norwegian owners while building and delivered to the USSB on February 26, 1918. Operated by the Shipping Board for service to France.

Westlake—5,852 tons, steel, 11.5 knot coal or oil-burning freighter, 216044, LJSP. Keel laid on November 9, 1917. Requisitioned from Norwegian owners while building and delivered to the USSB on March 19, 1918. Operated by the Shipping Board for service to France.

Canoga—5,575 tons, steel, 11.5 knot coal-burning freighter, 216095, LJVS. USSB contract ship. Keel laid on December 1, 1917, and delivered on March 23, 1918, 112 building days. Allocated to Williamson & Barnes for service to the United Kingdom.

Ossineke—5,577 tons, steel, 11.5 knot coal-burning freighter, 1918, 216155, LKMH. USSB contract ship. Keel laid on December 26, 1917, and delivered on April 14, 1918, 107 building days. Allocated to the Earn Line for service to France.

Western Queen—5,853 tons, steel, 11.5 knot coal or oil-burning freighter, 216224, LKPN. Keel laid on January 2, 1918. Requisitioned from Norwegian owners while building and delivered to the USSB on April 25, 1918. Allocated to Luckenbach as manager for service to the United Kingdom by the Texas Transport & Terminal Co.

West Lianga—6,222 tons, steel, 11.5 knot coal or oil-burning freighter, 216274, LKQR. USSB contract ship. Keel laid on February 14, 1918, and delivered on May 4, 1918, 80 building days. Allocated to J.H.W. Steele for service to the United Kingdom. Commissioned *West Lianga* in the Naval Overseas Transportation Service on August 19, 1918.

West Durfee—6,087 tons, steel, 11.5 knot coal or oil-burning freighter, 216342, LKRN. USSB contract ship. Keel laid on January 25, 1918, and delivered on May 10, 1918, 111 building days. Allocated to S. Sgitcovich for service to the United Kingdom.

West Alsek—5,573 tons, steel, 11.5 knot coal-burning freighter, 216415, LKTW. USSB contract ship. Keel laid on March 4, 1918, and delivered on June 4, 1918, 93 building days. Commissioned *West Alsek* in the Naval Overseas Transportation Service on June 4, 1918.

West Apaum—5,573 tons, steel, 11.5 knot coal-burning freighter, 216500, LMBC. USSB contract ship. Keel laid on March 19, 1918, and delivered on June 19, 1918, 93 building days. Commissioned *West Apaum* in the Naval Overseas Transportation Service on June 20, 1918.

West Ekonk—6,086 tons, steel, 11.5 knot coal or oil-burning freighter, 216520, LMFW. USSB contract ship. Keel laid on April 16, 1918, and delivered on June 22, 1918, 88 building days. Allocated to the American Line for the European Food Relief mission. Commissioned *West Ekonk* in the Naval Overseas Transportation Service on July 13, 1918.

West Cohas—6,087 tons, steel, 11.5 coal or oil-burning freighter, 216549, LMCP. USSB contract ship. Keel laid on April 2, 1918, and delivered on June 29, 1918, 89 building days. Commissioned *West Cohas* in the Naval Overseas Transportation Service on July 29, 1918.

West Gambo—6,087 tons, steel, 11.5 knot coal-burning freighter, 216664, LMHD. USSB contract ship. Keel laid on April 25, 1918, and delivered on July 20, 1918, 86 building days. Commissioned *West Gambo* in the Naval Overseas Transportation Service on July 20, 1918.

West Gotomska—6,087 tons, steel, 11.5 knot coal or oil-burning freighter, 216737, LMKD. USSB contract ship. Keel laid on May 17, 1918, and delivered on August 7, 1918, 83 building days. Commissioned *West Gotomska* in the Naval Overseas Transportation Service on August 7, 1918.

West Hobomac—6,087 tons, steel, 11.5 knot coal or oil-burning freighter, 216786, LMPD. USSB contract ship. Keel laid on May 29, 1918, and delivered on August 17, 1918, 80 building days. Commissioned *West Hobomac* in the Naval Overseas Transportation Service on August 17, 1918.

West Hosokie—6,087 tons, steel, 11.5 coal or oil-burning freighter, 216848, LMSH. USSB contract ship. Keel laid on June 11, 1918, and delivered on August 29, 1918, 79 building days. Commissioned *West Hosokie* in the Naval Overseas Transportation Service on August 29, 1918.

West Humhaw—6,087 tons, steel, 11.5 knot coal or oil-burning freighter, 216925, LMWF. USSB contract ship. Keel laid on June 27, 1918, and delivered on September 10, 1918, 80 building days. Commissioned *West Humhaw* in the Naval Overseas Transportation Service on September 16, 1918. Operated during World War II by the American West African Line. Sunk on November 8, 1942, by *U-161* off the west coast of Africa.

West Lashaway—6,087 tons, steel, 11.5 knot coal or oil-burning freighter, 217002, LNDR. USSB contract ship. Keel laid on July 8, 1918, and delivered on September 30, 1918, 83 building days. Commissioned *West Lashaway* in the Naval Overseas Transportation Service on September 30, 1918. Operated during World War II by the American West African Line. Sunk on August 30, 1942, by *U-66* in the South Atlantic.

West Loquassuck—6,087 tons, steel, 11.5 knot coal or oil-burning freighter, 217070, LNHR. USSB contract ship. Keel laid on July 20, 1918, and delivered on October 15, 1918, 76 building days. Commissioned *West Loquassuck* in the Naval Overseas Transportation Service on October 15, 1918.

West Elcasco—6,087 tons, steel, 11.5 knot coal or oil-burning freighter, 217128, LNKV. USSB contract ship. Keel laid on July 8, 1918, and delivered on October 23, 1918, 109 building days. Commissioned *West Elcasco* in the Naval Overseas Transportation Service on October 23, 1918. Operated during World War II by the Army as the *Maj. Gen. Henry Gibbins*. Sunk on June 23, 1942, by *U-158* in the Gulf of Mexico.

West Madaket—6,087 tons, steel, 11.5 knot coal or oil-burning freighter, 217176, LNQF.

USSB contract ship. Keel laid on July 31, 1918, and delivered on October 30, 1918, 91 building days. Commissioned *West Madaket* in the Naval Overseas Transportation Service on October 30, 1918. Operated during World War II by Waterman. Sunk on May 5, 1943, by *U-707* in the North Atlantic.

West Mahomet—6,087 tons, steel, 11.5 knot coal or oil-burning freighter, 217248, LNVG. USSB contract ship. Keel laid on August 20, 1918, and delivered on November 13, 1918, 84 building days. Commissioned *West Mahomet* in the Naval Overseas Transportation Service on November 13, 1918.

West Eldara—6,087 tons, steel, 11.5 knot coal or oil-burning freighter, 217282, LNWV. USSB contract ship. Keel laid on July 15, 1918, and delivered on November 22, 1918, 131 building days. Commissioned *West Eldara* in the Naval Overseas Transportation Service on November 23, 1918. Operated during World War II by A.H. Bull as the *Mae*. Sunk on September 17, 1942, by *U-515* off British Guiana.

Edenton—6,992 tons, steel, 11.5 knot coal and oil-burning freighter, 217336, LPDM. USSB contract ship. Keel laid on August 31, 1918, and delivered on December 5, 1918, 96 building days. Commissioned *Edenton* in the Naval Overseas Transportation Service on December 5, 1918.

West Cressy—6,091 tons, steel, 11.5 knot coal or oil-burning freighter, 217382, LPGK. USSB contract ship. Keel laid on June 26, 1918, and delivered on December 17, 1918, 174 building days. Commissioned *West Cressy* in the Naval Overseas Transportation Service on December 17, 1918.

Edgecombe—6,897 tons, steel, 11.5 knot coal and oil-burning freighter, 217412, LPHV. USSB contract ship. Keel laid on September 16, 1918, and delivered on December 24, 1918, 102 building days. Commissioned *Edgecombe* in the Naval Overseas Transportation Service on December 27, 1918.

Edgefield—6,990 tons, steel, 11.5 knot coal and oil-burning freighter, 217433, LPKG. USSB contract ship. Keel laid on September 25, 1918, and delivered on December 31, 1918, 98 building days. Bareboat chartered to the Atlantic Transport Co.

Sloan Shipyards Corp.
552 Colman Bldg., Olympia, Washington

In 1915 Philip D. Sloan was president and lived in Seattle with his wife Isabelle at 1125 23rd Ave. N. John O. Pierce was secretary-treasurer, and John C. Arnold was office manager and superintendent of the 8-way yard at the foot of Washington. In 1918 Jesse D. Kuhns was business manager, and Edward W. Heath superintendent.

Coolcha—2,403 tons, wood motor freighter, 1918, 219279, LTVF. J.E. Chilberg.

Cethana—2,341 tons, wood motor freighter, 1918, 219330, LVBG. Launched on February 17 for the Australian government. Owned in 1919 by J.E. Chilberg.

Challamba—2,400 tons, wood motor freighter, 1918, 219331, LVBH. Owned in 1919 by J.E. Chilberg.

Culburra—2,353 tons, wood motor freighter, 1918, 219540, LVMC. Launched on March 13 for the Australian government. Owned in 1919 by J.E. Chilberg.

Southwestern Shipbuilding Co.
224 H.W. Hellman Building,
356 S. Spring St., Los Angeles

Marco H. Hellman was president of the Hellman Commercial Trust & Savings Bank and president of Southwestern Ship. James A. Talbot was 1st vice-president and general manager, William F. Howard was 2nd vice-president, Emanuel Cohen, the cashier at Hellman's bank, was secretary, and Irving Hellman, vice-president of Hellman's bank, was treasurer. William Lambie was the naval architect and David Hollywood the works manager.

Work on the 65-acre yard in East San Pedro began on March 23, 1918, and was finished on June 20. The company built 22 ships, the last one being the big freighter *Papoose* in 1921. In 1920 they built the *Hollywood* in honor of David Hollywood who died on the evening of January 16, 1920, when the Cadillac he was driving collided with a street car in San Pedro.

West Carnifax—6,174 tons, steel, 10.5 knot freighter, 217373, LPGC. USSB contract ship. Keel laid on July 4, 1918, and delivered on December 31, 1918, 180 building days. Commissioned *West Carnifax* in the Naval Overseas Transportation Service on December 31, 1918.

Standifer-Clarkson Shipyard, Inc.
Portland, Oregon

A 4-way yard built in 1916 on 25 acres by contractors Guy Standifer and James Clarkson.

W.F. Burrows—1,580 tons, wood motor freighter, 1917, 215139, LHFS. Libby, McNeill & Libby.

James Timpson—2,016 tons, wood motor freighter, 1918, 216016, LJRV. Resolved Corp.

Libby Maine—1,811 tons, wood motor freighter, 1918, 216452, LKVQ. Libby, McNeill & Libby.

Supple-Ballin Shipbuilding Corp.
Portland, Oregon

Joseph Supple built boats at the foot of Belmont St. on the west side and had a dock at the foot of Yamhill on the east side of the Willamette River.

Fred A. Ballin was a naval architect, consulting engineer, and president of the Ballin Water Tube Boiler Co. in Portland. James B.C. Lockwood was secretary of Ballin Water Tube.

In January 1917 they opened a marine railway and repair yard on 8 acres at SE Oak at Water St. as Supple Ballin & Lockwood. Lockwood left the company and in 1918 Supple-Ballin Shipbuilding Corp. was formed with Supple as president, attorney Arthur Langguth vice-president, and Ballin secretary-treasurer.

Ballin came up with Shipping Board Design No. 1011, a well deck ship with two forward and two after holds with a single midship house and an after deckhouse. The company got a contract for 12 Ballin ships. They employed around 450.

Mount Hood—2,909 tons, wood motor freighter, 1918, 216058, LJSW. Gaston, Williams & Wigmore. Burned in 1919.

Mount Shasta—2,904 tons, composite-hull motor freighter, 1918, 216217, LKPB. Gaston, Williams & Wigmore.

Calala—2,799 tons, wood, 10.5 knot coal-burning Ballin freighter, 216748, LMKP. USSB contract ship. Keel laid on December 6, 1917, and delivered on August 13, 1918, 250 building days. Allocated to Callaghan, Atkinson as manager for operation by the Northland Steamship Co. to the east coast of South America.

Dalana—2,799 tons, wood, 10 knot coal-burning Ballin freighter, 216833, LMRJ. USSB contract ship. Keel laid on January 9, 1918, and delivered on September 3, 1918, 236 building days. Managed and operated by the USSB.

Airlie—2,799 tons, wood, 10.5 knot coal-burning Ballin freighter, 216918, LMVW. USSB contract ship. Keel laid on February 25, 1918, and delivered on October 22, 1918, 241 building days. Owned in 1919 by Nacirema Steamship.

Wallowa—2,799 tons, wood, 10 knot coal-burning Ballin freighter, 216922, LMWC. USSB contract ship. Keel laid on August 3, 1917, and delivered on September 28, 1918, 421 building days. Allocated to Callaghan, Atkinson as manager for service to France by Northland Steamship.

Thala—2,799 tons, wood, 10 knot coal-burning Ballin freighter, 217131, LNMF. USSB contract ship. Keel laid on August 2, 1917, and delivered on December 13, 1918, 498 building days. Managed and operated by the USSB.

Tacoma Shipbuilding Co.
Tacoma, Washington

The Foundation Co. of New York acquired 50 acres at Tacoma in September 1917 with plans to construct a plant and ten to twelve 300' ways to build ships for the French government. Franklin Remington was president in New York and Henry A. Rhodes was president of Tacoma Ship. John S. Baker was vice-president, Joseph S. Whitehouse secretary, and Arthur G. Prichard treasurer. The property was developed into a 6-way yard on 35 acres under manager James E. Bonnell.

Beloit—2,531 tons, wood, 10 knot coal-burning Ferris freighter, 216894, LMVB. USSB contract ship. Keel laid on August 1, 1917, and delivered on August 13, 1918, 378 building days. Allocated to American Shipping for service to the United Kingdom.

Coloma—2,526 tons, wood, 10 knot coal-burning Ferris freighter, 217167, LNPR. USSB contract ship. Keel laid on September 6, 1917, and delivered on December 28, 1918, 480 building days. Allocated to Callaghan, Atkinson for service to the Netherlands.

Todd Dry Dock & Construction Corp.
Tacoma, Washington

The Todd Dry Dock Engineering & Repair Corp. was headquartered in Brooklyn. Clarence W. Wiley was president at Tacoma, James A. Eves vice-president and general manager, and Carl S. Holmes assistant secretary-treasurer. The company had a foundry, a floating drydock, a boiler shop, and offered engineering services at their 8-way, 105-acre Hylebos Waterway plant. The company employed 6,600 workers.

Chebaulip—5,769 tons, steel, 10.5 knot coal-burning freighter, 216589, LMFD. Keel laid on September 17, 1917. Requisitioned from British owners while building and delivered to the USSB on June 30, 1918. Commissioned *Chebaulip* in the Naval Overseas Transportation Service on July 11, 1918.

Anacortes—4,829 tons, steel, 10.5 knot coal burning freighter, 216846, LMSF. Keel laid on September 24, 1917. Requisitioned from British owners while building and delivered to the USSB on August 31, 1918. Allocated to National Shipping for service to the east coast of South America.

Puget Sound—4,836 tons, steel, 10.5 knot coal-burning freighter, 217015. LNFG. Keel laid on November 23, 1917. Requisitioned from British owners while building and delivered to the USSB on September 30, 1918. Allocated to James W. Elwell as manager for service to the east coast of South America by the New York & Cuba Mail Steamship Co.

Willimantic—5,237 tons, steel, 10.5 knot oil-burning freighter, 217129, LNKW. USSB contract ship. Keel laid on November 24, 1918, and delivered on October 31, 1918, 241 building days. Allocated to the Oriental Navigation Co. for service on Swiss accounts.

Bellingham—4,584 tons, steel, 10.5 knot coal-burning knot freighter, 217150, LNMP. Keel laid on March 30, 1918. Requisitioned from British owners while building and delivered to the USSB on October 30, 1918. Commissioned *Bellingham* in the Naval Overseas Transportation Service on October 30, 1918.

Yukon—4,843 tons, steel, 10.5 knot coal-burning freighter, 217332, LPDJ. Keel laid on May 25, 1918. Requisitioned from British owners while building and delivered to the USSB on November 30, 1918. Allocated to the Atlantic Transport Company of West Virginia for service to the United Kingdom.

Deranof—5,245 tons, steel, 10.5 knot oil-burning freighter, 217406, LPHN. USSB contract ship. Keel laid on December 31, 1917, and delivered on December 24, 1918, 358 building days. Allocated to the American Line for service to the United Kingdom.

W.F. Stone
Oakland, California

In 1913, naval architects Lester F. and William F. Stone opened an office in William's home at 2134 42nd Ave. The next year William moved to 1534 E. Park St. In 1916 they were known as W.F. Stone and had a 2-way shipyard at the foot of Kennedy St.

William was president, general manager, and naval architect. Lester was vice-president and secretary, and accountant William A. Barlage, of San Francisco, was treasurer. Charles White was superintendent and ship joiner Edward H. Boyce was chief engineer.

Ryder Hanify—1,363 tons, wood steam freighter, 1917, 215007. J.R. Hanify Co.

Robert C. Sudden—1,363 tons, wood steam freighter, 1917, 215234. Sudden & Christenson.

Washington Shipping Corp.
409 Alaska Building, Seattle

Samuel H. Hedges was president of the Puget Sound Bridge & Dredging Co., Frank E. Burns was vice-president and general manager of the Puget Sound Navigation Co., Robert R. Fox was manager of the Simonds Manufacturing Co. (saw, knives, and mill supplies), and Bert Farrar sold real estate. The partners got the idea to build six auxiliary schooners for the French government so in 1916 they built a yard on Harbor Island. Hedges was president, Burns vice-president, Fox secretary, and Farrar treasurer. The ships were equipped with twin 240-hp Mietz & Weiss engines.

Portland—1,594 tons, 4-masted auxiliary schooner, 1917, 215463, LHRD. Pacific Motorship Co., Portland, Oregon. Burned at sea on June 3, 1918, with 21 on board. There were no casualties

Remittent—1,616 tons, 4-masted auxiliary schooner, 1917, 215652, LHWT. Washington Shipping Corp. Purchased by Alexander Prebensen of New York.

Barleux—1,607 tons, 4-masted auxiliary schooner, 1917, 215675, LJBV. Completed on a French account. The first ship ever built on Puget Sound for a foreign power.

Risor—1,604 tons, 4-masted auxiliary schooner, 1917, 215676, LJBW. Rosario Navigation Corp. Purchased by Alexander Prebensen of New York.

Western Pipe & Steel Company of California
South San Francisco, California

Henry Davies Thomson was born in San Antonio, Texas, on October 5, 1871. James L. Boyle was born in New Orleans on August 31, 1866. In 1894 James married Henry's sister, Catherine and in May 1896, Henry, James, and Catherine came out to California. That same year Henry married a California girl named Laura. In 1898 James and Catherine had a son, James T. and in 1901 Henry and Laura had a daughter, Laura E.

The two went into business and in August 1898 incorporated the Thomson & Boyle Co. at 1101 N. Main in Los Angeles to sell plumbing supplies. Henry was president, James vice-president, James A. Talbot, of Los Angeles, was secretary, and Allen C. Dunham, of Pasadena, was the bookkeeper. Henry was also president of the Phillips Sanitary Grate and Heating Co. The Thomsons lived at 1207 W. 6th and the Boyles at 507 Carondelet.

The partners were very successful and by 1907 they were manufacturing iron and steel pipe, well casings, steam boilers, and oil and cyanide tanks. At that time, William F. Howard was a laundryman who lived at 225 S. San Pedro St. in Los Angeles.

The exact chain of events is unknown, but in 1908 three things happened: Thomson went back to San Antonio and became a mine operator, James Boyle took a job as 2nd vice-president of the Los Angeles Nevada Mining Stock Exchange at 256 Spring St., and their business was reorganized as the Western Pipe & Steel Co. at 1101 N. Main. Howard was president, Talbot vice-president, and Dunham secretary. The company continued to manufacture iron and steel pipe, well casings, irrigation supplies, and oil and cyanide tanks. In 1909 the company moved to 1758–1764 Downey Ave. in Los Angeles.

In 1910 the company opened an office in San Francisco at 444 Market St. to manufacture sheet iron, steel pipe, tanks, and irrigation supplies.

In 1911 Talbot was president and Howard vice-president in Los Angeles. The company had plants in San Francisco, Richmond, and Los Angeles. They built three steel Shipping Board freighters homeported at San Francisco.

Nantahala—6,176 tons, steel, 10.5 knot oil-burning freighter, 217209, LNSD. USSB contract ship. Keel laid on December 29, 1917, and delivered on October 31, 1918, 302 building days. Allocated to Moore & McCormick for service to the River Plate. Commissioned *Nantahala* in the Naval Overseas Transportation Service on November 16, 1918.

Oskaloosa—6,162 tons, steel, 10.5 knot oil-burning freighter, 217391, LPGS. USSB contract ship. Keel laid on January 10, 1918, and delivered on December 20, 1918, 346 building days. Commissioned *Oskaloosa* in the Naval Overseas Transportation Service on December 20, 1918.

Wilson Bros.
300 Taylor Ave., Astoria, Oregon

Charles W. Kankonen was born in Gamla Carleby, Finland, on May 7, 1861, built boats, and on April 17, 1880, boarded the steamer *Bravo* at Gothernburg, Sweden, for Hull, England, with the ultimate destination of Muskegon, Michigan. He arrived at New York on May 4, 1880, on the Anchor Line steamer *Anchoria,* worked in lumbering, and at some point changed his surname to Wilson. In 1883 came to Astoria, Oregon, helped build a cannery, and then took up fishing. He helped organize the Fisherman's Co-operative Packing Co. and in 1887 built fishing boats. His brothers Frans and Frithiof joined him as general contractors and in 1906 they began building steamers as the Wilson Bros. at Smith's Point on 2 ways. In 1907 they delivered the *General Washington.*

In 1917 the brothers got a Shipping Board contract and expanded to 18 acres and reorganized their business with a capital of $100,000 from investors as the Wilson Shipbuilding Co.

Peter J. Brix was vice-president of the Fisher Bros. Co., a groceries, hardware, and ship chandlery at 546 Bond St. and had an interest in the Knappton Mills & Lumber Co. at Knappton, Washington. He became president. Everell Stanton Collins, co-owner of the Ostrander Railway & Timber Co. at Ostrander, Washington, was vice president. Lumberman John A. Byerly, of Portland, was treasurer, and Frithiof Kankonen secretary. Charles Wilson was manager of the 2-way yard. The company had 350 employees.

Wahkeena—1,053 tons, steam schooner, 1916, 214880, LGTW. McCormick Steamship Co.

Ernest H. Meyer—1,057 tons, steam schooner, 1917, 215333, LHMR. Broughton & Wiggins Navigation Co.

Quoque—2,540 tons, wood, 10 knot coal-burning Ferris freighter, 216917, LMVT. USSB contract ship. Keel laid on July 19, 1917, and delivered on September 21, 1918, 431 building days. Allocated to the Pendleton Bros. for service to France.

Lonoke—2,540 tons, wood, 10 knot coal-burning Ferris freighter, 217014, LNFD. USSB contract ship. Keel laid on July 26, 1917, and delivered on October 30, 1918, 461 building days. Allocated to W.J. Grandfield & Son for service to the Netherlands.

Bonifay—2,531 tons, wood, 10 knot freighter, 1918, 217037, LNGJ. USSB. Allocated to the Columbia Pacific Shipping Co. for service to the United Kingdom.

Winslow Marine Railway & Shipbuilding Co., Inc. Winslow, Washington

This 1-way yard on Bainbridge Island was leased to Donald W. Hartzell, treasurer of the Northwest Trading Co. and the Valdez Packaging Co. in Seattle, holder of a contract to install engines and auxiliary equipment in Shipping Board freighters. The company didn't build any Shipping Board vessels.

Anyox—1,287 tons, freighter, 1917, 215804, LJGT. Coastwise Steamship & Barge Co.

Wright Ship Yards Co. Tacoma, Washington

In early 1917, Theodore Brent, a representative from the Shipping Board, was scouring the West Coast looking for potential contractors to build wooden ships, "on account of the large growth of timber available on the Pacific coast and in the Northwest."

George P. Wright was a general contractor in Tacoma who lived with his wife Susan at 824 S. L St. On April 6, 1917, Brent called on Wright, Wright agreed, and built a 4-way yard just north of Seaborn's shipyard on the City Waterway as a private enterprise. Wright's son Horace was assistant manager. The company employed 300 to build the two freighters.

Yakima—2,523 tons, wood, 10 knot coal-burning Ferris freighter, 217249, LNVH. USSB contract ship. Keel laid on June 29, 1917, and delivered on November 22, 1918, 510 building days. Allocated to Luckenbach as manager for service to the Netherlands by Gailey, Davis & Co.

Eyota—2,529 tons, wood, 10 knot coal-burning Ferris freighter, 217393, LPGV. Keel laid on July 12, 1917, and launched on June 10, 1918, 333 days. Delivered sometime after May 31, 1919. Allocated to Struthers & Dixon for service from the Pacific to the United Kingdom.

The Great Lakes and Western Rivers

The American Shipbuilding Co. 2017 Superior Ave. NW, Cleveland, Ohio

In 1885, George Presley was president of Presley & Co. in Cleveland and lived at 144 Clinton St. His partners were John F. Pankhurst, Henry D. Coffinberry, John B. Cowle, and Robert Wallace. The firm operated Globe Dry Docks at 96 Elm St.

Cowle, Wallace, Pankhurst, and Coffinberry also operated Globe Iron Works, a maker of marine engines and boilers at Elm and Spruce Sts.

In 1886 Coffinberry, Wallace, and Cowle began building ships as the Globe Ship Building Co. at the foot of St. Paul on the Old River. Henry was president, Wallace secretary, Cowle treasurer, and John H. Smith was superintendent.

The next year, 1887, they changed the name of their firm to the Cleveland Ship Building Co. and Globe Iron started building iron and steel steamships. In 1890 Cleveland Ship's office was at 120 Viaduct St. and Wallace was vice-president, William M. Fitch secretary. In 1895 James C. Wallace was vice-president and general manager. Henry Coffinberry retired and in 1898 their machine and boiler shop and foundry were in Cleveland and they opened a new yard at Lorain. Ship repairs could be made in Cleveland or Lorain but shipbuilding and drydock was at Lorain. Robert Wallace was president, James C. Wallace vice-president and general manager, William Fitch was secretary. Globe Iron had become a major shipbuilder.

In 1900 a major shakeup occurred whereby Cleveland Ship and Globe Iron were merged into the American Shipbuilding Co. at 120 Viaduct. William L. Brown, of Chicago, was president, Robert L. Ireland, of Glenville, vice-president, Russell C. Wetmore secretary-treasurer, and James C. Wallace general manager. Robert Wallace stayed on but had no management role.

William L. Brown was vice-president of Picklands, Brown & Co., 1315 McCormick Building, marine engineers and naval architects.

Robert Ireland had been vice-president, treasurer, and general manager of the Ship Owners' Dry Dock Co. at W. Center & Spruce Sts. and lived at 661 Prospect.

The company grew and expanded and in 1914 the headquarters were at 2017 Superior Ave. NW in Cleveland. James C. Wallace was president and James L. Wallace was the local general manager. Both lived at 13526 Detroit Ave. Vice-president Edward N. Smith was running the company's Buffalo Dry Dock yard in Buffalo, New York. Ora J. Fish was secretary-treasurer, and Alfred G. Smith was the company general manager. He lived at 6710 Franklin Ave. NW.

By 1918, American Ship operated yards in Chicago; Brunswick, Georgia; Cleveland and Lorain, Ohio; Detroit and Wyandotte, Michigan; Superior and Milwaukee, Wisconsin; and Buffalo, New York. The company employed over 20,000 workers.

Ships built in the yards of the Great Lakes intended for ocean service had to pass through the Welland Canal. Around 448 lakers of 4,000 deadweight tons were built during the war years.

CLEVELAND

Howard M. Hanna, Jr.—6,204 tons, laker, 1914, 212353. W.C. Richardson & Co.

Socony 82—2,320 tons, schooner, 1916, 214518, LGKS. Standard Oil of New York.

Sioux—2,048 tons, freighter, 1916, 214569, LGNC. Atlantic Gulf & West Indies. Allocated to the Navy.

Kiowa—2,039 tons, freighter, 1917, 215375, LHNS. Atlantic Gulf & West Indies. Commissioned *Kiowa* in the Naval Overseas Transportation Service on February 26, 1918.

Lake Port—2,028 tons, steel, 9 knot coal-burning freighter, 215679, LJCF. Keel laid on June 2, 1917. Requisitioned from British owners while building and delivered to the USSB on October 8, 1917. Allocated to the New York & Cuba Mail Steamship Co. for service to the West Indies. Commissioned *Lake Port* in the Naval Overseas Transportation Service on January 21, 1918.

Lake Michigan—2,020 tons, steel, 9.5 knot coal-burning freighter, 215772, LJFV. Keel laid on June 20, 1917. Requisitioned from British owners while building and delivered to the USSB on November 17, 1917. Commissioned *Lake Michigan* in the Naval Overseas Transportation Service on January 30, 1918.

Lake Lillian—2,020 tons, steel, 10.5 knot coal-burning freighter, 215901, LJNH. Keel laid on September 6, 1917. Requisitioned from British owners while building and delivered to the USSB on January 31, 1918. Allocated to A.H. Bull for the phosphate service. Commissioned *Lake Lillian* in the Naval Overseas Transportation Service on October 2, 1918.

Lake Otisco—2,015 tons, steel, 9.5 knot coal-burning freighter, 216040, LJSM. Keel laid on September 26, 1917. Requisitioned from British owners while building and delivered to the USSB on February 25, 1918. Allocated to Munson for service to the West Indies. Commissioned *Lake Otisco* in the Naval Overseas Transportation Service on October 5, 1918.

Lake Lemando—2,370 tons, steel, 9.5 knot coal-burning freighter, 216156, LKMJ. USSB contract ship. Keel laid on November 27, 1917, and delivered on May 11, 1918, 155 building days. Commissioned *Lake Lemando* in the Naval Overseas Transportation Service on November 21, 1918.

Lake Edon—2,371 tons, steel, 9.5 knot coal-burning freighter, 216232, LKPT. USSB contract ship. Keel laid on December 18, 1917, and delivered on May 21, 1918, 153 building days. Allocated to the Army.

Lake Narka—2,367 tons, steel, 9.5 knot coal-burning freighter, 216399, LKTJ. USSB contract ship. Keel laid on March 19, 1918, and delivered on June 11, 1918, 48 building days. Allocated to the Coastwise Transportation Co. for service to the West Indies.

Lake Duncan—2,367 tons, steel, 9.5 knot coal-burning freighter, 216539, LMCH. USSB contract ship. Keel laid on March 19, 1918, and delivered on July 3, 1918, 106 building days. Allocated to the Atlantic Chartering Co. for service to the West Indies.

Lake Blanchester—2,346 tons, steel, 9.5 knot coal-burning freighter, 216657, LMGW. USSB contract ship. Keel laid on March 26, 1918, and delivered on July 31, 1918, 126 building days. Commissioned *Lake Blanchester* in the Naval Overseas Transportation Service on August 10, 1918.

Lake Gedney—2,333 tons, steel, 9.5 knot coal-burning freighter, 216775, LMNP. USSB contract ship. Keel laid on May 4, 1918, and delivered on August 21, 1918, 109 building days. Allocated to the Hasler Bros. for service to the West Indies. Commissioned *Lake Gedney* in the Naval Overseas Transportation Service on September 6, 1918.

Lake Hewes—2,330 tons, steel, 9.5 knot coal-burning freighter, 216841, LMRT. USSB contract ship. Keel laid on May 22, 1918, and delivered on August 31, 1918, 101 building days. Allocated to Crowell & Thurlow for service to the West Indies.

Lake Alvada—2,329 tons, steel, 9.5 knot coal-burning freighter, 216926, LMWG. USSB contract ship. Keel laid on June 8, 1918, and delivered on September 19, 1918, 103 building days. Allocated to Munson for service to the West Indies.

Lake Govan—2,343 tons, steel, 9.5 knot coal-burning freighter, 217042, LNGQ. USSB contract ship. Keel laid on July 4, 1918, and delivered on October 13, 1918, 101 building days. Allocated to the New York & Cuba Mail Steamship Co. for service to the West Indies.

Lake Fluvanna—2,349 tons, steel, 9.5 knot coal-burning freighter, 217117, LNKH. USSB contract ship. Keel laid on July 25, 1918, and delivered on October 29, 1918, 96 building days. Allocated to the Earn Line for service to the West Indies.

Lake Farge—2,624 tons, steel, 9.5 knot coal-burning freighter, 217237, LNTP. USSB contract ship. Keel laid on August 18, 1918, and delivered on November 20, 1918, 94 building days. Allocated to the West Indies Steamship Co. for service to the West Indies.

Buffalo, New York

The Union Dry Dock Co. came under the management of American Ship in 1899 and was renamed Buffalo Dry Dock Co. Buffalo's president was Edward N. Smith, vice-president of American Ship. The *Lake Bledsoe* and *Lake Cathcoon* were documented as being built by Buffalo Dry Dock.

Lake Bledsoe—2,316 tons, steel, 9.5 knot coal-burning freighter, 216669, LMHJ. USSB contract ship. Keel laid on January 24, 1918, and delivered on July 30, 1918, 187 building days. Allocated to Shawmut for the New England coal service.

Lake Delancey—2,316 tons, steel, 9.5 knot coal-burning freighter, 216903, LMVG. USSB contract ship. Keel laid on February 20, 1918, and delivered on September 20, 1918, 212 building days. Allocated to the Coastwise Transportation Co. for the New England coal service.

Lake Galera—2,316 tons, steel, 9.5 knot coal-burning freighter, 217057, LNHC. USSB contract ship. Keel laid on March 6, 1918, and delivered on October 22, 1918, 229 building days. Allocated to the Earn Line for service to the West Indies.

Lake Otsquago—2,316 tons, steel, 9.5 knot coal-burning freighter, 217139, LNMJ. USSB contract ship. Keel laid on May 18, 1918, and delivered on November 23, 1918, 102 building days. Allocated to Crowell & Thurlow for service to the West Indies by the New York & Cuba Mail Steamship Co.

Lorain, Ohio

Robert L. Ireland—6,387 tons, laker, 1914, 212078. Wilson Line.

William D. Crawford—6,385 tons, laker, 1914, 212197. Virginia Steamship Co.

W.F. White—7,180 tons, laker, 1915, 213555. Limestone Transportation Co.

Henry G. Dalton—7,810 tons, laker, 1916, 214051. Interlake.

D.G. Kerr—7,756 tons, laker, 1916, 214147. U.S. Steel.

Emory L. Ford—7,986 tons, laker, 1916, 214318. Franklin Steamship Co.

Herbert F. Black—6,262 tons, laker, 1916, 214466. Donner Steamship Co.

D.M. Clemson—7,278 tons, laker, 1916, 214599. Pittsburgh Steamship Co.

Eugene W. Pargny—7,724 tons, laker, 1917, 214747. Pittsburgh Steamship Co.

Carl D. Bradley—7,203 tons, laker, 1917, 215018. Bradley Transportation Co.

Homer D. Williams—7,742 tons, laker, 1917, 215159. Pittsburgh Steamship Co.

Louis W. Hill—7,038 tons, laker, 1917, 215518. Producers Steamship Co.

Carmi A. Thompson—7,038 tons, laker, 1917, 215614. Producers Steamship Co.

William A. Amberg—7,031 tons, laker, 1917, 215708. Producers Steamship Co.

Lake Como—2,018 tons, steel, 10 knot coal-burning freighter, 216039, LJSK. Keel laid on September 15, 1917. Requisitioned from British owners while building and delivered to the USSB on April 22, 1918. Allocated to Crowell & Thurlow as managers for service to the West Indies by United Fruit.

Lake Charles—2,013 tons, steel, 10 knot coal-burning freighter, 216091, LJVP. Keel laid on September 18, 1917. Requisitioned from British owners while building and delivered to the USSB on April 23, 1918. Allocated to Crowell & Thurlow as managers for service to the West Indies by the Munson Line.

Lake Jessup—2,013 tons, steel, 10 knot coal-burning freighter, 216092, LJVQ. Keel laid on September 26, 1917. Requisitioned from British owners while building and delivered to the USSB on April 30, 1918. Allocated to Crowell & Thurlow for service to the West Indies.

Lake Ogden—2,013 tons, steel, 10 knot coal-burning freighter, 216093, LJVR. Keel laid on September 27, 1917. Requisitioned from British owners while building and delivered to the USSB on April 30, 1918. Allocated to Crowell & Thurlow as manager for service to the West Indies by the Raporel Steamship Co.

William P. Cowan—5,429 tons, tanker, 1918, 216311. Standard Oil of Indiana. Cowan was president of the company and died on August 14, 1918.

Lake Benton—2,018 tons, steel, 10 knot coal-burning freighter, 216329, LKRH. Keel laid on December 31, 1917. Requisitioned from British owners while building and delivered to the USSB on June 10, 1918. Allocated to the Atlantic Fruit Co. for service to the West Indies under Capt. Frederick S. Walker.

Lake Weir—2,018 tons, steel, 10 knot coal-burning freighter, 216330, LKRJ. Keel laid on December 19, 1917. Requisitioned from British owners while building and delivered to the USSB on May 29, 1918. Allocated to the New England Fuel & Transportation Co. as manager for service to the West Indies by United Fruit.

Lake Fernwood—2,013 tons, steel, 10 knot coal-burning freighter, 216400, LKTM. Keel laid on December 4, 1917. Requisitioned from British owners while building and delivered to the USSB on June 29, 1918. Allocated to United Fruit for service to the West Indies. Commissioned *Lake Fernwood* in the Naval Overseas Transportation Service on September 27, 1918.

Lake Harney—2,013 tons, steel, 10 knot coal-burning freighter, 216537, LMCF. Keel laid on January 17, 1918. Requisitioned from British owners while building and delivered to the USSB on July 13, 1918. Commissioned *Lake Harney* in the Naval Overseas Transportation Service on July 27, 1918.

Lake Wimico—2,013 tons, steel, 10 knot coal-burning freighter, 216538, LMCG. Keel laid on February 14, 1917. Requisitioned from British owners while building and delivered to the USSB on July 25, 1918. Allocated to A.H. Bull for the phosphate service. Commissioned *Lake Wimico* in the Naval Overseas Transportation Service on August 6, 1918.

Lake Eckhart—2,337 tons, steel, 9.5 knot coal-burning freighter, 216717, LMJS. USSB contract ship. Keel laid on March 2, 1918, and delivered on August 8, 1918, 158 building days. Commissioned *Lake Eckhart* in the Naval Overseas Transportation Service on August 20, 1918.

Lake Eliko—2,330 tons, steel, 9.5 knot coal-burning freighter, 216790, LMPK. USSB contract ship. Keel laid on March 16, 1918, and delivered on August 21, 1918, 158 building days. Commissioned *Lake Eliko* in the Naval Overseas Transportation Service on September 4, 1918.

Lake Elsinore—2,330 tons, steel, 9.5 knot coal-burning freighter, 216816, LMQR. USSB contract ship. Keel laid on March 18, 1918, and delivered on August 31, 1918, 166 building days. Commissioned *Lake Elsinore* in the Naval Overseas Transportation Service on September 13, 1918.

Lake Garza—2,331 tons, steel, 9.5 knot coal-burning freighter, 216870, LMST. USSB contract ship. Keel laid on April 15, 1918, and delivered on September 9, 1918, 148 building days. Commissioned *Lake Garza* in the Naval Overseas Transportation Service on September 24, 1918.

Lake Deval—2,329 tons, steel, 9.5 knot coal-burning freighter, 216891, LMTS. USSB contract ship. Keel laid on July 11, 1918, and delivered on September 21, 1918, 72 building days. Allocated to John S. Emery & Co. for service to the West Indies.

Lake Gaspar—2,331 tons, steel, 9.5 knot coal-burning freighter, 216978, LNCB. USSB contract ship. Keel laid on April 25, 1918, and delivered on September 25, 1918, 156 building days. Allocated to the Warren Transportation Co. for the New England coal service. Commissioned *Lake Gaspar* in the Naval Overseas Transportation Service on October 10, 1918.

Lake Yelverton—2,332 tons, steel, 9.5 knot coal-burning freighter, 217053, LNBH. USSB contract ship. Keel laid on May 8, 1918, and delivered on October 9, 1918, 157 building days. Allocated to the Merchants Navigation Co. for Atlantic coastwise service.

Lake Zaliska—2,332 tons, steel, 9.5 knot coal-burning freighter, 217074, LNHV. USSB contract ship. Keel laid on May 23, 1918, and delivered on October 16, 1918, 146 building days. Allocated to the New York & Cuba Mail Steamship Co. for service to the West Indies.

Lake Pickaway—2,339 tons, steel, 9.5 knot coal-burning freighter, 217116, LNKG. USSB contract ship. Keel laid on June 6, 1918, and delivered on October 26, 1918, 144 building days. Allocated to the New York & Cuba Mail Steamship Co. for service to the West Indies.

Lake Copley—2,339 tons, steel, 9.5 knot coal-burning freighter, 217154, LNPC. USSB contract ship. Keel laid on July 1, 1918, and delivered on November 2, 1918, 124 building days. Allocated to the Caribbean Steamship Co. for service to the West Indies.

Lake Cahoon—2,339 tons, steel, 9.5 knot coal-burning freighter, 217155, LNPD. USSB contract ship. Keel laid on July 25, 1918, and delivered on November 15, 1918, 112 building days. Allocated to the New York & Cuba Mail Steamship Co. for service to the West Indies.

Lake Frampton—2,622 tons, steel, 9.5 knot coal-burning freighter, 217288, LPBD. USSB contract ship. Keel laid on August 5, 1918, and delivered on November 30, 1918, 117 building days. Allocated to the West India Steamship Co. as managers for service to the United Kingdom by the Texas Transport & Terminal Co.

Lake Franconia—2,622 tons, steel, 9.5 knot coal-burning freighter, 217308, LPCH. USSB contract ship. Keel laid on August 14, 1918, and delivered on November 30, 1918, 108 building days. Allocated to Lykes Bros. for service to the United Kingdom.

Chicago Ship Building Co.
Chicago, Illinois

A subsidiary of Globe Iron Works, of Cleveland, established in 1890 in South Chicago at 101st St. and the Pittsburgh, Fort Wayne & Chicago Railway tracks. William I. Babcock was manager. In 1891 the company built the *Marina*, first steel ship on Lake Michigan.

In 1900 the yard was taken over by the American Ship Building Co. of Cleveland. Globe continued under its own name. Kellogg Fairbanks was president in 1916, Charles W. Frey assistant treasurer of the 20-acre yard at 101st on the Calumet River.

Manta—2,081 tons, freighter, 1916, 214554, LGMJ. Atlantic Gulf & West Indies Steamship Lines. Allocated to the Navy.

Choctaw—2,056 tons, freighter, 1917, 215580, LHTP. Atlantic Gulf & West Indies Steamship Lines. Commissioned *Choctaw* on March 4, 1918, in the Naval Overseas Transportation Service.

Lake Worth—2,080 tons, steel, 9 knot coal-burning freighter, 215716, LJDG. Keel laid on May 8, 1917. Requisitioned from British owners while building and delivered to the USSB on October 31, 1917. Allocated to Crowell & Thurlow for the New England coal service. Commissioned *Lake Worth* in the Naval Overseas Transportation Service on February 26, 1918.

Lake Huron—2,080 tons, steel, 9.5 knot coal-burning freighter, 215788, LJGH. Keel laid on July 25, 1917. Requisitioned from British owners while building and delivered to the USSB on November 19, 1917. Allocated to C.H. Sprague as manager for service to the United Kingdom by the Kerr Steamship Co. Commissioned *Lake Huron* in the Naval Overseas Transportation Service on January 11, 1918.

Lake Crescent—2,054 tons, steel, 10.5 knot coal-burning freighter, 216098, LJVW. Keel laid on September 22, 1917. Requisitioned from British owners while building and delivered to the USSB on April 19, 1918. Allocated to A.H. Bull for Atlantic coastwise service. Commissioned *Lake Crescent* in the Naval Overseas Transportation Service on October 19, 1918.

Lake Clear—2,054 tons, steel, 9.5 knot coal-burning freighter, 216199, LKNJ. Keel laid on October 27, 1917. Requisitioned from British owners while building and delivered to the USSB on May 6, 1918. Allocated to the Baltimore & Carolina Steamship Co. for Atlantic coastwise service. Commissioned *Lake Clear* in the Naval Overseas Transportation Service on November 15, 1918.

Lake Yemasee—2,319 tons, steel, 9.5 knot coal-burning freighter, 216540, LMCJ. USSB contract ship. Keel laid on February 4, 1918, and delivered on July 5, 1918, 151 building days. Allocated to C.H. Sprague & Son for the New England coal service.

Lake Yahara—2,338 tons, steel, 9.5 knot coal-burning freighter, 216698, LMHV. USSB contract ship. Keel laid on February 12, 1918, and delivered on July 29, 1918, 167 building days. Allocated to the Hasler Bros. for service to the West Indies and France. Commissioned *Lake Yahara* in the Naval Overseas Transportation Service on December 28, 1918.

Lake Bloomington—2,342 tons, steel, 9.5 knot coal-burning freighter, 216741. USSB contract ship. Keel laid on February 28, 1918, and delivered on August 26, 1918, 179 building days. Commissioned *Lake Bloomington* on September 12, 1918, in the Naval Overseas Transportation Service. Stranded at Ponta de la Coubre, France, on December 11, 1918, with 49 on board. There were no casualties.

Lake Berdan—2,342 tons, steel, 9.5 knot coal-burning freighter, 216814, LMQN. USSB contract ship. Keel laid on March 11, 1918, and delivered on September 9, 1918, 182 building days. Allocated to the Earn Line as manager for service to Belgium by the Triangle Steamship Co. Commissioned *Lake Berdan* on September 24, 1918, in the Naval Overseas Transportation Service.

Lake Buckeye—2,338 tons, steel, 9.5 knot coal-burning freighter, 216915, LMVS. USSB contract ship. Keel laid on April 30, 1918, and delivered on September 27, 1918, 149 building days. Allocated to the Coastwise Transportation Co. for service to the West Indies.

Lake Fondulac—2,338 tons, steel, 9.5 knot coal-burning freighter, 216998, LNDG. USSB contract ship. Keel laid on April 1, 1918, and delivered on October 18, 1918, 200 building days. Allocated to Munson for service to the West Indies.

Lake Beacon—2,338 tons, steel, 9.5 knot coal-burning freighter, 216999, LNDM. USSB contract ship. Keel laid on May 28, 1918, and delivered on October 25, 1918, 149 building days. Allocated to Munson for service to the West Indies.

Lake Desha—2,338 tons, steel, 9.5 knot coal-burning freighter, 217119, LNKP. USSB contract ship. Keel laid on June 18, 1918, and delivered on November 4, 1918, 139 building days. Allocated to Munson for service to the West Indies.

Lake Mattato—2,338 tons, steel, 9.5 knot coal-burning freighter, 217183, LNQP. USSB contract ship. Keel laid on July 6, 1918, and delivered on November 15, 1918, 131 building days. Allocated to the New York & Cuba Mail Steamship Co. for service to Cuba.

Lake Gradan—2,632 tons, steel, 9.5 knot coal-burning freighter, 217266, LNWC. USSB contract ship. Keel laid on August 7, 1918, and delivered on November 26, 1918, 111 building days. Allocated to the Texas Transport & Terminal Co. for service to the United Kingdom.

Lake Grafton—2,630 tons, steel, 9.5 knot coal-burning freighter, 217295, LPBM. USSB contract ship. Keel laid on September 5, 1918, and delivered on November 30, 1918, 86 building days. Allocated to the Moore & McCormick Steamship Co. for service to the United Kingdom.

Detroit Shipbuilding Co.
Detroit, Michigan

Hugh McMillan was president of the Detroit Dry Dock Co. and the Dry Dock Engine Works. Alexander McVittle was vice-president and general manager and Gilbert N. McMillan was secretary-treasurer. The company's metal ship building works was located at Wyandotte about 11 miles downriver from Detroit.

In 1899, Detroit Dry Dock was reorganized and expanded with a capitalization of $1,428,000. McVittle was president and general manager, William C. McMillan vice-president, Merton E. Farr secretary-treasurer, Charles B. Calder general superintendent, and John D. Langell superintendent of drydocks. The offices, drydock and repair plant was at the foot of Orleans St.

The Clark Dry Dock Division was at the foot of Clark Ave. and the sheet metal works, foundry, and brass, piping, and lifeboat department was at 72 Orleans. The engine works, boiler and joiner shops were at Atwater and Orleans, and a foundry was on the southwest corner of Guoin and Dequindre Ave.

In 1913, the American Shipbuilding Co. acquired a controlling interest in the company. In 1918 Merton Farr was president and treasurer, Ernest Ketcham secretary, Frank Jeffrey general superintendent, and John D. Langell superintendent of drydocks and repairs. Frederick J. Pilon was superintendent of the shipyard at Wyandotte.

Adrian Iselin—2,130 tons, laker, 1914, 212089, LKPF. George Hall Coal & Transportation Co.

George L. Eaton—2,008 tons, laker, 1916, 214023. George Hall Coal & Transportation Co. Foundered on June 26, 1918, off Yeu Island, France. All 50 on board were rescued by HMS *Coronado.*

Levisa—2,118 tons, freighter, 1916, 214484, LGKB. United Fruit Co. Commissioned *Levisa* in the Naval Overseas Transportation Service on November 19, 1918.

Ozama—2,118 tons, freighter, 1916, 214502, LGMR. Atlantic Gulf & West Indies Steamship Lines. Commissioned *Ozama* in the Naval Overseas Transportation Service on January 9, 1918.

Carib—2,118 tons, freighter, 1916, 214561, LGMQ. Atlantic Gulf & West Indies Steamship Lines. Requisitioned by the Navy.

War Song—2,534 tons, steel, 9 knot coal-burning freighter, 215596, LHVD. Keel laid on April 10, 1917, for the Cunard Steamship Co. Requisitioned while building and delivered to the USSB on September 22, 1917. Returned to Great Britain before documentation. Torpedoed by *U-93* on January 15, 1918, off Brest, France, with the loss of 16 crew.

Lake Placid—2,534 tons, steel, 9 knot coal-burning freighter, 215656, LJBD. Keel laid on April 12, 1917. Requisitioned from British owners while building and delivered to the USSB on September 9, 1917. Commissioned *Lake Placid* in the Naval Overseas Transportation Service on January 12, 1918. Hit a mine 20 miles off Gothenberg, Sweden, on May 19, 1919, and sank. There were no casualties from the 19 on board.

Lake Erie—1,948 tons, freighter, 1917, 215747, LJFD. Keel laid on June 8, 1917. Requisitioned from British owners while building and delivered to the USSB on November 1, 1917. Allocated to the Army then commissioned *Lake Erie* in the Naval Overseas Transportation Service on January 17, 1918. Sunk in a collision off Cardiff, Wales, on January 16, 1919.

Lake Forest—1,948 tons, steel, 9.5 knot coal-burning freighter, 215784, LJGD. Keel laid on July 28, 1917. Requisitioned from British owners while building and delivered to the USSB on November 24, 1917. Allocated to the Raporel Steamship Co. for service to the West Indies. Commissioned *Lake Forest* in the Naval Overseas Transportation Service on January 15, 1918.

Lakewood—1,948 tons, steel, 9.5 knot coal-burning freighter, 215785, LJGF. Keel laid on June 27, 1917. Requisitioned from British owners while building and delivered to the USSB on November 14, 1917. Allocated to the New York & Porto Rico Steamship Co. for service to the West Indies then transferred to the Navy.

Lake Duane—1,948 tons, steel, 10.5 knot coal-burning freighter, 215819, LJKR. Keel laid on August 14, 1917. Requisitioned from British owners while building and delivered to the USSB on February 14, 1918. Allocated to the Coastwise Transportation Co. for service to the West Indies by Munson.

Lake Conway—1,948 tons, steel, 10.5 coal-burning freighter, 215940, LJPG. Keel laid on September 25, 1917. Requisitioned from British owners while building and delivered to the USSB on February 14, 1918. Allocated to the Atlantic Fruit Co. for service to the West Indies.

Lake Arthur—1,948 tons, steel, 9.5 knot coal-burning freighter, 216036, LJSF. Keel laid on November 7, 1917. Requisitioned from British owners while building and delivered to the USSB on April 30, 1918. Allocated to the United Fruit Co. for service to the West Indies. Commissioned *Lake Arthur* in the Naval Overseas Transportation Service on September 17, 1918.

Lake Butler—1,948 tons, steel, 10 knot coal-burning freighter, 216037, LJSD. Keel laid on October 13, 1917. Requisitioned from British owners while building and delivered to the USSB on February 27, 1918. Allocated to Coastwise Transportation for service to the West Indies.

Lake Weston—1,948 tons, steel, 9.5 knot coal-burning freighter, 216075, LJVF. Keel laid on November 12, 1917. Requisitioned from British owners while building and delivered to the USSB on April 30, 1918. Commissioned *Lake Weston* in the Naval Overseas Transportation Service on November 27, 1918.

Lake Stirling—1,948 tons, steel, 9.5 knot coal-burning freighter, 216135, LJWQ. Keel laid on December 10, 1917. Requisitioned from British owners while building and delivered to the USSB on April 30, 1918.

Lake Felicity—1,948 tons, steel, 9 knot coal-burning freighter, 216254, LKQC. Keel laid on January 5, 1918. Requisitioned from British owners while building and delivered to the USSB on May 20, 1918. Allocated to the New York & Cuba Mail Steamship Co. for service to the West Indies.

Lake Ennis—2,371 tons, steel, 9.5 knot coal-burning freighter, 216366, LKSD. USSB contract ship. Keel laid on January 10, 1918, and delivered on May 31, 1918, 141 building days. Allocated to W.H. Randall for the New England coal service.

Lake Larga—2,371 tons, steel, 9.5 knot coal-burning freighter, 216389, LKTB. USSB contract ship. Keel laid on January 29, 1918, and delivered on June 13, 1918, 135 building days. Allocated to the Earn Line as managers for service to the West Indies by the Kerr Steamship Co. Commissioned *Lake Larga* in the Naval Overseas Transportation Service on November 21, 1918.

Lake Lasang—2,371 tons, steel, 9.5 knot coal-burning freighter, 216390, LKTC. USSB contract ship. Keel laid on February 19, 1918, and delivered on June 25, 1918, 126 building days. Allocated to the France & Canada Steamship Co. for service to the West Indies. Commissioned *Lake Lasang* in the Naval Overseas Transportation Service on September 10, 1918.

Lake Daraga—2,371 tons, steel, freighter, 216483, LKWQ. USSB contract ship. Keel laid on March 16, 1918, and delivered on July 8, 1918, 114 building days. Commissioned *Lake Daraga* in the Naval Overseas Transportation Service on July 27, 1918.

Lake Damita—2,371 tons, steel, 9.5 knot coal-burning freighter, 216556, LMCR. USSB contract ship. Keel laid on April 10, 1918, and delivered on July 22, 1918, 103 building days. Commissioned *Lake Damita* in the Naval Overseas Transportation Service on August 2, 1918.

Lake Benbow—2,368 tons, steel, 9.5 knot coal-burning freighter, 216600, LMFJ. USSB contract ship. Keel laid on April 24, 1918, and delivered on July 31, 1918, 97 building days. Allocated to Allocated to the New York & Porto Rico Steamship Co. for Atlantic coastwise service.

Lake Gakona—2,368 tons, steel, 9.5 knot coal-burning freighter, 216663, LMHC. USSB contract ship. Keel laid on April 30, 1918, and delivered on August 10, 1918, 102 building days. Commissioned *Lake Gakona* in the Naval Overseas Transportation Service on August 22, 1918.

Lake Ormoc—2,368 tons, steel, 9.5 knot coal-burning freighter, 216716, LMJR. USSB contract ship. Keel laid on May 14, 1918, and delivered on August 21, 1918, 104 building days. Allocated to the New York & Cuba Mail Steamship Co. for service to the West Indies.

Lake Akkra—2,368 tons, steel, 9.5 knot coal-burning freighter, 216776, LMNQ. USSB contract ship. Keel laid on May 25, 1918, and delivered on August 31, 1918, 98 building days. Allocated to Crowell & Thurlow for service to the West Indies.

Lake Licking—2,368 tons, steel, 9.5 knot coal-burning freighter, 216844, LMSC. USSB contract ship. Keel laid on June 13, 1918, and delivered on September 12, 1918, 91 building days. Allocated to the National Shipping Co. for service to the West Indies.

Lake Ypsilanti—2,368 tons, steel, 9.5 knot coal-burning freighter, 216874, LMTC. USSB contract ship. Keel laid on June 13, 1918, and delivered on September 23, 1918, 101 building days. Allocated to the West India Steamship Co. for service to the West Indies. Commissioned *Lake Yipsilanti* in the Naval Overseas Transportation Service on January 11, 1919.

Goodspeed—2,368 tons, steel, 9.5 knot coal-burning freighter, 216971, LNBR. USSB contract ship. Keel laid on June 29, 1918, and delivered on September 30, 1918, 93 building days. Allocated to A.H. Bull for service to the west coast of South America.

Goree—2,368 tons, steel, 9.5 knot coal-burning freighter, 216972, LNBS. USSB contract ship. Keel laid on June 29, 1918, and delivered on October 9, 1918, 101 building days. Allocated to A.H. Bull for service to the West Indies.

Lake Gorin—2,368 tons, steel, 9.5 knot coal-burning freighter, 217017, LNFJ. USSB contract ship. Keel laid on July 6, 1918, and delivered on October 16, 1918, 102 building days. Allocated to the New York & Cuba Mail Steamship Co. for service to the West Indies.

Lake Gormania—2,368 tons, steel, freighter, 217029, LNFW. USSB contract ship. Keel laid on July 16, 1918, and delivered on October 24, 1918, 100 building days. Allocated to Hasler Bros. for service to the West Indies.

Lake Grandon—2,611 tons, steel, 9.5 knot coal-burning freighter, 217204, LNRT. USSB contract ship. Keel laid on July 30, 1918, and delivered on November 11, 1918, 104 building days. Allocated to A.H. Bull as managers for service to the United Kingdom by the Gulf & Southern Steamship Co.

Lake Gratis—2,611 tons, steel, 9.5 knot coal-burning freighter, 217216, LNSM. USSB contract ship. Keel laid on August 13, 1918, and delivered on November 27, 1918, 106 building days. Allocated to Munson for service to the United Kingdom.

Lake Graphite—2,611 tons, steel, 9.5 knot coal-burning freighter, 217217, LNSP. USSB contract ship. Keel laid on July 30, 1918, and delivered on November 21, 1918, 113 building days. Allocated to A.H. Bull as manager for service to the west coast of South America by the American Metal Transportation Co.

Lake Gravett—2,611 tons, steel, 9.5 knot coal-burning freighter, 217245, LNVC. USSB contract ship. Keel laid on September 17, 1918, and delivered on November 30, 1918, 86 building days. Allocated to the Phelps Bros. as managers for service to the United Kingdom by John S. Emery & Co.

Globe Shipbuilding Co.
N. 1st & Hughitt, Superior, Wisconsin

Benjamin C. Cooke was president, Chester A. Massey vice-president, Frank A. Miller secretary, and Joseph E. Adams treasurer. Ben Cooke's daughter, Miriam, 22, was assistant treasurer.

Massey was also president of the Massey Steamship Co. at the Globe yards. The company employed about 500.

Lake Washburn—2,121 tons, steel, 10 knot coal-burning freighter, 216437, LKVG. Keel laid on September 25, 1917. Requisitioned from Norwegian owners while building and delivered to the USSB on June 28, 1918. Allocated to C.H. Sprague & Son as manager for service to the West Indies by Capt. Arthur L. Crowley.

Lake Borgne—2,100 tons, steel, 10 knot coal-burning freighter, 216751, LMKS. Keel laid on February 19, 1918. Requisitioned from Norwegian owners while building and delivered to the USSB on August 17, 1918. Commissioned *Lake Borgne* on September 2, 1918, in the Naval Overseas Transportation Service.

Lake Medford—2,084 tons, steel, 10 knot coal-burning freighter, 216924, LMWD. Keel laid on February 25, 1918. Requisitioned from Canadian owners while building and delivered to the USSB on September 23, 1918. Allocated to the Warren Transportation Co. for service to the West Indies.

Lake Arline—2,084 tons, steel, 10 knot coal-burning freighter, 217059, LNHD. Keel laid on April 16, 1918. Requisitioned from Canadian owners while building and delivered to the USSB on October 18, 1918. Allocated to the West India Steamship Co. for service to the West Indies.

Contoocook—2,344 tons, steel, 10 knot coal-burning freighter, 217267, LNWD. USSB contract ship. Keel laid on July 30, 1918, and delivered on November 11, 1918, 104 building days. Allocated to New England Fuel & Transportation as manager for service to the West Indies by the West Indies Steamship Co.

Great Lakes Engineering Works
Ecorse, Michigan

In 1901, Harry S. Hodge was president and treasurer of Samuel F. Hodge & Co., a foundry and machine shop at 308 Atwater E. in Detroit. The Riverside Iron Works was also at the same location. John R. Russel was secretary-treasurer of George H. Russel's Russel Wheel & Foundry Co., a wheel manufacturer and foundry in Detroit. Henry W. Hoyt, of Chicago, was 2nd vice-president of the Allis-Chalmers Co.

Henry Hoyt and John Russel obtained financing, bought out Hodge, and organized the Great Lakes Engineering Works at Rivard St. and Detroit River. Hodge then became vice-president of the Detroit Twist Drill Co.

Hoyt and Russel apparently felt it was necessary to hire a president with some connections and through trade associations they knew of Antonio C. Pessano. Pessano was vice-president and general manager of the George V. Cresson Co., a maker of power transmitting machinery, machinists, and founders, in Philadelphia and was a very highly regarded officer in the American Foundrymen's Association. He was offered the position of president and in mid–June 1902 he resigned his position to become president and general manager of Great Lakes. Pessano went to Great Lakes and the next year the company established the shipyard at Ecorse.

Ecorse

Huron—4,810 tons, laker, 1914, 212080. Wyandotte Transportation Co.

South American—2,662 tons, passenger, 1914, 212244. Chicago, Duluth & Georgian Bay Transit Co. Homeported at Erie, Pennsylvania.

International—1,815 tons, freighter, 1914, 212420, LDNB. Atlantic Coast Steamship.

International—1,709 tons, freighter, 1915, 213378, LFKQ. Atlantic Coast Steamship.

Yaque—1,414 tons, freighter, 1915, 213693, LFPN. Clyde Steamship Co.

Inca—1,414 tons, freighter, 1915, 213730, LFQH. Clyde Steamship Co.

Yuna—1,414 tons, freighter, 1915, 213734, LFQC. Clyde Steamship Co.

Conneaut—4,749 tons, laker, 1916, 214061. Wyandotte Transportation.

William A. Mcgonagle—7,811 tons, laker, 1916, 214234. Pittsburgh Steamship Co.

Munisla—1,607 tons, freighter, 1916, 214291, LGDV. Munson Steamship Corp.

Norfolk—2,482 tons, freighter, 1916, 214564, LGMV. Clyde Steamship Co.

Philadelphia—2,482 tons, freighter, 1916, 214565, LGMW. Clyde Steamship Co.

Pontiac—8,200 tons, laker, 1917, 215048. Grande Isle Steamship.

Frank H. Goodyear—8,243 tons, laker, 1917, 215204. Buffalo Steamship Co.

Florence H—3,819 tons, steel, 11 knot coal-burning freighter, 215643, LHWM. Keel laid on February 8, 1917, as *Souk Ahras* for Compagnie Algérienne Navigation. Requisitioned from French owners while building and delivered to the USSB on October 24, 1917. Named *Florence H* on November 19, 1917.

Lake City—2585 tons, steel, 9.5 knot coal-burning freighter. 215743, LJFB. Keel laid on April 21, 1917, as *Sidi Mabrock*. Requisitioned from French owners while building and delivered to the USSB on October 25, 1917. Sank in collision with the steamer *James McGee* off Key West, Florida, on October 3, 1918. Of 32 aboard, only Navy Engineman 3c John Terry Mathers was lost.

Lake Tahoe—3,819 tons, steel, 11 knot coal-burning freighter, 215745, LJFC. Keel laid as *S.N.A. 4* on April 20, 1917, for the Société Nationale D'affrétements. Requisitioned from French owners while building and delivered to the USSB on December 1, 1917. Allocated to the Navy on December 24, 1917, and commissioned *Lake Tahoe* on February 16, 1918, in the Naval Overseas Transportation Service. Renamed *Evansville* on November 25, 1918.

Lakeland—2,364 tons, steel, 9.5 knot coal-burning freighter, 215766, LJFS. Keel laid on March 29, 1917. Requisitioned from British owners while building and delivered to the USSB on November 11, 1917. Allocated to Munson for service to the West Indies.

Lake Dunmore—2,396 tons, steel, 9.5 knot coal-burning freighter, 215815, LJHC. Keel laid on April 10, 1917. Requisitioned from Norwegian owners while building and delivered to the USSB on November 22, 1917. Allocated to John S. Emery & Co. as managers for service to the River Plate by F.D. Dimmick.

August Ziesing—8,271 tons, laker, 1917, 215870. Pittsburgh Steamship Co.

Lake Elizabeth—2,308 tons, steel, 9.5 coal-burning freighter, 216050, LJSR. Keel laid on August 16, 1917. Requisitioned from Norwegian owners while building and delivered to the USSB on April 30, 1918. Allocated to the Merchants Navigation Co. for service to the West Indies. Commissioned *Lake Elizabeth* in the Naval Overseas Transportation Service on September 26, 1918.

Lake Houghton—2,308 tons, steel, 10 knot coal-burning freighter, 216051, LJST. Keel laid on November 1, 1917. Requisitioned from British owners while building and delivered to the USSB on April 30, 1918. managed and operated by the USSB. Wrecked on Cranberry Isle, Nova Scotia, on June 16, 1918, with 48 aboard. There were no casualties.

Lake St. Clair—2,308 tons, steel, 10 knot coal-burning freighter, 216052, LJSV. Keel laid on September 10, 1917. Requisitioned from British owners while building and delivered to the USSB on April 30, 1918. Allocated to W.J. Grandfield & Co. for the New England coal service. Commissioned *Lake St. Clair* in the Naval Overseas Transportation Service on October 31, 1918.

Lake Crystal—2,308 tons, steel, 10 knot coal-burning freighter, 216168, LKMV. Keel laid on December 15, 1917. Requisitioned from British owners while building and delivered to the USSB on May 15, 1918. Allocated to New England Fuel & Transportation for service to the West Indies.

Lake Owens—2,308 tons, steel, 10 knot coal-burning freighter, 216169, LKMW. Keel laid on November 8, 1917. Requisitioned from British owners while building and delivered to the USSB on May 10, 1918. Managed and operated by the USSB.

Lake Winona—2,308 tons, steel, 9.5 knot coal-burning freighter, 216170, LKNB. Keel laid on November 23, 1917. Requisitioned from British owners while building and delivered to the USSB on May 6, 1918. Allocated to Shawmut as manager for service to the West Indies by the New York & Porto Rico Steamship Co.

Lake Allen—2,308 tons, steel, 10 knot coal-burning freighter, 216194, LKNH. Keel laid on December 26, 1917. Requisitioned from British owners while building and delivered to the USSB on May 20, 1918. Allocated to Munson for service to the West Indies.

Lake Hemlock—2,308 tons, steel, 10 knot coal-burning freighter, 216261, LKQH. Keel laid on January 2, 1918. Requisitioned from British owners while building and delivered to the USSB on May 31, 1918. Allocated to A.H. Bull for Atlantic coastwise service.

Lakehurst—2,308 tons, steel, 10 knot coal-burning freighter, 216262, LKQJ. Keel laid on January 26, 1918. Requisitioned from British owners while building and delivered to the USSB on June 12, 1918. Allocated to Crowell & Thurlow as managers for service to the West Indies by the Panama Railroad Co.

Lake Mary—2,292 tons, steel, 10 knot coal-burning freighter, 216481, LKWP. Keel laid on February 20, 1918. Requisitioned from British owners while building and delivered to the USSB on June 22, 1918. Allocated to New England Fuel & Transportation for the New England coal service. Commissioned *Lake Mary* in the Naval Overseas Transportation Service on November 29, 1918.

Lake Conesus—2,292 tons, steel, 10 knot coal-burning freighter, 216501, LMBD. Keel laid on February 20, 1918. Requisitioned from British owners while building and delivered to the USSB on June 30, 1918. Commissioned *Lake Consesus* in the Naval Overseas Transportation Service on November 14, 1918.

Lake Janet—2,292 tons, steel, 9.5 knot coal-burning freighter, 216569, LMDF. Keel laid on April 6, 1918. Requisitioned while building and delivered to the USSB on August 7, 1918. Allocated to Atlantic Chartering for service to the West Indies.

Lake Silver—2,292 tons, steel, 10 knot coal-burning freighter, 216570, LMDG. Keel laid on March 23, 1918. Requisitioned from British owners while building and delivered to the USSB on July 31, 1918. Allocated to John S. Emery for the New England service. Commissioned *Lake Silver* in the Naval Overseas Transportation Service on August 15, 1918.

Lake Pearl—2,292 tons, steel, 9.5 knot coal-burning freighter, 216609, LMFR. Keel laid on April 23, 1918. Requisitioned while building and delivered to the USSB on August 24, 1918. Allocated to the Coastwise Transportation Co. for service to the West Indies.

Crawl Keys—2,308 tons, steel, 9.5 knot coal-burning freighter, 216727, LMJW. USSB contract ship. Keel laid on July 11, 1918, and delivered on August 14, 1918, 34 building days. Allocated to Eastern Steamship for service to the West Indies.

Lake Gardner—2,292 tons, steel, 9.5 knot coal-burning freighter, 216811, LMQK. Keel laid on May 13, 1918. Requisitioned while building and delivered to the USSB on August 31, 1918. Allocated to Munson for service to the West Indies under Capt. Albert L. Warner.

Lakeville—2,203 tons, steel, 9.5 knot coal-burning freighter, 216890, LMTR. Keel laid on May 25, 1918. Requisitioned while building and delivered to the USSB on September 16, 1918. Allocated to the National Shipping Corp. for service to the West Indies.

Crain Creek—2,292 tons, steel, 9.5 knot coal-burning freighter, 216948, LMWV. USSB contract ship. Keel laid on June 12, 1918, and delivered on September 30, 1918, 110 building days. Allocated to Warren Transportation as manager for service to the West Indies by the Panama Railroad Co.

Cranenest—2,292 tons, steel, 9.5 knot coal-burning freighter, 217024, LNFS. USSB contract ship. Keel laid on July 8, 1918, and delivered on October 8, 1918, 92 building days. Allocated to the New York & South American Line for service to the west coast of South America.

Craycroft—2,292 tons, steel, 9.5 knot coal-burning freighter, 217055, LNHB. USSB contract ship. Keel laid on July 8, 1918, and delivered on October 21, 1918, 105 building days. Allocated to the West India Steamship Co. for service to the West Indies.

Costilla—2,351 tons, steel, 9.5 knot coal-burning freighter, 217203, LNRS. USSB contract ship. Keel laid on July 30, 1918, and delivered on November 11, 1918, 104 building days. Allocated to the Coastwise Transportation Co. as manager for service to the West Indies by the New York & Cuba Mail Steamship Co.

Corydon—2,351 tons, steel, 9.5 knot coal-burning freighter, 217236, LNTM. USSB contract ship. Keel laid on July 9, 1918, and delivered on November 20, 1918, 133 building days. Managed and operated by the USSB.

Cote Blanche—2,351 tons, steel, 9.5 knot coal-burning freighter, 217269, LNWG. USSB contract ship. Keel laid on September 3, 1918, and delivered on November 29, 1918, 87 building days. Sold in December 1918 to New York banker John E. Dockendorff.

Cotopaxi—2,351 tons, steel, 9.5 knot coal-burning freighter, 217270, LNWH. USSB contract ship. Keel laid on August 29, 1918, and delivered on November 30, 1918, 93 building days. Allocated to the Earn Line for service to the east coast of South America.

Ashtabula, Ohio

Great Lakes acquired the Ashtabula yard in 1912. Frank C. Pahlow was superintendent and lived with his wife Mayme at 29 Kingsville Ave. The yard was on Great Lakes Ave. near Lake.

William H. Donner—6,311, laker, 1914, 212354. Mahoning Steamship Co.

Morris Adler—2,481, laker, 1915, 213747. Passolt & Co. Lost December 31, 1915.

Edith J. Osgood—2,483 tons, freighter, 1916, 214085, LGBM. Corona Coal Co. Sold alien June 30, 1916.

Clinchwood—2,532 tons, laker, 1916, 214541. Clinchfield Navigation Co.

Begna—2,436 tons, freighter, 1916. Built for Norwegian owners. Became the *Condor* and then the American post-war *Thomas Tracy* (225122, MFTH).

Midvale—8,271 tons, laker, 1917, 215153. Johnstown Steamship Co.

Lake George—2,486 tons, steel, 9.5 knot coal-burning freighter, 215765, LJFR. Keel laid on April 25, 1917. Requisitioned from French owners while building and delivered to the USSB on November 15, 1917. Allocated to the New Orleans & South American Steamship Co. for service to the west coast of South America.

Sekstant—2,468 tons, coal-burning freighter, 1916. Became the Norwegian *Nordstrand* and then the American *Maurice Tracy* (225917, MGFN).

Lake Maurepas—2,242 tons, steel, 10 knot coal-burning freighter, 216184, LKMR. Keel laid on October 15, 1917. Requisitioned from British owners while building and delivered to the USSB on May 16, 1918. Allocated to the New England Fuel & Transportation Co. as manager for service to the United Kingdom by the West Indies Steamship Co.

Lake St. Regis—2,238 tons, steel, 10 knot coal-burning freighter, 216185, LKND. Keel laid on September 10, 1917. Requisitioned from British owners while building and delivered to the USSB on April 30, 1918. Allocated to West India Steamship for service to the West Indies. Commissioned *Lake St. Regis* in the Naval Overseas Transportation Service on October 17, 1918.

Lake Louise—2,242 tons, steel, 10 knot coal-burning freighter, 216406, LKTR. Keel laid on November 27, 1917. Requisitioned from British owners while building and delivered to the USSB on June 10, 1918. Allocated to New England Fuel & Transportation as managers for service to the West Indies by United Fruit.

Lake Charlotte—2,239 tons, steel, 10 knot coal-burning freighter, 216424, LKVC. Keel laid on January 11, 1918. Requisitioned from British owners while building and delivered to the USSB on June 20, 1918. Allocated to W.J. Grandfield & Co. Commissioned *Lake Charlotte* in the Naval Overseas Transportation Service on October 17, 1918.

Lake Harris—2,217 tons, steel, 10 knot coal-burning freighter, 216646, LMGQ. Keel laid on February 11, 1918. Requisitioned from British owners while building and delivered to the USSB on July 29, 1918. Allocated to J.H.W. Steele for service to the United Kingdom. Commissioned *Lake Harris* in the Naval Overseas Transportation Service on August 8, 1918, at Montreal.

Lake Pleasant—2,221 tons, steel, 10 knot coal-burning freighter, 216738, LMKF. Keel laid on March 20, 1918. Requisitioned from British owners while building and delivered to the USSB on August 24, 1918. Commissioned *Lake Pleasant* in the Naval Overseas Transportation Service on September 8, 1918.

Lake Marion—2,230 tons, steel, 10 knot coal-burning freighter, 216912, LMVQ. Keel laid on April 24, 1918. Requisitioned while building and delivered to the USSB on September 26, 1918. Allocated to Munson for service to the West Indies.

Connersville—2,232 tons, steel, 9.5 knot coal-burning freighter, 217071, LNHS. USSB contract ship. Keel laid on July 8, 1918, and delivered on November 9, 1918, 124 building days. Allocated to the Raporel Steamship Co. for service to the Netherlands.

Lake Sanford—2,232 tons, steel, 9.5 knot coal-burning freighter, 217086, LNJH. Keel laid on May 28, 1918. Requisitioned while building and delivered to the USSB on October 17, 1918. Allocated to the New York & Porto Rico Steamship Co. for service to the west coast of South America.

Crathorne—2,241 tons, wood, 9.5 knot coal-burning freighter, 217189, LNSV. USSB contract ship. Keel laid on August 1, 1918, and delivered on November 23, 1918, 114 building days. Allocated to the Lind Navigation Co. for service to the West Indies.

McDougall-Duluth Co.
715 Sellwood Building, Duluth, Minnesota

Donald B. McDonald was president of the Northern Cold Storage & Warehouse Co. in Duluth. Alexander McDougall was vice-president and A. Miller McDougall was vice-president and treasurer. In 1917 Alexander and A. Miller organized the McDougall-Duluth Co. to build ships. Their office was at. Alexander was president, A. Miller secretary, treasurer, and general manager.

R.L. Barnes—1,914 tons, freighter, 1917, 215115, LHFB. Barnes-Ames Co. Requisitioned while building and delivered to the USSB on June 29, 1918. Commissioned *Robert L. Barnes* in the Naval Overseas Transportation Service on October 19, 1918, and remained in naval service.

Lake Moor—1,955 tons, steel, 10 knot coal-burning freighter, 215770, LJFT. Keel laid on February 4, 1917. Requisitioned from Norwegian owners while building and delivered to the USSB on November 19, 1917. Allocated to the Navy.

Lake Traverse—1,993 tons, steel, 10 knot coal-burning freighter, 216150, LKMF. Keel laid on July 31, 1917. Requisitioned from British owners while building and delivered to the USSB on April 30, 1918. Allocated to A.H. Bull for the phosphate service. Commissioned *Lake Traverse* in the Naval Overseas Transportation Service on October 19, 1918.

Lake Portage—1,993 tons, steel, 10 knot coal-burning freighter, 216409, LKTS. Keel laid on October 20, 1917. Requisitioned from French owners while building and delivered to the USSB on June 13, 1918. Sunk by submarine attack on August 3, 1918, off Brest, France.

Lake Markham—1,993 tons, steel, 10 knot coal-burning freighter, 216587, LMFB. Keel laid on October 20, 1917. Requisitioned from French owners while building and delivered to the USSB on July 16, 1918. Allocated to the Caribbean Steamship Co. for service to the West Indies.

Lake Pepin—1,993 tons, steel, 10 knot coal-burning freighter, 216718, LMJT. Keel laid on November 14, 1917. Requisitioned from French owners while building and delivered to the USSB on August 22, 1918. Allocated to the South Atlantic Maritime Corp. for service to the West Indies. Commissioned *Lake Pepin* in the Naval Overseas Transportation Service on September 4, 1918.

Lake Geneva—1,993 tons, steel, 10 knot coal-burning freighter, 216827, LMRD. Keel laid on November 14, 1917. Requisitioned from French owners while building and delivered to the USSB on August 31, 1918. Allocated to the Western Transportation Co. for the New England coal service. Commissioned *Lake Geneva* in the Naval Overseas Transportation Service on September 21, 1918.

Lake Helen—1,998 tons, 10 knot coal-burning freighter, 216892, LMTV. Keel laid on February 27, 1918. Requisitioned from French owners while building and delivered to the USSB on September 18, 1918. Commissioned *Lake Helen* in the Naval Overseas Transportation Service on October 10, 1918.

Lake Indian—1,991 tons, steel, 10 knot coal-burning freighter, 216990, LNCS. Keel laid on March 14, 1918. Requisitioned while building and delivered to the USSB on September 30, 1918. Allocated to the Hasler Bros. for service to the West Indies.

Lake Orange—1,991 tons, steel, 10 knot coal-burning freighter, 217151, LNMS. Keel laid on March 30, 1918. Requisitioned while building and delivered to the USSB on October 29, 1918. Allocated to Munson for service to the West Indies.

Cedar Springs—2,343 tons, steel, 10 knot coal-burning freighter, 217299, LPBT. USSB contract ship. Keel laid on June 24, 1918, and delivered on November 29, 1918, 158 building days. Allocated to Coastwise Transportation Co. for the coastwise New England coal trade.

Manitowoc Ship Building & Dry Dock Co. Manitowoc, Wisconsin

In 1902, three employees of the Chicago Ship Building Co. at 101st and Calumet River in Chicago, superintendent Elias Gunnell, naval architect Charles C. West, and chief clerk Lynford E. Geer decided to go into business for themselves and purchased the Burger yard on Lueps Island in Manitowoc for $110,000. Gunnell was president, West vice-president, and Geer secretary-treasurer. In 1903 they delivered their first boat, the small wooden passenger vessel *Chequamegon*. In 1905 they built their first steel ship, the passenger vessel *Maywood*, homeported at Marquette, Michigan. In June 1917 the Hill Transportation Co. sold the *Maywood* to the French government for conversion to a minesweeper. Their last ship was the towboat *Harry B. Williams* in 1931 for the Great Lakes Dredge & Dock Co. in Duluth.

The company continues today as the Manitowoc Co., Inc., in the materials handling, crane, and food service industries.

Nevada—2,122 tons, freighter, 1915, 213782, LFFR. Sold Russian December 31, 1917.

Lake Oneida—2,124 tons, steel, 9 knot oil-burning motor freighter, 215630, LHVW. Keel laid on October 31, 1916. Requisitioned from Norwegian owners while building and delivered to the USSB on October 8, 1917. Unallocated.

Lake Mohonk—2,124 tons, steel, 9 knot oil-burning motor freighter, 215702, LJCS. Keel laid on November 29, 1916. Requisitioned from Norwegian owners while building and delivered to the USSB on October 24, 1917. Allocated to Mallory.

Lakeview—2,003 tons, steel, 9.5 knot coal-burning freighter, 215773, LJFW. Keel laid on June 30, 1917. Requisitioned from British owners while building and delivered to the USSB on November 17, 1917. Allocated to Raporel for service to the West Indies. Commissioned *Lake View* in the Naval Overseas Transportation Service on February 23, 1918.

Lake Ontario—2,003 tons, steel, 9 knot coal-burning freighter, 215816, LJHD. Keel laid on January 19, 1917. Requisitioned from British owners while building and delivered to the USSB on November 24, 1917. Allocated to Kerr for service to the United Kingdom. Commissioned *Lake Ontario* in the Naval Overseas Transportation Service on February 19, 1918.

Lake Pewaukee—2,167 tons, steel, 9 knot coal-burning freighter, 216226, LKPQ. Keel laid on August 11, 1917. Requisitioned from British owners while building and delivered to the USSB on April 30, 1918. Allocated to the New York & Porto Rica Steamship Co. for service to the West Indies. Commissioned *Lake Pewaukee* in the Naval Overseas Transportation Service on September 7, 1918.

Lake Lida—2,177 tons, steel, 9 knot coal-burning freighter, 216277, LKQT. Keel laid on October 25, 1917. Requisitioned from Norwegian owners while building and delivered to the USSB on May 25, 1918. Allocated to the Coastwise Transportation Co. as managers for service to the West Indies by the West India Steamship Co.

Lake Shawano—2,177 tons, steel, 9 knot coal-burning freighter, 216278, LKQV. Keel laid on October 6, 1917. Requisitioned from Norwegian owners while building and delivered to the USSB on May 16, 1918. Allocated to the Coastwise Transportation Co. as manager for service to France by the Raporel Steamship Co.

Lake Annette—2,177 tons, steel, 9 knot coal-burning freighter, 216403, LKTQ. Keel laid on December 6, 1917. Requisitioned from Norwegian owners while building and delivered to the USSB on June 13, 1918. Allocated to Munson for service to the West Indies.

Lake Greenwood—2,150 tons, steel, 9 knot coal-burning freighter, 216616, LMGC. Keel laid on February 27, 1918. Requisitioned from British owners while building and delivered to the USSB on July 24, 1918. Allocated to Crowell & Thurlow for the New England coal service.

Lake Monroe—2,150 tons, steel, 9 knot coal-burning freighter, 216739, LMKG. Keel laid on January 21, 1918. Requisitioned from British owners while building and delivered to the USSB on August 10, 1918. Allocated to Shawmut for service to the West Indies.

Lake Winthrop—2,150 tons, steel, 9 knot coal-burning freighter, 216780, LMPG. Keel laid on March 27, 1918. Requisitioned from British owners while building and delivered to the USSB on August 26, 1918. Allocated to the Coastwise Transportation Co. for service to the West Indies.

Lake Linden—2,150 tons, steel, 9 knot coal-burning freighter, 216805, LMQD. Keel laid on February 18, 1918. Requisitioned from British owners while building and delivered to the USSB on August 31, 1918. Allocated to Munson for service to the West Indies.

Lake Wilson—2,150 tons, steel, 9 knot coal-burning freighter, 216959, LNBJ. Keel laid on April 6, 1918. Requisitioned from British owners while building and delivered to the USSB on September 28, 1918. Allocated to M.H. Tracy for service to the West Indies by the Panama Railroad Co.

Corsicana—2,157 tons, steel, 9.5 knot coal-burning freighter, 217011, LNDW. USSB contract ship. Keel laid on July 10, 1918, and delivered on October 19, 1918, 97 building days. Allocated to Munson for service to the West Indies.

Lake Kyttle—2,157 tons, steel, 9.5 knot coal-burning freighter, 217087, LNJK. USSB contract ship. Keel laid on May 1, 1918, and delivered on October 31, 1918, 182 building days. Allocated to the Independent Steamship Co. for service to the West Indies.

Corrales—2,157 tons, steel, 9.5 knot coal-burning freighter, 217197, LNRH. USSB contract ship. Keel laid on June 1, 1918, and delivered on November 20, 1918, 172 building days. Allocated to M.H. Tracy & Co. for service to the West Indies.

Rieboldt, Wolter & Co.
Sturgeon Bay, Wisconsin

August Rieboldt was born in Germany in 1849 and came with his parents to Sheboygan, Wisconsin, in 1853. He became a ship carpenter and in 1876 married Maria Davidson.

Joseph Wolter was a ship carpenter foreman who lived with his wife Regina at 384 Grove St. in Milwaukee. In 1885 the Wolters moved to Sheboygan, Joseph joined forces with Rieboldt, and they formed the Rieboldt & Wolter Co. In 1895 they moved their yard to Sturgeon Bay.

Conrad Starkey was a lake captain who lived at 340 Hanover in Milwaukee. In 1888 he joined the firm and they became Rieboldt, Wolter & Co.

In early 1917 the company signed a contract to build a wood freighter for the Lake & Ocean Navigation Co. at 208 S. La Salle St. in Chicago. The plan for the *Sturgeon Bay* was approved by the Shipping Board as Design No. 1007, a three-hold, well deck freighter with the house and machinery spaces aft. The boiler was oil-fired and had a 13' diameter sectional propeller made by the Sheriffs Manufacturing Co. at Milwaukee.

Gustavus A. Huck was born in Germany in 1877, came to America in June 1892, and moved in with his older sister and her husband, Herman Sempke, at 754 1st St. in Milwaukee. He worked as a barber with Joseph W. Kirschstein, a boarder in the same household. Gustav was naturalized on November 5, 1900, and in 1908 he married Ola May Gelhaar, 23. They went to live with her parents, Otto and Luisa, in Hartford, Wisconsin, and Gus worked on their farm.

By 1914 Gus was vice-president of the Palace Mining & Leasing Co. at 379 3rd in Milwaukee and in 1916 partnered with Alf Bues in the ice cream freezer business at 612 Grand Ave.

By 1917 Alf was selling insurance while restless Gus was in the meat business and that summer he happened to visit Sturgeon Bay when noticed the Rieboldt & Wolter shipyard, a modest operation with about 300 employees. With the country now at war and business prospects looking very good, he decided the yard could be significantly expanded and made much more profitable with a year-round workforce of 700–1,000. He approached Rieboldt and Wolter about selling. Reiboldt was pushing 70 and they no doubt were closely watching the frenzy of shipbuilding going on around them. They knew the shipping business very well and they also knew that every boom has a bust.

On December 27, 1917, the partners sold Gus a $5,000 option to purchase their shipyard on or before February 1, 1918. Huck left his meat business and went after capital. The idea attracted big money investors and in January 1918 the Universal Shipbuilding Co. was incorporated with a capital stock of $1.5 million to take over Huck's option on February 1, which they did. Gustav Huck stayed on as a director.

The deal required Reiboldt & Wolter to complete boat No. 61, the *Sturgeon Bay*, which was expected to launch around May 1, 1918. The new owners anticipated that 600 additional workers would be needed by August 1 but things don't always work out. Universal Ship built only three tugboats for the Shipping Board, Gus and Ola divorced, and Lake & Ocean Navigation was in receivership by 1920. Gus died in 1948.

Sturgeon Bay—2,383 tons, wood, 10 knot oil-burning freighter, 216276, LKQS. Lake & Ocean Navigation Co. Keel laid on July 1, 1917, and launched on April 25, 1918. Requisitioned while building and delivered to the USSB on November 23, 1918, 509 building days. Retained by the Shipping Board for recruiting service.

Saginaw Shipbuilding Co.
Saginaw, Michigan

Chesley Wheeler was born in New York on June 17, 1823, started out in farming, married Eliza Haselton, then decided to become a ship carpenter. Their kids were Fred, Sarah, Emma, Franklin, and Hattie. Franklin Willis Wheeler was born in Chaumont, New York, on March 2, 1853, and would loom large in Great Lakes shipping.

In late 1865 the family moved to Saginaw where Chesley set up his own shipyard. In 1870 he was 48, Eliza was 45 and kept house, Fred, 23, was a ship carpenter, Sarah, 21, helped at home, Emma, 20, was a music teacher, and Franklin and Hattie were in school.

In 1875 Franklin married Eva Armstrong, of Saginaw, and in 1876 they had a daughter May. That same year the families moved to West Bay City, Michigan, where Chesley and Franklin set up shop as Wheeler & Son, shipbuilders in Wenona. They lived on the southeast corner of Van Buren between 3rd & 4th. Frank engaged in ship building and operation, obtaining his master's license on the lakes. Frank also opened his own small yard where he did repair work, got into building, and in 1877 built the steamer *Mary Martini*.

In 1883 Frank partnered with Albert A. Crane, a lumberman in Bay City, as Wheeler & Crane, shipbuilders and dry dock operators on SE Washington.

In 1886 Chesley was partnered with Capt. Charles M. Averell, of Bay City, in a dry dock business as C. Wheeler & Co. at his same location and Frank had a yard adjacent to his father. Frank lived at 307 Litchfield. Ship carpenter George F. Williams was superintendent and brother Fred was the foreman.

In 1887 Frank established F.W. Wheeler & Co. with a capital stock of $500,000. Frank was president, George Williams vice-president, lumberman Frederick L. Gilbert secretary, and John R. Goodfellow, sister Emma's husband, was treasurer. Fred became superintendent. On January 11, 1889, the company was incorporated by Frank, William Forbes, and George Williams. Facilities for building steel ships were put up and in June 1890 the passenger steamer *City of Chicago* was launched. The company then had the largest wood shipbuilding yard in the United States.

Frank was said to be the first Great Lakes shipbuilder to seek work from the Navy and he apparently was irritated by a provision in a treaty with Great Britain that prevented him from securing work from British companies. That apparently was his motivation for running for Congress.

In 1888 Frank ran, defeated his Democratic opponent, and on March 4, 1889, was sworn in as a Republican representative and served until March 3, 1891.

One aspect of Wheeler's time in Congress was written about in the Buffalo Courier on Saturday, July 18, 1891:

Free Ships for Privileged Persons.

Nice customs courtesy to great kings, and stiff Protectionists temporarily forget their principles when their personal interests point in the other direction. People who keep abreast of lake shipping news will remember the two steel ships, the *Mackinaw* and the *Keweenaw*, which were built at Bay City by Congressman Frank W. Wheeler. One of them was cut in two in Buffalo, taken through the Welland Canal, and then put together, while the other was launched in halves, the two parts being towed down the lakes and through the canal and then put together. They were built for a company which is said to be represented by Timothy Hogan & Sons, stevedores, of New York City. Congressman Wheeler is a staunch Republican and a sturdy opponent of British goods of all sorts, but when the Hogans asked him to secure quietly an American register for two British-built steamships, the *Mineola* and the *Montauk*, his scruples were laid aside, and he introduced in the House a bill to accomplish this object, that other staunch Protectionist, Senator Evarts, attending to the Senate and of the business. The bill was duly passed and signed, and the *Mineola*, built by Laing in Sunderland, England, and having Liverpool for her hailing port, arrived in New York the other day flying American colors. The law requires the master of an American ship to be an American, and this matter was easily arranged, the old captain of the *Mineola* having taken out naturalization papers three years ago. It is said to be a common thing for British captains and mates, whose families remain on the other side, to qualify themselves for the American service by becoming naturalized in the United States. sometimes they temporarily take a room in a New York hotel, which they give as their home, and sometimes pay a broker $10 and let him attend to the perjury part of the transaction.

The most interesting part of the business in this case, however, is that the *Mineola* and *Montauk* are not to engage in the foreign trade, but are to enter into competition with the Bath-built clipper ships which ply between New York and San Francisco. They will be able to make three round trips between these ports while sailing vessels are making one. It is possible, however, that, owing to the large amount of coal they must carry on so long a voyage, the steamers may not be able to make this business pay, in which case it is not improbable that they will engage in trade between American ports on the Pacific Coast.

Of course if the owners of American sailing vessels had known anything about the bill which the two Protectionists, Wheeler and Evarts, were working through Congress there would have been a lively outcry from their direction, but the arrival in New York of the *Mineola* under American colors was the first announcement to the trade of the admission of herself and her sister ship, the *Montauk*, to an American register. Free ships would be an excellent thing for the American people, though the advocates of free ships have intended that they should be employed in the foreign and not the domestic trade, but free ships from those persons alone who happen to have friends among Republican Congressmen, while their competitors are restricted to the purchase of vessels built in the United States can hardly be regarded as a fair deal.

The *Mackinaw* and the *Keweenaw* were put together in Montreal. In 1891 the company began building marine engines and by mid–1896 the yard was fully-engaged in steel shipbuilding with 2,500 feet of frontage on the Saginaw River.

In 1893 Frank was the general manager of the Bay City & Alpena Line of Steamers. Frank also helped organize the Hopkins Transportation Co. in Detroit with Mark Hopkins of St. Clair, E.C. Recor, a St. Clair shipbuilder, and John Mitchell, a shipowner in Cleveland. Frank was then building the "immense new freight boat" *Centurion*, valued at $275,000, and the company was formed to manage it from East China near St. Clair. Frank named the *Centurion* as it was the 100th ship from his yard. The keel was laid on his 40th birthday.

In March 1894 Frank testified at a murder trial in Troy, New York, in which Republicans John McGough and John H. Boland were charged with shooting and killing William Ross in a "factional dispute" on election day. Frank was a Republican of the 2nd District, 13th Ward, and witnessed the event.

In 1899 Frank lived at 900 Center St. and early that year removed to Detroit. In June he sold his yard to the American Shipbuilding Co. trust. Rumor had it that the deal was forced by the "Rockefeller syndicate to absorb the Wheeler yard and drop the idea of forcing him out of business." The yard sold for $700,000 and after Wheeler paid off his creditors he netted $300,000.

In Detroit he became president of the Shelldrake Lumber Co., a wholesale lumber business at 42 Campus Bldg. Dr. George F. Clark, his son-in-law, husband of daughter May, was secretary-treasurer. The Wheelers and Clarks were celebrated in Detroit society. Wheeler lived at 40 Joy St. In 1902 Wheeler and Clark owned the steamer *Iroquois* which went ashore on Victoria Island in Lake Superior, and was the subject of some litigation in Buffalo in August 1903.

In 1906 he moved to Buffalo, New York, and became president and general manager of the Niagara & St. Lawrence Navigation Co.

When war came in 1917 he went back to Saginaw and established the Saginaw Shipbuilding Co. to build wood Shipping Board freighters. William J. Wickes was chairman of the board, Frank was president, George H. Hannum vice-president, Walter C. Hill treasurer, Charles W. Stiver secretary, and Charles W. Watson general manager. The yard was at SE Carrolton near Hickory. Frank lived at 1100 N. Michigan.

Chesley died at Bay City on June 4, 1911, leaving a widow, Imogene. In 1920 the Wheelers and Clarks lived at 1100 Michigan Ave. in Saginaw. Frank died in Saginaw on August 9, 1921.

Lake Pachuta—2,416 tons, steel, 10 knot coal-burning freighter, 216758, LMKV. USSB contract ship. Keel laid on January 16, 1918, and delivered on August 29, 1918, 220 building days. Allocated to C.H. Sprague & Son. for service to the West Indies.

Lake Osweya—2,416 tons, steel, 10 knot coal-burning freighter, 216914, LMVR. USSB contract ship. Keel laid on February 21, 1918, and delivered on September 26, 1918, 220 building days. Allocated to the Atlantic Chartering Co. for service to France. Commissioned *Lake Osweya* in the Naval Overseas Transportation Service on October 17, 1918. Operated during World War II by the Ford Motor Co. Sunk on February 20, 1942, in the North Atlantic by *U-96*.

Lake Winooski—2,416 tons, steel, 10 knot coal-burning freighter, 217045, LNGR. USSB contract ship. Keel laid on April 1, 1918, and delivered on November 7, 1918, 202 building days. Allocated to the Phelps Bros. as manager for service to the West Indies by the Merchants Navigation Co. Commissioned *Lake Winooski* in the Naval Overseas Transportation Service on November 25, 1918.

Lake Belnona—2,350 tons, steel, 10 knot coal-burning freighter, 217230, LNTF. USSB contract ship. Keel laid on May 17, 1918, and delivered on November 29, 1918, 195 building days. Allocated to Allocated to the New York & Porto Rico Steamship Co. for service to the West Indies.

The Superior Shipbuilding Co.
Superior, Wisconsin

In 1899 the American Steel Barge Co. at Lamborn Ave. and Howard's Pocket was taken over by The Superior Ship Building Co. Capt. Augustus B. Wolvin became president and Daniel E. Ford was manager. The new company was apparently taken over almost immediately by the American Shipbuilding Co. at Cleveland.

Capt. Wolvin went to sea at 12 as a cabin boy on his father's ship, the *Dean Richmond*. His first command was the steamer *Anne Smith* in 1879 and in 1882 he left sailing. In 1889 he and Capt. Fred N. LaSalle became steamship agents in Duluth, Minnesota, as LaSalle & Wolvin but LaSalle died in 1901 leaving a widow, Jennie, of 1226 E. 1st St.

In 1902, U.S. Steel acquired 112 vessels and organized the Pittsburgh Steamship Co. to operate them. Capt. Wolvin was appointed general manager and managed other vessels.

A large fire at Superior on November 8, 1907, badly damaged Superior Ship's buildings.

In 1914 he moved to the Thornton Hotel in Butte, Montana, and was president of the Duluth Mining Co. He was back and forth between there, Duluth, and Superior and in 1917 he had an office in Virginia, Minnesota.

Superior's first ship was the freighter *D.G. Kerr* in 1903. John Sinclair was the superintendent in 1918.

Sir Trevor Dawson—7,215 tons, laker, 1917, 214499. American Interlake Co.

Cleveland—2,945 tons, freighter, 1917, 214709, LGQR. Sold to Norwegian owners.

Lake Champlain—2,046 tons, steel, 9 knot coal-burning freighter, 215642, LHWJ. Keel laid on April 21, 1917. Requisitioned from British owners while building and delivered to the USSB on September 22, 1917. Allocated to the United Steamship Co. for service to the West Indies. Commissioned *Lake Champlain* on January 24, 1918, in the Naval Overseas Transportation Service.

Lakeside—2,046 tons, steel, 9 knot coal-burning freighter, 215715, LJDC. Keel laid on May 17, 1917. Requisitioned from British owners while building and delivered to the USSB on November 2, 1917. Allocated to Moore & McCormick for service to the United Kingdom. Commissioned *Lake Side* in the Naval Overseas Transportation Service on January 25, 1918.

Lake Tulare—2,005 tons, steel, 9.5 knot coal-burning freighter, 216066, LJTD. Keel laid on August 20, 1917. Requisitioned from British owners while building and delivered to the USSB on April 30, 1918. Allocated to the Seaboard & Gulf Steamship Co. for Atlantic coastwise service. Commissioned *Lake Tulare* in the Naval Overseas Transportation Service on October 10, 1918.

Lake Capens—2,026 tons, steel, 9 knot coal-burning freighter, 216295, LKJP. Keel laid on August 30, 1917. Requisitioned from British owners while building and delivered to the USSB on May 20, 1918. Commissioned *Lake Capens* in the Naval Overseas Transportation Service on October 16, 1918.

Lake Ledan—2,353 tons, steel, 9.5 knot coal-burning freighter, 216522, LMBQ. USSB contract ship. Keel laid on January 19, 1918, and delivered on July 2, 1918, 163 building days. Allocated to Edward M. Raphael & Co. for service to the West Indies.

Lake Lesa—2,352 tons, steel, 9.5 knot coal-burning freighter, 216678, LMHK. USSB contract ship. Keel laid on March 2, 1918, and delivered on July 31, 1918, 150 building days. Allocated to the Hasler Bros. for service to the West Indies.

Lake Dymer—2,357 tons, steel, 9.5 knot coal-burning freighter, 216808, LMQH. USSB contract ship. Keel laid on March 14, 1918, and delivered on August 26, 1918, 165 building days. Allocated to the Pensacola Shipping Co. for service to the United Kingdom. Commissioned *Lake Dymer* in the Naval Overseas Transportation Service on September 16, 1918.

Lake Aurice—2,353 tons, steel, 9.5 knot coal-burning freighter, 216866, LMSR. USSB contract ship. Keel laid on April 30, 1918, and delivered on September 14, 1918, 137 building days. Allocated to the France & Canada Steamship Co. for service to the West Indies.

Lake Dancey—2,323 tons, steel, 9.5 knot coal-burning freighter, 216981, LNCG. USSB contract ship. Keel laid on March 3, 1918, and delivered on September 30, 1918, 197 building days. Allocated to the New York & Cuba Mail Steamship Co. for service to the West Indies. Commissioned *Lake Dancey* in the Naval Overseas Transportation Service on October 29, 1918.

Lake Agomak—2,339 tons, steel, 9.5 knot coal-burning freighter, 217095, LNJR. USSB contract ship. Keel laid on March 28, 1918, and delivered on October 22, 1918, 207 building days. Allocated to the West India Steamship Co. for service to the West Indies.

Lake Fostoria—2,618 tons, steel, 9.5 knot coal-burning freighter, 217235, LNTK. USSB contract ship. Keel laid on July 18, 1918, and delivered on November 21, 1918, 125 building days. Allocated to Munson for service to the United Kingdom.

Lake Fouche—2,615 tons, steel, 9.5 knot coal-burning freighter, 217304, LPCD. USSB contract ship. Keel laid on July 18, 1918, and delivered on November 30, 1918, 134 building days. Allocated to the Moore & McCormick Steamship Co. for service to the United Kingdom.

Toledo Shipbuilding Co. Toledo, Ohio

Alexander McVittle was vice-president and general manager of the Detroit Dry Dock Co. and the Dry Dock Engine Works in Detroit, Michigan.

In 1899, Detroit Dry Dock was reorganized with McVittle as president and general manager and Charles B. Calder general superintendent.

In 1906, McVittle incorporated Toledo Ship. He was president, Horace S. Wilkinson was secretary, Charles Calder general manager, and Archibald D. Black purchasing agent. The yard was at Front & Craig Sts. The company's first ship was the freighter *Sierra.*

In 1916 Horace Wilkinson was president and treasurer, Charles Calder was vice-president and general manager, Archibald Black secretary and assistant treasurer. William G. Henderson was the general superintendent, and Edward Hopkins the naval architect. The company's last ship was the freighter *Windsor* in 1930.

Limoges—1,984 tons, steel, 10 knot coal-burning freighter. Keel laid on November 28, 1916. Requisitioned from French owners while building and delivered to the USSB on August 3, 1917. Later returned to France.

Moonlite—1,930 tons, motor tanker, 1916, 214217, LGDB. Standard Oil of New Jersey.

Starlite—1,930 tons, motor tanker, 1916, 214218, LGDC. Standard Oil of New Jersey.

Twilite—1,955 tons, motor tanker, 1916, 214440, LGJF. Standard Oil of New Jersey.

Dawnlite—1,976 tons, motor tanker, 1916, 214491, LGKH. Standard Oil of New Jersey.

Sunlite—1,976 tons, motor tanker, 1916, 214557, LGMK. Standard Oil of New Jersey.

Daylite—1,976 tons, motor tanker, 1916, 214568, LGND. Standard Oil of New Jersey.

Horace S. Wilkinson—8,338 tons, laker, 1917, 215122. Great Lakes Steamship Co.

Lake Shore—1,984 tons, steel, 10 knot coal-burning freighter, 215651, LHWS. Keel laid on April 23, 1917. Requisitioned while building and delivered to the USSB on October 4, 1917. Allocated to Crowell & Thurlow for service to the east coast of South America. Commissioned *Lake Shore* in the Naval Overseas Transportation Service on January 11, 1918.

Lake Superior—1,977 tons, steel, 10 knot coal-burning freighter, 215756, LJFM. Keel laid on May 20, 1917. Requisitioned from British owners while building and delivered to the USSB on November 6, 1917. Allocated to the West India Steamship Co. for service to the West Indies. Commissioned *Lake Superior* in the Naval Overseas Transportation Service on January 30, 1918.

Lake Bridge—1,984 tons, steel, 10 knot coal-burning freighter, 215796, LJGO. Keel laid on July 9, 1917. Requisitioned from British owners while building and delivered to the USSB on November 16, 1917. Allocated to A.H. Bull for Atlantic coastwise service. Commissioned *Lake Bridge* in the Naval Overseas Transportation Service on February 16, 1918.

Lake Sunapee—2,009 tons, steel, 10 knot coal-burning freighter, 216260, LKQG. Keel laid on June 4, 1917. Requisitioned from British owners while building and delivered to the USSB on April 30, 1918. Allocated to Raporel for service to the West Indies. Commissioned *Lake Sunapee* in the Naval Overseas Transportation Service on November 19, 1918.

Lake Sebago—2,010 tons, steel, freighter, 216355, LKRV. Keel laid on September 5, 1917. Requisitioned from British owners while building and delivered to the USSB on May 25, 1918. Allocated to the Coastwise Transportation Co. for service to the West Indies.

Lake Cayuga—2,010 tons, steel, 10 knot coal-burning freighter, 216365, LKSC. Keel laid on October 4, 1917. Requisitioned from British owners while building and delivered to the USSB on June 6, 1918. Allocated to Shawmut for service to the West Indies.

Lake Chelan—2,010 tons, steel, 10 knot coal-burning freighter, 216454, LKVR. Keel laid on October 23, 1917. Requisitioned from British owners while building and delivered to the USSB on June 21, 1918. Allocated to Crowell & Thurlow as managers for service to the West Indies by Munson.

Lake Catherine—2,016 tons, steel, 10 knot coal burning freighter, 216528, LMBV. Keel laid on December 31, 1917, for the Magnolia Petroleum Co. Requisitioned while building and delivered to the USSB on July 6, 1918. Allocated to the West Indies Steamship Co. for phosphate service. Commissioned *Lake Catherine* in the Naval Overseas Transportation Service on October 4, 1918.

Lake Frances—2,016 tons, steel, 10 knot coal-burning freighter, 216571, LMDK. Keel laid on March 14, 1918. Requisitioned while building and delivered to the USSB on July 20, 1918. Commissioned *Lake Frances* in the Naval Overseas Transportation Service on August 1, 1918.

Calaveras—2,379 tons, steel, 10 knot coal-burning freighter, 216804, LMQF. USSB contract ship. Keel laid on April 22, 1918, and delivered on September 30, 1918, 161 building days. Allocated to A.H. Bull for Atlantic Coastwise trade.

Calicorock—2,379 tons, steel, 10 knot coal-burning freighter, 217079, LNJC. USSB contract ship. Keel laid on April 30, 1918, and delivered on October 18, 1918, 171 building days. Allocated to A.H. Bull for service to the West Indies.

Calispell—2,379 tons, steel, 10 knot coal-burning freighter, 217143, LNMQ. USSB contract ship. Keel laid on May 15, 1918, and delivered on October 31, 1918, 168 building days. Allocated to Atlantic Charter for service to the West Indies.

Lake Calistoga—2,379 tons, steel, 10 knot coal-burning freighter, 217207, LNSB. USSB contract ship. Keel laid on May 25, 1918, and delivered on November 16, 1918, 174 building days. Allocated to Shawmut for service to the West Indies.

Lake Calicoon—2,379 tons, steel, 10 knot coal-burning freighter, 217264, LNWB. USSB contract ship. Keel laid on June 15, 1918, and delivered on November 23, 1918, 161 building days. Allocated to the United Steamship Co. for service to the West Indies.

The Atlantic Coast

American Car & Foundry Co.
Wilmington, Delaware

In 1863, Job Hays Jackson, 30, a tinsmith from Pennsylvania, and Jacob F. Sharp, a carpenter, started building railroad cars, streetcars, and ships as Jackson & Sharp at Wilmington, Delaware, and in 1875 they acquired the Christina River Shipyard. Charles S. Gawthorp was the private secretary at Jackson & Sharp. Jacob Sharp died on August 2, 1888.

In 1899, 13 railroad car builders merged to form American Car & Foundry and on May 23, 1901, Job died. American Car & Foundry leased the Jackson & Sharp shipyard at the foot of 8th St. for ten years. Charles Gawthorp was district manager. In 1911 American Car purchased the works. Their first vessel was the *Ethel* in 1902. During the war the company built six schooner barges for Francis P. Dougherty, 407 Warren Ave. in Baltimore, who was in the tug and barge business as P. Dougherty Co., Inc.

Wilmington—1,371 tons, schooner barge, 1915, 213593, LFNS.
Montauk—1,371 tons, schooner barge, 1915, 213744, LFQJ.
Delaware—1,371 tons, schooner barge, 1916, 214209, LGCS.
Maryland—1,371 tons, schooner barge, 1917, 214687, LGQJ.
Providence—1,371 tons, schooner barge, 1917, 215749, LJFH.
Annapolis—1,371 tons, schooner barge, 1918, 216084, LJVM.

American International Shipbuilding Corp.
Hog Island, Philadelphia, Pennsylvania

In May 1917 a proposal was made to construct a large shipyard on Hog Island near Philadelphia to rush ships into the war effort using prefabrication methods. The expertise of the American International Corp. was immediately involved.

American International signed a contract on September 13, 1917, to build a yard with a wet basin, 20 ways for building 7,500-ton ships, 30 ways for building 8,000-ton ships, fitting-out piers for 28 ships at a time, and shops on 846 acres. The contract called for 180 ships at a cost of $267,825,000. Frederick Holbrook, a vice-president of American International, was in charge of the work and George O. Muhlfeld was construction manager.

Someone had to build the ships so on September 27 the company formed a subsidiary, the American International Shipbuilding Corp. Electrical engineers Charles A. Stone and Edwin F. Webster, contractors in business as Stone & Webster in Boston were early major stockholders of the American International Corp. so Charles was made president and Edwin Webster a director.

On March 29, 1918, Frederick Holbrook was elected president. Harris D.H. Connick, of San Francisco, another vice-president at American International, became vice-president and a director of American International Shipbuilding.

Alfred R. Patterson, general auditor of Stone & Webster, became treasurer and a director on April 16, 1918. Since the American International Corp. had little to no people with shipbuilding experience in their boardroom, they brought in many officers from shipbuilders who served with minimal or no compensation as officers and directors of American International Ship under Emergency Fleet Corporation contracts to aid the war effort.

Marvin A. Neeland, president of New York Shipbuilding Corp. at Camden, New Jersey, and Luther D. Lovekin, formerly chief engineer at New York Ship became vice-presidents of American International Shipbuilding. Others came from Sparrows Point, Maryland Steel, and International Terminals. James B. Andrew was manager of construction.

The total cost was around $65 million for the yard, 250 buildings, and 80 miles of standard gauge railroad track. Four ships were delivered in 1918 and the yard built 118 more ships through 1921. The famous "Hog Islanders" would form a good portion of the merchant fleet at the beginning of World War II.

The whole operation came under a scathing review from Congress and many people, lawmakers and taxpayers alike, considered the whole scheme a huge boondoggle of enormous proportions.

Quistconck—5,735 tons, steel, 11.5 knot oil-burning freighter, 217144, LNMR. USSB contract ship. Keel laid on February 12, 1918, and delivered on December 3, 1918, 296 building days. The ship was sponsored by President Wilson's wife. Allocated to J.H.W. Steele for service to the United Kingdom.

Saccarappa—5,735 tons, steel, 11 knot oil-burning freighter, 217328, LPDF. USSB contract ship. Keel laid on March 26, 1918, and delivered on December 31, 1918, 280 building days. Commissioned *Saccarappa* in the Naval Overseas Transportation Service on December 31, 1918.

The Atlantic Coast Corp.
131 State St., Boston

The Atlantic Coast Co. was a subsidiary of the Crowell & Thurlow Steamship Co. The company built schooners in Maine at Boothbay Harbor and Thomaston. They built a total of 16 schooners on their own account during the war and homeported them at Boston. The last one at Thomaston was the schooner *Elizabeth Freeman* in 1920. The last of six ships at Boothbay was the schooner *Josiah B. Chase* in 1921.

Boothbay

Anna Laura Mckenney—1,021 tons, schooner, 1917, 215873, LJMD.
Francis L. Taussig—1,070 tons, schooner, 1918, 216927, LMWH.

Thomaston

Jessie G. Noyes—1,376 tons, schooner, 1917, 215670. LJBS.
Augusta G. Hilton—1,562 tons, schooner, 1918, 216434, LKVF.
Ida S. Dow—1,411 tons, schooner, 1918, 217434, LPKH.

The Baltimore Dry Dock & Ship Building Co. Foot of E. 4th Ave., Baltimore

The Baltimore Ship Building & Dry Dock Co. opened in 1903 under president John Q. Lovell and secretary-treasurer J. Triplett Haxall. The yard with a 504' drydock was at the foot of Martin St. at Locust Point. Their first vessel was the ferryboat *Queen Caroline* in 1903 for the Queen Anne Ferry Co. followed by the *Anne Arundel* in 1904. In 1905 the company was in liquidation when they delivered the *Geo. F. Randolph* in 1906.

In 1906 the company was succeeded by the Skinner Ship Building & Dry Dock Company of Baltimore City. Harry G. Skinner was president and treasurer, Decatur H. Miller, Jr., secretary.

Skinner had been president of the William Skinner & Sons Ship Building & Dry Dock Co. at the foot of E. Cross St. in Baltimore. Miller had been secretary of the Merchants & Miners Transportation Co. and William Skinner & Sons.

The yard was expanded to an Upper Plant at the foot of E. Cross and a Lower Plant and main office at the foot of E. Fort Ave. at Locust Point. The company's first ship was the 308-ton *Ohio*.

In 1914 Harry was president, Edwin W. Poe secretary, and William C. Seddon treasurer. The next year Harry retired and Thomas H. Bowles became president and Holden A. Evans vice-president and general manager,

In 1916 Skinner Ship Building became the Baltimore Dry Dock & Ship Building Co. Thomas Bowles was president, Holden Evans vice-president and general manager, John M. Willis assistant general manager, and William C. Seddon, treasurer.

In 1917 Holden Evans was president and general manager, Clement C. Smith vice-president, John Willis, II, vice-president, George Allison treasurer, and Edwin Poe secretary.

When war came the plant expanded and the company had to buy out the Baltimore Yacht Club. The club put its clubhouse on scows and towed it to a new home at the foot of Light St. Their last ship was the freighter *Warwick* in 1921.

Bramell Point—3,250 tons, motor tanker, 1916, 214444, LGJK. Vacuum Oil Co.

Pennant—3,253 tons, motor tanker, 1916, 214597, LGNQ. Pierce Oil Corp.

Holden Evans—3,253 tons, 32,000-barrel motor tanker, 1917, 214673, LGVP. Empire Transportation & Oil Co.

Charles Braley—3,253 tons, tanker, 1917, 214817, LGSQ. Sinclair Navigation Co. Braley was a director of the Sinclair Gulf Corp. Sold to the American Italian Commercial Corp. for $1,533,000.

Albert Watts—3,302 tons, tanker, 1917, 215064, LHCW. Sinclair Navigation Co. Watts was a vice-president of the Sinclair Gulf Corp. Sold to the American Italian Commercial Corp. for $1,533,000.

Joseph Cudahy—3,302 tons, tanker, 1917, 215310, LHFN. Sinclair Navigation Co. Cudahy was a vice-president of the Sinclair Gulf Corp. Sold to the American Italian Commercial Corp. for $1,533,000.

William Isom—3,321 tons, 10.5 knot oil-burning, 33,000-barrel tanker, 215654, LJBC. Keel laid on March 3, 1917, for the Sinclair Navigation Co. Sold to the American Italian Commercial Corp. for $1,533,000. Requisitioned while building and delivered to the USSB on November 12, 1917. Commissioned *William Isom* in the Naval Overseas Transportation Service on May 1, 1918. William Isom, of Chicago, was vice-president of the Sinclair Gulf Corp. His daughter Evelyn sponsored the ship at launching on June 9.

Elinor—4,295 tons, steel, 10.5 knot oil-burning freighter, 215982, LJQN. Keel laid on May 29, 1917. Requisitioned from French owners while building and delivered to the USSB on March 20, 1918. Commissioned *Elinor* in the Naval Overseas Transportation Service on March 20, 1918.

Luella—4,152 tons, steel, 11.5 knot coal-burning refrigerated freighter, 216147, LKMD. Keel laid on June 18, 1917. Requisitioned from British owners while building and delivered to the USSB on May 31, 1918. Allocated to the Army then commissioned *Luella* in the Naval Overseas Transportation Service on September 16, 1918.

Polar Sea—4,120 tons, steel, 11.5 coal-burning refrigerated freighter, 216530, LMCB. Keel laid on October 11, 1917, for the Garland Steamship Corp. Requisitioned while building and delivered to the USSB on August 30, 1918. Allocated to the Oriental Navigation Co. for service to the east coast of South America and the United Kingdom. Commissioned *Polar Sea* in the Naval Overseas Transportation Service on October 23, 1918.

Polar Land—4,130 tons, steel, 11.5 knot coal-burning refrigerated freighter, 216982, LNCH. Keel laid on November 4, 1917. Requisitioned from French owners while building and delivered to the USSB on October 31, 1918. Commissioned *Polar Land* in the Naval Overseas Transportation Service on November 5, 1918.

Naiwa—6,240 tons, steel, 12 knot oil-burning freighter, 217066, LNHP. USSB contract ship. Keel laid on December 22, 1917, and delivered on October 31, 1918, 310 building days. Commissioned *Naiwa* in the Naval Overseas Transportation Service on November 4, 1918.

South Pole—4,130 tons, steel, 11.5 knot coal-burning refrigerated freighter, 217268, LNWF. Keel laid on May 5, 1918. Requisitioned while building and delivered to the USSB on November 29, 1918. Commissioned *South Pole* in the Naval Overseas Transportation Service on December 2, 1918.

Polar Bear—4,297 tons, steel, 11.5 knot coal-burning freighter, 217303, LPCB. Keel laid on June 11, 1918, for Garland Steamship. Requisitioned while building and delivered to the USSB on November 30, 1918. Commissioned *Polar Bear* in the Naval Overseas Transportation Service on December 3, 1918.

Fort Wayne—6,245 tons, steel, 12 knot oil-burning freighter, 217319, LPCS. USSB contract ship. Keel laid on December 30, 1917, and delivered on December 12, 1918, 358 building days. Commissioned *Fort Wayne* in the Naval Overseas Transportation Service on December 27, 1918.

North Pole—4,130 tons, steel, 11.5 knot coal-burning refrigerated freighter, 217401, LPHF. Keel laid on February 21, 1918. Requisitioned from French owners while building and delivered to the USSB on December 31, 1918. Allocated to the Red Star Line for service to Belgium. Commissioned *North Pole* in the Naval Overseas Transportation Service on January 14, 1918.

Polar Star—4,297 tons, steel, 11.5 knot coal-burning freighter, 217407, LPHQ. Keel laid on June 20, 1918. Requisitioned from British owners while building and delivered to the USSB on December 31, 1918. Commissioned *Polar Star* in the Naval Overseas Transportation Service on January 4, 1919.

Bethlehem Shipbuilding Corp., Ltd.
Bethlehem, Pennsylvania

Charles Michael Schwab and Joseph Wharton founded Bethlehem Steel in 1904 at Bethlehem, Pennsylvania, with Schwab as president. In 1913 the company got into shipbuilding under president Eugene G. Grace and in 1917 incorporated the Bethlehem Shipbuilding Corp., Ltd.

During the Great War the company delivered ships from yards at Alameda and San Francisco, California; Elizabeth, New Jersey; Sparrows Point, Maryland; and Wilmington, Delaware. The company was a major World War II builder.

Elizabeth, New Jersey

Plainfield—2,771 tons, steel, freighter, 216822, LMQW. Keel laid on May 12, 1917. Requisitioned while building and delivered to the USSB on September 11, 1918. Allocated to United Fruit for service to the West Indies.

Garfield—2,948 tons, steel, 11.5 knot coal-burning freighter, 217145, LNMT. Keel laid on May 12, 1917. Requisitioned while building and delivered to the USSB on October 31, 1918. Allocated to A.H. Bull for service to the east coast of South America.

Sparrows Point, Maryland

Bethlehem Steel opened an office in 1917 on the 15th floor of the Continental Building in Baltimore. Jesse A. Davis was the sales agent.

The *Cornelia* through the *Hatteras* were documented as built by the Bethlehem Steel Co. but beginning with the *Hatteras* and the 216000 official numbers the ships were documented as Bethlehem Shipbuilding Corp.–built ships. Sparrows Point was a major World War II shipbuilder.

Cornelia—3,216 tons, freighter, 1916, 214382, LGHK. Bull-Insular Steamship Co.

Munrio—3,868 tons, freighter, 1916, 214467, LGJT. Munson. Commissioned *Munrio* in the Naval Overseas Transportation Service on November 14, 1918.

Helen—3,216 tons, freighter, 1916, 214503, LGKP. Bull-Insular Steamship Co.

James McGee—10,667 tons, tanker, 1917, 214816, LGSP. Standard Oil of New Jersey.

Beatrice—3,450 tons, freighter, 1917, 214848, LGTD. A.H. Bull. Operated by Bull during World War II. Sunk on May 24, 1942, by *U-558* off Kingston, Jamaica.

W.C. Teagle—10,677 tons, tanker, 1917, 215284, LHKR. Standard Oil of New Jersey.

Cubore—7,151 tons, freighter, 1917, 215583, LHTR. Ore Steamship Co. Operated by the Army.

Walter D. Munson—4,088 tons, steel, 12 knot oil-burning freighter, 215621. LHVQ. Keel laid on November 15, 1916, for Munson. Requisitioned while building and delivered to the USSB on September 20, 1917. Commissioned *Walter D. Munson* in the Naval Overseas Transportation Service on April 19, 1918.

Feltore—7,117 tons, steel, 10 knot coal-burning freighter, 215723, LJDR. Keel laid on October 18, 1917, for Ore Steamship. Requisitioned while building and delivered to the USSB on October 30, 1918.

Hatteras—4,919 tons, steel, 10 knot coal-burning freighter, 215865, LJMC. Keel laid on May 24, 1917. Requisitioned from British owners while building and delivered to the USSB on December 21, 1917.

Santore—7,117 tons, steel, 10 knot coal-burning freighter, 216009, LJRN. Keel laid on March 1, 1917, for Ore Steamship. Requisitioned while building and delivered to the USSB on March 3, 1918. Operated during WWII. Sunk on June 17, 1942, by *U-701* off the East Coast.

Ampetco—8,301 tons, steel, 10.5 knot oil-burning tanker, 216233, LKPV. Keel laid on May 15, 1917, for the Standard Oil of New Jersey. Requisitioned while building and delivered to the USSB on May 6, 1918. Operated by Standard Oil. Operated during World War II by the Cities Service Oil Co. as the *Cities Service Empire*. Sunk on February 22, 1942, by *U-128* off the Atlantic Coast.

Cape Henry—4,992 tons, steel, 11.5 knot coal-burning freighter, 216327, LKRG. Keel laid on August 20, 1917. Requisitioned from British owners while building and delivered to the USSB on May 22, 1918. Allocated to Mallory. Commissioned *Cape Henry* in the Naval Overseas Transportation Service on October 25, 1918.

Mangore—4,066 tons, steel, 9.5 knot coal-burning freighter, 216413, LKTV. Keel laid on July 12, 1917, for Ore Steamship. Requisitioned while building and delivered to the USSB on June 11, 1918. Managed and operated by the company.

Cape Romain—4,992 tons, steel, 11.5 knot coal-burning freighter, 216466, LKWD. Keel laid on October 6, 1917. Requisitioned from British owners while building and delivered to the USSB on June 24, 1918. Commissioned *Cape Romain* in the Naval Overseas Transportation Service on June 26, 1918.

Firmore—7,117 tons, steel, 10 knot coal-burning freighter, 216565, LMDB. Keel laid on June 14, 1917, for Ore Steamship. Requisitioned while building and delivered to the USSB on July 8, 1918. Allocated to the company.

Cape Lookout—4,986 tons, steel, 11 knot coal-burning freighter, 216667, LMHG. Keel laid on December 18, 1917. Requisitioned from British owners while building and delivered to the USSB on July 25, 1918. Commissioned *Cape Lookout* in the Naval Overseas Transportation Service on July 26, 1918.

Berwyn—4,992 tons, steel, 11 knot coal-burning freighter, 216957, LNDH. USSB contract ship. Keel laid on April 19, 1917, and delivered on September 27, 1918, 159 building days. Commissioned *Berwyn* in the Naval Overseas Transportation Service on September 28, 1918.

Cape May—6,867 tons, steel, 12 knot coal-burning freighter, 217115, LNKJ. Keel laid on February 15, 1918. Requisitioned from British owners while building and delivered to the USSB on October 22, 1918. Commissioned *Cape May* in the Naval Overseas Transportation Service on October 24, 1918.

Wilmington, Delaware—Harlan Branch

Joseph B. Weaver was general manager, Samuel H. Smith assistant general manager. The company built ships and cars at the foot of West St.

W.H. Tilford—5,579 tons, steel, 10.5 knot coal and oil-burning tanker, 215856, LJKW. Keel laid on April 20, 1916, for Standard Oil of New Jersey. Requisitioned while building and delivered to the USSB on December 28, 1917. Wesley H. Tilford was an officer with Standard Oil.

Saetia—2,573 tons, steel, 9.5 coal-burning freighter, 215965, LJPW. Keel laid on October 14, 1916, for United Fruit. Requisitioned while building and delivered to the USSB on March 1, 1918. Commissioned *Saetia* in the Naval Overseas Transportation Service on March 1, 1918.

Guaro—2,873 tons, steel, 10 knot coal-burning freighter, 216079, LJVH. Keel laid on May 17, 1917, for the United Fruit Co. Requisitioned while building and delivered to the USSB on April 4, 1918. Allocated to Harriss, Magill for service to Belgium.

Biran—2,873 tons, steel, 10 knot coal-burning freighter, 216462, LKTP. Keel laid on December 21, 1917, for United Fruit. Requisitioned while building and delivered to the USSB on June 13, 1918. Allocated to the New York & Cuba Mail Steamship Co. for service to the west coast of South America.

Garibaldi—2,873 tons, steel, 10 knot coal-burning freighter, 216488, LKWR. Keel laid on March 15, 1918, for the New York & Porto Rico Steamship Co. Requisitioned while building and delivered to the USSB on June 28, 1918. Allocated to Gaston, Williams & Wigmore for service to Finland.

J.A. Bostwick—7,929 tons, steel, 10.5 knot oil-burning tanker, 216608, LMFQ. Keel laid on July 13, 1916, for Standard Oil of New Jersey. Jabez Abel Bostwick was treasurer of the company. Requisitioned while building and delivered to the USSB on August 6, 1918. Operated during World War II by the Cities Service Oil Co. as the *Cities Service Toledo*. Sunk on June 12, 1942, by *U-158* off the Atlantic Coast.

O.T. Waring—5,579 tons, steel, 10.5 knot coal or oil-burning tanker, 216770, LMNK. Keel laid on March 1, 1917, for Standard Oil of New Jersey. Requisitioned while building and delivered to the USSB on September 24, 1918. Orville Waring was a director. Operated by the company under Capt. Walter J. Healey.

Charles M. Everest—5,579 tons, 10.5 knot coal-burning tanker, 217130, LNMB. Keel laid on August 6, 1917, for Vacuum Oil. Requisitioned while building and delivered to the USSB on December 13, 1918. Operated by Vacuum Oil. Operated during WWII by Charles Kurz's Pennsylvania Shipping Co. as the *Naeco*. Sunk on March 23, 1942, by *U-124* off North Carolina.

Tipton—2,648 tons, steel, 11.5 coal-burning freighter, 217134, LNMC. Keel laid on May 7, 1918, for United Fruit. Requisitioned while building and delivered to the USSB on November 18, 1918. Allocated to United Fruit for service to the West Indies.

Norma—2,648 tons, steel, 11.5 coal-burning freighter, 217399, LPHC. Keel laid on June 21, 1918, for United Fruit. Requisitioned while building and delivered to the USSB on December 31, 1918. Allocated to United Fruit for service to the West Indies.

Chester Shipbuilding Co., Ltd.
Finance Building, Philadelphia

Richard Hallett Meredith Robinson was the general manager of the Lake Torpedo Boat Co. in Bridgeport, Connecticut. Charles B. Segar was vice-president of the Guaranty Trust Co. in New York. In 1915 they incorporated Chester Ship. Robinson was president, Segar vice-president and comptroller, Carl W. Hamilton vice-president, Anton D. Wagner treasurer, and James W. Kissick, a Philadelphia contractor, was secretary. The company acquired property in Bristol.

In August 1917, 1,500 International Brotherhood of Boilermakers, Iron Shipbuilders and Helpers of America went on strike for week. After the strike the company was taken over by the Merchants Shipbuilding Corp., controlled by William A. Harriman, chairman of the board. Chester Ship retained control of the yard at Chester and the property at Bristol was purchased by Merchants Ship. The last ship built at Chester was the freighter *Missourian*, delivered in 1922.

Hisko—5,665 tons, steel, 10.5 coal or oil-burning tanker, 215758, LJFN. Keel laid on June 5, 1916. Requisitioned from Norwegian owners while building and delivered to the USSB on December 6, 1917. Allocated to the Navy.

Sudbury—5,075 tons, 11 knot coal or oil-burning freighter, 215991, LJQV. Keel laid on May 25, 1916, for the Shawmut Steamship Co. Requisitioned while building and delivered to the USSB on March 5, 1918. Commissioned *Sudbury* in the Naval Overseas Transportation Service on March 5, 1918.

Overbrook—5,724 tons, steel, 10.5 knot coal or oil-burning tanker, 216157, LKMN. Keel laid on July 5, 1916. Requisitioned from British owners while building and delivered to the USSB on April 26, 1918. Allocated to Walker & Daly for service to Mexico.

Avondale—5,731 tons, steel, 10.5 knot coal or oil-burning tanker, 216368, LKSF. Keel laid on August 29, 1917. Requisitioned from British owners while building and delivered to the USSB on June 4, 1918. Allocated to Walker & Daly for Atlantic Coastwise trade.

Phoenix—5,731 tons, steel, 10 knot oil-burning tanker, 216824, LMRB. Keel laid on October 1, 1916. Requisitioned from British owners while building and delivered to the USSB on August 31, 1918. Allocated to Walker & Daly for service to Mexico.

Auburn—6,047 tons, steel, 10 knot oil-burning freighter, 217438, LPKN. Keel laid on April 28, 1917. Requisitioned from British owners while building and delivered to the USSB on December 31, 1917. Commissioned *Auburn* in the Naval Overseas Transportation Service on January 24, 1918.

Coastwise Shipbuilding Co.
Baltimore, Maryland

Thomas Martin Whedbee was born in Maryland on July 2, 1878, and became a civil engineeer, explorer, and railroad builder in Bolivia, Peru, Chile, and Argentina. He arrived at Baltimore on December 28, 1914, on the steamer *Denis* and incorporated Coastwise in May 1916 with a capital of $50,000. He was single and lived at 1408 Park Ave. in Baltimore. The yard was at the foot of Andre Street on Locust Point.

Rossiter S. Scott was assistant engineer at the Consolidated Gas, Electric Light & Power Co. and was vice-president. He lived at 605 Park Ave. Secretary-treasurer Frank A. Brandy was an accountant who lived in Cantonsville and Frank A. McDonald was general manager. The company built two schooner barges during the war for the Reading Co. They built 14 sailing vessels, the last being the 1,281-ton schooner *Carroll* in 1920.

Trenton—1,013 tons, schooner barge, 1917, 215835, LJHV. Reading Co.

Tioga—1,012 tons, schooner barge, 1918, 216038, LJSH. Reading Co.

Cobb, Butler & Co.
Rockland, Maine

In 1868 Francis Cobb, Hezekiah W. Wight, and John S. Case were grocers, lime manufacturers, and granite dealers in Rockland as Cobb, Wight & Case.

Cobb also built ships on the South End as Francis Cobb & Co. In 1882, William T. Cobb was a partner in Cobb, Wight & Co. Francis and William were lime manufacturers at 407 Main and wholesale grocers at 409 Main with Albert W. Butler, and Fred W. Wight as Cobb, Wight & Co. At some point Francis teamed up with Albert Butler to build ships as Cobb Butler & Co. Albert Butler was mayor of Rockland in 1897. Francis died on December 2, 1890, and William continued in the shipyard.

In 1917 the shipyard at Atlantic Wharf was the Francis Cobb Co. with William as president, Nathan T. Cobb treasurer, and Roy L. Knowlton business manager. They continued building ships as Cobb, Butler & Co.

Ella Pierce Thurlow—1,505 tons, schooner, 1918, 216015, LJRT. Atlantic Coast Co.
Paul E. Thurlow—1,590 tons, schooner, 1918, 216809, LMQJ. Atlantic Coast Co.

Crowninshield Shipbuilding Co.
Fall River, Massachusetts

Bowdoin Bradlee "B.B." Crowninshield, of the famous nautical family, was born in Marblehead on October 13, 1867, and became a noted naval architect of fine schooners and yachts.

Clarence T. Birkett was an insurance broker in New York and Harold W. Krogman was a ship broker in Boston.

In August 1917 the three incorporated the Crowninshield Shipbuilding Corp. with capital of $300,000. Crowninshield was president and general manager, Birkett vice-president, and Krogman secretary-treasurer.

The company's 8-way occupied the former Read yard on the Taunton River at South Somerset. Their first vessel was the *Hesper* and the last was the motorboat *Betty* in 1927.

Hesper—1,348 tons, schooner, 1918, 216630, LMGH. Charles F. Webb, Boston.

Cumberland Shipbuilding Co.
South Portland, Maine

James C. Hamlen and his son James C. Hamlen, Jr., ran a cooperage and lumber business in Portland. In 1918 they opened a shipyard at 162 Pickett St. in South Portland to build wooden Shipping Board freighters.

Cumberland—2,551 tons, wood, 10 knot coal-burning Ferris freighter, 217080, LNJD. USSB contract ship. Keel laid on January 2, 1918, and delivered on December 27, 1918, 361 building days. Allocated to General Navigation Co. service to the Netherlands.

Downey Shipbuilding Corp.
Arlington, Staten Island, New York

Wallace Downey went to work in a shipyard in 1880 and became a partner, ship owner, and operator of up to 50 ships. He obtained financing and purchased an abandoned yard on

Shooters Island and opened the Standard Shipbuilding Corp. He built steel ships for a Norwegian customer but the controlling interest in his company was bought out by "foreign capital" that had no experience in shipbuilding. Downey couldn't get along with the new owners so he sold his interest.

In 1912 he was president and general manager of the Downey Ship Yard & Marine Co. at the foot of 23rd St. His office was at 21 State St. in Manhattan and he lived at 819 Carroll St. in Brooklyn.

In 1916 Downey purchased the dilapidated Providence Engineering Co. in Providence, Rhode Island, which he converted into a shipyard and built ten, 150' ocean tugs for the Emergency Fleet Corp. In January 1917 he purchased the Milliken steel fabricating plant and established the Downey Shipbuilding Corp. On July 3, 1917, Downey signed the Emergency Fleet Corp. contract No. 30 S.C. to build ten steel freighters. He delivered the last one in December 1919.

Downey was accused of using John Cranor, an energetic, ambitious railroad man from Richmond, Virginia, to bribe Emergency Fleet Corp. officials Tucker K. Sands, R. Wilmer Bolling, and Lester Sisler to get the Shipping Board contract and to expedite the delivery of materials and supplies to his yard. During testimony, Downey at first said Cranor came looking for a job and then admitted he actually sought him out. The controversy caused the Emergency Fleet Corp. to suspend payments, which put his company into receivership. Downey's last project was the steamer *El Petrolero* in 1921.

Abron—5,383 tons, steel, 10.5 knot oil-burning freighter, 1918, 217381, LPGJ. USSB contract ship. Keel laid on November 24, 1917, and delivered on December 12, 1918, 398 building days. Allocated to W.R. Grace for the French-India trade.

Dunn & Elliot Co.
Thomaston, Maine

John Elliot, of Halifax, Nova Scotia, settled in Boston. John married Abilail Kneeland in 1767 they moved to Wiscasset, District of Maine, and had a daughter named Abigail. Thomas Dunn, of Liverpool, England, settled in Boston. Thomas Dunn and Abigail Elliot married and their son Thomas W. Dunn was born in Boston on October 10, 1821.

Abigail Elliot died when Thomas W. Dunn was six weeks old and he was raised by John and Abigail Elliot until he was 14. In 1837 he went to Thomaston to work with his uncle, Richard E. Elliot as Elliot & Dunn. On March 9, 1846, he married Eliza Giles and in 1856 he became a partner with Capt. William Tobey as Tobey & Dunn to manufacture sails and rigging. In 1859 Richard Elliot died and his son, Tom's cousin George Elliot, joined the firm as Tobey, Dunn & Co, sailmakers. In 1864 Thomas and George had an interest in a chandlery and general store on Wadsworth St. and in 1865 Tom and George Elliot teamed up as Dunn & Elliot to build ships.

They had seven children including Richard E. Dunn. Eliza died in 1878 and in 1880 he teamed up with George Elliot as Dunn & Elliot, sailmakers in Thomaston. That partnership lasted until 1902.

Richard O. Elliott was born on February 6, 1873, to George and Mary Ella Libby Elliot.

Thomas W. Dunn's son, Richard E., graduated from M.I.T. in 1896, became a sailmaker and shipbuilder, and then teamed up with Richard O. Elliot to build ships as Dunn & Elliot.

Their first 21st century schooner was the *E. Marie Brown* in 1903 followed by the *E. Starr Jones* and the *Helen E. Taft*. That apparently was the last big ship built until 1918 when the 4-masted schooner *Margaret Throop* was delivered on the account of Richard Elliot.

In 1918 Richard O. Elliot was president, A. Junkins Elliot vice-president, Frank D. Elliot, secretary-treasurer.

Margaret Throop—1,264 tons, schooner, 1918, 216315, LKRD.

E.G. Potter
Jacksonville, Florida

Ellis G. Potter was secretary of the American Push Button Telephone Co. at 136 Liberty St. in New York City and also owned the big 5-masted, 267' schooner *Louise B. Crary*, built in 1900. The *Crary* sank in a collision with the smaller 4-masted, jinxed schooner *Frank A. Palmer* on December 17, 1902, off Cape Ann.

Potter had his share of legal troubles and George H. Barnes was an attorney in New York at 120 Broadway. Potter built only the two ships.

De Witt Brown—1,325 tons, schooner, 1918, 216354, LKRT. Armour Fertilizer Works.

George H. Barnes—1,620 tons, motor freighter, 1918, 216662, LMHB. E.G. Potter, New York.

Federal Shipbuilding & Dry Dock Co.
Kearny, New Jersey

Judge Elbert Henry Gary, of Chicago, was chief counsel for the Illinois Steel Co. and in 1893 was made president of the Federal Steel Company. He was a founder of the U.S. Steel Corp. in 1901 and the namesake for Gary, Indiana.

Federal Ship was incorporated on July 24, 1917, as a subsidiary U.S. Steel with a capital of $3 million to build Shipping Board freighters on Hackensack meadows. Judge Gary was president of Federal Ship when the company signed a contract for 30 ships.

Liberty—6,868 tons, steel, 11 knot coal-burning freighter, 217061, LNHG. USSB contract ship. Keel laid on November 15, 1917, and delivered on October 5, 1918, 325 building days. Commissioned *Liberty* in the Naval Overseas Transportation Service on October 7, 1918. Operated during World War II by the Army. Sunk on January 11, 1942, by *I-166* off Bali.

Federal—6,868 tons, steel, 11-knot coal-burning freighter, 217147, LNMV. USSB contract ship. Keel laid on November 23, 1917, and delivered on November 11, 1918, 353 building days. Commissioned *Federal* in the Naval Overseas Transportation Service on November 16, 1918.

Piave—6,868 tons, steel, 11 knot coal-burning freighter, 217349, LPDW. USSB contract ship. Keel laid on November 28, 1917, and delivered on December 18, 1918, 385 building days. Commissioned *Piave* in the Naval Overseas Transportation Service on December 18, 1918.

The Fore River Shipbuilding Co.
Quincy, Massachusetts

Thomas Augustus Watson was born in Salem, Massachusetts, on January 18, 1854. He became a machinist and was hired by Alexander Graham Bell at Boston University. He worked at Bell Telephone until 1881 and on September 5, 1882, married Elizabeth Kimball.

Frank O. Wellington was the superintendent of Brayton Petroleum in Brockton, Massachusetts, and in 1883 moved to Boston where became acquainted with Watson and the next year the two opened the Fore River Engine Co. in East Braintree to build small marine engines. By 1898 they were building ships in Weymouth.

The naval appropriation act of May 4, 1898, authorized the construction of 12 torpedo boats and 16 torpedo-boat destroyers. Fore River Engine got contracts to build the *Lawrence* and the *Macdonough*. They laid the keels for the two destroyers on April 10, 1899, and the same year built the Diamond Shoal *Lightship No. 72*.

In 1900 they became the Fore River Ship & Engine Co. at 107 E. Howard Ave. in Quincy. Watson was president and Wellington treasurer and general manager. The families lived next door to each other in East Braintree.

In 1901 they incorporated in New Jersey and got another naval contract. On September 20, 1902, they launched the cruiser *Des Moines* then the 403', 7-masted collier schooner *Thomas W. Lawson*, the largest schooner in the world. It was chartered by the Sun Oil Co. and converted to carry petroleum. The ship made 12 knots under sail but was wrecked in 1907.

On December 6, 1902, the company executed a mortgage with the Adams Trust Co. to secure performance bonds to build the *Lawrence* and *Macdonough*. The *Lawrence* was commissioned on April 7, 1903, and the *Macdonough* on September 5, 1903.

Apparently all of the builders of the destroyers experienced financial difficulties with the Navy Department in executing these contracts and the company defaulted on interest payments to Adams. On April 1, 1904, the mortgage was foreclosed and Adams "conveyed all property and assets to attorney Henry Endicott, Jr., on behalf of a new company, the Fore River Shipbuilding Co., incorporated under the laws of Massachusetts. On October 10, 1904, all moneys due to the Fore River Ship & Engine Co. for the *Lawrence* and *Macdonough* were assigned to the Fore River Shipbuilding Co. on account of Fore River Engine "having failed to comply with certain provisions of the laws of the State of New Jersey" and from the U.S. government, "on account of said torpedo-boat destroyers *Lawrence* and *Macdonough*, together with all the rights, claims, and demands connected therewith or arising therefrom."

In early 1905 the Watsons moved to 35 Everett Ave. in Somerville, Massachusetts, and Thomas became the assistant yard master for the Boston & Albany Railroad. He retired in 1907. Wellington became assistant to the president at Fore River Shipbuilding.

On or about January 5, 1909, the governor of the state of New Jersey revoked the corporate charter of the Fore River Ship & Engine Co.

On January 12, 1912, a bill was introduced in the Senate for the relief of the Fore River Ship & Engine Co., for $329,033.98 on the *Lawrence* and *Macdonough*, "being the difference between the actual cost of said boats to their respective builders and the amount paid them under their contracts." On June 21, 1912, the claim was referred to the Court of Claims for fact-finding. The suit was dismissed on March 10, 1919.

In 1913 the shipyard was purchased by the Bethlehem Steel Co. with headquarters at 97 E. Howard. Thomas Watson died on December 13, 1934.

Amolco—3,074 tons, tanker, 1914, 211906, LDJF. Boston Molasses Co.

Atlantic—5,398 tons, freighter, 1914, 212387, LDMS. Union Sulphur Co.

Pacific—5,398 tons, freighter, 1914, 212560, LDPG. Union Sulphur Co.

Texas—6,678 tons, tanker, 1916, 213878, LFSD. The Texas Co.

New York—6,578 tons, tanker, 1916, 213937, LFTP. The Texas Co.

Cubadist—5,788 tons, tanker, 1916, 214053, LFWV. Cuba Distilling Co.

Sucrosa—5,788 tons, tanker, 1916, 214190, LGCQ. Cuba Distilling Co.

Edward Luckenbach—8,151 tons, freighter, 1916, 214500, LGMP. Luckenbach Co. Commissioned *Edward Luckenbach* in the Naval Overseas Transportation Service on August 29, 1918. Sunk during World War II on July 2, 1942, by an American mine off Key West, Florida.

Julia Luckenbach—8,151 tons, freighter, 1917, 214677, LGPW. Luckenbach Co. Commissioned *Julia Luckenbach* in the Naval Overseas Transportation Service on August 15, 1918.

Mielero—5,853 tons, tanker, 1917, 214740, LGRN. Cuba Distilling.

Pennsylvania—6,666 tons, tanker, 1917, 215286, LHMF. The Texas Steamship Co.

Virginia—6,666 tons, tanker, 1917, 215371, LHNP. The Texas Steamship Co.

F.J. Luckenbach—8,074 tons, steel, 13 knot oil-burning freighter, 215668, LJBR. Keel laid on January 29, 1917, for the Luckenbach Co. Requisitioned while building and delivered to the USSB on November 28, 1917. Commissioned *F.J. Luckenbach* in the Naval Overseas Transportation Service on January 9, 1918.

K.I. Luckenbach—8,074 tons, steel, 13 knot oil-burning freighter, 215800, LJGS. Keel laid on January 4, 1917, for Edgar F. Luckenbach. Requisitioned while building and delivered to the USSB on January 25, 1918. Commissioned *K.I. Luckenbach* in the Naval Overseas Transportation Service on August 9, 1918.

Katrina Luckenbach—8,074 tons, steel, 13 knot oil-burning freighter, 216090, LJVN. Keel laid on March 12, 1917, for Edgar F. Luckenbach. Requisitioned on March 11, 1918, at about 75 percent complete and completed with public funds to government specifications. The ship was then sold back to the Luckenbach Co. for $1,535,000. On May 18 the company signed a bareboat charter party agreement with the Navy Department and the ship was commissioned into the Naval Overseas Transportation Service as *Katrina Luckenbach* on May 18, 1918.

George W. Barnes—6,450 tons, steel, 10.5 oil-burning tanker, 216388, LKSW. Keel laid on June 27, 1917, for Pan American Petroleum & Transport. Requisitioned while building and delivered to the USSB on June 4, 1918.

W.L. Steed—6,450 tons, steel, 10.5 knot oil-burning tanker, 216877, LMTQ. Keel laid on November 9, 1917, for Standard Oil of New Jersey. Requisitioned while building and delivered to the USSB on September 18, 1918. Time chartered by the Shipping Board. Operated during World War II by Standard Oil. Sunk on February 2, 1942, by *U-103* off the East Coast.

Nantasket—5,769 tons, steel, 10.5 knot coal and oil-burning freighter, 216977, LNBW. Keel laid on November 13, 1917, for Munson. Requisitioned while building and delivered to the USSB on October 17, 1918. Allocated to the American Line for service to the United Kingdom.

Cohasset—5,769 tons, steel, 10.5 knot coal or oil-burning freighter, 217211, LNSQ. Keel laid on April 2, 1918. Requisitioned from British owners while building and delivered to the USSB on November 30, 1918. Commissioned *Cohasset* in the Naval Overseas Transportation Service on December 5, 1918.

The Foundation Co.
35 Nassau St., Rm 817, New York

Franklin Remington was born on November 16, 1866, in Utica, New York, the son of Samuel Remington, president of the Remington Arms Co. He graduated from Harvard in 1887, tried farming, then lost a lot of money in the Kansas City real estate market. In 1892 he and his brother, Carver Remington, became general contractors in Chicago. They built the water system in Hammond, Indiana, the elevated electric railway at the 1893 Chicago World's Fair, and the Chicago Drainage Canal.

Daniel S. Moran was a contractor in Jersey City, New Jersey, and Edwin Jarrett was a civil engineer. In 1902 Franklin, Moran, and Jarett established the Foundation Co. with Franklin as president, Moran, vice-president, Edwin Jarrett, treasurer, and Cora E. Jarrett, secretary.

Their plan was to specialize in difficult foundation work for buildings using compressed air tools to break up hard rock. Under the soil of Manhattan is a very large and extremely hard 25 million-year-old chunk of granite called the Manhattan Shift. Outcroppings are in Central Park.

In 1916 John W. Doty and Louis L. Brown were vice-presidents. Charles A.D. Bayley had succeeded Brown by 1918 and Frank Quilter was secretary-treasurer.

Franklin decided to get into shipbuilding during the war and had yards at Kearny, New Jersey; Portland, Oregon; Tacoma, Washington; and Brunswick and Savannah, Georgia.

Accoma—2,551 tons, wood, 10 knot coal-burning Ferris freighter, 216621, LMGD. USSB contract ship. Keel laid on October 20, 1917, at Kearny and delivered on August 31, 1918, 309 building days. Sold to Smith & Terry and foundered east of Florida on February 11, 1919. No lives were lost of the 40 on board.

Coyote—2,551 tons, wood, 10 knot coal-burning Ferris freighter, 216679, LMHN. USSB contract ship. Keel laid on October 26, 1917, and delivered on August 1, 1918, 279 building days. Sold to American Steamship.

Coweta—2,551 tons, wood, 10 knot coal-burning Ferris freighter, 216836, LMRP. USSB contract ship. Keel laid on August 30, 1917, and delivered on September 21, 1918, 387 building days. Allocated to Harriss, Magill for service to the Netherlands.

Congaree—2,551 tons, wood, 10 knot coal-burning Ferris freighter, 216888, LMTP. USSB contract ship. Keel laid on August 30, 1917, and delivered on October 1, 1918, 396 building days. Allocated to manager J.A. Elliott for service to France by the F.D. Dimmick Co.

Alanthus—2,551 tons, wood, 10 knot coal-burning Ferris freighter, 216889, LMTQ. USSB contract ship. Keel laid on March 22, 1917, and delivered on October 29, 1918, 221 building days. Allocated to Potter Transportation for service to the Netherlands.

Frye Flynn Co.
Harrington, Maine

Edwin M. Frye was born on November 27, 1870, in Columbia, Maine, to John F. and Helen Frye. John was a ship carpenter. Edwin went to sea, served in the Navy during the Spanish-American War, studied bookkeeping, and settled in Harrington.

On April 1, 1893, he married Mabel Walsh. In 1903 he went into the pulpwood business in Harrington with Otis S. Plummer as Plummer & Frye. In 1908 Edwin partnered with Charles A. Coffin as E.M. Frye & Co. but by 1910 Coffin was obliged to retire and Bartlett W. Brown, of Milbridge, became a partner. In 1911 the company became the Frye Pulp Wood Co. with Frye as treasurer and general manager.

In 1917 Edwin and sea captain Carl B. Flynn established the Frye-Flynn Co. to build ships but in 1918 Frye became sole owner. The company built six schooners. The first was the *Lizzie D. Peabody* in 1917 and the last was the *Mabel A. Frye* in 1920.

Sally Persis Noyes—1,034 tons, schooner, 1918, 217315, LPCN. Atlantic Coast Co.

G.G. Deering Co.
Bath, Maine

In 1915, Gardiner G. Deering was an 80-year-old ship builder, engineer, and president of the company at 226 Washington St. Harry G. was treasurer and Frank M. was a clerk. The three schooners were built on the company's own account. The company owned five schooners.

Jerome Jones—1,891 tons, schooner, 1916, 213944, LFTQ. Name changed to *Frank M. Deering*.
Maude M. Morey—1,364 tons, schooner, 1917, 215165, LHGF.
Harry G. Deering—1,342 tons, schooner, 1918, 216137, LJWR.

G.S. Baxter & Co. Ship Yards
Jacksonville, Florida

George S. Baxter lived in New York City and was a partner in a wholesale lumber business at 434 E. Bay St. in Jacksonville. His shipyard at Talleyrand and 14th was managed by John L. Douglass. The *Macdowell* was the only ship the company built.

Charles H. MacDowell—1,325 tons, schooner, 1917, 215662, LJBN. Armour Fertilizer.

George A. Gilchrist
Thomaston, Maine

George Alden Gilchrist was born in St. George, Maine, on May 27, 1851, the son of a sea captain. He went to work in a general store and in 1873 married Alice Robinson. In 1876 he owned a country store in Tenant's Harbor. In 1883 the couple lived in Rockland and he was a traveling salesman for a wholesale flour and grocery business in Portland.

They had two children before Alice died in 1886. The next year he married Annie Foster, of Belfast, and decided to quit traveling and get into the shipbuilding business. In 1889 rented property in Belfast and then bought a marine railway in Rockland.

In 1904 he built the steamer *Key West* and the schooner *Theoline* in 1905. When shipbuilding declined he sold everything but in 1916 opened a new yard in Thomaston. After the *Nancy Hanks,* he built the wood Shipping Board freighter *Utoka,* launched on December 17, 1918, and delivered it in 1919.

Nancy Hanks—1,162 tons, schooner, 1917, 215129, LHFK. Amos D. Carver.

The Gildersleeve Ship Construction Co. Gildersleeve, Connecticut

Oliver Gildersleeve was born on March 6, 1844, in Gildersleeve and after schooling went to work in his father Henry's shipyard, S. Gildersleeve & Sons in East Haddam, Connecticut. The yard built naval vessels during the Civil War.

On November 8, 1871, he married Mary Ellen Hall and they had a son Alfred on August 23, 1872, and then eight more children ending with Oliver Jr., on March 9, 1890.

Oliver's brother Sylvester opened a ship brokerage and commission house in New York at 1 Broadway where the Gildersleeve ships were sold or chartered.

In 1918 Alfred Gildersleeve was president and treasurer, Berkley C. Stone vice-president, and Oliver Gildersleeve secretary. The company built steam and schooner yachts, barges, car floats, tankers, and other craft. On July 12, 1917, they signed a contract for a wood Shipping Board freighter. Their last ship was the schooner *Allegheny* in 1921.

Batta Hatchie—2,551 tons, wood, 10 knot freighter, 1919, 217054, LNGW. USSB. Allocated to Nafra for service to the Netherlands.

Groton Iron Works Noank, Connecticut

Established in 1916 by notorious shipping, ice, and lumber tycoon Charles Wyman Morse. His sons ran the company: Erwin Albert was president, Harry Franklin vice-president, and Benjamin Wyman, secretary. The yard at Noank built six wooden ships. In 1918 they opened a yard in Groton to build steel Shipping Board freighters and delivered 11 ships between 1919 and 1921.

Francis J. Mcdonald—1,059 tons, schooner, 1917, 215905, LJNM. A.D. Cummins.

Hokah—2,551 tons, wood, 10 knot coal-burning Ferris freighter, 216803, LMQG. USSB contract ship. Keel laid on September 13, 1917, and delivered on August 31, 1918, 352 building days. Allocated to Potter Transportation for service to the Netherlands.

Balsto—2,551 tons, wood, 10 knot coal-burning Ferris freighter, 216842, LMRW. USSB contract ship. Keel laid on September 29, 1917, and delivered on October 29, 1918, 394 building days. Allocated to Smith & Terry for service to the west coast of South America.

Harlan & Hollingsworth Corp. Wilmington, Delaware

Mahlon Betts was born in Pennsylvania on March 16, 1795. In 1812 he moved to Wilmington, established a foundry, and set up the first stationary steam engine in the state.

Samuel Nichols Pusey was a machinist in Wilmington. Betts and Pusey came up with $5,000 and on March 1, 1836, established Betts & Pusey at Tenth & Market Sts. to build railroad cars and the trucks for them. They needed a woodworker and found Samuel Harlan, Jr., a cabinet maker in Wilmington and on January 1, 1837, he joined the firm as Betts, Pusey & Harlan.

In 1840 the company was in a 3-story rented building at First & Tatnall when they built the first fire engine in Delaware for the Water Witch Engine House. Shortly after, they moved onto property belonging to the Wilmington Whaling Co. at the foot of West St. on the Christiana River, right next to the Thatcher & Young shipyard, where they built a boiler shop, smithy, machine shop, and fabrication shop for marine work.

On August 28, 1841, Pusey sold his interest to machinist Elijah Hollingsworth, of Philadelphia, and the firm became Betts, Harlan & Hollingsworth. Pusey and Hollingsworth had worked together at the Baldwin Locomotive Works in Philadelphia. Hollingsworth's family was engaged in shipbuilding and after his arrival the company concentrated on more marine engineering work and then ship repair. They leased a wharf and a 200' way. The same year the company designed and built a new cylinder for the 160' passenger steamer *Sun* and that led to the firm establishing the first iron shipbuilding yard in the United States and the location for the famous Pusey & Jones.

In October 1843 they laid the keel for the Bangor Steam Navigation's 120', 231-ton propeller steamer *Bangor*, launched in May 1844 and delivered in June. According to Col. Charles S. Hill, national secretary for the American Shipping & Industrial League, the *Bangor* was the first seagoing steamer of its kind built in the United States.

On May 8, 1849, Betts retired and the company was Harlan & Hollingsworth until July 8, 1858, when John Taylor Gause, a longtime employee, became a partner and the name changed to Harlan & Hollingsworth & Co. On March 14, 1867, the company was incorporated as Harlan & Hollingsworth.

In 1914 the shipyard was at the foot of West St. on the Christina River, the company's original location. William G. Coxe was president, Samuel K. Smith, treasurer, and Henderson Weir secretary. The old company's last ships were delivered in 1917.

Francis Hanify—2,588 tons, freighter, 1914, 212805, LDWR. Name changed to *Freeport Sulphur No. 1*. Freeport Sulphur.

Silver Shell—5,605 tons, tanker, 1915, 213752, LFQM. Shell Line.

Gold Shell—5,614 tons, tanker, 1916, 213873, LFSB. Shell Line.

George E. Paddleford—4,787 tons, tanker, 1916, 214161, LGCJ. Pan American Petroleum & Transport.

Pearl Shell—5,614 tons, tanker, 1916, 214490, LGKD. Shell Line.

Georgeanna Weems—2,088 tons, freighter, 1916, 214587, LGNK. Baltimore & Carolina Steamship Co.

Benjamin Brewster—5,579 tons, tanker, 1917, 215002, LHBN. Standard Oil of New Jersey.

Brewster was an officer with Standard Oil. Sunk by *U-67* on July 9, 1942, 2½ miles off Grande Isle, Louisiana.

Olean—2,783 tons, tanker, 1917, 215362, LHNF. Vacuum Oil Co.

Hillyer-Sperring-Dunn Co.
Jacksonville, Florida

Charles E. Hillyer and William L. Sperring were general contractors at 23 Blum St. in Jacksonville. Joseph R. Dunn was a real estate broker at 501 West Building.

The three established a short-lived yard in 1916 at the foot of St. John's Ave. S. Hillyer was president, Sperring, vice-president, and Dunn secretary-treasurer. Their first vessel was the schooner *Lohocla*. The *Northern No. 30* was the last ship they built.

Northern No. 29—1,267 tons, schooner barge, 1918, 216265, LKQM. Northern Transportation.

Northern No. 30—1,640 tons, schooner barge, 1918, 216421, LKVB. Northern Transportation.

J.M. Murdock
Jacksonville, Florida

John Milton Murdock was born in West Virginia on February 10, 1870. He went to work as a young book keeper for William D. and Watson M. Murdock and Charles W. James, the real estate firm of Murdock, James & Co. at 17 N. Leslie in Chicago.

In 1907 John married a Kansas girl, Ardesta, 24, and they had a son Mearl. In 1915 they had a daughter Murdesta. Around that same time the Charlotte Harbor & Northern Railroad built a 100-mile line in Florida between Boca Grande on Gasparilla Island and Arcadia to carry phosphate. Around 1910, Murdock, James & Co. purchased 18,000 acres of swamp in the area and set up Murdock Station on the line just north of Port Charlotte. In 1912 they hired a firm from Arcadia to survey the land and design a drainage system. John then founded Murdock Colony Farms and invited settlers in. Sam E. Roberts was the first but not many followed. John and Ardesta built a hunting lodge on the property. John became a contractor and built a shipyard at the foot of E. Adams St.

John was general manager, Esther M. Johannesen was secretary, Robert H. Russell superintendent, and Henry L. Russell assistant superintendent. Murdock had contracts for eight Ferris Shipping Board freighters but built only three. The last one was the *St. George* delivered in 1919.

In October 1919 he left Key West and went to Cuba "to investigate storage facilities" and planned to build a $1 million cold storage plant in Havana. In January 1920 Ardesta was living with Murdesta in a rented house at 238 Lee St. in Tampa. John died in Jacksonville in 1925.

Dancey—2,589 tons, wood, 10 knot coal-burning Ferris freighter, 216908, LMVK. USSB contract ship. Keel laid on November 28, 1917, and delivered on August 31, 1918, 276 building days. Allocated to the American Metal Transportation Co. as manager for service to Belgium the by the American Shipping Corp.

Harish—2,551 tons, wood, 10 knot coal-burning Ferris freighter, 217228, LNTC. USSB contract ship. Keel laid on January 3, 1918, and delivered on October 31, 1918, 303 building days. Allocated to Charles T. Magee for trans–Atlantic service.

Johnson Shipyards Corp.
2941 Richmond Terrace, Richmond, Staten Island, New York

Established in 1916 by Joseph A. Johnson at Mariners Harbor to build scows, barges, tugs, and lighters. Robert Magruder was "a minor official" in the company. Sylvanus Bedell, cashier at the Mariners Harbor National Bank, purchased ten shares of stock in the company.

In 1917 Magruder approached the Shipping Board and in July the company signed contracts to build three wood hulls for $900,000. Magruder then opened "a small account" with Bedell's bank and the directors hired Bedell as treasurer at $10,000 a year. Bedell then acquired 50 shares of common stock and 300 shares of preferred stock, "most of which he did not pay for."

Army Lt. Col. Edward E. Hardin (ret.) then became president of the company but the next month went on active duty. Magruder was then elected president, general manager, and purchasing agent. Joseph Johnson was vice-president, and Hamline Magruder was secretary.

It was later alleged that Magruder said to Bedell, "Whenever we are short, pay any checks that are presented and endorsed by me because our credit is good on account of our contracts with the Shipping Board." That October, Magruder gave Bedell a $16,920 check to cover the payroll but the company didn't have that much in the Mariners Harbor Bank. Bedell mentioned that and was told, "If you don't pay it, we cannot go along." Two weeks later Magruder gave Bedell another check for $16,752.78. Bedell again hesitated and told Magruder, "You had better carry on the financing of the shipyard or we'll go broke." The directors of the bank immediately paid Magruder's overdraft and declared they were solvent.

On May 12, 1920, the bank's directors discovered that Bedell had "misappropriated" $3,300 from other accounts at the bank to cover Magruder's overdrawn checks. When Bedell as much as admitted it, he was fired and the Federal Banking Dept. was notified.

On Tuesday afternoon, June 28, 1921, indictments were returned against Magruder and Bedell. Magruder was arrested the next day and held on $25,000 bail on charges that around $300,000 had been taken from the bank over the previous four years by Bedell to cover checks. Bedell's doctor came forward with dire warnings concerning his patient's health but when the U.S. Marshals went to his home at 73 Townsend Ave. in Clifton, Long Island, he was not there. The next morning Magruder presented himself at the prosecutor's office with his attorney and pleaded not guilty before Judge Garvin and was then held for trial. Magruder said the company was owed a large sum of money for the hulls.

The company constructed a 350' pier and built five ships and a number of small craft. The *Aowa* was the only Shipping Board steamer they built. Their last ship was the schooner *Eastern* delivered in 1922.

Aowa—2,551 tons, wood, 10 knot freighter, 1919, 217018, LNFK. Allocated to Smith & Terry for service to Scandinavia.

The Kelley Spear Co.
Bath, Maine

John R. Kelley was a ship broker on Front St. and Daniel Howard Spear was a blacksmith. In 1886 they opened a shipyard at 91 Bowery to build wooden vessels as Kelley, Spear & Co. They incorporated in 1900 and primarily built schooners and schooner barges.

By 1902 Elijah F. Sawyer was president and Spear was treasurer. The company's first 21st century delivery was the schooner barge *Annie* in 1903.

In 1914 Fred R.E. Dean was president, Harry B. Sawyer was treasurer. Sawyer was a flour, grain, hay, feed, and seed dealer at 15 Broad St. The company's last ship was the schooner *Winnegance* in 1919. J. Edward Fuller was president and Harry Sawyer was vice-president.

Wollaston—1,378 tons, schooner barge, 1915, 213435, LFKW. Staples Transportation.

Atlantic—1,900 tons, schooner barge, 1915, 213708. LFPR. Staples Transportation.

Larimer—1,584 tons, schooner barge, 1915, 213806, LFRC. Westmoreland Coal Co.

Westmoreland—1,593 tons, schooner barge, 1916, 214344, LGFV. Westmoreland Coal Co.

Northern No. 6—1,602 tons, schooner barge, 1916, 214542, LGMF. Northern Transportation Co.

Northern No. 7—1,631 tons, schooner barge, 1917, 214785, LGSD. Northern Transportation Co.

Manor—1,603 tons, schooner barge, 1917, 215253, LHKP. Westmoreland Coal Co.

Samuel W. Fancher—1,972 tons, schooner barge, 1917, 215731, LJDP. J.B. King Transportation.

Osceola—1,621 tons, schooner barge, 1918, 215952, LJPQ. Westmoreland Coal Co.

L.H. Shattuck Co.
Newington, New Hampshire

Louis H. Shattuck was a general contractor at 208 Granite St. in Manchester, New Hampshire, and lived with his wife Mary at 969 Chestnut. He built a 110-acre shipyard in the Newington section of Portsmouth to build wood Shipping Board freighters. The company's last ship was delivered in 1921.

Roy H. Beattie—2,551 tons, wood, 10 knot coal-burning Ferris freighter, 217110, LNKD. USSB contract ship. Keel laid on December 19, 1917, and delivered on April 13, 1919, 450 building days. Reported burned in 1919.

Chibiabos—2,551 tons, wood, 10 knot coal-burning Ferris freighter, 217111, LNKF. USSB contract ship. Keel laid on November 14, 1917, and delivered on November 14, 1918, 365 building days. Allocated to Lowrance & Co. for service to Scandinavia.

M.M. Davis & Sons
Solomons, Maryland

Marcellus M. Davis was born in August 1857 in Maryland and became a ship carpenter. He married his wife Emma in 1880 and they had sons Clarence in July 1883 and Raymond in July 1885. That same year he began shipbuilding in Solomons just north and west on the harbor side of Stoney's Solomons Pier Restaurant. In 1913 he moved the yard a half a mile across the harbor to the foot of present day Rousby Hall Road and used the original site for repair until around 1918. His last ship was the wood motor freighter *State Port* built in 1932 for the North Carolina Line at Wilmington.

Sarah Weems—1,521 tons, wood freighter, 1917, 215839, LJKB. Baltimore & Carolina Steamship Co.

Maryland Steel Co.
Sparrows Point, Maryland

In 1887 the Pennsylvania Steel Co., 208 S. 4th St., Philadelphia, opened an extension plant in Maryland on land first settled in the 1600s by Thomas Sparrow. Sparrow's Point was ideal for deepwater shipping. Walter S. Franklin was appointed manager.

In 1891 Pennsylvania Steel reorganized and expanded their Sparrow's Point plant as the Maryland Steel Co. with Frederick W. Wood as president and Rufus K. Wood as cashier. They made pig iron, railroad track rails, billets, and built iron and steel steam and sailing vessels, tugs, marine engines, boilers, and did marine repair work.

Charles Michael Schwab and Joseph Wharton founded Bethlehem Steel in 1904 at Bethlehem, Pennsylvania, with Schwab as president. In 1913 the company got into shipbuilding under president Eugene G. Grace and in 1916 purchased Maryland Steel. The company expanded to building steel floating dry docks with Frederick Wood continuing as president. The last ship built under a Maryland Steel building certificate was the *Munsomo*.

Washingtonian—6,649 tons, passenger, 1914, 211887. LDHW. American-Hawaiian.

Iowan—6,649 tons, freighter, 1914, 212144, LDKR. American-Hawaiian.

Ohioan—6,649 tons, freighter, 1914, 212314, LDMK. American-Hawaiian. Commissioned *Ohioan* in the Naval Overseas Transportation Service on August 7, 1918.

Ulysses—10,910 tons, collier, 1915, 213028, LFHM. Panama Canal Commission.

Achilles—11,081 tons, collier, 1915, 213174, LFJP. Panama Canal Commission.

Edith—3,628 tons, freighter, 1915, 213472, LFMD. A.H. Bull. Commissioned *Edith* in the Naval Overseas Transportation Service on October 1, 1918. Operated during World War II by Bull. Sunk on June 7, 1942, by *U-159* in the South Atlantic.

Floridian—6,930 tons, lumber freighter, 1915, 213813, LFRG. American-Hawaiian.

Munamar—3,477 tons, freighter, 1915, 213814, LFRH. Munson.

Clare—3,455 tons, freighter, 1915, 213833, LFRJ. A.H. Bull. Commissioned *Clare* in the Naval Overseas Transportation Service on September 19, 1918. Operated by Bull during World War II. Sunk on May 20, 1942, by *U-103* near Cuba.

Artisan—5,825 tons, lumber freighter, 1916, 213885, LFSH. American-Hawaiian. Sold Norwegian in October 1916 for $170,000.

Arborean—5,825 tons, lumber freighter, 1916, 213931, LFTM. American-Hawaiian. Sold Norwegian for $170,000 in October 1916.

Margaret—3,372 tons, freighter, 1916, 213988, LFVS. A.H. Bull. Commissioned as *Chatham* in the Naval Overseas Transportation Service on March 25, 1918. Operated by Bull during World War II. Sunk on April 14, 1942, by *U-571* off the East Coast.

Munplace—3,315 tons, freighter, 1916, 214042, LFWR. Munson. Commissioned *Munplace* in the Naval Overseas Transportation Service on August 31, 1918.

Munsomo—3,315 tons, freighter, 1916, 214087, LGBM. Munson. Commissioned *Munsomo* in the Naval Overseas Transportation Service on June 28, 1918.

Merchant Shipbuilding Corp. Harriman (Bristol), Pennsylvania

In 1915, Richard Hallett Meredith Robinson and Charles B. Segar incorporated the Chester Shipbuilding Co., Ltd., at Chester, Pennsylvania. Robinson was president, Segar vice-president and comptroller, Carl W. Hamilton vice-president, Anton D. Wagner treasurer, and James W. Kissick, a Philadelphia contractor, was secretary. The company acquired property in Bristol.

William Averell "W. Averell" Harriman was born in New York on November 15, 1891, to railroad magnate Edward and Mary Williamson Harriman. He graduated from Yale, inherited millions, and set up shop as W.A. Harriman & Co., an investment bank concentrating on railroads and steamship lines. He lived with his wife Kitty and their children at 1155 Park Ave.

In February 1917, the Harriman interests incorporated Merchant Ship to acquire the Bristol property. Richard Robinson was president, Charles Segar vice-president and comptroller, Carl W. Hamilton vice-president, and William M Simpson secretary-treasurer.

On September 17, 1917, the company contracted with the Emergency Fleet Corp. to build a 12-way yard in north Bristol with fitting out piers, and shops that could produce about 15 percent of the prefabricated steel parts for the ships. The estimated cost was $12 million. The contract called for 60 ships at a cost of $82,165,000. The first keel was laid on February 16, 1918.

The housing development for the workers came to be known as Harriman. The company's last ship at Harriman was the 5,976-ton *Naamok.*

Watonwan—6,107 tons, freighter, 1919, 217103, LNJW.

Wathena—6,107 tons, freighter, 1919, 217392, LPGT.

Merrill-Stevens Shipbuilding Corp. 614–616 E. Bay St., Jacksonville, Florida

Brothers James E. and Alexander R. Merrill were blacksmiths, shipsmiths, and boilermakers on E. Bay St. as J.E. Merrill & Bro. In 1888 they went into business with engineer Arthur D. Stevens as the Merrill-Stevens Engineering Co., mechanical and electrical engineers.

Stevens was superintendent of the Citizens Gas Co. and secretary-treasurer of the American Lighting Co. Stevens was president, engineer Harry C. Pike was secretary, and William L. Lampkin treasurer. Lampkin was assistant cashier of the Savannah, Florida & Western Railway Co. The company was the state agent for Westinghouse engines and operated a foundry, built marine and stationary boilers, mining equipment, light plants, iron jails, sold steam engines and new and second-hand machinery and metals.

In 1891 they started building wood, composite, and steel boats and by 1894 they had built 12 steamers. By 1904 they were specializing in designing and building light draft steamers, operated a marine railway, and engaged in wrecking and diving.

When the war started, Stevens was still president, Alexander Merrill was vice president, Archibald M. Main, vice president and general manager, William J. Payne, vice president, and John T. Manier, secretary-treasurer. The company had acquired a dry dock,

Naval architect and marine engineer Thomas Mullan Cornbrooks, 41, of Baltimore was hired to oversee their Shipping Board contracts. The company's yard was in confined waters and the largest vessel they had launched was the side wheel steamer *Jose Pedro* in 1906.

The company would be obliged to launch the Shipping Board freighters sideways. The sideways launching of vessels had been used in England since 1904 and was common on the Great Lakes and in the Mississippi Valley but not on the Atlantic or Gulf Coasts and no one at Merrill-Stevens was familiar with the process.

Cornbrooks asked Edward N. Hurley at the Shipping Board office in Washington if he knew anyone who specialized in launching ships sideways. He needed an experienced person to supervise several anticipated launchings between May 20 and the 25th.

Hurley publicized Cornbrooks' request in the various shipping publications and in early May he got a letter from Walter W. Johnston, a specialist in the launching process who was then employed at the Stewart Construction Co. in Baldwinsville, New York. He had been in the shipbuilding business since 1898. Johnston outlined his expertise in the launching of vessels sideways and said he had perfected a device for accomplishing the job.

The method Johnston had perfected resulted in less expensive construction for the ways, the launching process was safer, and the cost of each launching was much less. The process could also be used for traditional stern first launchings. The system employed a cradle that slid over the ways with trip levers that released blocking and packing onto the ground so it didn't float away.

Hurley was intrigued and invited Walter to Washington for a meeting with him and Cornbrooks. At the meeting, Cornbrooks asked Johnston to come to Jacksonville.

Johnston agreed and met with Archibald Main and William C. McGowan, the Shipping Board's Fourth District construction supervisor at Jacksonville. McGowan had control of all shipyards since they had been requisitioned by the Shipping Board.

McGowan invited Johnston to look over the company's plans for launching the ships. Johnston found them unacceptable, explained his reasons, and outlined what he would do instead. McGowan then requested Johnston to construct his own sliding ways and packing for the imminent launching of the *Red Cloud*. He did, and on or about May 30 the ship went into the water sideways without a hitch. Johnston was then put in charge of future launchings with the understanding that he would be compensated for his special device by the Shipping Board. He discussed fixing the amount of pay by contract with McGowan on several occasions but no agreement was concluded. In the meantime he was put on Merrill's payroll at $65 a week.

McGowan eventually wrote out a proposal of terms for Johnston to sign, which he did, but without reading it. Johnston asked for a copy of the agreement several times but never got one. McGowan said he would draw up a formal contract on the terms of the signed proposal the next time he was in Washington.

During the time Johnston was at Merrill-Stevens he reconstructed the ways and oversaw the launching of eight other ships. He also made at least three trips to the Merrill-Stevens yard at Slidell, Louisiana, and many trips to other shipyards in the region at the behest of the Shipping Board. These included the Morey & Thomas and J.N. Murdock yards in Jacksonville; Pensacola Shipbuilding at Pensacola; American Shipbuilding at Brunswick, Georgia; and the Oscar Daniels yard at Tampa. None of these companies paid Johnston anything. He was on the payroll of Merrill-Stevens at all times and was never paid anything by the Shipping Board.

In July 1918 the St. Johns River Shipbuilding Corp. at Jacksonville was operating shipyards acquired by the Emergency Fleet Corporation. Around July 4, 1918, the company attempted to launch the *Baxley* sideways on ways constructed by the Emergency Fleet Corporation but the ship got hung up on the ways and stuck there. McGowan contacted Johnston and requested he go to the shipyard and report back to him. Johnston did, reported the problem, and told McGowan how to fix it. McGowan then put Johnston in full charge of the operation and the ship was launched successfully shortly after. McGowan then directed Johnston to inspect all seven ways at the shipyard, which he did, and reported to McGowan that all the ways were unsafe and improperly constructed. McGowan transferred Johnston to St. Johns where he was put on the payroll at $75 a week with instructions from McGowan to redo all the ways to accommodate his launching device. Eight vessels were successfully launched from the ways through November 1919 and two were launched after that with Johnston's methods.

The sliding ways Johnston had constructed at Merrill-Stevens were towed to the A. Bentley & Sons yard at Jacksonville at the direction of the Emergency Fleet Corporation for the launching of two concrete ships, the *Dinsmore* and *Moffitt*.

All the ships constructed for the Shipping Board at St. Johns were on a fixed-price basis. Johnston worked in the Shipping Board's Jacksonville district for about 18 months. During this time he earned approximately $5,495 plus expenses. When the war ended on November 18, 1918, Johnston sought out McGowan but during an official visit to the Shipping Board's regional office in Philadelphia that month McGowan died. Neither the original proposal McGowan had written out for Johnston or any copy of it was ever found. Johnston believed the reasonable compensation for his services was $50,000 and he felt his only recourse was a lawsuit.

At trial, the government argued that Johnston had no contract with the Shipping Board nor had he provided any service to them. The court stated:

> It is urged on behalf of defendant that under all of the contracts involved in this case the duty was cast upon the builders of the vessels to launch them, and that the launching therefore was no concern of the Shipping Board, and that it was not likely or reasonable that they would have any understanding or expectation that the Government was to pay for any part of the work connected with the launching. Under ordinary circumstances this argument would have much force, but the circumstances were not ordinary. The vessels were needed for war purposes. It is a matter of common knowledge that the Government was in pressing and dire need of them. If they could not be obtained quickly, they might be of no use whatever, and the interest of the Government in obtaining them quickly was far greater than that of the shipbuilding companies. The Government was putting immense sums into its shipbuilding program, and the expenditure of a few thousands for the purpose of getting the ships quicker was, under the circumstances, quite reasonable.
>
> The evidence does not show that the plaintiff had any contract with the Shipping Board. On the contrary, it appears that the contract was never completed, but under the circumstances we think there was an implied contract on the part of the Shipping Board to pay the plaintiff.

On April 30, 1930, Walter Johnston was awarded $20,000 minus the wages he had already been paid making his total $14,505.

The company got a Shipping Board contract for eight composite-hull wood freighters designed by consulting engineer Norman Emile McClelland, in business as N.E. McClelland & Co., Ltd., 286 St. James St., Montreal—naval architects, marine engineers, surveyors, yacht and vessel brokers. The company had a branch office in New York at 2 Stone St.

Shipping Board Design No. 1009 was a well deck freighter with two holds forward and two aft, a single midship house, and an after deckhouse.

W.M. Tupper—1,852 tons, freighter, 1917, 215841, LJKD. Gulf & Southern Steamship.

Red Cloud—2,474 tons, wood, 10 knot coal-burning McClelland freighter, 216815, LMQP. USSB contract ship. Keel laid on September 29, 1917, and delivered on August 28, 1918, 333 building days. Sold to the Red Cloud Steamship Co., 18 Broadway, New York.

Apalachee—2,474 tons, 10 knot coal-burning McClelland freighter, 216860, LMSJ. USSB contract ship. Keel laid on October 11, 1917, and delivered on August 31, 1918, 324 building days. Allocated to Lykes Bros. for Atlantic coastwise service.

Botsford—2,474 tons, wood, 10 knot coal-burning McClelland freighter, 1918, 216928, LMWJ. USSB contract ship. Keel laid on November 27, 1917, and delivered on September 30, 1918, 308 building days. Allocated to Callaghan, Atkinson for service to the West Indies.

Kanabec—2,474 tons, wood, 10 knot coal-burning McClelland freighter, 216989, LNCR. USSB contract ship. Keel laid on November 27, 1917, and delivered on December 31, 1918, 393 building days.

Morey & Thomas Shipyard Co. Jacksonville, Florida

Richard Morey was born in Missouri in 1873, became a civil engineer, and in 1910 was a partner with Ernest A. Faulhaber in the Morey-Faulhaber Construction Co., general contractors in St. Louis. He lived at 5236 Washington Ave. with his wife Mary, son Richard, Jr., and a servant, Christine Sanders.

Charles Sewell Thomas was a partner with Ira G. Harmon, engineering contractors in St. Louis as the Thomas-Harmon Co., and lived at 611 Olive. He was born in Denver, Colorado, on February 24, 1884, graduated from Columbia University School of Mines in 1906, and worked as a mine examiner in Nevada, Mexico, and then back in Colorado. He married Marie Louise Wade in St. Louis, Missouri, on April 10, 1912. He got into general contracting in Texas as Thomas & Jones and in 1915 built the Missouri-Kansas-Texas Railroad terminal in San Antonio and in 1916–17 a supply depot at Fort Sam Houston.

Morey and Thomas decided to build a shipyard at Jacksonville, Florida, on the St. Johns River to build wood Shipping Board freighters. They contracted with the Emergency Fleet Corp. and work started on the 4-way yard on September 1, 1917. On October 5 they incorporated as the Morey & Thomas Shipyard Co. with a capital stock of $150,000. Morey was president and general manager and Thomas vice president.4-way yard. They subsequently signed a contract to build four ships for $1,200,000.

William L. Sperring, a general contractor with Hillyer & Sperring in Jacksonville, was treasurer. Howard L. Parsons, a carpenter, was assistant general manager and purchasing agent. John H. Thacker was superintendent. The last of their four ships was the *Kusdeca* delivered in 1919.

Bedminster—2,589 tons, wood, 10 knot coal-burning Ferris freighter, 217001, LNDQ. USSB contract ship. Keel laid on December 12, 1917, and delivered on September 26, 1918, 288 building days. Operated by the Shipping Board.

Bagoso—2,551 tons, wood, 10 knot coal-burning Ferris freighter, 217294, LBPK. USSB contract ship. Keel laid on December 14, 1917, and delivered on November 12, 1918, 333 building days. Allocated to American Shipping for service to the United Kingdom.

Nelson Ingalls
Wilmington, North Carolina

In July 1916, the Washington Iron Works, of Wilmington, started building the 240', 4-masted schooner *Hauppauge* on Fagle's Island in the Cape Fear River. Some difficulties arose and on May 18, 1917, the job was taken over by a well-known marine insurance man, Arthur Penrhyn Stanley Naul, of 39 Bow St., Forest Hills, Queens, New York.

Naul incorporated the Naul Shipbuilding Co. in early September 1917 at Baltimore with a capital of $125,000 and hired Capt. Nelson Ingalls as manager and Andrew Gilmour as the naval architect. The ship was launched on October 9 and after it was delivered and accepted, Naul went back to the marine insurance business in New York and Capt. Ingalls took over.

Capt. Ingalls built the *Commack* and then went back to his wife Margaret and son George in their rented home at 11 Union St. in Belfast, Maine.

Hauppauge—1,394 tons, schooner, 1918, 215997, LJQH. East Coast Transportation Co.

Commack—1,446 tons, schooner, 1918, 216356, LKRW. Smith Navigation Corp.

Newburgh Shipyards, Inc.
Newburgh, New York

Thomas C. Desmond was a construction engineer in New York at 110 W. 34th St., Room 505, and lived at 17 Gramercy Park. He wanted to construct wood freighters for the Emergency Fleet Corp. and chose Newburgh, 60 miles north of New York City.

Desmond was vice-president in New York and Edwin C. Bennet was vice-president, naval architect, and general manager in the yard. Eugene W. Brown was assistant treasurer, and Harry A. Marvel general superintendent. Marvel was the commodore of the Newburgh Yacht Club.

The company delivered only one freighter during the war but completed 16 in all, the last one being the *Comayagua* in 1921.

Newburgh—6,106 tons, steel, 11 knot coal-burning freighter, 217324, LPCW. USSB contract ship. Keel laid on March 14, 1918, and delivered on December 31, 1918, 292 building days. Commissioned *Newburgh* in the Naval Overseas Transportation Service on December 31, 1918.

Newport News Shipbuilding & Dry Dock Co.
Newport News, Virginia

Founded in 1880 by railroad "Big Four" magnate Collis P. Huntington. The company built their first vessel in 1891, the tugboat *Dorothy* for James Sheffield. In 1914 Albert L. Hopkins was president, Homer L. Ferguson vice-president and general manager, and Leroy F. Boggs assistant secretary. Newport News continued as a major shipbuilder.

John D. Archbold—8,374 tons, tanker, 1914, 211990, LDJR. Standard Oil of New Jersey. John Dustin Archbold was a competitor of Rockefeller who became vice-president of Standard Oil.

John D. Rockefeller—8,374 tons, tanker, 1914, 212317, LDNQ. Standard Oil of New Jersey. John Davison Rockefeller founded the Standard Oil Company of Ohio.

Neches—5,426 tons, freighter, 1914, 212464, LDNH. Mallory. Collided with a British patrol boat on May 14, 1918, off Sixpenny Point, England, and sank. None of the 91 aboard were lost.

Medina—5,426 tons, freighter, 1914, 212561, LDPH. Mallory.

Edward Pierce—4,387 tons, collier, 1914, 212756, LDVR. Crowell & Thurlow.

Mariana—3,063 tons, freighter, 1915, 213385, LFKT. New York & Porto Rico. Commissioned *Mariana* in the Naval Overseas Transportation Service on July 28, 1918. Operated during World War II by A.G.W.I. Lines. Sunk on March 5, 1942, by *U-126* in the South Atlantic.

Walter D. Noyes—4,387 tons, freighter, 1915, 213479, LFMG. Crowell & Thurlow.

Stephen R. Jones—4,387 tons, freighter, 1915, 213740, LFQD. Crowell & Thurlow. Commissioned *Stephen R. Jones* in the Naval Overseas Transportation Service on May 3, 1918.

Charles Pratt—9,959 tons, tanker, 1916, 213899, LFSR. Standard Oil of New Jersey.

Edgar F. Luckenbach—8,061 tons, freighter, 1916, 214040, LFWN. Luckenbach. Commissioned *Edgar F. Luckenbach* in the Naval Overseas Transportation Service on July 11, 1918.

H.H. Rogers—10,050 tons, tanker, 1916, 214066, LGBH. Standard Oil of New Jersey. Henry H. Rogers was an officer with Standard Oil.

Munalbro—4,293 tons, freighter, 1916, LGBJ, 214074, Munson. Commissioned *Munalbro* in the Naval Overseas Transportation Service on September 17, 1918.

Antwerpen—7,955 tons, tanker, 1916, 214341, LGFT. Built for the Standard Oil Company of New Jersey and sold to the American Petroleum Co. of Rotterdam. Left New York for London on August 3, 1916, on Voyage No. 1 under the Dutch flag. Sunk on September 12.

Henry R. Mallory—6,063 tons, cargo liner, 1916, 214458, LGJQ. Mallory. Requisitioned by the Shipping Board on a bareboat charter basis to the Navy for operation by the Army. Operated during World War II by A.G.W.I. Lines. Sunk on February 2, 1943, by *U-402* southwest of Iceland.

William A. Mckenney—6,256 tons, freighter, 1916, 214583, LGNH. Crowell & Thurlow. Commissioned on June 5 as *William A. McKenney*. Operated during World War II by the Eastern Gas & Fuel Associates. Sunk on October 5, 1942, by *U-175* off Venezuela.

Wm. G. Warden—10,289 tons, tanker, 1917, 214684, LGQF. Standard Oil of New Jersey.

Felix Taussig—6,253 tons, freighter, 1917, 214726, LGRC. Crowell & Thurlow. Commissioned *Felix Taussig* in the Naval Overseas Transportation Service on August 29, 1918.

F.Q. Barstow—10,289 tons, tanker, 1917, 214873, LGTQ. Standard Oil of New Jersey.

Mundelta—4,746-to tonsn freighter, 1917, 214916, LGVM, Munson. Commissioned *Mundelta* in the Naval Overseas Transportation Service on April 20, 1918.

Torres—4,943 tons, tanker, 1917, 215233, LHJV. Southern Pacific Co.

El Almirante—5,216 tons, freighter, 1917, 215411, LHPR. Southern Pacific Co.

El Capitan—5,215 tons, steel, freighter, 215587, LHTV. Keel laid on March 17, 1917, for the Southern Pacific Co. Requisitioned while building and delivered to the USSB on September 18, 1917. Commissioned *El Capitan* in the Naval Overseas Transportation Service on March 21, 1918.

O.B. Jennings—10,289 tons, 10.5 oil-burning tanker, 215588, LHTW. Keel laid on January 12, 1917, for Standard Oil of New Jersey. Requisitioned while building and delivered to the USSB on October 26, 1918. Oliver Burr Jennings was an officer with Standard Oil.

Munindies—5,200 tons, steel, 11 knot oil-burning freighter, 215748, LJFG. Keel laid on May 21, 1917, for the Munson. Requisitioned while building and delivered to the USSB on December 6, 1917. Commissioned *Munindies* in the Naval Overseas Transportation Service on December 17, 1918.

J.C. Donnell—10,241 tons, steel, 10.5 knot oil-burning tanker, 215852, LJKT. Keel laid on March 23, 1917, for Atlantic Refining. Requisitioned while building and delivered to the USSB on January 19, 1918.

Munairies—5,095 tons, steel, 11 knot oil-burning freighter, 215853, LJKV. Keel laid on June 21, 1917, for Munson. Requisitioned while building and delivered to the USSB on January 10, 1918. Commissioned *Munairies* in the Naval Overseas Transportation Service on January 14, 1918.

H.M. Flagler—8,207 tons, steel, 10.5 knot oil-burning tanker, 216456, LKVT. Keel laid on September 4, 1917, for Standard Oil of New Jersey. Requisitioned while building and delivered to the USSB on July 16, 1918. Henry M. Flagler was a co-founder of Standard Oil of Ohio.

Agwidale—5,080 tons, steel, 11 knot oil-burning freighter, 217184, LNQR. Keel laid on May 2, 1918, for Atlantic, Gulf & West Indies. Requisitioned while building and delivered to the USSB on November 16, 1918. Commissioned *Agwidale* in the Naval Overseas Transportation Service on November 16, 1918.

F.D. Asche—8,293 tons, steel, 10.5 knot oil-burning tanker, 217246, LNVD. Keel laid on December 11, 1917, for Standard Oil of New Jersey. Requisitioned while building and delivered to the USSB on December 9, 1918. Frederick Asche was a vice president of Standard Oil.

New York Shipbuilding Corp. Camden, New Jersey

Incorporated in October 1899 as the New York Shipbuilding Co., Inc., to build ships, marine engines, and boilers. Henry G. Morse was president, William G. Randle treasurer,

George L. Brown secretary, and R.L. Newman general superintendent. The yard was on 141 acres with 4,200 feet of Delaware River frontage. Their first ship was the steamer *J.M. Gufley* in 1902. They delivered the sidewheel steamer *Robert Fulton* in 1909.

In 1910, DeCourcey May became president, Henry A. Magoun vice-president, Samuel M. Knox secretary-treasurer, George M. Andrews general manager, and William G. Randle marine superintendent. The new team's first delivery was the steamer *Coastwise.*

In 1914 Samuel Knox was chairman of the board, DeCourcey May was president, Henry Magoun vice-president, J. Townsend Wickersham secretary-treasurer, and George Andrews general manager. New York Ship was a major shipbuilder through World War II.

Hampden—4,725 tons, collier, 1914, 211888, LDJB. Coastwise Transportation.

Gulflight—5,185 tons, tanker, 1914, 212524, LDNR. Gulf Refining.

Gulfstream—5,185 tons, tanker, 1914, 212676, LDTH. Gulf Refining.

William O'Brien—5,483 tons, tanker, 1915, 213085, LFHQ. Pierce Oil Corp.

Gulfcoast—5,188 tons, tanker, 1915, 213099, LFHV. Gulf Refining.

Franklin—5,266 tons, collier, 1915, 213624, LFPD. Coastwise Transportation.

Plymouth—5,266 tons, tanker, 1915, 213670, LFPM. American Italian. Operated by the Nafra Co. Commissioned *Plymouth* in the Naval Overseas Transportation Service on August 2, 1918.

Virginia—3,289 tons, freighter, 1915, 213761, LFQN. Pocahontas Fuel. Sold alien June 30, 1917.

Jonancy—3,289 tons, freighter, 1915, 213803, LFRB. Pocahontas Fuel.

Bylayl—3,289 tons, freighter, 1916, 213834, LFRK. Pocahontas Fuel.

Bristol—3,971 tons, freighter, 1916, 213859, LFRQ. Coastwise Transportation.

Standard Arrow—7,794 tons, tanker, 1916, 214196, LQCR. Standard Transportation. Commissioned *Standard Arrow* in the Naval Overseas Transportation Service on January 9, 1918.

Royal Arrow—7,794 tons, tanker, 1916, 214581, LGNF. Standard Transportation.

Tidewater—5,266 tons, tanker, 1917, 214699, LGPT. American Italian Commercial Corp. Operated by Nafra.

Gulfmaid—5,225 tons, tanker, 1917, 214827, LGSV. Gulf Refining.

Edward L. Doheny, Jr.—7,006 tons, tanker, 1917, 214951, LGWP. Pan American Petroleum & Transport.

Gulf of Mexico—7,807 tons, tanker, 1917, 215404, LHPQ. Gulf Refining.

Deepwater—7,610 tons, 10.5 knot coal-burning collier, 215535, LHSM. Keel laid on July 22, 1916, for the Darrow-Mann Coal Co. Sold in August 1917 to the American Italian Steamship Co. Requisitioned while building and delivered to the USSB on October 10, 1917. Allocated to Coastwise Transportation. Operated during World War II by the Eastern Gas & Fuel Associates as the *Lemuel Burrows.* Sunk on March 14, 1942, by *U-404* off Atlantic City, New Jersey.

Sylvan Arrow—6,669 tons, steel, 10.5 knot coal or oil-burning tanker, 215840, LJKC. Keel laid on March 22, 1917, for the Standard Transportation Co. Requisitioned while building and delivered to the USSB on January 2, 1918. Commissioned *Sylvan Arrow* in the Naval Overseas Transportation Service on July 25, 1918.

Fairmont—5,400 tons, steel, 10.5 knot coal-burning collier, 215957, LJPS. Keel laid on

February 13, 1917, for the Coastwise Transportation Co. Requisitioned while building and delivered to the USSB on February 15, 1918. Commissioned *Fairmont* in the Naval Overseas Transportation Service on December 19, 1918. Operated during World War II by the Sprague Steamship Co. as the *Black Point*. Sunk on May 5, 1945, by *U-853* off Point Judith, Rhode Island.

Broad Arrow—7,796 tons, steel, 11 knot coal or oil-burning tanker, 215988, LJQT. Keel laid on April 26, 1917, for Standard Transportation. Requisitioned while building and delivered to the USSB on March 12, 1918. Commissioned *Broad Arrow* in the Naval Overseas Transportation Service on March 17, 1918. Operated during World War II by the Socony–Vacuum Oil Co. Sunk on January 9, 1943, by *U-124* off Dutch Guiana.

Freeman—3,350 tons, steel, 20.5 knot coal-burning freighter, 216032, LJSC. Keel laid on July 7, 1917, for Pocahontas Fuel. Requisitioned while building and delivered to the USSB on April 8, 1918.

Sewalls Point—5,413 tons, steel, 10.5 knot coal-burning freighter, 216101, LJWB. Keel laid on June 30, 1917, for Castner Curran & Bullitt. Requisitioned while building and delivered to the USSB on March 28, 1918.

Tuckahoe—3,372 tons, steel, 10.5 knot coal-burning freighter, 216286, LKQW. Keel laid on April 8, 1918, for National Shipping. Requisitioned while building and delivered to the USSB on May 15, 1918. Allocated to National Steamship for service to the United Kingdom.

Gulfland—5,276 tons, steel, 11 knot coal or oil-burning tanker, 216299, LKJQ. Keel laid on August 2, 1917, for Gulf Refining. Requisitioned while building and delivered to the USSB on May 29, 1918.

Santa Tecla—2,576 tons, steel, 10 knot oil-burning freighter, 216380, LKSR. Keel laid on August 9, 1917, for W.R. Grace Requisitioned while building and delivered to the USSB on July 19, 1918.

Absecon—3,372 tons, 10.5 knot coal-burning freighter, 216513, LMBJ. Keel laid on October 2, 1917, for National Shipping. Requisitioned while building and delivered to the USSB on June 28, 1918. Operated by National Ship for service to Italy.

Mineola—2,576 tons, steel, 10 knot oil-burning freighter, 216514, LMBK. Keel laid on August 30, 1917, for the Atlantic Transport Line. Requisitioned while building and delivered to the USSB on July 12, 1918.

Glen White—5,438 tons, steel, 10.5 knot coal-burning freighter, 216529, LMBT. Keel laid on October 17, 1917, for Castner Curran & Bullitt. Requisitioned while building and delivered to the USSB on July 10, 1918. Commissioned *Glen White* in the Naval Overseas Transportation Service on July 22, 1918.

Winding Gulf—5,438 tons, steel, 10.5 knot coal-burning freighter, 216767, LMNH. Keel laid on November 10, 1917, for Castner, Curran & Bullitt. Requisitioned while building and delivered to the USSB on August 15, 1918. Commissioned *Winding Gulf* in the Naval Overseas Transportation Service on August 19, 1918.

M.J. Scanlon—5,672 tons, steel, 11 knot oil-burning lumber carrier, 216821, LMQV. Keel laid on December 15, 1917, for East Coast Transportation. Requisitioned while building and delivered to the USSB on September 20, 1918. Commissioned *M.J. Scanlon* in the Naval Overseas Transportation Service on September 23, 1918. Operated during World War II by Waterman as the *Minotaur*. Sunk on January 9, 1943, by *U-124* in the Atlantic.

William N. Page—5,438 tons, steel, 10.5 knot coal-burning freighter, 217126, LNKS. Keel laid on February 19, 1918, for Castner, Curran & Bullitt. Requisitioned while building and delivered to the USSB on November 30, 1918. Commissioned *William N. Page* in the Naval Overseas Transportation Service on December 18, 1918.

E.L. Doheny Third—7,753 tons, 11 knot oil-burning tanker, 217173, LNQB. Keel laid on December 24, 1917, for Pan American Petroleum & Transport. Requisitioned while building and delivered to the USSB on November 19, 1918. Commissioned *Edward L. Doheny III* the Naval Overseas Transportation Service on November 25, 1918. Operated during World War II by the Standard Oil of New Jersey as the *E.G. Seubert*. Sunk on February 23, 1944, by *U-510* in the Gulf of Aden.

North Carolina Shipbuilding Co.
Morehead City, North Carolina

Lumberman Gordon D. Canfield was president and general manager, attorney Edmund H. Gorham vice-president, David Stoneman treasurer, Joseph C. Taylor secretary and purchasing agent, and J. Banks Bailey superintendent. The company built only the two schooners on their own account.

A.W. Coburn—2,197 tons, wood schooner barge, 1918, 222401, MDRS.
James Blight—2,197 tons, wood schooner barge, 1918, 222353, MLKN.

Pendleton Brothers, Inc.
Bath, Maine

Brothers Edwin and Fields Seely Pendleton came from a long line of Maine and Brooklyn shipowners, brokers and builders. Edwin Pendleton owned the schooners *Gen'l Adelbert Ames* and *Susie H. Davidson*, homeported at New York.

In 1916 they got into shipbuilding for the war effort and built ships in Bath, Maine, and Noank and Mystic, Connecticut, under various business names. The last ship the family built was the schooner *Blanche C. Pendleton* in 1920 at their yard in Belfast, Maine.

Noank, Connecticut

Asta—1,965 tons, wooden gas motor freighter, crew of 10, 1917, 215430, LNSW. Astral Shipping.

Bath, Maine

In 1919 they delivered the second and last ship they built at Bath, the schooner *Anna G. Lord*.

Brina P. Pendleton—1,513 tons, schooner, 1918, 217397, LPHB. Fields S. Pendleton.

The Pendleton Co.—Mystic, Connecticut

Kingsway—1,272 tons, schooner, 1918, 217090, LNJT. Russell G. Morris, New York.

Pennsylvania Shipbuilding Co.
Gloucester, New Jersey

Hankon E. Norbom was president of the Norbom Engineering Co. in Philadelphia. George S. Hoell was vice-president and secretary, Harry Kaupp treasurer.

Norbom and Norwegian speculator Christopher Hannevig incorporated Pennsylvania Ship on February 6, 1916, as a subsidiary and built only one ship before the yard was taken over by Pusey & Jones. Norbom remained president, mechanical engineer Charles H. Moyer was vice-president, George Hoell was secretary-treasurer, and Henry V. Lysholm general manager.

Chestnut Hill—4,935 tons, steel, 11 knot coal or oil-burning tanker, 215976, LJQG. Keel laid on September 9, 1916, as the *Desdemona* for Cunard. Requisitioned from British owners while building and delivered to the USSB on March 14, 1918. Commissioned *Chestnut Hill* in the Naval Overseas Transportation Service on March 14, 1918.

Percy & Small, Inc.
Bath, Maine

In the mid–1880s, Samuel R. Percy was a sea captain and Frank A. Small was clerk for James B. Drake, a ship and insurance broker in Bath.

The two decided to build schooners and by 1897 they were in business as Percy & Small at Center & Front Sts. In 1902 they moved to 52 Front St. and in 1903 delivered their first ship, the big schooner *Elizabeth Palmer*. They built 23 schooners altogether and in 1917 built their one and only steamer, the *Sam C. Mengel*. Their last delivery was the schooner *Cecilia Cohen* in 1920.

Dunham Wheeler—1,926 tons, schooner, 1917, 215323, LHMK. East Coast Transportation.

St. Johns, N.F.—2,046 tons, schooner, 1918, 216231, LKPS. St. Johns, N.F., Shipping Corp.

Portland Ship Ceiling Co.
132 Commercial St., Portland, Maine

Horatio B. and Clinton T. Swett were grocers as the C.T. Swett Co. at 132 Commercial. In 1917 they decided to try shipbuilding. Horatio was president, Clinton, treasurer and manager, and civil engineer Loyal L. Alden was superintendent. The yard was at 27 Berwick St. and the ship's machinery was installed at Brown's Wharf. The three Shipping Board freighters were the only ships the company built.

Andra—2,550 tons, wood, 10 knot coal-burning Ferris freighter, 216726, LMJV. USSB contract ship. Keel laid on August 29, 1917, and delivered on August 31, 1918, 367 building days. Allocated to the Pendleton Bros. for service to France.

Basaan—2,499 tons, wood, 10 knot coal-burning Ferris freighter, 216865, LMSQ. USSB contract ship. Keel laid on September 7, 1917, and delivered on September 19, 1918, 377 building days. Allocated to Lowrence & Co. for service to the United Kingdom.

Okesa—2,503 tons, wood, 10 knot coal-burning Ferris freighter, 216904, LMVH. USSB contract ship. Keel laid on October 20, 1917, and delivered on October 23, 1918, 368 building days. Allocated to Rogers & Webb for service to the east coast of South America.

The Pusey & Jones Co. Wilmington, Delaware

In 1848, Joshua L. Pusey and John Jones began a manufacturing and boatbuilding business in Wilmington, Delaware. They made papermaking and pulp machines, engines, boilers, autoclave diffusion cells, and welded steel digesters. The boatyard was at the foot of Poplar Ave. where they built steam and gasoline propelled yachts and tugs.

In 1851, Edward Betts and Joseph Seal bought into the business and they became Betts, Pusey, Jones & Seal. Betts and Seal retired in 1859 and Edward's place was taken by Alfred Betts and the firm became Pusey, Jones & Betts. In 1860 Alfred Betts retired and was replaced by William G. Gibbons. The name then became Pusey, Jones & Co.

In 1866 John Jones retired and was replaced by Thomas H. Savery but the company name remained the same and was incorporated as the Pusey & Jones Co. in 1879. When Joshua Pusey retired, William Gibbons became president but when he died in October 1886, Joshua was brought back as president. Thomas Savery was vice-president, Samuel C. Biddel secretary, William W. Pusey treasurer, and Charles W. Pusey general manager.

Joshua died on February 8, 1891, and Charles became president, Savery vice-president, Biddle secretary, and William Pusey treasurer.

One notorious individual in the Pusey & Jones story was a Norwegian named Christoffer Hannevig. Hannevig was born in Norway on September 2, 1884, came to America in 1915, and opened an office at 139 Broadway in New York to deal in foreign exchange and to purchase ships. He got into shipbuilding as Christopher Hannevig, Inc., as a resident alien, and by 1918 he owned the controlling interest in the Pusey & Jones through stock ownership.

When Germany threw in the towel in November 1918 the Shipping Board cancelled a large number of contracts leaving shipbuilders holding the bag and causing bankruptcies. A huge number of claims were then filed against the Shipping Board for breach of contract. Settlements involving foreign nationals often involved existing treaties and, if necessary, disputes were settled in the Permanent Court of Arbitration at The Hague.

In June 1921, a U.S. attorney involved in the United States-Norway Arbitration hearings stated, in part:

> This group of claims is differentiated from the other Norwegian claims in which settlement has been made by several circumstances of which perhaps the most important are the history of their ownership and the physical status of the hulls involved. These claims have been latterly referred to in the negotiations between the claimants and the Board, and in the diplomatic correspondence as the claims of the "Christiana Group." The original name of the Group, however, appears to have been the "Hannevig Group," and the original name was well chosen because most, if not all, of these claims owe their very existence to Christopher Hannevig, whose meteoric and ill-fated career as a shipbuilder, shipowner, insurance magnate and banker forms one of the minor though dramatic footnotes to the history of the great war.
>
> Mr. Hannevig is a citizen and subject of Norway. He came to the United States in 1915 with one sound idea and the audacity to stake everything upon it. To Mr. Hannevig belongs the merit of seeing before most of his business contemporaries that the German submarine campaign and the demands of the World War would inevitably greatly increase the price of tonnage, and the merit of acting boldly upon this conception. Shortly after coming to the United States in 1915 he entered into a contract for the purchase of a ship, made

the first payment, and sold his contract for a sufficient profit to buy two more. He continued this process successfully until his paper profits ran into the millions. Unfortunately, like most other speculators, he was not content to limit his transactions within the bounds of reason.

Not content with selling the ships which other men built, he essayed the role of both broker and builder in order that there might be more ships to sell.

He acquired in the latter part of 1915, in the furtherance of his gigantic speculative schemes, practically the entire stock of the Pusey & Jones Company, of Wilmington, Delaware, a company which did a general foundry work and was engaged in shipbuilding to a limited extent. He not only undertook to increase the facilities of this company but in quick succession organized two other shipbuilding companies the Pennsylvania Shipbuilding Company on February 6, 1916, and the New Jersey Shipbuilding Company on May 3, 1917. The record is explicit that Mr. Hannevig was the sole owner of the stock of the Pennsylvania Shipbuilding Company, and the controlling owner of the New Jersey Shipbuilding Company, and everything indicates that he was also the sole beneficial owner of the stock of the New Jersey Shipbuilding Company.

In the meantime he commenced the process of organizing other companies for "building and operating vessels," the first of these being the Bulk Oil Transports, Inc., incorporated June 6, 1916. This was followed by the Manss Steamship Corporation, incorporated January 24, 1917, ostensibly for the purpose of operating vessels, and the Baltimore and South American Navigation Company, incorporated January 25, 1917, for the same object.

The beneficial ownership of the entire stock of these operating companies was in Christopher Hannevig.

Having created these "operating companies" Mr. Hannevig proceeded to cause them to contract with his shipbuilding companies for the construction of vessels, and personally, or in one of his personal-corporate capacities, for he was incorporated as Christopher Habbevig, Inc., in the United States and as Christopher Hannevig A/S in Norway, he acted as broker to sell the contracts which he had thus caused to be made.... In each case Christopher Hanevig was, as a matter of fact, both the buyer and the seller.

Hannevig, like so many others, lost his shirt in shipbuilding and bankruptcy proceedings were brought against him personally on February 11, 1921, and against Christopher Hannevig, Inc., on the 23rd.

In 1918 Sterling H. Thomas was president, John M. Mendenhall, vice president, and Clarence B. Lynch, secretary-treasurer. The yard was on seven acres with 1,400 feet of wharfage.

All ships built at Wilmington were wood Shipping Board steamers. The company opened a second yard at Gloucester City, New Jersey, to build steel ships for the war effort. The company remained a major builder through World War II.

Piqua—2,585 tons, steel, 10 knot coal-burning freighter, 216246, LKPW. Keel laid on January 1, 1917. Requisitioned from British owners while building and delivered to the USSB on May 3, 1918. Allocated to Mallory for service to the West Indies.

Waukesha—2,585 tons, steel, 10 knot coal-burning freighter, 216371, LKSH. Keel laid on April 1, 1917. Requisitioned from British owners while building and delivered to the USSB on May 31, 1918. Allocated to A.H. Bull for service to the Netherlands.

Middlebury—2,585 tons, steel, 10 knot coal-burning freighter, 216680, LMHP. Keel laid on May 16, 1917. Requisitioned from British owners while building and delivered to the USSB on July 31, 1918. Bareboat chartered to the Panama Railroad Co.

Aurora—2,585 tons, steel, 10 knot coal-burning freighter, 216820, LMQT. Keel laid on December 19, 1917. Requisitioned from British owners while building and delivered to the USSB on September 7, 1918. Allocated to Shawmut for service to Scandinavia.

Lynchburg—2,585 tons, steel, 11 knot coal-burning freighter, 217200, LNRM. Keel laid on May 26, 1917. Requisitioned from British owners while building and delivered to the USSB on November 22, 1918. Allocated to Mallory for Atlantic coastwise service.

Gloucester City, New Jersey

The yard built 19 ships. The last delivery was the Shipping Board motor freighter *William Penn* delivered in 1920.

John M. Connelly—4,939 tons, steel, 11 knot coal or oil-burning 45,000-barrel tanker, 216213, LKNT. Keel laid on October 14, 1916, for the Cities Service Co. Owned in 1917 by the subsidiary Empire Transportation & Oil Co. Requisitioned while building and delivered to the USSB on May 7, 1918. Commissioned John M. Connelly in the Naval Overseas Transportation Service on May 9, 1918.

Allentown—4,938 tons, steel, 10-knot oil-burning tanker, 216610, LMFS. Keel laid on September 8, 1917. Requisitioned from British owners while building and delivered to the USSB on September 12, 1918. Allocated to Atlantic Refining for service to Mexico.

Indianapolis—8,164 tons, steel, 11 knot oil-burning freighter, 217165, LNPQ. Keel laid on January 9, 1917. Requisitioned from British owners while building and delivered to the USSB on November 30, 1918. Commissioned *Indianapolis* in the Naval Overseas Transportation Service on December 12, 1918.

Brandywine—4,969 tons, steel, 11 knot coal or oil-burning freighter, 217210, LNSF. Keel laid on November 15, 1918. Requisitioned from British owners while building and delivered to the USSB on November 30, 1918. Allocated to Walker & Daly for service to Mexico.

R.L. Bean
Camden, Maine

Holly Marshall Bean was born on July 9, 1837, in Chatham, New Hampshire, became a ship carpenter, married Emaline Pottle, and settled in Tennents Harbor, Maine. Mary E. was born in 1858, Miller E. in 1864, Edith L. in 1870, and Robert L. in 1873.

In 1872, Holly built the 3-masted schooner *Earl H. Potter* for Fred Sheerer and in 1875 moved to Camden, Maine, and built eight 5-masted schooners followed by the *George W. Wells*, the first 6-masted schooner. Robert worked in the shipyard and in 1906, after they had delivered their 70th vessel, the yard came under Robert's management. Robert's first vessel was the schooner *Frank L. Swain* delivered in 1909 and his last was the schooner *T.N. Barnsdall* in 1920.

Percy R. Pyne II—1,341 tons, auxiliary schooner, 1916, 214658, LGPM. Oneida Navigation Corp.

Blue Peter—1,234 tons, schooner, 1917, 215501, LHSB. Blue Peter Navigation Corp.

Edna M. McKnight—1,326 tons, schooner, 1918, 216979, LNCD. Lewis K. Thurlow.

Read Bros. Co.
846 President Ave., S. Somerset, Massachusetts

William V. and Henry R. Read lived in Fall River and built boats in Somerset.

Luther Little—1,234 tons, schooner, 1917, 215845, LJKG. Charles F. Webb, Boston.

Richard T. Green Co.
211 Marginal St., Chelsea, Massachusetts

Green was a shipwright, caulker, and marine railway operator who incorporated in 1908. He was president and treasurer, William R. Green vice-president, and Richard T. Green, Jr., assistant treasurer. They delivered the schooner *Barbara* in 1909 and built five ships total. The last was the steamer *Noodle Island* in 1921.

Herdis—1,220 tons, barkentine, 1917, 214685, LGQH. American Star Line.
Richard T. Green—1,467 tons, schooner, 1918, 216161, LKMT. Pittsburgh Coal Co.

Riter-Conley Manufacturing Co.
3255 Preble Ave., Fairfield, Maryland

Thomas B. Riter was a coppersmith at 114 Elm St. in Pittsburgh, Pennsylvania, and William Henry Conley was a bookkeeper who lived at 81 Webster St. in Allegheny.

Conley came to Pittsburgh and in 1874 they were building boilers at 114 Elm as Riter & Conley. The business flourished and expanded, they moved to 56 Water St., and on July 25, 1897, Conley died. The company then became the Riter-Conley Co. with Riter as president and John S. Craig secretary. In 1899 the firm became the Riter-Conley Manufacturing Co. specializing in steel construction. John Craig was secretary-treasurer.

In 1903 they built one vessel at a yard in Leetsdale, Pennsylvania, the *Anton Wilbert*. By 1912 they were building natural gas machinery, iron buildings, cement kilns, oil refining plants, steel plant ladles, and other assorted heavy industrial products.

In 1915, Fletcher J. Gilmore was president and Cyrus Gray secretary. They built only the one wartime ship in Maryland.

Monongahela—1,677 tons, schooner barge, 1918, 215916, LJNT. Gulf Refining.

Robert Palmer & Son Shipbuilding
and Marine Railway Co.
Noank, Connecticut

An old firm established in 1850 by two brothers, John and Robert Palmer, the sons of shipbuilder John Palmer, Sr. John, Jr., died in 1879 and the next year Robert's son, Robert, Jr., joined the firm. They built small schooners and steamers. Their last ship was the schooner *Julia Loft* in 1917 when the business was purchased that year by the Charles W. Morse combination and the name changed was to Groton Iron Works.

Exeter—1,780 tons, schooner barge, 1914, 212127, LDKN. Reading Co.

Russell Shipbuilding Co.
44 Water St., Portland, Maine

The four Russell brothers, Eben, George, Jonas, and William were ship carpenters in Canada. In the 1860s they moved to the East Westbrook area of Portland and by 1873 they were building ships in East Deering where they were joined by another brother, Stuart.

The brothers married, raised families, and most lived at 20, 29, and 30 Veranda St. By 1899, George had retired, Jonas was gone, but Eben and William were still working. By 1904 Stuart was the only one left working. Stuart died on December 20, 1915, leaving his wife, Marcena, and the old 3-way yard to history, none of the family apparently having any interest in the business.

Frank A. Rumery, like the Russells, was from an established Portland family. He was a lumber dealer and contractor who was able the use the Russell's yard and on June 11, 1917, signed a Shipping Board contract for three wood freighters. Frank was president and treasurer, Charles W. Robinson superintendent, and Leo P. Hinds purchasing agent. the *Allison* was the company's first ship of the three they built. The *Brevard* and *Briarcliff* were delivered in 1919. The last ship the company built was the schooner *Hartford* in 1921.

Allison—2,551 tons, wood, 8-knot freighter, 1919, 217194, LNRD. Allocated to Charles T. Megee for service to France.

St. Johns River Shipyard Co.
South Jacksonville, Florida

John H. Fahey was born in Manchester, New Hampshire, on February 19, 1873, and became a newspaperman and resident of Boston. He apparently was able to raise a capital of $1,250,000 and the company was incorporated on October 7, 1918. Fahey was president, Arthur P. Gifford assistant general manager, August C. Bergh purchasing agent, and John D. Zimmerman general superintendent at the 7-way yard. Philip Davis was the employment manager. The *Baxley* was the first of eight ships and the last two were the Shipping Board freighter *Wayhut* and the 1,313-ton schooner *Ormond*, both delivered in 1919.

Baxley—2,551 tons, wood, 10 knot coal-burning Ferris freighter, 217410, LPHS. USSB contract ship. Keel laid on November 7, 1917, and delivered on September 30, 1918, 328 building days. Allocated to the Naval Overseas Transportation Service and then the Texas Transportation & Terminal Co. for service to France.

Smith & Terry, Inc.
11 Broadway, New York

Shipbrokers and operators Edwin Smith and Howard A. Terry established a 2-way, 215' shipyard on Broad Creek at Bethel, Delaware, a little east of the Nanticoke River, which flows in a generally southern direction to the Atlantic. Howard was manager, William H. Smith was chief of the hull department.

Smith & Terry No. 1—1,156 tons, schooner barge, 1917, 215546, LHST. Smith & Terry.
Smith & Terry No. 2—1,251 tons, schooner barge, 1918, 216876, LMTF. Smith & Terry.

Standard Shipbuilding Corp.
Shooters Island, New York

Wallace Downey was born in Nova Scotia on May 11, 1862, came to Brooklyn in 1880, and worked as a shipwright for James A. Townsend and George Edgett's company Townsend

& Edgett. Downey became the manager and on May 1, 1891, became a partner but the company name remained the same. On November 30, 1891, Edgett retired and the firm became Townsend & Downey. In 1900 the company built piers at Mariners Harbor and a shipyard.

In 1901 the Townsend & Downey Shipbuilding & Repair Co. delivered the 325-ton steamer *Kingfisher*, built some famous yachts, and in 1904 completed their last vessels, two 4,167-ton schooners, the *S.O. Co. No. 94* and *S.O. Co. No. 95*. Legal issues intervened concerning the previous partnership and in 1916 Wallace Downey and a group of investors purchased Shooters Island as the Shooters Island Shipyard Co.

The Guantanamo & Western Railroad was incorporated at Portland, Maine, on September 25, 1909, and on March 1, 1910, the company acquired the assets and property of the Cuba Eastern Railroad, the Northeastern Cuba Railroad, the Cuba Eastern Terminal Co., and the Eastern Railroad of Cuba with a capital stock of $2,750,000.

Montgomery H. Lewis was president of the Santa Cecilia Sugar Co. with holdings in Cuba and lived at 2350 Broadway in New York. He became president of Guantanamo & Western at the 82 Beaver St. office in New York. José Bosch was 1st vice-president in Santiago, Cuba; Irvin McD. Garfield was 2nd vice-president and treasurer in Boston; Robert L. Dean was secretary and assistant treasurer; Robert H. Caplan was assistant secretary in New York; and Edward M. Wise was general manager at Guantanamo, Cuba. The directors were American and Cuban. José Marimon was a director in Havana, Cuba, and Antonio B. Caragol was a director in New York. Caragol was president of Manuel Caragol & Son, sugar importers at 127 Water St.

When war came the Cubans saw an investment opportunity in America and incorporated the Standard Shipbuilding Corp. in 1916 at 44 Whitehall St., Room 421, in New York. José Marimon was president, Gabriel Juvé vice-president; Emilio S. Godoy vice-president and general manager; D.C. Mulheman secretary, Antonio Caragol treasurer; John Hunter was the naval architect and superintendent of hulls; G. Norman Scott was the engineer; and Juane A. Cortada was the purchasing agent.

They leased facilities on Shooters Island, built a plant and marine railway at Mariners Harbor, and purchased additional land from the Shooters Island Shipyard Co. with mortgages and a $1,337,000 loan from the Emergency Fleet Corp.

The company's last ship was delivered in 1921 and like so many shipbuilders of the day, Standard Ship ran into financial difficulties and defaulted on two annual payments of $267,000 to the Emergency Fleet Corp. for 1921 and 1922. The EFC sued and Shooters Island Ship sued to foreclose the mortgage. Standard Ship declared bankruptcy on March 7, 1922, and John J. Fitzpatrick, Albert Conway, William A. Young, and Alfred A. Stein were appointed receivers in 1925.

Jupiter—4,430 tons, steel, 10 knot coal-burning freighter, 215822, LJHK. Keel laid on November 11, 1916. Requisitioned from French owners while building and delivered to the USSB on November 5, 1917. Allocated to International Freighting for service to France. Commissioned *Democracy* in the Naval Overseas Transportation Service on October 25, 1918.

Muscatine—4,730 tons, steel, 10.5 knot coal-burning refrigerated freighter, 216220, LKPH. Keel laid on November 11, 1916. Requisitioned from Norwegian owners while building and delivered to the USSB on May 2, 1918. Commissioned *Muscatine* in the Naval Overseas Transportation Service on May 2, 1918. Operated during World War II by A.H. Bull & Co. as the *Elizabeth*. Sunk on May 20, 1942, by *U-103* south of Cuba.

Passaic—4,451 tons, steel, 10.5 knot coal-burning refrigerated freighter, 216290, LKJH. Keel laid on November 11, 1916. Requisitioned from British owners while building and delivered to the USSB on June 19, 1918. Commissioned *Ice King* in the Naval Overseas Transportation Service on July 3, 1918.

Montclair—4,773 tons, steel, 10.5 knot coal-burning refrigerated freighter, 216607, LMFP. Keel laid on June 30, 1917, for Cunard. Requisitioned while building and delivered to the USSB on June 28, 1918. Allocated to the New York & Cuba Mail Steamship Co. for service to the River Plate. Commissioned *Montclair* in the Naval Overseas Transportation Service on August 19, 1918.

Morristown—5,138 tons, steel, 10.5 knot coal-burning freighter, 216777, LMNR. Keel laid on November 17, 1917. Requisitioned from Russian owners while building and delivered to the USSB on October 15, 1918. Commissioned *Morristown* in the Naval Overseas Transportation Service on October 18, 1918.

Galesburg—5,138 tons, steel, 10.5 knot coal-burning freighter, 216778, LMNS. Keel laid on May 26, 1917. Requisitioned from Russian owners while building and delivered to the USSB on September 10, 1918. Allocated to J.W. Ellwell as manager for service to the United Kingdom by Strachan Shipping.

Englewood—5,138 tons, steel, 10.5 knot coal-burning freighter, 216779, LMNT. Keel laid on May 26, 1917. Requisitioned from French owners while building and delivered to the USSB on August 31, 1918. Allocated to Cosmopolitan Shipping for service to the Netherlands.

Hickman—5,121 tons, steel, 10.5 knot coal-burning freighter, 217170, LNPT. Keel laid on January 18, 1918. Requisitioned from British owners while building and delivered to the USSB on November 13, 1918. Commissioned *Hickman* in the Naval Overseas Transportation Service on November 16, 1918.

Monmouth—5,125 tons, steel, 10.5 knot coal-burning freighter, 217318, LPCR. Keel laid on April 6, 1918. Requisitioned from British owners while building and delivered to the USSB on December 13, 1918. Allocated to American Line as manager for service to the United Kingdom by Barber.

A case involving the Consorzio Veneziano di Armamento e Navigazione's relationship with Standard Ship illustrates some of the complexities of the day. The Consorzio Veneziano di Armamento e Navigazione was an Italian shipping consortium. Carlo de Luca of Venice was their assignor in the United States.

The first Italian Parliament created the Kingdom of Italy on March 1, 1861, whereupon Victor Emmanuel changed his title from King of Sardinia to the King of Italy. With help from the Prussians, the Austrians were driven out of Venice and the city was annexed in 1866. Prussia's war with France in 1870–71 pulled French troops out of the city and territory of Rome. Trieste and Trento remained in Austria's hands until the World War drove them out. Austria, Hungary, and Germany were joined by Bulgaria, East Prussia, and the Ottoman Empire, which was then Turkey.

The United States entered into a treaty with His Majesty, the King of Italy, on November 23, 1871. Article XXIII of the treaty, dealing with commerce and navigation, allowed

Italian subjects full access to U.S. courts, and vice versa, to "defend their rights without any conditions, reservations, or taxes other than such as are imposed upon the natives."

On April 23, 1917, Carlo de Luca signed two contracts with the Standard Shipbuilding Corp. to build two single screw, two-deck, steel cargo steamships of approximately 7,300 deadweight tons for $150 per deadweight ton, or about $1,095,000 per ship. Payments would be made in installments and all materials delivered to the shipyard would become the property of the actual purchaser. In case the job did not progress on schedule, de Luca had the right to employ any number of his own workers in the shipyard and bring in any necessary equipment, supplies, or tools, and purchase materials at his discretion to get the job done.

Standard Ship was required to place $328,500 in escrow on the ships—15 percent of the purchase price for both ships—until completion of the work. The money was deposited with the Guaranty Trust Company of New York and Guaranty Trust purchased a $328,500 performance bond from the Globe Indemnity Co. on July 3 with themselves as benefactor. The bond would be released to Guaranty Trust and the money paid to Standard Ship on completion of the ships.

The first vessel, Hull No. 10 was to be completed on March 31, 1918, and Hull No. 11 on April 30, 1918.

Sometime prior to July 25, 1917, de Luca paid $635,100, the first of several installment payments, and on that date he formally assigned the contracts to the Consorzio Veneziano di Armamento e Navigazione. He was then paid $1,000,000 plus $491,731 for fees, commissions, and other charges related to the purchase of the contracts and he was reimbursed the $635,100 for the first installment.

On August 3, 1917, Standard Ship received a telegram advising them that all 13 ships the company was then building were being requisitioned by the United States Shipping Board Emergency Fleet Corporation. Two were for the Russian Volunteer Fleet. A formal letter arrived the next day from Eads Johnson, the Shipping Board District Officer, requesting the company to furnish copies of all the contracts and all details concerning the construction of the vessels. The balance due on Hulls No. 10 and 11 was on that date $1,401,600.

All the ships Standard was building were a 7,300 deadweight ton standard design. Hull No. 1 had been launched, steelwork was 90 percent complete; carpenter work, 75 percent; joiner work, 15 percent; main engines, 95 percent; boilers, 90 percent; piping, 30 percent with a delay due to late delivery of valves and fittings. The other hulls were in lesser stages of completion. Hull No. 10 was 9.4 percent complete and Hull No. 11 3.5 percent complete.

Standard ship also had a Navy contract for two minesweepers and the keel blocks had just been laid in readiness for the keel laying.

The Shipping Board ordered all the ships to be completed to their specifications if the work in each particular area was not too far along. Some of the specifications included allowing only two ventilators at each end of a cargo hold; no special cargo handling equipment; and no watertight doors in 'tween deck bulkheads between holds, just in the forward and aft end of the machinery spaces. Additional watertight transverse bulkheads were to be added to create a "two-compartment ship," one where any two compartments could flood and the ship would remain afloat. Fuel oil capacity sufficient for an 8,000-mile cruising radius for the vessel's cruising speed and tonnage was required.

One item, totally foreign to commercial vessels was called for:

> Accommodations for gun crew to be provided for, consisting of room in vicinity of officers' quarters for lieutenant and an additional room for two gunners and further quarters for fifteen gun crew with berths, outfit, and furnishings similar to firemen's or sailors' quarters, and the gun crew quarters provided for as separate quarters for either sailors or firemen.

Gun foundations and magazines for one gun forward and one aft was to be provided. The Navy Bureau of Ordnance would send an inspector to view each ship and determine its armament and structural integrity.

Other requirements included an 18-inch searchlight; a camouflage paint scheme; and accommodations for two wireless operators adjoining the wireless room. The Shipping Board would supply and install a 1 kw. Radio, having a radius of 100–150 miles, on vessels of less than 5,000 deadweight tons, and a 2 kw. Radio, radius 300–400 miles, on vessels over 5,000 deadweight tons.

All vessels being built for foreign owners were required to conform to U.S. Steamboat Inspection rules and laws for crew accommodations and lifesaving equipment unless construction was so far advanced as to make changes impractical.

On August 16, 1917, Adm. Washington L. Capps, the Emergency Fleet Corporation General Manager, sent a letter to Eads which said, in part:

> You will immediately notify the shipbuilding company that you are instructed to take charge, for the corporation, of the completion of these vessels, and you are authorized hereby to take over into the employ of the corporation, temporarily, the owner's local inspecting officers at the present rates of compensation, so far as they can comply with existing instructions and those hereafter issued in regard to citizenship, being careful that they take the usual oath of office.
>
> Report promptly the name, compensation, the date of taking the oath, and the former employers of those whom you take over.

After the information was furnished, Capps wrote to de Luca saying he was required to complete the construction of his consortium's ships "with all practical dispatch" and that compensation for the ships, materials, and contracts purchased would be determined later.

Standard Ship's management sent many letters and telegrams to the Shipping Board requesting details of the requisitioning process. The company's First Vice President, Gabriel Juve, was particularly worried about the labor situation in a September 10 letter to Eads Johnson:

> We note what you say regarding the method of requisitioning vessels and we will accordingly proceed, without delay, the completion of these vessels following the instructions given. Of course, we must make this reservation, that unless we are paid a sum equal to the sums set out in the contract there will, of necessity, result in delay in construction; the responsibility of that, however, will rest with your principals, not with us.
>
> We would respectfully request that you give consideration to the labor situation now pending. It has been stated in meetings of the old Shipping Board that in the event of any increase in the cost of labor this increase will be added to the price of the vessels as set out in the contracts.
>
> Are we correct in assuming that if the Labor Adjustment Board now in session increases the coast of labor on any of these vessels that your board will add to the price of the vessels as determined in the contract the added cost of labor so fixed by such board?

Charles S. Bookwalter, a Shipping Board district officer, wrote back to Standard Ship's office at 44 Whitehall St. informing them that the Board would pay the amount "equal to the total amount unpaid on your contracts with the former owners" and they figured the total amount was $13,552,550 and that as of September 29, 1917, the company had been paid $6,611,224.04 and was owed a balance of $6,941,325.96. The company paid commissions of

$170,988 which Bookwalter said would be withheld, leaving a balance of $6,770,338. He told the company not to make any further payments for commissions or brokerage charges and to forward all "process served" on the company from brokers claiming commissions.

Bookwalter said the Emergency Fleet Corporation would open a $500,000 account at the Columbia Trust Co. into which Standard Ship could deposit payments or withdraw funds on Bookwalter's approval.

Consorzio Veneziano di Armamento e Navigazione's Hull No. 10 was completed and delivered on January 15, 1919, and Hull No. 11 on February 18, 1919, and they filed a $1,226,400 claim for just compensation plus 6 percent interest from August 3, 1917. The Emergency Fleet Corporation sent a check for $816,651 on July 23, 1920, which the company rejected. In accordance with the law, the check was returned and the company was paid $612,488.25 on August 27, 1920, which was 75 percent of the amount claimed, which the company accepted reserving the right to sue for the balance in federal court.

The total cost to build the Consortium's ships was $2,919,442.64, of which $506,854.12 was for overhead. Due to government changes, the cost was $729,422.64 over the original contract price of $2,190,000. Standard ship was paid in full less the $788,400 received before August 3, 1917.

The accounting officers were not amenable to further negotiations so the Consortium sued and was awarded $1,165,182.34. This included the full amount of their just compensation claim and $227,657.60 for 6 percent interest from August 3, 1917, to August 27, 1920. Seventy-five percent of the August 27 award was deducted, leaving a balance due of $839,569.35 and they were awarded 6 percent interest on this balance, $325,612.99, from August 27 to February 14, 1927, the date of the judgement. The court stated:

> The record in this case abundantly establishes the existence of a demand for ship contracts during the years 1916 and 1917. Such contracts were in demand not only among domestic shipowners but with foreign governments as well. It arose by reason of the destruction of ships under the German submarine campaign, and it was immediate and urgent. Ships already in process of construction, or those merely under contract for which materials had been delivered or ordered, were preferred as offering the first promise of relief. Shipowners gave consideration not merely to the immediate needs pending the war but were influenced in making such purchases by the natural expectation of an unusual demand for world tonnage after the war. Many ship contracts were purchased prior to the requisition date, and a few were purchased thereafter. It was an active and rising market, and reached its highest activity in March and April, 1917. Prices for ship contracts took a wide range, but in all such transactions the prices paid were in excess of the original contract prices.

Apparently Carlo de Luca was an investor who contracted for the ships and his contracts were purchased by the Consortium, which in turn were purchased or appropriated by the Shipping Board Emergency Fleet Corporation under the Act of June 15, 1917.

De Luca was awarded $1,592,464 plus interest on $777,064.64 from December 7, 1936, until paid for two other 7,300 deadweight ton ships he contracted for at $1,277,500 each on June 25, 1917, which were taken under the same circumstances.

Staten Island Shipbuilding Co.
Port Richmond, New York

Established in 1914 with William J. Davidson president and general manager; John Tracey vice-president, and George H. Bates secretary-treasurer. The main office was on Staten

Island. The company built three vessels at Mariners Harbor, the steamer *Wm. J. Davidson* in 1914, the *Sag Harbor*, and the steamer *York Harbor* in 1919. They also had another yard in New York City and one at Port Richmond. The yard in New York built 32 vessels between 1918 and 1929.

Snug Harbor—2,388 tons, steel, 9.5 knot coal-burning freighter, 215910, LJNR. Keel laid on October 6, 1916. Requisitioned from British owners while building and delivered to the USSB on January 22, 1918. Allocated to Coastwise Transportation for service to the West Indies.

Mariners Harbor—2,431 tons, steel, 9.5 knot coal-burning freighter, 216395, LKTF. Keel laid on April 4, 1917. Requisitioned from British owners while building and delivered to the USSB on June 3, 1918. Allocated to the New York & Porto Rico Steamship Co. for service to the West Indies.

Bar Harbor—2,431 tons, steel, 9.5 knot coal-burning freighter, 217273, LNWM. Keel laid on February 22, 1917. Requisitioned from British owners while building and delivered to the USSB on December 3, 1918. Allocated to the New York & Cuba Mail Steamship Co. for service to the West Indies.

Mariners Harbor, New York

Sag Harbor—2,431 tons, steel, 9.5 knot coal-burning freighter, 216845, LMSD. Keel laid on June 23, 1917. Requisitioned from British owners while building and delivered to the USSB on August 31, 1918. Allocated to M.H. Tracy for service to the West Indies.

Submarine Boat Corp.
Newark, New Jersey

The Electric Boat Co. was established in 1899 by financier Isaac Rice to build a 54' submarine designed by John P. Holland. In 1914 the company organized the Submarine Boat Co. to build Shipping Board freighters and signed a contract on September 14, 1917, for a 28-way yard to build 5,000-ton ships. Total cost was around $17 million. The contract called for 150 ships at a cost of $144,900,000. The first keel was laid on September 20, 1917. The company built 149 ships, the last being the *Suphenco* delivered in 1921.

Agawam—3,658 tons, steel, 10.5 knot oil-burning freighter, 216996, LNDF. USSB contract ship. Keel laid on December 20, 1917, and delivered on December 17, 1918, 362 building days. Allocated to Mississippi Shipping for service to the United Kingdom.

The Sun Shipbuilding Co.
Chester, Pennsylvania

Joseph Newton Pew and Edward O. Emerson were produced natural gas as the Pew & Emerson Co., 116 Market St. in Pittsburgh. Pew was also secretary-treasurer of the Penn Fuel Co. at 516 Market St. and in 1884 Pew & Emerson moved to Room 10, 516 Market. In 1885, Penn Fuel moved to 102 4th Ave. and Pew, Emerson & Co., Ltd. Was established at 102 4th Ave. to purchase a lease on an oil field in Lima, Ohio.

In 1886 Pew & Emerson were at 91 5th Ave. In 1891 Joseph Pew was president of the

Peoples Natural Gas Co. at 531 Wood St. Pew & Emerson, the Sun Oil Co., and the Sun Line Oil Co. were in the same office.

In 1899 Edward Emerson was president of the Sun Oil Co. and Pew was president of the Peoples Natural Gas Co., both at 1347 Tradesmens Building.

Around 1901 the company decided to operate its own ships to move oil from the fields in Texas to its refinery at Marcus Hook, Pennsylvania, to "guarantee a stable rate of cost for each barrel of oil, irrespective of accelerated cost per barrel due to the supply and demand of tankships in periods of crisis and accelerated business." The company also found advantage in having "stable and reliable personnel employed by the company itself."

Sun's first ship was the 242', 18,000-barrel, 9-knot *Paraguay*, converted from a Great Lakes ore carrier, as was the company's second ship in 1902, the *Toledo*. The company next chartered the big, 7-masted schooner collier *Thomas W. Lawson*, converted to carry oil. The ship made 12 knots under sail, faster than the steamers of the day.

In 1904 the office was moved to Philadelphia. The company had a wharf and foreign trade was established. J. Howard Pew was president and Joseph N. Pew, Jr., was vice-president.

In 1907 the 60,000-barrel *Lawson* was wrecked in England and the company decided to have two tankships built from scratch, the world's first such vessels, one in England and the other at Newport News, Virginia. The English-built ship was the *British Sun* and was operated by a Sun subsidiary, the British Sun Co., Ltd., under the British flag. The world's biggest tanker was the American-built 51,600-barrel *Sun*.

Sun Shipbuilding & Dry Dock was incorporated at Chester on May 23, 1916, as a subsidiary of Sun Oil. The yard was at the foot of Morton Ave. in Chester to build its own ships and steel Shipping Board freighters, marine boilers and engines, Corliss engines, special machinery, and to provide machine work and iron castings. Joseph N. Pew, Jr., was chairman of the board. John G. Pew was president, J. Howard Pew, Robert Haig, and John K. Graham were vice-presidents, and Hervey Schumacher was secretary-treasurer. He was succeeded soon after by Frank S. Reitzel.

Sun Ship pioneered the use of all-welded construction and was the innovator of individual cabins for crew members, now no longer packed into the forecastle. The rooms came to be known as fo'c'sles. The yard became the Sun Shipbuilding & Dry Dock Co. around 1923 and remained a major shipbuilder through World War II.

Chester Sun—7,799 tons, steel, 11 knot coal or oil-burning 81,000-barrel tanker, 215878, LJMK. Keel laid on January 27, 1917, for Sun Oil. Requisitioned while building and delivered to the USSB on December 29, 1918. Operated by Sun Oil.

Sabine Sun—6,728 tons, steel, 10.5 knot coal or oil-burning tanker, 216033, LJSD. Keel laid on February 8, 1917, for the Sun Co. Requisitioned while building and delivered to the USSB on March 28, 1918. Operated by Sun Oil.

Radnor—7,699 tons, steel, 10.5 knot coal or oil-burning freighter, 216293, LKJN. Keel laid on April 19, 1917. Requisitioned from British owners while building and delivered to the USSB on May 13, 1918. Allocated to the Army.

Lancaster—7,699 tons, steel, 10.5 knot coal or oil-burning freighter, 216451, LKVP. Keel laid on May 3, 1917. Requisitioned from British owners while building and delivered to the USSB on June 19, 1918. Commissioned *Lancaster* in the Naval Overseas Transportation Service on June 19, 1918.

Neponset—7,622 tons, steel, 10.5 knot oil-burning refrigerated freighter, 217019, LNPM. Keel laid on November 7, 1917. Requisitioned while building and delivered to the USSB on October 7, 1918. Commissioned *Neponset* in the Naval Overseas Transportation Service on October 28, 1918.

Deerfield—7,622 tons, steel, 10.5 knot oil-burning freighter, 217030, LNGB. Keel laid on February 8, 1918, for Shawmut. Requisitioned while building and delivered to the USSB on October 19, 1918. Operated during WWII by the as the *Mathew Luckenbach*. Sunk on March 19, 1942, by *U-527* and *U-523* in the North Atlantic.

The Texas Steamship Co., Inc.
17 Battery Place, Rm 711, New York

The marine branch of the Texas Co. was organized by William Andrew Thompson, Jr., in April 1912. Thompson was president, William W. Bruce secretary, and Guy Stevens treasurer. The company decided to get into the shipbuilding business at the foot of Front St. in Bath, Maine. George B. Drake was general manager of the shipyard.

They built 24 ships, the last one being the tanker *New Jersey* in 1921 for themselves. The Shipping Board spent $1,449,968.99 on housing for workers.

Maine—6,457 tons, steel, 10.5 knot oil-burning freighter, 215581, LHTQ. Keel laid on June 5, 1916. Requisitioned while building and delivered to the USSB on September 27, 1917. Allocated to Robert Ramsay for service on the northern European civilian food relief mission.

Woonsocket—6,457 tons, steel, 11 knot oil-burning freighter, 215832, LJHS. Keel laid on July 25, 1916. Requisitioned while building and delivered to the USSB on December 1, 1917. Allocated to Harris, Magill for the European Food Relief mission.

Sagadahoc—6,846 tons, 11 knot oil-burning freighter, 216584, LMDV. Keel laid on June 15, 1917. Requisitioned while building and delivered to the USSB on July 9, 1918. Commissioned *Sagadahoc* in the Naval Overseas Transportation Service on July 9, 1918. Operated during WWII by the American South African Line. Sunk on December 3, 1941, by *U-124* in the South Atlantic.

Canibas—6,846 tons, steel, 11 knot coal or oil-burning freighter, 216864, LMSP. Keel laid on July 26, 1917, as the *New Jersey*. Requisitioned while building and delivered to the USSB on September 10, 1918. Commissioned *Canibas* in the Naval Overseas Transportation Service on September 10, 1918.

Traylor Shipbuilding Corp.
Cornwells, Pennsylvania

Samuel W. Traylor, Sr., was president of the Traylor Engine & Manufacturing Co. in Allentown, Pennsylvania, a maker of mining machinery. The Traylor family had numerous business holdings in Allentown. In 1917 Samuel decided to get into the wood shipbuilding business and built a 10-way shipyard in Cornwells Heights off State Road. Samuel, Sr., was president and manager, T.W. Hopkins and Samuel W. Traylor, Jr., were vice presidents; Charles G. Davis was chief of the hull department, and Ruland R. Shafter was chief of the engine department.

Alvada—2,573 tons, wood, 10 knot coal-burning Ferris freighter, 216796, LMPS. USSB contract ship. Keel laid on September 20, 1917, and delivered on August 31, 1918, 345 building days. Allocated to W.R. Weist for service to Italy.

Alapaha—2,573 tons, wood, 10 knot coal-burning Ferris freighter, 216930, LMWK. USSB contract ship. Keel laid on November 13, 1917, and delivered on September 24, 1918, 316 building days. Allocated to A.D. Cummings for service to France.

Buhisan—2,573 tons, wood, 10 knot coal-burning Ferris freighter, 217036, LNGH. USSB contract ship. Keel laid on December 20, 1917, and delivered on October 29, 1918, 312 building days. Allocated to Charles T. Magee as manager for service to the west coast of South America by M.E. Kinsley.

Bulana—2,573 tons, wood, 10 knot coal-burning Ferris freighter, 217212, LNSH. USSB contract ship. Keel laid on December 25, 1917, and delivered on November 27, 1918, 338 building days. Allocated to A.D. Cummins for service to France.

Daram—2,573 tons, wood, 10 knot coal-burning Ferris freighter, 217213, LNSJ. USSB contract ship. Keel laid on April 17, 1918, and delivered on December 28, 1918, 255 building days. Allocated to Charles T. Magee as manager for service to France by the F.D. Dimmick Co.

In December 1918, John A. Donald, acting chairman of the Shipping Board, sent telegrams to operators and managers of the wood freighters to see how well they were performing. He contacted Charles T. Magee regarding the *Buhisan*. Magee wired back:

> Your telegram 7th. *Buhisan* built by Traylor Shipbuilding Co. is only Ferris wooden steamer under our management. Hull well and strongly built. Does not leak. Total dead-weight cargo capacity 3,200 tons. Bunker capacity 385 tons. Speed 11½ knots light; 10 knots loaded. Captain states ships seaworthiness excellent. Imperfect construction of condensers permitted salt water to enter boilers. Feed pumps were too tightly packed. Controlling steam valve on steering gear gave trouble. Hoisting winches improperly fastened to deck. These matters now fully adjusted and ship is in satisfactory condition. Operated carefully, overcoming determination and delays, these ships should give good results.

Union Ship Building Co.
Fairfield, Baltimore, Maryland

A 4-way yard established in 1917. Willis M. Sterrett was employed by the McClintic-Marshall Co. in Pittsburgh as manager of their plant in Pottstown then ran the company's office in Washington. He was a highly-regarded steel engineer and was elected president of Union Ship. Henry Starkey Sanford was general manager; William W. Darsie purchasing agent; and naval architect H.A. Everett was hull and engine chief. William H. Blakeman was the second manager at the company's office at 601 E. Lombard. The company built eight ships, the last one being the *Gulfprince* in 1921.

Ohio—1,677 tons, schooner tank barge, 1918, 216378, LKSQ. Gulf Refining.

William Cramp & Sons' Ship and Engine Building Co.
Philadelphia, Pennsylvania

William Cramp was born in the Kensington district in Philadelphia on September 22, 1807, went to work in shipbuilding for Samuel Grice in Kensington at age 15, and married

Sophia Miller. In 1830 he purchased property on the Delaware River and built wooden sailing vessels and steamboats. He had five sons, Charles H. in 1828; William M. in 1832; Samuel H. in 1833, Jacob C. in 1835, and Theodore H. in 1840. Daughter Susan was born in 1830 and five more girls would follow.

By 1850 Samuel was a machinist. In 1857 Charles and William joined the firm and in 1863 Sam, Jacob, and Theodore joined. Ten years later, when the Civil War started, William, William M., and Charles were at the foot of Otis St. as C.H. & William M. Cramp building iron ships. William, Samuel, a shipwright, and Theodore, a clerk, lived at 1033 Palmer, Charles H. at 1035 Palmer, and William M. at 421 Richmond.

In 1864 the firm became W. Cramp & Sons run by William and Charles at 814 Richmond. In 1867 they had a plant at 1415 Beach. By 1870, Charles, William M., Samuel, Jacob, and Theodore were partners as patriarch William's health began to fail.

On March 6, 1872, the company was incorporated as William Cramp & Sons' Ship and Engine Building Co. to produce ships, engines, and ordnance.

William M. died on July 6, 1879, and the boys carried on. In 1882, Andrew became the first grandson to join the management of the firm. In 1884 the company became the William Cramp & Sons Ship & Engine Building Co. at Beach and Norris. In 1885 Theodore died, the first of the sons to go, leaving a widow, Mary E. at their home at 1714 Master Street. Theodore had worked as a clerk in the office, the only son to not become a ship carpenter.

Over the decades the company thrived, Cramp family members came and went, and by the start of the World War Henry S. Grove was president, Harry W. Hand vice-president and general manager, Charles T. Taylor secretary-treasurer, and Roland L. Howe and George H. Carey assistant treasurers. In June 1917, Grove resigned as president and was succeeded by Harry Hand.

The company's last ship was the 5,943-ton steamer *Yarmouth*, delivered in 1927. The yard was revived to become a major shipbuilder during World War II.

Great Northern—8,255 tons, passenger liner, 1914, 212719. Army Quartermaster Corps. Crew of 18 officers and 280 enlisted men.

Henry M. Flagler—2,699 tons, car ferry, 1914, 212826, LFBN. Florida East Coast Car Ferry.

Northern Pacific—8,255 tons, passenger liner, 1915, 212926. Army Quartermaster Corps with a crew of 18 officers and 280 enlisted men.

Santa Barbara—6,621 tons, freighter, 1916, 213845, LFRN. Grace. Commissioned *Santa Barbara* in the Naval Overseas Transportation Service on April 15, 1918. Operated in 1942 by American-Hawaiian as the *American* and sunk by *U-504* on June 11 off Honduras.

Jalisco—2,585 tons, freighter, 1916, 213928, LFTJ. Sold Mexican on March 31, 1916.

J.M. Danziger—6,487 tons, tanker, 1916, 214311, LGFK. Pan American Petroleum & Transport.

Sunoil—7,157 tons, tanker, 1916, 214459, LGJR. Sun Co.

Joseph R. Parrott—2,406 tons, freighter, 1916, 214493, LGKM. Florida East Coast Car Ferry.

Wm. Rockefeller—7,157 tons, tanker, 1916, 214582, LGNJ. Standard Oil of New Jersey. On May 19, 1917, an explosion occurred on board seriously injuring seaman Louis G.

Scaros. Commissioned *William Rockefeller* in the Naval Overseas Transportation Service on January 9, 1918.

Santa Rosa—6,415 tons, freighter, 1917, 214725, LGRB. Grace. Allocated to the Army Transport Service on August 29, 1917, transferred to the Navy on March 10, 1919, and returned to Grace Line on October 27.

Santa Paula—6,415 tons, freighter, 1917, 214901, LGVF. Grace. Commissioned *Santa Paula* in the Naval Overseas Transportation Service on August 15, 1918. Operated during WWII by American-Hawaiian as the *Montanan*. Sunk on June 3, 1943, by *I-27* south of Masirah Island.

Harold Walker—6,784 tons, tanker, 1917, 215152, LHGC. Pan American Petroleum & Transport.

William Green—6,784 tons, tanker, 1917, 215348, LHMV. Pan American Petroleum & Transport.

Santa Ana—4,942 tons, steel, 14 knot coal or oil-burning passenger ship, 215880, LJMQ. Keel laid on February 1, 1917, for W.R. Grace. Requisitioned while building and delivered to the USSB on January 1, 1918.

Siboney—7,652 tons, steel, 17 knot coal or oil-burning passenger ship, 216082, LJVK. Requisitioned while building and delivered to the USSB on April 18, 1918. Allocated to the Navy as manager for operation by the Army.

Santa Luisa—4,944 tons, steel, 14 knot coal or oil-burning passenger ship, 216215, LKNV. Keel laid on March 26, 1917, for W.R. Grace. Requisitioned while building and delivered to the USSB on May 18, 1918. Commissioned *Santa Luisa* in the Naval Overseas Transportation Service on August 12, 1918.

Orizaba—7,652 tons, steel, 17 knot coal or oil-burning passenger ship, 216294, LKJM. Keel laid on December 6, 1916. Requisitioned while building and delivered to the USSB on June 5, 1918. Allocated to the Army and the Navy.

Santa Olivia—6,421 tons, steel, 12 knot coal or oil-burning passenger ship, 216438, LKJV. Keel laid on May 7, 1917, for W.R. Grace. Requisitioned while building and delivered to the USSB on July 2, 1918. Commissioned *Santa Olivia* in the Naval Overseas Transportation Service on July 1, 1918.

Santa Teresa—5,102 tons, steel, 14 knot oil-burning passenger ship, 216969, LNBP. Keel laid on October 22, 1917, for W.R. Grace. Requisitioned while building and delivered to the USSB on November 15, 1918. Allocated to the Navy as manager for operation by the Army.

William E. Woodall & Co.
Baltimore, Maryland

John T. Fardy & Bros. had a shipyard at the east end of Hughes in Baltimore and ran a steam marine railway with Philip Auld as Fardy & Auld. On May 27, 1851, John married Emily J. Kone. Years later John became ill and he went to Berkeley Springs, West Virginia, to take the medicinal waters and on Sunday, July 22, 1867, died "from a protracted case of consumption."

William E. Woodall was a shipbuilder born in England in 1837 and was a relative of the Fardys and he arrived on the steamer *City of Baltimore* on February 17, 1868, to help Emily run the yard, a chandlery at 33 Hughes, Fardy's Wharf, and the marine railway at the foot of Montgomery St. as Fardy & Woodall.

Charles A. Witler was a ship carpenter who lived at 110 Cross. In 1873 Woodall went into partnership with Witler and William's younger ship carpenter brother, James E. Woodall, as William E. Woodall & Co. They opened a marine railway at 46 Covington and a 270' floating drydock and side basin at the foot of Allen on Locust Point.

After William died in 1884, James and Charles ran the company. James died in 1915 and the company was controlled by James' estate. They built five ships delivering the first one, the 219-ton steamer *Charles A. Witler*, in 1904. Their last ship was the *Baltimore*.

Georgia—1,318 tons, schooner barge, 1915, 212943, LFDV. Foundered off Chincoteague, Virginia, on December 9, 1917, with 4 aboard. There were no casualties.

Baltimore—1,307 tons, schooner barge, 1916, 214479, LGJW. P. Dougherty Co.

Wilmington Wooden Ship Construction Co. Wilmington, North Carolina

Capt. J.F. Cushman and Capt. George W. McKown, of Philadelphia, decided to build barges and coastal schooners as Cushman & McKown and on Saturday, June 30, 1917, they leased 225 feet of property at the foot of Surry between Queen and Wooster St. in Wilmington from the Chadbourn Lumber Co. The ways were done by July 29.

Richard T. Harriss and his wife Belle grew cotton in Greenville, Texas. On September 4, 1900, their daughter Isabel C. Harriss was born in Greenville and by 1916 Richard was a successful cotton merchant in New York City where the family lived at 324 W. 103rd St. He wanted his own ship and ordered a schooner from Cushman & McKown.

In 1898, John J. Turney went to work for William J. Grandfield as a ship broker in Philadelphia and in 1913 became a partner. Grandfield died the same year and Turney took over the firm and expanded into coastwise barge service. He formed the Eastern Lighterage Co., the Turney Transportation Co., the Philadelphia Barge Co., and got into shipbuilding as the Delaware Shipbuilding & Repair Co. at Cooper's Point in Camden, New Jersey.

In March 1918, Turney Transportation took over Cushman & McKown. The schooner was built of North Carolina yellow pine and launched on June 25, 1918.

Isabel C. Harriss—1,078 tons, 4-masted schooner, 1918, 216506, LMJH. Pyne Co., Inc., New York.

The Gulf Coast

Alabama & New Orleans Transportation Co. Violet, Louisiana

Incorporated in New York in 1913 to bring Alabama coal from Gilmore, 370 miles above Mobile at Lock 13 on the Black Warrior River, to New Orleans for 20 cents a ton. John H. Bernhard was vice-president and general manager. The company built a yard on the Lake Borgne Canal at Violet, 12 miles below New Orleans, to build and 15 self-propelled, twin screw barges. It was necessary to hire and train black laborers from the nearby sugar cane fields to build the barges, no other source of labor being available.

On August 16, 1918, the Shipping Board purchased six barges, the collier *New Orleans*, and two gantry coal towers from the company for $322,500.

The two, 260' tankers cost $250,000 each and were built to bring crude oil from Tampico, Mexico, to New Orleans with a crew of 15. In 1917 Bernhard was president of the Delta Gulf & Pacific Transportation Co. at 32 Broadway.

Mexoil—1,370 tons, tanker, 1918, 216397, LKTG. Pan American Petroleum & Transport Co. A sistership, the *Panoil*, was built in 1919, the last ship the company built.

Alabama Dry Dock & Shipbuilding Co. Mobile, Alabama

David R. Dunlap was president of Alabama Iron Works, George H. Dunlap was president of the Gulf Dry Dock Co., William M. Evans was president of Gulf City Boiler Works, and William M. Marshall was a businessman at 63 N. Commerce.

On January 1, 1917, the four consolidated into Alabama Ship and set up shop at the foot of Eslava St. with David Dunlap as president. William Evans and William Marshall were vice presidents and George Dunlap was secretary-treasurer.

They built the 483-ton schooner *Rena A. Murphy* in 1918 and their last ship, the wood freighter *Alta* was delivered in 1919. The company was revived for World War II construction.

Banago—2,551 tons, wood, 10 knot coal-burning Ferris freighter, 216887, LMTN. USSB contract ship. Keel laid on October 20, 1917, and delivered on September 26, 1918, 341 building days. Allocated to the Naval Overseas Transportation Service and then to W.R. Weist for service to Scandinavia.

The American Lumber Co. Millville, Florida

Only one vessel was delivered in 1919 but the story is interesting. In 1898 a French-Canadian named Henry Bovis built a sawmill in the dense pine forests around St. Andrews Bay in southeast Florida and organized the St. Andrews Bay Lumber Co.

Kaiser Wilhelm II apparently wanted a supply of American timber. Prince Furstlich Schaumber von Hoofkoner studied the prospects and around 1900 sent Freiherr (Baron) Julius Schreyer to America to purchase property in Florida. Schreyer contacted broker Boykin P. Jones at 24 E. Government St. in Pensacola, about 175 miles away, to purchase the Bovis operation and 1,000 acres of southern white pine around St. Andrews Bay. St. Andrews Bay was an ideal site for a deep-water port, secluded from storms, and became the closest harbor to the Panama Canal.

On June 15, 1901, the German-American Lumber Co. was incorporated at Pensacola with a capitalization of $1 million. Herman H. Boyer was the German vice-consul in Pensacola and acted as the company's manager. The books of the company were kept in Bremen, all the funds were held in the Deutsche Bank of Berlin, and all the stock was owned by the Kaiser.

In 1907 the town of Millville was set up as a modern company town and as time went on, more timbered and un-timbered parcels were added. In 1911, Herman G. Kulenkampff, the Kaiser's nephew, was president. He lived with his wife Flora at 1400 N. 12th Ave. Boyer was 1st vice-president, Schreyer 2nd vice-president, and J. Gerhard Rolfs, the German consul in Pensacola, was secretary-treasurer. He lived with his wife Gila at 1550 N. 12th.

On July 11, 1914, the company shipped 308,000 feet of lumber to Bremen via St. Andrews on the steamer *Port Inglis* and on November 18, 1,583,000 feet to London on the *Antigua*.

When war came, A. Mitchell Palmer, the Alien Property Custodian, investigated the company. One officer had an expense account of $50,000 but not one cent was in an American bank. Huge amounts of German propaganda literature were found in the homes of all the officers. It was learned that Samuel J. Carpenter, of Jacksonville, and Marc L. Fleishel, a lumber dealer in Jacksonville, were involved with the company. In all, there were about 150,000 acres, 80 miles of logging roads, and two sawmills and Millville had about 1,500 inhabitants.

Palmer seized the property, changed the name to the American Lumber Co. and appointed new directors: attorney William A. Blount, of Pensacola; William L. Wilson, of Millville; bank president Edgar R. Malone, of Pensacola; Charles S. Hebard, of Chestnut Hill, Philadelphia; and Hollins N. Randolph, of Atlanta.

Mitchell urged the Emergency Fleet Corp. to use the site for shipbuilding. On July 18, 1918, a contract was awarded for eight 2,500-ton seagoing schooner barges to carry coal from Norfolk, Virginia, to New England. The first keel was laid on September 28, 1918. The company was sold at public auction on March 10, 1919, to Minor C. Keith and Walter C. Sherman.

Millville—1,213 tons, schooner barge, 1919, 218755, LSTB.

American Shipbuilding Co.
2017 Superior Ave. NW in Cleveland, Ohio

James C. Wallace was president. Michael Brown was vice-president in Brunswick and John H. Leo was general manager at the yard at Newcastle & Bay. Brown was with the Oglethorpe Land & Investment Co., the Manufacturers Timber Co., and the American Tie & Timber Co.

American Ship built three ships at Brunswick. The third was the big schooner *Tenas* delivered in 1919.

BRUNSWICK, GEORGIA

Alabat—2,551 tons, wood, 10 knot coal-burning Ferris freighter, 217227, LNTB. USSB contract ship. Keel laid on August 31, 1917, and delivered on October 31, 1918, 426 building days. Retained for recruiting and training purposes at Jacksonville, Florida.

Beaumont Shipbuilding & Dry Dock Co.
Beaumont, Texas

John H. Cooke and Gustave T. Nelson were boat builders as Cooke & Nelson on the river between Elizabeth & Mulberry. They decided to build wood Shipping Board freighters and delivered four, the last was the *Angelina*. The schooners *Marie F. Cummins* and *Albert D. Cummins* were delivered after the war.

Oneco—2,571 tons, wood, 10 knot coal-burning Ferris freighter, 216765, LMNF. USSB contract ship. Keel laid on September 1, 1917, and delivered on October 31, 1918, 425 building days. Allocated to J.H.W. Steele for service to the United Kingdom.

Brunswick Marine Construction Corp.
Brunswick, Georgia

Willard U. Taylor was president, R. Lawrence Smith vice-president, and E.W. Brown secretary-treasurer. The company built eight ships under superintendent Walter S. Nathan and general manager Wilson E. Skinner at Bay and Dartmouth Sts. The last ship delivered was the steamer *Dan F. Hanlon* in 1920.

William Taylor—1,358 tons, schooner, 1917, 215717, LJDH. Apex Navigation.
Myron C. Taylor—1,358 tons, schooner, 1918, 215987, LJQS. Apex Navigation.
Sir Thomas J. Lipton—1,358 tons, schooner, 1918, 217405, LPHM. Apex Navigation.

Clooney Construction & Towing Co.
Westlake, Louisiana

Company founder John Francis "Jack" Clooney, Jr., was born in Westlake, Louisiana, on May 24, 1880. The company delivered the schooner *Ankla* in 1915 and built 19 ships total. The last delivery was the sidewheel steamer *Panuco* in 1920.

James M. Hudson—1,011 tons, schooner barge, 1917, 214928.
A.H. Olwine—1,077 tons, schooner barge, 1917, 215737, LJDV. Eastern Transportation Co.
Col. Bowie—1,363 tons, wood freighter, 1917, 216212, LKNS. Gulf Export & Transportation Co.

Dierks-Blodgett Shipbuilding Co.
Pascagoula, Mississippi

Hans, Herman, and Peter Dierks came from Germany and built a lumber, railroad, and paper empire that stretched across several states. Hans Dierks was president of the Dierks Lumber & Coal Co. in Kansas City, Missouri. Herman Dierks was vice-president, and Harry Dierks was secretary-treasurer. Herman had a daughter named Mae. Robert Sutherland, Jr., was in the lumber business in Kansas City and on September 14, 1911, Robert and Mae married.

Laurence G. Blodgett was vice-president of the A.M. Blodgett Construction Co. in Kansas City. The company built bridges, foundations, and steel structures.

When war came, the sons of the Dierks founders and cousins Fred H. Dierks and DeVere Dierks teamed up with Laurence Blodgett and Robert Sutherland, Jr., to form Dierks-Blodgett Ship. Fred was president, Sutherland vice-president, DeVere secretary, Blodgett treasurer and general manager. The company built five wood Shipping Board freighters. The last was the *Panga* delivered in 1919. The company had a contract for 10 steel freighters that was cancelled.

Pascagoula—2,551 tons, wood, 10 knot coal-burning Ferris freighter, 216991, LNCT. USSB contract ship. Keel laid on September 1, 1917, and delivered on October 1, 1918, 394 building days. Commissioned *Pascagoula* in the Naval Overseas Transportation Service on October 4, 1918.

Belair—2,551 tons, wood, 10 knot coal-burning Ferris freighter, 217062, LNHJ. USSB contract ship. Keel laid on August 22, 1917, and delivered on October 31, 1918, 435 building days. Allocated to Lykes as managers and operated by Dale Universal on time charter.

Cawker—2,469 tons, wood, 10 knot coal-burning Ferris freighter, 217279, LNWR. USSB contract ship. Keel laid on August 31, 1917, and delivered on December 30, 1918, 487 building days. Allocated to New Orleans & South American for service to France.

F.H. Swails
Westlake, Louisiana

Frederick Hostler Swails was born on January 24, 1876, and became a boat captain in Galveston, Texas. He acquired property in Westlake, Louisiana, and in 1910 built the little steamers *John Sealy* and *R.P. Clark*. In 1913 he was a ship carpenter foreman for the Bowers Southern Dredging Co. in Galveston and at some point he acquired a shipyard on Harbor Island at Orange, Texas, which was purchased by Henry Piaggio in 1916.

In 1916 he lived at 1019 Fern St. in New Orleans with his wife Claudie and signed a contract with Piaggio to build the *City of Orange, City of Houston*, and the smaller *City of Pensacola* at his old yard in Westlake. The steam tug *Robert P. Clark* in 1917 was his last delivery before the yard was bought by Piaggio. In September 1918 he was working for Jancke Shipbuilding and in 1920 he was president of the Swails, Hance & Rawlings Construction Co. in Port Arthur, Texas.

City of Orange—1,632 tons, motor freighter, 1917, 214819, LGST. Henry Piaggio.

City of Houston—1,519 tons, motor freighter, 1917, 215246. Henry Piaggio. Converted to a schooner and renamed *Jessie Bounds* at Gulfport, Mississippi.

Gulf City Boiler Works Co.
307 E. Water St., Mobile, Alabama

In 1903, William M. Evans was a boilermaker at the Louisville & Nashville Railroad. William T. Hieronymus was a partner in Hieronymous Bros., lumber dealers. Felix A. Cirlot and his wife Annie ran the People's Furniture Co. and in 1904 he was an engineer with the Sullivan Timber Co. On November 30, 1904, the three incorporated Gulf City Boiler with a capital of $25,000 to manufacture boilers, oil and water tanks, dredge buckets, iron and steel masts, standpipes, etc. This company built one ship and on January 1, 1917, merged into the Alabama Dry Dock & Shipbuilding Co.

Susquehanna—1,381 tons, steel composite hull schooner tank barge, 1917, 215387, LHNT. Gulf Refining.

Hodge Ship Co., Inc.
Moss Point, Mississippi

Otis E. Hodge was in the lumber business in Ruston, Louisiana, and lived with his wife Mary and their five kids at 507 N. Vienna St. Hodge was president and general manager, E.H. Allen vice-president, attorney Joseph D. Barksdale secretary, and John S. Hunt treasurer. The *Banaran*, *Boonsborough*, and *Nika* were delivered in 1919.

Alpaca—2,551 tons, wood, 10 knot coal-burning Ferris freighter, 217088, LNJM. USSB contract ship. Keel laid on October 2, 1917, and delivered on October 31, 1918, 363 building days. Allocated to the Naval Overseas Transportation Service but found unfit and turned over to Page & Jones for service to the United Kingdom.

International Shipbuilding Co.
Orange, Texas

In 1861, Enrico Piaggio was a commission merchant and importer of wines and brandies at 60 Old Levee Rd. in New Orleans. During the Civil War he was a captain in the 6th Regiment, European Brigade, Louisiana Militia, Italian Guards Battalion. At some point Enrico and his wife moved back to Italy and on March 17, 1874, their son Enrico was born in Genoa.

On December 21, 1891, Angelina Piaggio, 45, Idalia, 25, and Enrico, 17, arrived at New York on the steamer *La Gascogne* with the destination of Pensacola, Florida. Angelina was the widow of Dario Piaggio, a ship broker, commission merchant, and vice-consul for France and Italy at Pensacola. Angelica, Enrico, now Henry, and his sisters Emma, Idalia, Irma, and Amalia lived at 509 N. Barcelona St. in Pensacola. By 1898 Henry was a traveling salesman with the family business.

In 1898 Theodore D. Baars was a student growing up in the Baars, Dunwoody & Co., a lumber exporting business. The family lived at 114 W. La Rua in Pensacola.

Robert A. Hyer was a partner with William and Knowles Hyer as the Hyer Bros., lumber merchants and ship brokers at 18 E. Government. Knowles Hyer lived at 501 N. Barcelona.

Henry decided to follow in Dario's footsteps, made the acquaintance of Theodore Baars and Robert Hyer and the three formed H. Piaggio Co., ship brokers in Pensacola at 307 S. Palafox. Theodore was president, Robert vice-president, and Henry secretary-treasurer. The venture was short-lived. By 1905 Theodore was treasurer of H. Baars & Co., lumber exporters at 804 S. Palafox and Robert was back at Hyer Bros.

Henry apparently went abroad and spent some time organizing a venture and while there met Margaret Muldowned, of Ohio, and they married. Arriving home, he purchased a property on Harbor Island in Orange, Texas, from shipbuilder Frederick H. Swails and in 1916 organized International Ship in 1916. Piaggio also had yards at Beaumont, Texas, and Pascagoula, Mississippi. His intention was to build a fleet of sailing vessels to export lumber. He built ten ships in Orange and four in Pascagoula. Two of the ships built at Orange were 5-masted barkentines, the *Macerata* and *Monfalcone*, both in 1919. He built two 5-masted barkentines at Pascagoula, the *Marsala* in 1919 and the *Molfetta* in 1920. The Pascagoula yard built four ships, all after the war.

City of Gulfport—1,844 tons, motor freighter, 1918, 215942, LJPH. Burned in 1919.

City of Mobile—1,975 tons, motor freighter, 1918, 216350, LKRS. Burned in 1919.
City of Beaumont—2,014 tons, motor freighter, 1918, 216473, LKWM.
City of Austin—2,231 tons, 1918, motor freighter, 217160, LNPJ. Sold alien on May 22, 1919.

Beaumont, Texas

City of Dallas—1,977 tons, 1918, gas motor freighter, 216909, LMVN.

Jahncke Shipbuilding Co.
Madisonville, Louisiana

Ernest Lee Jahncke was president and general manager of the Jahncke Navigation Co. in New Orleans. Paul F. Jahncke was vice-president, Walter F. Jahncke secretary-treasurer. The company operated towboats and dredging barges and incorporated Jahncke Ship to build wood Shipping Board freighters at Madisonville 30 miles north of New Orleans on the opposite side of Lake Pontchartrain. The wood *Abbeville*, *Pontchartrain*, and *Balabac* freighters were delivered in 1919. They subsequently got a $10 million contract to construct a shipyard and build 12 steel freighters in New Orleans but the steel ships were never built.

Bayou Teche—2,551 tons, wood, 10 knot coal-burning Ferris freighter, 217320, LPCT. USSB contract ship. Keel laid on October 31, 1917, and delivered on October 31, 1918, 365 building days. Allocated to Page & Jones for service to France.

The Johnson Iron Works, Ltd.
Algiers, Louisiana

Lewis Johnson's fabricating works were in New Orleans and he built and repaired ships at Algiers. Henry D. Stearns was secretary-treasurer. The company built six small ships between 1902 and 1919. Wilmer H. Johnson was president during the war, Warren Johnson, a consulting engineer, was secretary-treasurer.

Juniata—1,432 tons, steel schooner tank barge, 1917, 215238, LHKC. Gulf Refining.

Lone Star Ship Building Corp.
111 Broadway, Rm 2116, New York

Louis Duncan, McKee Duncan, and S. Marsh Young were consulting engineers as Duncan Young & Co. at 50 Church St. in New York. George B. Leighton and Isaac F. Woodbury were building contractors in Boston as Woodbury & Leighton. They got the idea to build government ships in Beaumont, Texas, and in 1917 organized Lone Star Ship in Maryland. Leighton was president, James S. Bailey vice-president, Henry W. Brooks, Jr., secretary, and Frederick E. Hazler treasurer. Edward J. Barber, Young, and the Duncans were directors.

They leased land in Beaumont but maintained accounts and conducted business in New York. They got a contract for eight wood Shipping Board freighters and built four. The *Batanes* was delivered in 1919. When their contract was cancelled they went bankrupt and filed claims against the government for breach of contract.

Lone Star—2,584 tons, wood, 10 knot coal-burning Ferris freighter, 216843, LMSB. USSB contract ship. Keel laid on November 9, 1917, and delivered on October 14, 1918, 341 building days. Allocated to J.H.W. Steele for service to the West Indies.

Arado—2,551 tons, wood, 10 knot coal-burning Ferris freighter, 217281, LNWT. USSB contract ship. Keel laid on November 10, 1917, and delivered on April 7, 1919, 467 building days. Allocated to Robert Hasler for service to France.

Arambi—2,551 tons, wood, 10 knot coal-burning Ferris freighter, 217444, LPKW. USSB contract ship. Keel laid on November 5, 1917, and delivered on May 22, 1919, 563 building days. Allocated to Texas Transportation & Terminal for service to the United Kingdom.

McBride & Law
209 Keith Building, Beaumont, Texas

Richard J. McBride and Charles F. Law were general contractors in Beaumont as McBride & Law. They signed the very first contract with the Emergency Fleet Corp. to build wood freighters. The contract called for four Ferris freighters but they built two.

Beaumont—2,551 tons, wood, 10 knot coal-burning Ferris freighter, 217313, LPCM. USSB contract ship. Keel laid on August 25, 1917, and delivered on April 28, 1918, 246 building days. Allocated to Lykes Bros. for service to the United Kingdom.

Quapaw—2,551 tons, wood, 10 knot coal-burning Ferris freighter, 217446, LPMC. USSB contract ship. Keel laid on September 22, 1917, and delivered on April 28, 1918, 218 building days. Allocated to New Orleans & South American Steamship for service to the Netherlands.

Midland Bridge Co.
504 Railway Exchange Building, Kansas City, Missouri

An old company run in 1917 by engineers Henry Freygang and Albert A. Trocon. They designed and built bridges, viaducts, foundations, buildings, and steel structures of all kinds. In 1916 they built bridges in Tucson, Arizona, on St. Mary's Road over the Santa Cruz River, one in Sabino Canyon, and one on Oracle Rd. over the Rillito River.

They decided to build wood ships in Houston, Texas, and built six wood Shipping Board freighters and the schooner *Aransas* in 1919. Their last ship was the wood freighter *Wiscasta* also delivered in 1919. The keel for the *Katonah* was laid on October 4, 1917, and launched on July 13, 1918, after 283 days, but as of May 31, 1919, the ship had not been delivered.

Batino—2,559 tons, 10 knot coal-burning Ferris freighter, 217429, LPKB. USSB contract ship. Keel laid on November 15, 1917, and delivered on April 11, 1918, 512 building days. Mothballed at Beaumont, Texas.

Mobile Shipbuilding
Mobile, Alabama

William Lee Kelly was born in Waukegan, Illinois, on March 12, 1864, grew up on the family farm, worked as a railroad section hand then in a boiler shop out West. He came back to Chicago, became a contractor in 1892, and on September 21 married Mary McGary. In 1894 he teamed up with Charles E. Atkinson as Kelly & Atkinson.

When they incorporated The Kelly-Atkinson Construction Co., bridge builders at 189 W. Madison, Kelly was president, Atkinson vice-president, and Mac R. Fife secretary-treasurer. By 1912 Atkinson had apparently left the firm and moved to Champaign-Urbana where he was superintendent of grounds at the University.

Mobile Ship was formed in 1917 with Kelly as president, Fife vice-president and treasurer, and Frank Casper secretary. Casper was also secretary-treasurer of the Mobile Ship Chandlery Co. run by Capt. Irving T. Newbury.

They signed a contract for 18 composite hull freighters and built 14 between 1918 and 1920 at the foot of Texas St. Their last was the *Oklahoma City* in 1920.

Balino—2,551 tons, wood, 10 knot coal-burning McClelland freighter, 217137, LNMH. USSB contract ship. Keel laid on December 18, 1917, and delivered on October 31, 1918, 316 building days. Allocated to Page & Jones as managers and operated by Merchants Navigation for service to the West Indies.

Morganza—2,469 tons, composite wood, 10 knot coal-burning McClelland freighter, 217280, LNWS. USSB contract ship. Keel laid on December 27, 1917, and delivered on December 4, 1918, 342 building days. Allocated to Texas Transport & Terminal for service to the United Kingdom.

National Shipbuilding Co.
1608 Woolworth Building, New York

Incorporated to build ships at Orange, Texas. Patrick J. Reilly was president, John F. Penrose vice-president, Guy Chase Van Alstyne secretary and purchasing agent, and Frederick Straith-Miller treasurer. William A. Ebsen was the naval architect and chief engineer and Albert A. "Silent Daugherty" was the general manager.

Ebsen and Daugherty modified the Ferris design that bore Daugherty's name as Design No. 1006. The well deck freighter had three holds with machinery spaces and house aft. Daugherty was born on October 3, 1878, in Saltsburg, Pennsylvania, and became a lumber salesman in Dayton, Ohio. In 1902 he married Mayme and they had a son and two daughters. He died on January 8, 1962, in Saltsburg where he was a lumber dealer.

The company built 16 American ships all told including the big schooner *W.A. Ebsen* in 1919. The last ship was the steamer *P.J. Reilly* in 1920. The company had a shop at Pier 41 in Galveston where machinery was installed in the hulls.

They built the 880' wood ships *War Mystery* and *War Marvel* for Cunard, launched on February 27, 1918, reported to be the largest wood steamships in the world. The company got a contract for 28 Daugherty ships.

Bonham—3,349 tons, wood, 10 knot coal-burning Daugherty freighter, 216964, LNBM. USSB contract ship. Keel laid on September 20, 1917, and delivered on October 28, 1918, 464 building days. Allocated to Lykes Bros. for service to South America.

Boreta—3,361 tons, wood, 10-knot coal-burning Daugherty freighter, 217311, LPCJ. USSB contract ship. Keel laid on September 20, 1917, and delivered on December 11, 1918, 447 building days. Converted into a tanker and burned in 1920.

Beechland—3,355 tons, 10 knot coal-burning Daugherty freighter, 217443, LPKV. USSB contract ship. Keel laid on September 20, 1917, and delivered on April 19, 1918, 211 building days. Allocated to Nacirema for service to France.

Savannah Engineering & Construction Co. Savannah, Georgia

Wayne Cunningham was president of the Appraisal Company of the South and lived with his wife Kate at 118 E. 36th. He built just the two ships on Lathrop Ave.

Chaska—1,347 tons, schooner, 1917, 215707, LJCT. East Coast Transportation. Foundered on March 13, 1919.

Sherewog—1,353 tons, 1918, wood motor freighter, 216386, LKST. East Coast Transportation.

Slidell Dry Dock & Shipbuilding Co., Inc. Slidell, Louisiana

Fritz Salmen was a German-Swiss born in August 1854. In 1867 his family settled in Hansboro near Gulfport, Mississippi, and in 1880 he was a clerk in a grocery store. On November 23 1882, he married Ella Rose Bell Liddle and they moved to Slidell, Louisiana, where Fritz got involved in the brickmaking and lumber business. In May 1885 they had a daughter Ella and in June 1887 a son, Frederick W.

Fritz purchased more forestland and had a thriving lumber and brickmaking business. He is considered a pioneer of the state's lumber industry as the Salmen Brick & Lumber Co.

In 1916 Frederick organized the Slidell Dry Dock & Shipbuilding Co. Frederick was president, Capt. L.T. Miles was 1st vice-president, and Edward F. Hailey secretary-treasurer. The company built just the three American ships. On December 30, 1916, the company launched the 230' lumber carrier *Baltic I*, built of Louisiana yellow pine and cypress, woods heavily invested in by Salmen. About 200 ship experts from all over the country were invited to inspect the ship and the *Baltic II*, *Baltic III*, and *Balltic IV* on the ways being built for the Norwegian company Damskibs-Aktieselskabet Baltic. Several thousand local residents attended as well. Part of the Norwegian crews were already in town helping with the installation of machinery and becoming familiar with their ship's unique construction. Frederick was also president of the Slidell Brick & Lumber Co. and the Gulf Coast Transportation Co.

In May 1917, the Continental Transportation & Oil Co. purchased Slidell for $187,000 and on January 1, 1918, sold the property but not the capital stock to the Louisiana Shipbuilding Co.

Flush—1,474 tons, freighter, 1917, 214995. Bulk Oil Transports.

Flirt—1,491 tons, freighter, 1917, 215025. Bulk Oil Transports.

Maple—1,098 tons, freighter, 1918, 216003, LJRG. C.L. Dimon, New York.

The Southern Dry Dock & Shipbuilding Co., Inc.
Orange, Texas

Solomon M. White was president of the White Implement Co. and the East Texas Investment Co. White Implement sold farm implements, harnesses, saddlery, and vehicles at 107B 5th St. He lived with his wife Inez at 610 7th St.

Frederick H. Farwell was secretary-treasurer of the Yellow Pine Paper Mill Co. and general manager of the Lutcher-Moore Lumber Co. He lived with his wife Fannie at 812 Green Ave.

William H. Tippin was a partner with Max Rowland as Tippin & Rowland, steamship agents and brokers in the Stark Building. Tippin boarded at 312 5th St.

Dr. Tom F. Coyle was a dentist in the Stark Building. He lived with his wife Grace at 604 Pine Ave. The four incorporated Southern Dry Dock with White as president, Farwell vice-president, Tippin secretary, and Coyle treasurer.

Work started on the 5-way yard on August 20, 1917, and the company signed a contract on August 9, 1918, for 16 Daugherty type wood freighters. They also had facilities at Galveston to install machinery. The company built only three ships: the 784-ton motor vessel *Elmir Roberts* in 1918, the *Gonzalis*, and the wood Shipping Board freighter *Arenas* in 1919.

Drinking had become a serious problem for shipyard managers around the country and on May 1, 1918, a nationwide prohibition went into effect. All the shipyards were contacted by telegram to get feedback and on June 21 the company provided a fairly common reply:

> Your wire 20th: Unquestionably there has been decided improvement in working conditions at our plant since saloons were closed. Larger percentage of our men make full time. We would regret to see any compromise and absolutely can not agree that prohibition has reduced shipbuilding output in our locality.

In addition to showing up sober and putting in a full days' work, all of the employees donated their entire salary from one Sunday a month to the Red Cross for war relief work.

Gonzalis—2,654 tons, wood freighter, 1919, 218911, LSVJ.

Tampa Dock Co.
Foot of 16th St., Tampa, Florida

Andrew J. and L. Esli Knight were fire insurance, real estate, and rental agents in Tampa. Dr. John C. Vinson was a physician. The three formed Tampa Dock with Andrew Knight as president, Dr. Vinson secretary, and L. Esli treasurer. Tampa Dock built four wood freighters. the *Coulter* and *Hoosac* were delivered in 1919.

Namecki—2,525 tons, wood, 10 knot coal-burning Ferris freighter, 216873, LMTB. USSB contract ship. Keel laid on October 27, 1917, and delivered on August 31, 1918, 309 building days. Allocated to West India Steamship for service to the West Indies.

Agria—2,560 tons, wood, 10 knot coal-burning Ferris freighter, 217012, LNFB. USSB contract ship. Keel laid on October 29, 1917, and delivered on September 30, 1918, 337 building days. Allocated to Philip Shore for the French trade.

Tampa Shipbuilding & Engineering Co.
19th & Grant, Tampa, Florida

Ernest Kreher was president and general manager of the Tampa Foundry & Machine Co., Paul Kreher was assistant manager of Tampa Foundry, Max Kreher was secretary-treasurer of Tampa Foundry, and Peter O. Knight was an attorney.

Ernest was president and general manager of Tampa Ship, Knight was vice-president, Paul was assistant manager, and Max secretary-treasurer. Their last ship was Shipping Board freighter *Everglades*, delivered in 1919.

Poughkeepskie—1,566 tons, steel, passenger, 1917, 215041. Central Hudson Steamboat Co. Homeported at Poughkeepskie, New York, to provide local service on the river. Former New York governor Benjamin B. Odell was president of Central Hudson. The boat arrived at 3:30 p.m. on Monday, May 28, 1917, to great fanfare and the presentation of flags to the company.

Lithopolis—2,362 tons, steel, freighter, 1918, 217289, LPBF. Keel laid on June 5, 1917, and delivered on February 6, 1919, 610 building days. Allocated to A.H. Bull for the phosphate trade.

Universal Shipbuilding Co., Inc.
513 Union National Bank Building, Houston, Texas

James Craven was a partner with Rufus Cage as Cravens & Cage, fire, tornado, accident, liability, motorboat, automobile, marine, and life insurance agents.

Craven was president, James T. Taylor and J. D Hall vice-presidents, Edward J. McCullough secretary-treasurer. Work on the yard began on July 7, 1917, to build wood ships and they got a contract on July 25, 1918, for eight ships. Universal built nine ships total, six were Shipping Board freighters, and the last was the steamer *Lieutenant* in 1919. The *Nacogdoches* was the first Ferris freighter built on the Gulf Coast and the launching was a big affair.

Nacogdoches—2,628 tons, wood, Ferris freighter, 1918, 216401, LKTN. Keel laid on August 24, 1917, and launched on April 6, 1918.

14. *The Naval Overseas Transportation Service*

In 1917, merchant seamen signed articles 506,941 times to make one or more voyages from an American port to one or more overseas ports and back to an American port. In 1918 that figure was 457,248, which reflects the workload taken up by the Naval Overseas Transportation Service. The figure for 1919, after the war ended, rose to 485,796.

In 1917, the Navy had no cargo ships of its own to speak of yet during the Great War, the bulk of trans–Atlantic cargo was transported in Navy-manned merchant ships

Supplies needed in France were first sent across in existing American cargo or passenger ships. Added to the very inadequate existing fleet were the interned German ships seized early in the war, purchased Austrian ships, and chartered Chinese, Cuban, Danish, French, and Norwegian sailing vessels and steamers. New construction under private and Shipping Board contracts created an ever greater need for merchant seamen, but finding enough seafarers to fill the berths on all these ships soon became difficult, even with the new training programs and the sea had not been a particularly sought after vocation for American boys before the war. There was no real labor shortage during the war. Workers merely sought out the best deal they could get and no one had to go to sea if they didn't want to.

Secondly, and somewhat more troubling at the time, was the fact that very few, if any, ships left port with all-American crews. Prior to the war that was of little consequence and, in fact, it had always been the rule rather than the exception in the world's merchant fleets. But after the war started and German plots and intrigues popped up everywhere, the presence of foreign nationals sometimes created tension and suspicion on board ships. The large numbers of merchant seamen needed produced "an element among them sullen and suspiciously neutral."

A typical example was the *Orleanian*, a ship that was lost. In the crew of 36 were 6 Americans, 1 Canadian, 4 Danes, 1 Dutch, 1 English, 3 Finns, 6 Norwegians, 5 Russians, 7 Spaniards, and 1 Swede.

One ship kept developing frequent engine problems that caused the ship to straggle from a convoy. When ground glass was found in the lubricating oil, a member of the Armed Guard was stationed in the engine room night and day. It's impossible to know how much deliberate sabotage occurred versus how much ordinary trouble was suspected sabotage but these incidents caused concern.

In addition, several instances of alleged or perceived cowardice, disobedience to convoy commanders, and lax discipline about showing lights at night aboard merchant ships were brought to the Shipping Board's attention. Secretary of Labor Redfield addressed a conference on manning ships and stated, in part:

> It has not been always the case that merchant captains crossing the North Atlantic have either known of or have been willing when they were told to avoid submarines. In several cases that have come to my knowledge they have refused to take very simple precautions for the protection of their vessels, themselves, and their crew. I am making no sweeping statement. I say "in a few cases." Let me instance a case of a captain of a ship who was asked by the naval commander of the gun crew on board to take a zig-zag course, and he refused to do it on the ground that it was all a matter of luck anyhow and he had crossed eight or nine times without being hit and he thought he could cross again. So he did cross once more and yet once again, but the next time his ship went to the bottom and some of his crew.
>
> In another case, a merchant captain, being requested to tell his crew to go to sleep in the danger zone with a life preserver handy by and have a considerable amount of their heavy clothing on—this was in the colder part of the season—the captain refused to give the orders. The crew went unwarned to their beds. They were torpedoed that night and, fortunately, all rescued, but one of those poor fellows had on only a flannel shirt and was twenty-four hours in an open boat in the latter part of March in the North Atlantic because the captain had refused to give the order.
>
> There are other captains who have at last seen there was a reason when we issued a public statement that no captain's license would be taken if he ran ashore or had an accident. You remember we issued such a rule and the reason was that certain captains refused to run without lights in the danger zone for fear they would lose their license.
>
> I think one of the worst things in this whole war is unjustified suspicion... those captains really thought we would take their licenses away if they obeyed instructions to run without lights in the danger zone, so we had to issue a definite statement that they may run safely without their lights.

The two main solutions to these crewing problems that emerged were the militarization of the merchant marine or the creation of a shipping division within the Navy Department. It was soon realized that bringing merchant seamen into the Armed Forces under the existing age and physical fitness requirements would be impossible and there was still the problem of numbers. The only practical solution was a naval shipping division. The Navy could readily get the manpower needed, either through voluntary enlistment or draft.

Requisitioned merchant ships could be bareboat chartered through the Shipping Board, steamers built under Shipping Board contracts could be allocated to the Navy, or the Navy Department could negotiate with owners and purchase or charter ships on its own account.

The Naval Overseas Transportation Service was established on January 9, 1918, with 73 Navy, Shipping Board, and requisitioned American or seized foreign merchant ships. Each ship was commanded by a Naval Reserve Force officer and manned by regular and reserve enlisted men. Any Armed Guard crews aboard simply became part of the ship's company.

By the end of January, 16 of the original 73 ships were under charter to the War Department to supply A.E.F. forces, 52 carried supplies to naval forces, and five were awaiting commission. By July, 198 ships were in commission. The service was soon termed "The Ferry to France."

The major ports of operation were New York, Baltimore, and Newport News where around 33,000 people were employed in office and shoreside operations.

It was originally intended for NOTS to operate 490 ships but when the war ended rather abruptly there were 378 in operation with 4,672 officers and 29,175 sailors. Of the 378 ships, 213 were operated on the Army's account.

Besides carrying war supplies and munitions during the war, the service also carried Food Administration cargo for "distressed regions of Europe where relief was urgently needed," according to Secretary of the Navy Josephus Daniels. Almost all of the privately-owned ships were returned to their owners in 1919.

This section lists all the privately-owned merchant ships that served in NOTS including those that were in the Navy prior to January 9, 1918. The ships are arranged in order of the date of commissioning. Ships built under Shipping Board contracts and delivered between 1914 and 1918 and transferred to NOTS are noted in the Merchant Ship Construction section.

The name is followed by gross tonnage, type of service, year built, official number, call sign, owner, and date commissioned. Most ships underwent some conversion work to accommodate guns, additional crew, or troops.

Ships Commissioned Before January 9, 1918, and Transferred to NOTS

Sterling—freighter purchased from the Black Diamond Transportation Co. in 1898 and commissioned on April 16, 1898. Transferred to NOTS on January 9, 1918.

Gargoyle—4,433 tons, tanker, 1902, 212825, LFBK, Vacuum Oil Co. Commissioned on August 8, 1917, as *Gargoyle.*

Nahunta—1,213 tons, 2-masted schooner-barge, 1866, 81350, KJWN. Luckenbach. Commissioned on August 8, 1917, as *Nahunta* to carry coal in the 5th Naval District out of Norfolk and transferred to NOTS on August 8, 1918.

Los Angeles—6,872 tons, tanker, 1916, 213962, LFTS. Union Oil. Commissioned at New Orleans on August 9, 1917, as *Los Angeles.*

Standard Arrow—7,794 tons, tanker, 1916, 214196, LGCR. Standard Transportation. Commissioned on August 22, 1917, as *Standard Arrow* and transferred to NOTS on January 9, 1918.

Gen. Knox—1,587 tons, 2-masted schooner barge, 1881, 85609, JVRM. Luckenbach. Purchased on October 18, 1917, and commissioned on November 8, 1917, as the *General Knox.* Carried coal between Boston and Bar Harbor. Transferred to NOTS on August 8, 1918, and carried coal from Norfolk to New England and Key West.

Old Dominion—1,802 tons, 3-masted schooner-barge, 1872, 19350, JLKN. Luckenbach. Purchased on October 18, 1917, and entered NOTS on August 8, 1918, as *Old Dominion* to carry coal in the 5th Naval District out of Norfolk.

W.F. Babcock—2,128 tons, 3-masted schooner-barge, 1882, 80935, JWRK. Luckenbach. Commissioned on November 8, 1917, as *W.F. Babcock* to carry coal from Norfolk to New England.

Manta—2,081 tons, freighter, 1916, 214554, LGMJ. Atlantic Gulf & West Indies. Commissioned on December 8, 1917, as *Manta.*

Sioux—3,048 tons, freighter, 1916, 214569, LGNC. Atlantic Gulf & West Indies. Commissioned on December 1, 1917, as *Sioux.*

Peter H. Crowell—3,101 tons, freighter, 1912, 210618, LCRN. Crowell & Thurlow. Commissioned on December 20, 1917, as *Peter H. Crowell.*

Iowan—6,649 tons, freighter, 1914, 212144, LDKR. American-Hawaiian. Commissioned on December 23, 1917, as *Iowan.*

Mexican—8,414 tons, freighter, 1907, 203971, KWBC. American-Hawaiian. Commissioned on December 23, 1917, as *Mexican.*

Ozama—2,118 tons, freighter, 1916, 214562, LGMR. Atlantic Gulf & West Indies. Commissioned on December 24, 1917, as *Ozama.*

Carib—2,118 tons, freighter, 1916, 214561, LGMQ. Atlantic Gulf & West Indies. Commissioned *Carib* on December 27, 1917.

Middlesex—4,727 tons, freighter, 1912, 210644, LCQD. Coastwise Transportation. Commissioned on January 2, 1918, as *Middlesex.*

Carrie Clark—1,298 tons, 2-masted schooner barge, 1874, 128309, JQLP. Luckenbach. Purchased in 1917 to carry coal in the 5th Naval District. Transferred to NOTS on August 8, 1918, as *Carrie Clark.*

Josephus—1,406 tons, 3-masted schooner-barge, 1876, 75888, JSBQ. Luckenbach. Purchased on October 18, 1917, and commissioned on November 8. Transported coal in the 3rd Naval District in New York Harbor. Transferred to NOTS. in August 1918.

Gold Shell—5,614 tons, tanker, 1916, 213873, LFSB. Shell Line. Commissioned *Gold Shell* on August 8, 1917.

Hatteras—4,919 tons, freighter, 1917, 215865, LJMC. USSB. Commissioned *Hatteras* on October 23, 1917.

Hisko—5,665 tons, tanker, 1917, 215758, LJFN. USSB. Commissioned *Hisko* on December 6, 1917.

Lake Moor—5,665 tons, tanker, 1917, 215758, LJFN. USSB. Commissioned *Lake Moor* on January 4, 1918.

Lakewood—1,948 tons, freighter, 215785, LJGF. USSB. Commissioned *Lake Wood* on December 15, 1917.

Ships Built Before 1914 and Requisitioned for NOTS in 1918

Winnebago—4,362-ton freighter tons, 1900, 13528, LFMV. American Transatlantic. Commissioned on March 6 as *Winnebago.*

Corozal—3,063-ton freighter tons, 1911, 208330, LBSV. New York & Porto Rico. Commissioned on March 18 as *Corozal.*

Alaskan—5,671-ton freighter tons, 1902, 107736, KRNV, American-Hawaiian. Commissioned on March 23 as *Alaskan.*

Texan—8,615-ton freighter tons, 1902, 145944, KSBG, American-Hawaiian. Commissioned on March 23 as *Texan.*

The *Zeelandia*

The *Zeelandia* was a passenger ship built in 1910 by Alexander Stephen & Son, Ltd., in Glasgow, Scotland, for Koninklijke Hollandsche Lloyd (Royal Holland Lloyd) in Amsterdam.

The ship had accommodations for 120 first class passengers, 120 second class passengers, and 800 third class, or steerage, passengers. The steerage berths were removable so cargo could be carried in those spaces. The normal crew consisted of six head officers, 14 junior officers, and 145 unlicensed sailors and passenger attendants.

Holland was neutral during the war and engaged in passenger and freight service between Holland, New York City, and South America. The British required all Royal Holland ships to stop at Halifax, Nova Scotia, for inspection by British authorities.

On September 28, 1917, the *Zeelandia* left Bahia under Capt. Jacob Haas with coffee, cocoa, wool, tallow, and paper rings for cigars and arrived at New York on October 16. Also on the cargo manifest were 11 boxes of personal effects consigned to the Netherlands Overseas Trust Co., an organization formed to guarantee that goods consigned to it would not be requisitioned by any belligerent power. There was also a quantity of mail for the United States.

There were about 55 passengers, including 41 U.S. citizens and 21 of those were naval officers. Cdr. David E. Theleen, of Kenosha, Wisconsin, and Lt. John H. Culin, of Ellensberg, Washington, were the senior officers overseeing 19 ensigns.

Capt. Haas was not planning to take on any fuel or stores unless additional passengers coming aboard at New York required extra provisions. In cases where no additional passengers were coming aboard, a ship would normally discharge passengers and mail onto tugs for transport into the harbor in order to avoid paying an export license, but a company official had been assured by customs authorities that no export fee would be levied so the ship came into port.

The next day a Holland Lloyd official wired the director of the Shipping Board's Export License Bureau to say the ship had not intended to take on fuel or stores but there were a number of Dutch citizens who were stranded in the United States who had sought passage on the ship and they now needed an export license to purchase additional stores for these passengers.

On the 19th the Shipping Board official in Washington wired his New York agent to forward copies of the ship's manifest and these were received the same day. The company's officials waited until the morning of the 22nd and, not having heard from Washington, submitted their manifest to Treasury Department customs officials at the Port of New York and said they were no longer planning to transport the Dutch citizens and would like to leave immediately. The request was denied on the grounds that the company failed to obtain an export license for the bunkers, cargo, and stores already on board.

The company applied for the license on the same day, under protest, claiming no license was necessary since they never discharged any cargo, bunkers, or stores at New York and requiring the license was illegal since none of the items had been purchased in the United States or in any port controlled by the United States.

The ship sat in New York until March 21, 1918, by order of the War Trade Board to prevent the cargo from reaching the Netherlands. The ship was also being held for possible requisitioning and on the 21st, company officials were notified the vessel would be seized. The ship was then bareboat chartered from the company at $66,458.44 per month based on a market value of $8.3125 per gross ton per month but no compensation would be paid for the period from October 22, 1917, to March 21, 1918. The bareboat rate for this period would have been $332,299.20 plus expenses the company incurred while sitting in the harbor.

The company used 7.73 tons of coal per day for a total of 1,159.5 tons. On October 22, the *Zeelandia* had 1,810.5 tons of coal, which had been purchased at Port Castries for $18.85 a ton. The company had to purchase 1,149 tons of additional coal at $7.90 per ton on February 25, 1918, while sitting at New York, for a total of $9,077.10. On March 20, the ship had 1,800 tons of coal on board.

Crew wages were $327.24, a total of $49,086; food was $127.50 a day, a total of $19,125; water was $850.50; repairs and surveys and for deck and engine supplies cost $7,500; port expenses like towage and refuse disposal cost $4,192.50; insurance for the period was $21.18 per day, a total of $3,177; and general overhead costs to Royal Holland were $7,500. Royal Holland's costs allotted to the *Zeelandia* for advertising the company's services was $18.27 a day, or $2,740.50. The company was also obliged to borrow money in order to maintain operations at an average interest rate of 5.088 percent per year. General demurrage costs for the vessel itself was $332,292.20 for a grand total of $446,826.22 for the period of detention.

The ship was surveyed and given SP #2507, converted into a troopship, and commissioned on April 3, 1918, as *Zeelandia* under Cdr. Robert Henderson. The ship made five voyages to France during the war, seven more after the Armistice bringing troops home, and was decommissioned on October 6, 1919, and turned over to the Shipping Board.

Much diplomatic correspondence ensued between the United States and Holland over the seizing of this ship. On November 20, 1925, President Coolidge authorized Secretary of State Frank B. Kellogg to notify the minister of the Netherlands that a court had been appointed to hear the company's claim and on March 3, 1927, Congress referred the company's claim to a federal court to determine damages and compensation. The company filed suit on June 15, 1927.

Royal Holland claimed that officers of the United States had seized a vessel belonging to a friendly, neutral nation. In international relations, a government is responsible for the tortuous actions of its administrative officials and officers of the armed forces.

The government argued in court the *Zeelandia* was held under the Espionage Act of June 15, 1917, which allowed customs officials, on approval of the Secretary of Commerce, to refuse clearance to any vessel carrying any type of cargo designated by the President as contraband. The court found no evidence of contraband and ruled the vessel was retained illegally for the sole purpose of using it for the needs of the United States.

On March 7, 1932, the court awarded Royal Holland $446,826.22 plus 5 percent interest per year from June 15, 1927, to the date of judgement.

Calamares—7,782 tons, passenger with refrigerated cargo space, 1913, 212654, LDSQ. United Fruit. Commissioned on April 10 as *Calamares*.

Isabela—3,063 tons, freighter, 1911, 208911, LCBM. New York & Porto Rico. Commissioned on April 12 as *Isabela*.

Norlina—4,596 tons, freighter, 1909, 212840, LFBS. Garland. Commissioned *Norlina* on May 1, 1918.

Craster Hall—4,319 tons, freighter, 1909, 212886, LDQC. U.S. Steel Products Co. Commissioned on May 9 as *Craster Hall.*

Californian—5,707 tons, freighter, 1900, 127446, KQDV. American-Hawaiian. Commissioned on May 13 as *Californian.* Hit a mine in the Bay of Biscay on June 22, 1918, with 44 on board and sank. Everyone was rescued by the *Corsair.*

Antilla—3,652 tons, freighter, 1904, 212936, LFBT. New York & Cuba. Commissioned on May 20 as *Antilla.*

Santa Rosalia—5,409 tons, freighter, 1911, 212509, LDPN. U.S. Steel Products. Commissioned on May 20 as *Santa Rosalia.*

Guantanamo—3,292 tons, freighter, 1910, 212946, LSGB. New York & Cuba. Commissioned on May 21 as *Guantanamo.*

American—5,591 tons, freighter, 1900, 107591, KQJN. American-Hawaiian. Commissioned on May 25 as *American.*

Santiago—3,490 tons, freighter, 1906, 212906, LFCV. New York & Cuba. Commissioned on June 11 as *Santiago.*

Charlton Hall—5,631 tons, freighter, 1907, 212853, LDQB. U.S. Steel Products. Commissioned on June 14 as *Charlton Hall.*

Winifred—2,551 tons, tanker, 1898, 81626, KNRC. Gulf Refining. Commissioned on June 21 as *Winifred.*

Tivives—5,017 tons, passenger with refrigerated cargo space, 1911, 212565, LDPJ. United Fruit. Commissioned on July 5 as *Tivives.*

Buena Ventura—4,786 tons, freighter, 1913, 212747, LDPW. U.S. Steel Products. Commissioned on July 26 as *Buena Ventura.* Torpedoed by *U-46* on August 16 in convoy to Philadelphia with the loss of 15 out of 93 on board.

El Oriente—6,008 tons, freighter, 1910, 206080, LBRC. Southern Pacific. Commissioned on July 29 as *El Oriente.*

El Sol—6,008 tons, freighter, 1910, 207751, LBPK. Southern Pacific. Commissioned on August 3 as *El Sol.*

Kermanshah—4,947 tons, freighter, 1910, 215401, LHPM. Kerr Navigation. Commissioned on August 3 as *Kermanshah.*

Newton—5,194 tons, freighter, 1911, 209388, LCGB. New England Coal & Coke. Commissioned *Newton* on August 3.

J.B. Walker—2,136 tons, 3-masted schooner barge, 1879, 760002, JTMC. Luckenbach. Commissioned on August 8 as *J.B. Walker* to transport coal between Norfolk and New England.

Panaman—6,649 tons, passenger, 1913, 211629, LDGF. American-Hawaiian. Commissioned on August 8 as *Panaman.*

San Joaquin—1,551 tons, 2-masted schooner-barge, 1876, 115504, JSCB. Luckenbach. Entered service on August 8 as a coal barge.

Hawaiian—5,597 tons, freighter, 1900, 90534, KQMW. American-Hawaiian. Commissioned on August 10 as *Hawaiian.*

Solitaire—1,515 tons, 2-masted schooner-barge, 1879, 115654, JTHF. Luckenbach. Purchased on October 18, 1917, and commissioned on August 8 as *Solitaire* to carry coal from Norfolk to New England.

Arizonan—5,671 tons, freighter, 1903, 107823, KSGM. American-Hawaiian. Commissioned on August 14 as *Arizonan*.

Oregonian—5,597 tons, freighter, 1901, 155406, KQSC. American-Hawaiian. Commissioned on August 17 as *Oregonian*.

Cacique—6,202 tons, freighter, 1910, 212852, LDWK. Grace. Commissioned on August 19 as *Cacique*.

City of Macon—5,311 tons, passenger, 1903, 127750, KBLD. Barber. Commissioned on August 19 as *Macona*.

Severance—4,927 tons, freighter, 1909, 215795, LGHD. Union Sulphur. Commissioned on August 19 as *Severance*.

Alabama—2,621 tons, tanker, 1901, 130907, KQRJ. The Texas Co. Commissioned on August 21 as *Amabala*.

Minnesotan—6,655 tons, passenger, 1912, 210634, LCNV. American-Hawaiian. Commissioned on August 23 as *Minnesotan*.

Howick Hall—4,922 tons, freighter, 1910, 212093, LDQF. U.S. Steel Products. Commissioned on August 24 as *Howick Hall*.

Pleiades—3,753 tons, freighter, 1900, 150656, KPSL. Edgar F. Luckenbach. Commissioned on August 26 as *Pleiades*.

El Occidente—3,747 tons, freighter, 1910, 208166, LBRM. Southern Pacific. Commissioned *El Occidente* on August 27.

Jean—3,125 tons, freighter, 1909, 206563, LBFG. A.H. Bull. Allocated to NOTS and commissioned on August 30 as *Jean*.

Robert M. Thompson—2,458 tons, freighter, 1912, 210287, LCMP. Girard Trust. Commissioned on August 30 as *Robert M. Thompson*.

Sherman—4,881 tons, freighter, 1893, 214146, LFWS. Sherman Steamship. Commissioned on August 30 as *Durham*.

Topila—5,125 tons, tanker, 1913, 211469, LDCM. East Coast Oil. Commissioned on September 8 as *Topila*.

Pennsylvanian—6,649 tons, freighter, 1913, 211297, LDBH. American-Hawaiian. Commissioned on September 13 as *Scranton*.

Carrillo—5,012 tons, passenger ship, 1911, 212652, LDSN. Carrillo Steamship. Commissioned on September 16 as *Carrillo*.

The *Sara Thompson*

The 321' tanker *Gut Heil* was built in 1888 by William Armstrong, Mitchell & Co. at Newcastle-upon-Tyne, England, for the Deutsch-Amerikanische Petroleum Gesellschaft.

Capt. Albert Permien left Tampico, Mexico, with 20,000 barrels of Mexican crude and arrived at Baton Rouge, Louisiana, on April 25, 1913. His destination was the Standard Oil refinery but conditions were dicey owing to a heavy spring thaw that broke several levees on the Mississippi River, caused flooding and strong currents, and Capt. Permien was directed to an anchorage to await a berth.

On the evening of the 27th, Capt. Permien received word that he could come in and around 6 p.m. he weighed anchor but a strong current took hold of the ship and they began drifting down toward the *Willkommen*, another Deutsch-Amerikanische Petroleum tanker anchored further out. The collision opened the seams in the *Gut Heil*'s after cargo tank and the ship sank, covering the water with 5,000 barrels of oil. Four engineers were feared lost but they turned up later. Several Standard Oil vessels came out to assist and the ship's papers were floating in the oil in a leather pouch.

The owners made several unsuccessful attempts to raise the ship and on February 3, 1914, they notified the local U.S. Army Engineers that they were abandoning the vessel. The Army replied that no government salvage effort would be undertaken and the ship was left in the mud.

Around January 1916, John S. Thompson contacted the War Department to see about salvaging the vessel. Thompson was a 68-year-old widowed packing plant owner who lived at 16 Presidents Place in Kingston City, New York, with his single daughter Sara, 39, and his married daughter Rae, 28, and her husband George W. Potter, 22.

He was told no permit was necessary and this was reiterated by the local Army Engineer's office. In November 1916, Thompson signed a contract with the Atlantic Towing Co. and King & Wotherspoon to refloat the wreck. For their trouble, Atlantic Towing would get half ownership in the vessel and Thompson advanced the company $60,000 for the work.

On April 1, 1917, five days before the United States declared war on Germany, Atlantic Towing made Thompson an offer to sell their half-interest in the ship. Thompson agreed and paid Atlantic Towing $428,000 in cash on the 13th, released them from the $60,000 he had advanced them, and received a formal bill of sale.

Thompson applied to the Shipping Board for permission to sell the ship to a citizen of an allied government and change the flag and on April 24 the Board agreed on condition the allied government assured the United States that the vessel "would be confined during the period of the war to essential war trades."

That April, while the ship was still on the bottom, the State of Louisiana decided to assess Thompson all the taxes due on the ship. Thompson argued the ship had no value until it was raised. The state wouldn't budge so Thompson sued. The Louisiana Supreme Court sustained the tax, the Supreme Court bowed out for lack of jurisdiction, and Thompson paid up.

The *Gut Heil* was refloated on December 5, 1917, and towed to New Orleans.

In mid–January 1918 Atlantic Towing filed a salvage claim on the ship and the U.S. marshal seized the vessel. The court permitted it to be dry docked at Atlantic Towing's expense and authorized them to remove salvage gear that was still aboard.

Thompson then went to see Cdr. Edward C. Hammer, the industrial manager at the New Orleans Naval Station, to find out if the yard could put the ship in condition to pass the highest standards of the Bureau Veritas. Hammer said they could and received permission from the Navy Department to do the work. The plans for the necessary repairs were drawn up under Hammer's supervision and on April 29 the ship was drydocked.

On August 7, when the work was about 90 percent complete, the Secretary of the Navy's office informed Thompson that the ship would be requisitioned and on August 12 it was formally taken. By then Thompson had paid $140,000 in various installments for the work. The same day, two naval officers went aboard to assess the value of the consumable stores on board and they arrived at a value of $25,230.05. Thompson submitted a bill to the Navy for the stores but was never paid. Title to the *Gut Heil* was transferred to Commodore Valentine

S. Nelson, representing the Navy Department. At Thompson's request, the ship was renamed *Sara Thompson.*

In February 1918, Deutsche-Amerikanische filed suit claiming they still owned the ship. Thompson and Atlantic Towing countersued together saying the company was "an alien enemy residing in enemy territory" and that unless restoration moved forward the ship would deteriorate. The German company's suit was dismissed.

On October 2, 1918, Secretary of the Navy Josephus Daniels wrote to Thompson stating the value of the ship was determined to be $700,000 and he would pay that amount plus the $25,230.05 for the stores. Thompson replied that $700,000 was too low. After a bureaucratic runaround between Secretary Daniels, Capt. Alexander Halstead, of the Board of Appraisal of Merchant and Private Vessels in New York, who Thompson visited personally, and the Solicitor of the Navy, Thompson filed suit in federal court. No money had yet been paid when he filed suit.

On November 29, 1918, with the war just over, the Navy's solicitor wrote to the Chief of Naval Operations saying that if Thompson refused the original offer he should be paid 75 percent of the offer, which would allow him, if he accepted it, to file suit in the federal Court of Claims to recover the full amount. A check for $700,000 was then delivered to Thompson's agent on December 29. Thompson was not aware of the 75 percent provision then and had told his agent not to sign for any settlement, pending his lawsuit. By the time Thompson got the check in St. Louis, he had heard about the 75 percent provision and wrote to Secretary Daniels offering to refund one-fourth of the money pending a court decision. There was no provision for this so the Secretary told Thompson he would place the offer on file.

The court determined that in August 1918, the market value of tankers was anywhere from $175 to $375 per deadweight ton and the *Gut Heil*'s assessed value was found to be $250 per deadweight ton, which amounted to $1,100,000.

While the case was being adjudicated, John Thompson died and Sara E. Thompson and Marcella Thompson Berkley, of St. Louis, were appointed administers of his estate by the probate court in St. Louis. The heirs continued the suit in federal court to recover for the estate.

The government argued they owned the ship pursuant to an Act of Congress on May 12 1917, subsequent to her ownership by an enemy alien, and that Thompson was entitled only to costs of salvaging the vessel and repairs to it prior to August 12, 1918, the date it was seized. The court did not agree. The government had recognized years before that the ship had been abandoned, refuted interest in the vessel, and sanctioned Thompson's ownership.

In part, the court said:

> It seems well settled that when a vessel is derelict and abandoned in the navigable waters of the United States or anywhere else it belongs to that person who finds it and reduces it to possession. Congress could undoubtedly provide that the proceeds of derelicts and abandoned vessels in the navigable waters of the United States be paid into the Treasury; but no such law has been passed, and until it is the principles of natural law must prevail.
>
> The claim that the vessel was alien property and became the property of the United States by virtue of the resolution of May 12, 1917, is obviously not well taken. The German owner had abandoned the property in the year 1914 and notified the United States authorities that it had done so. The vessel was no longer the property of the German owner, and the United States could not acquire title to property from a person who had no title.
>
> The defendant also contends that if the *Gut Hiel* was not the property of the United States, but was the

> property of the decedent, he has received a sum from the United States largely in excess of just compensation, and that the United States is entitled to a judgement for the excess over just compensation paid for the vessel. We do not think the evidence sustains this contention. The facts show the contrary to be true.

The court awarded Thompson's estate $1,100,000, less the $700,000, and interest on the difference of $400,000 from December 18, 1918, to July 1, 1926, the date of the judgement, which came to $180,733.34. The total judgement was $580,733.34.

The *Sara Thompson* was commissioned on September 17, 1918, at New Orleans under Lt. Cdr. Frederick S. Hayes.

Keresan—4,507 tons, freighter, 1912, 215395, LHPC. Kerr Navigation. Commissioned on September 18 as *Keresan.*

Sixaola—5,017 tons, passenger with refrigerated cargo space, 1911, 212585, LDQJ. United Fruit. Commissioned on September 19 as *Sixaola.*

Herman Frasch—3,803 tons, freighter, 1910, 207246, LBJP. Union Sulphur. Commissioned on September 20 as *Herman Frasch.* Collided with the NOTS freighter *George G. Henry* just after midnight on October 4, 1918, about 150 miles southeast of Nova Scotia and sank in seven minutes. Of 33 on board, 17 were lost. The *Henry* rescued several survivors the next day.

Katharine W. Cullen—703 tons, 3-masted steel schooner barge, 1903, 166529. Cullen Transportation. Purchased on September 23, 1918, and commissioned on September 28 as *Katherine W. Cullen* to transport coal from Norfolk to New York. Sank on December 30, 1918, 15 miles southeast of Boston Light.

Kerkenna—3,621 tons, freighter, 1900, 215394, LHPB. Kerr Navigation. Commissioned on September 28 as *Kerkenna.*

Kermoor—4,795 tons, freighter, 1907, 215398, LHPG. Kerr Navigation. Commissioned on November 1 at Cardiff, Wales, as *Kermoor.*

Kerwood—3,651 tons, freighter, 1911, 215397, LHPF. Kerr Navigation. Commissioned on November 5 as *Kerwood.* Hit a mine and sank March 19, 1919.

Ausable—3,153 tons, freighter, 1901, 213697, LFNR. American Transatlantic. To War Department on February 2, 1918. Commissioned on October 1 as *Ausable.*

Satsuma—4,204 tons, freighter, 1901, 212922, LFBV. New York & Oriental. Commissioned on October 1 as *Satsuma.*

Cushing—6,894 tons, tanker, 1903, 212766, LDVT. Standard Oil of New Jersey. Commissioned on October 2 as *Chinampa.*

Frederick Luckenbach—2,926 tons, freighter, 1888, 110972, KQCJ. Edgar F. Luckenbach. Commissioned on October 5 as *Frederick Luckenbach.*

Carolinian—4,170 tons, freighter, 1906, 212803, LFCM. Garland. Commissioned October 5 as *Carolinian.*

Frieda—3,365 tons, freighter, 1912, 210837, LCRT. Union Sulphur. On August 12, 1918, John Joseph Ciechowski, an American wiper was accidentally shot and killed on board by a "military guard." His mother, Julia, lived at 22 Windermere Ave. in Rockville, Connecticut. Commissioned on October 9 as *Frieda.*

Santa Clara—6,309 tons, freighter, 1913, 211480, LDCQ. Atlantic & Pacific. Commissioned October 12 as *Santa Clara.*

J.M. Guffey—2,520 tons, tanker, 1902, 77519, KRNL. Gulf Refining. Commissioned on October 14 as *J.M. Guffey.*

Kerowlee—3,350 tons, freighter, 1901, 215400, LHPK. Kerr Navigation. Commissioned on October 17 as *Kerowlee*.

Moldegaard—2,852 tons, freighter, 1906, 212643, LDNV. Bull-Insular. Commissioned on October 18 as *Moldegaard*.

Dochra—4,309 tons, freighter, 1906, 212798, LDTJ. Barber & Co. Commissioned on October 21 as *Dochra*.

Hilton—3,102 tons, freighter, 1911, 208880, LCBH. A.H. Bull. Commissioned on November 6 as *Hilton*.

Lewis K. Thurlow—3,173 tons, freighter, 1913, 211738, LDHG. Crowell & Thurlow. Commissioned on November 25 at Barry Docks, Wales, as *L.K. Thurlow*.

Montoso—3,063 tons, freighter, 1911, 208367, LBTC. New York & Porto Rico. Commissioned on December 11 as *Montoso*.

Ships Built Overseas Between 1914 and 1918

Saint Francis—5,102 tons, freighter. North Ireland Ship Building Co, Londonderry, 1914, 212786, LDPV. U.S. Steel Products. Commissioned on June 19 as *St. Francis*.

Tanamo—3,298 tons, passenger. Newcastle-on-Tyne, England, 1914, 214666, LGPR. Sarnia. Commissioned on August 17 as *Tanamo*.

Argonne—5,754 tons, freighter. Kawasaki Dockyards, Kobe, Japan, 1916, 214640, LGPH. Argonne Steamship. Commissioned on October 19 as *Argonne*.

Munwood—3,190 tons, freighter. Greenock, Scotland, 1914, 212924, LFDM. Munson. Commissioned on October 26 as *Munwood*.

American Merchant Ships Converted into Army Troopships on Which Merchant Seamen Served or Were Commissioned in the Navy but Not in NOTS

Ancon—9,332 tons, passenger, 1902, 117125, KRNT, Panama Railroad.

Dakotan—6,660 tons, freighter, 1912, 210753, LCQR, American-Hawaiian.

General G.W. Goethals—4,833 tons, freighter, 1911, 215106, LHDT. Panama Railroad.

General W.C. Gorgas—4,546 tons, freighter, 1902, 215110, LHDV. Panama Railroad.

Havana—6,391 tons, passenger, 1906, 203791, KVST. New York & Cuba. Served as an Army troopship and commissioned on March 14, 1918, as the hospital ship *Comfort*.

Iowan—6,716 tons, freighter, 1914, 212144, LDKR. American-Hawaiian.

Kentuckian—6,582 tons, freighter, 1910, 207489, LBMV. American-Hawaiian.

Santa Cecilia—6,462 tons, freighter, 1913, 211720, LDGW. Nafra.

Santa Rosa—6,415 tons, freighter, 1917, 214725, LGRB. Grace.

Santa Teresa—5,102 tons, passenger, 1918, 216969, LNBP. USSB.

Saratoga—6,391 tons, passenger, 204160, KWDB. New York & Cuba. Served as an Army troopship and commissioned on January 24, 1918, as the hospital ship *Mercy*.

Scranton—2,015 tons, freighter, 1888, 116235, KJFH. Sydney C. McLouth, Port Huron, Michigan.

Troy—3,655 tons, freighter, 145786. Great Lakes Transportation Corp.

Virginian—7,914 tons, freighter, 193269, KSJF. American-Hawaiian.

Yale—3,731 tons, passenger, 204047, KWBR. Delaware River Steamboat Co.

PART SIX

Shutting Up the Hornets

15. The End at Sea

The end, oddly enough, starts at the beginning. To get from their North Sea bases, the subs either had to go southwest through the Straits of Dover or go north and up around Scotland. In contemplating ways of "Shutting up the Hornets in Their Nests," the use of mine fields was immediately proposed in the Admiralty.

Mine warfare was nothing new, in fact more British warships, 44 to be exact, were lost to mines in the Great War—more than from submarine torpedoes—including three battleships in the Dardenelle offensive. But there was an element of mine warfare expressed later by First Lord Sir Eric Geddes in a speech he gave in New York: "Before the war, minelaying was considered unpleasant work for a naval man, an occupation like that of rat-catching, and not attractive." Nevertheless, the idea went ahead out of desperation and in 1915 the Royal Navy tried sealing off the Dover Straits with mines. When that proved to be impractical, the French and Italians advocated a barrage of mines from Santa Maria di Leuca, Italy, to Fano Island using British minelayers and French mines but again that idea was abandoned as impractical. In time, the Dover Strait was guarded by the Dover Patrol.

These developments were watched closely by Adm. Ralph Earle, head of the Navy's Bureau of Ordnance, who thought the idea was worth considering, and he got to work. On April 15, 1917, his team presented a plan "urging that we stop the submarines at their source" using mines.

The General Board felt the idea would work but it could not be implemented without the approval of the Admiralty and the endorsement of Adm. Sims in London, Commander of Naval Forces in Europe. The primary operational aspect was where exactly to put the mines but getting the British to approve the plan would not be easy since they had already dismissed the idea.

On April 16, Secretary of the Navy Daniels cabled Adm. Sims:

> Is it not practicable to blockade German coast efficiently and completely, thus making practically impossible the egress and ingress of submarines? The steps attempted or accomplished in this direction are to be reported at once.

On the 18th, Sims cabled back:

> To absolutely blockade the German and Belgian coast against the entrance and departure of submarines has been found quite unfeasible.

The next day Sims wrote an elaborate letter "amplifying the difficulties and reporting against any such barrier."

The Ordnance Bureau then suggested that "barriers" of mines be laid across the North Sea, the Adriatic, and the Dardanelles. The northern barrier would extend from the mid-eastern coast of Scotland to the Norwegian coast, a distance of about 250 miles and the southern barrier would extend from the southeast coast of England and to a point on the French coast near the Belgian frontier, a distance of about forty miles.

On May 9, 1917, the Bureau sent an enthusiastic memo of their plan to be submitted to the Admiralty, "thoroughly convinced the project was feasible."

On May 11, Secretary Daniels cabled Sims stating that the prevailing opinion was that it would work. Sims replied that the notion had been contemplated by the Admiralty since the beginning of the war and ultimately abandoned as impractical. Sims stated, in part:

> As may well be imagined, this whole subject has been given the most earnest consideration, as it is, of course, realized that if submarines could be kept from coming out, the whole problem would at once be solved. But I cannot too strongly emphasize the fact that during nearly three years of active warfare this whole question had been the most serious subject of consideration by the British Admiralty which had concluded that no barrier can be completely effective.

The Bureau persevered but on May 31 Sims reported that "the British Admiralty consider we can more profitably concentrate on other work." In other words, "Forget it."

But the Bureau kept working and by July 31 they had developed a new type of mine "particularly adapted to deep waters" that a ship did not have to actually hit to explode. A key part of this design was an ingenious electrical device invented by Ralph C. Browne, 17 Beckford St., Salem, Massachusetts, an X-ray technician at Salem Hospital. He worked with a group of officers at the Bureau to adapt his device for mines and on August 17 Sims was notified that a practical mine had been developed, the ingenious Mark 6 antisubmarine mine that could hover deep underwater at any set depth while anchored to the seabed.

An Allied Naval Conference was scheduled for September 14 in London to discuss the idea. The British were still in doubt the plan would work and at that time the American delegates could not guarantee an adequate supply of mines. But the American's presentation had apparently struck a nerve and on October 9 Sims reported the Admiralty apparently had undergone a change of heart and was "thoroughly investigating the question." The Bureau then let contracts for 100,000 mines and on October 20 word was received from the Admiralty approving the project. Secretary Daniels endorsed it on the 29th and President Wilson, already annoyed at the months' long delay, approved it immediately.

The North Sea Mine Barrage

The project was called the Northern Mine Barrage, laid in the North Sea from Norway almost to the Orkney Islands. Adm. Joseph Strauss, in his flagship *San Francisco*, was appointed commander of mining operations in the North Sea. The Navy had no suitable mine carrying or minelaying ships and had to turn to the merchant fleet.

The Admiralty was informed that mine planters were being fitted out and the first mines would be delivered by January 1, 1918. A total of 23 merchant freighters were converted to mine carriers to bring the mines over. Mine Squadron One, Atlantic Fleet, was formed at Hampton

Roads on Wednesday, April 10, 1918, on board the cruiser *San Francisco* under Capt. Reginald R. Belknap. The *San Francisco* and the *Baltimore* were fitted for carrying 350 mines.

The eight ships of the new squadron were coastwise vessels purchased outright from the Eastern Steamship Line, the Southern Pacific Co., and the Old Dominion Steamship Co.

The Squadron operated out of Inverness, Scotland, as Naval Base 18 where they arrived on May 26. The Squadron was designated as the Second Minelaying Squadron in His Britannic Majesty's Grand Fleet. The Squadron also had four seagoing tugs.

Eastern Steamship Lines

Eastern Steamship's 387' passenger liners were gutted and completely rebuilt inside to carry up to 352 mines. They were armed with one 5"/51 gun aft, three 3" anti-aircraft guns. The crew was 20 officers and 346 sailors.

Bunker Hill—4,779 tons, 1907, 204264, KWDT. Commissioned on December 17, 1917, as the minelayer *Aroostook*.

Massachusetts—4,779 tons, 1907, 204012, KWBM. Commissioned on January 7, 1918, as the minelayer *Shawmu*.

Southern Pacific Co.

The Southern Pacific Steamship Co., the "Morgan Line" was a subsidiary of the Southern Pacific Co., a rail carrier, and as such was under the exclusive jurisdiction of the Interstate Commerce Commission. The Shipping Board had no control over the "port-to-port rates, regulations, or practices of that carrier."

The company's freighters were purchased within a month after the Eastern Steamship vessels. The 405-foot coastwise freighters for service between New York, New Orleans, and Galveston. They were armed with one 5"/51 gun aft, two 3" anti-aircraft guns forward, and were fitted to carry up to 860 mines each. The crew was 21 officers and 400 sailors. The ships were known as the Big Four.

El Dia—4,613 tons, 1901, 136916, KRDQ. Commissioned on January 25, 1918, as the mine planter and troop transport *Roanoke* under Capt. Clark D. Stearns. Embarked 250 men for the mine bases and returned 5,500 troops.

El Rio—4,604 tons, 1899, 136731, KPJW. Commissioned on January 25, 1918, as the mine planter and troop transport *Housatonic*.

El Cid—4,608 tons, 1899, 136771, KPMH. Commissioned on March 2, 1918, as the mine planter *Canonicus*.

El Siglo—4,616 tons, 1901, 136951, KRHL. Commissioned on March 2, 1918, as the mine planter *Canandaigua*.

The Old Dominion Steamship Co.

Old Dominion's 375' cargoliners ran between New York and Norfolk and were purchased on December 6, 1917. They were armed with one 5"/51 gun aft, two 3" anti-aircraft guns forward, and were fitted to carry up to 642 mines. The crew had 18 officers and 392 sailors.

Jefferson—3,723 tons, 1899, 77356, KPCM. Commissioned on March 28, 1918, as the mine planter *Quinnebaug*. Planted 6,040 mines.

Hamilton—3,723 tons, 1899, 96439, KNWT. Commissioned on April 9, 1918, as the mine planter *Saranac*. Planted 4,822 mines.

The first mine was laid on June 8, 1918, and the first casualty occurred on July 9 when a sub was disabled and had to return to Germany.

Capt. Belknap estimated that 10 subs had been destroyed up to the middle of October 1918 and he believed Germany was becoming desperate. But when First Lord of the Admiralty Sir Eric Geddes paid a visit to Secretary Daniels, he said a major German U-boat offensive was on the horizon. England was building as many merchant and naval ships as quickly possible and Geddes said he anticipated another two years of war.

Daniels then summoned every "leading shipbuilder" to Washington for a conference. At that time over 200 destroyers were being built or had been contracted for and Daniels had hoped to spur production by urging a round-the-clock, seven-day work schedule. The builders said they would do all they could to speed up destroyer and "anti-submarine craft" production, "if labor could be secured."

While Geddes was on his way home, an astounding event occurred. The Germans recalled all of their submarines, even as hundreds of new boats were coming off the ways.

The German people were tired of the war and the sailors had mutinied against their officers. New technological advances in underwater listening devices, surface and air patrols, and especially the mine barrage were taking a huge toll on the submarines.

A total of 70,263 mines were planted on the surface and at depths from 90 to 240 feet. The operation was completed on October 26, 1918, at a cost of $80 million. *U-156*, one of the famous *Deutschland*-class subs was lost in the Barrage leaving only 21 survivors who reported about conditions in Germany.

The mines were impossible for a submarine to locate or evade and these new weapons had so demoralized the crews that they refused to sail. A mutiny was reported to have occurred on October 29, 1918, throughout the fleet.

During the war the Germans laid 43,000 mines, the British 129,000, and the United States 71,000 in the Northern Mine Barrage.

All told, the Germans lost 62 combat surface ships, 40 U-boats, and 42 naval auxiliaries to mines. The British lost 225 naval auxiliaries and the Allies lost 586 merchant ships to mines.

While the exact number might never be known for certain, the Admiralty's final accounting of U-boats lost to the Northern Mine Barrage was 23. That is out of approximately 200 total boats lost. A German stated, "the barrage caused no small amount of panic in some of the submarine flotillas." The Germans threw in the towel before the planned Adriatic and Aegean Sea barrages were started.

Adm. Strauss expressed the opinion that the Northern Mine Barrage could have been built in 1917. The United States possessed "antenna mines" that would have been suitable but it would have taken two or three as many to do the same job as the Mark 6 and the mine planters were not then available but could have been brought into service. Strauss and Secretary Daniels both admitted after the war that not starting earlier was "the greatest naval error of the war." Daniels later stated:

If the British had erected it early in the war, and put a similar effective barrier across the Straits of Dover and Otranto, the Germans would have been so restricted that widespread U-boat warfare, with its terrible destruction of life and shipping, would have been impossible.

Two Eyewitnesses to the End

Two Americans from the Navy transport *Ticonderoga* were being held as prisoners aboard Kapitäteutnant Adolf Franz's *U-152*: Lt. (jg) Frank Muller, the executive officer, and Junius Harris Fulcher, the 1st assistant engineer.

At this time peace negotiations had commenced and Germany was in a state of revolution. The Kaiser's monarch was being replaced with the new Weimar Republic.

On October 11, when the *U-152* was about 1,000 miles off Newfoundland, Franz received a radio message: "Engage men of war only. The merchant war has ended." Franz told his prisoners, "This is the first act of our new government." But the next day Franz stopped the Norwegian bark *Stifinder*, en route from New York to Australia, looted all the stores, and sank the boat. On the 15th he sank a steamer. More trouble was fomenting in the country and on the 20th another radio message was received: "All submarines return to Kiel."

The German High Seas Fleet had been ordered out to face England's Grand Fleet but on October 28 a mutiny started in the fleet. When Franz got the news, he and his crew had a choice to make. Submarine crews loyal to the Kaiser were setting a course for Sweden to be interned. Franz received another message from the commandant at Kiel ordering him to return there. Franz then called a meeting of all 80 hands to take a vote—Sweden or Germany? Ten said Kiel, 70 said Sweden. Many radio messages were being received "concerning mutinies and chaos ashore in Germany." Franz told the Americans, "You two gentlemen are now free. You are no longer prisoners of war. I don't know whether we shall finally reach Kiel, or Sweden, or Denmark, but rest assured that in any case you will be safe. I shall protect you at all hazards. Whatever we do, you will be well off." The sub's crew confided to Muller and Fulcher, who they regarded as friends, that they were upset that Germany's surface fleet had contributed almost nothing to the war and they were taking all the risks and doing all the work.

At 4 a.m. on November 12, the crew learned an Armistice had been signed and Franz headed for Kiel. Muller and Fulcher later reported the crew seemed happy at the news. One German officer admitted they had been fighting a losing battle since the United States entered the war. The sub entered the Skagerack that evening and tied up next to the *U-53* at Kiel at 5 p.m. on November 15 along with five other boats.

The Americans were taken to the sub tender *Kronprinz Heinrich* where they had a compartment to sleep in but they continued taking their meals on the *U-152*. The executive officer of the *Heinrich* formally released them as prisoners but advised them not to go ashore as there was considerable anti–American feelings since America was being blamed for the collapse of Germany by entering the war. They elected to remain aboard.

The next morning, the 16th, they were told by an officer they would be leaving for Copenhagen at 10 a.m. on the 17th but that didn't happen. Apparently all decisions were being made by the "soldiers' and workingmen's representatives aboard" the tender.

Franz, the first officer, several warrant officers, the chief engineer, and about half the crew had left the *U-152* at Kiel and had either gone home or got jobs on merchant ships at Hamburg.

The terms of the Armistice required ships to be surrendered so the crews who remained at Kiel decided to take 28 submarines and transports to Harwich, England. Lt. Wilhelm Wille was elected captain of the *U-152* for the voyage. Fulcher and Muller were to go on a transport but the crew of the *152* wanted them to stay aboard to insure their safety. At 2 p.m. on the 20th they got underway from Kiel in two columns, one led by the *U-152*, the other by the *U-155*, the ex–*Deutschland*, and arrived at Brunsbuttel the next afternoon. They arrived at Heligoland at 11 p.m. that night and at 7:30 the next morning, November 22, left for Harwich, arriving there without incident on Sunday, the 24th. Eventually, 114 U-boats were surrendered to the British.

The soldiers and sailors on the *Kronprinz Heinrich*, mutineers known as the "soldation" were enthusiastic about the new republic. One said, Take all our submarines and battleships and battle cruisers. Sink them in the Atlantic. We don't want them anymore. What we do want from now on are ploughs and picks and shovels, to get back to work again." Not everyone felt that way, especially a highly-decorated Army corporal named Adolphus Hitler.

16. American Ship Losses

This section lists all of the American vessels and the merchant seamen on them known or presumed to have been lost from contact with the enemy after the United States entered the war.

The number of vessels lost from normal, peacetime causes during the war years was no doubt higher than average due to steaming in blacked out conditions in groups, a much larger number of ships at sea, blacked out aids to navigation, and being required to maintain a schedule or course regardless of weather. It is impossible now to differentiate any given non enemy-caused loss from the incidental war conditions, the ordinary perils of the sea, an unsafe ship, or the state of navigation at the time.

The identification of specific U-boats presents a problem since there are discrepancies between what survivors observed or were told, contemporary historical accounts, and postwar investigation by historians. Many U-boats had no numbers painted on their conning towers and some commanders were known to paint out one number and replace it with another number. Survivors of sinkings who were taken aboard a sub might never have known the identity of the boat and in at least one instance the sailors wore hats with two different numbers on them.

In the case of mines, it was sometimes unknown whose mine it was since mines were known to drift from their original location due to storms. On rare occasions someone actually saw the mine and was able to identify the owner.

The Navy Department designated 40° West Longitude as roughly the dividing line between North American and European operations. At 40° North Latitude, 40° West Longitude is about mid-way between New York and Lisbon. The sections are arranged in the order of East from 40° Longitude to France and the South Atlantic; The British Isles and English Channel; The Mediterranean; West from 40° Longitude to Atlantic Coast and South to Caribbean Waters; and The West Coast and Pacific.

East from 40° Longitude to France and the South Atlantic

The French established lookout posts and signaling stations all along their coast with special telegraph lines to district headquarters and each station was equipped with a powerful telescope.

The general procedure when any vessel sighted a submarine was to immediately broadcast at full power the radio message A.L.L.O., the French word for "hello," three times or S.O.S. with along any pertinent details.

Percy Birdsall—In early 1917, Capt. Conrad Emil Petersen, 31, took the Foreign & Domestic Transportation Co.'s 4-masted auxiliary schooner to Sabine, Texas, to load sulphur for Bordeaux, possibly under charter to Edward L. Holmes. Petersen had a crew of nine and his wife, Agnes, 24, was aboard.

The cargo was delivered without incident and they headed home for New York in ballast. On April 22, when they were about 25 miles south of Cordouan Light in the Bay of Biscay, the *U-21* surfaced and commenced firing at them. Everyone got off into boats while the sub was still firing. Seven out of the ten shots fired hit the schooner. The survivors were picked up about 90 minutes later when a British patrol boat arrived and they were landed at Bordeaux. Apparently the only casualty was the engineer, Albert Day, 38, who suffered from "exposure." On May 1, all left Bordeaux on the steamer *Chicago* and arrived at New York on the 13th. The Petersens lived at 916 Moody St. in Lowell, Massachusetts.

Frances M—Charles V. Minnot, Jr.'s 4-masted schooner, of Philadelphia, was en route from New York to Oran, Algeria, with case oil. They were stopped in the approaches to Gibraltar on May 18, 1917, by *U-47* and burned. There were no casualties except one sailor from the sub.

Harwood Palmer—France & Canada Steamship's big 5-masted schooner was sunk by gunfire from *UC-21* in the Bay of Biscay on May 23, 1917, about six miles southwest of Le Blanche Island off Basse-Loire, France. The crew of 15 was taken aboard the French naval vessel *Cobra*. There were no casualties.

Barbara—Archibald M. Ostrum's schooner left Port Arthur, Texas, for La Spezia, Italy, with petroleum products. The schooner was burned on May 24, 1917, by *U-47* about 90 miles off Gibraltar. There were no casualties. Ostrum was a loan clerk with the Trust Company of America in New York.

Magnus Manson—Capt. George Torrey took the McIntyre Lumber & Export Co.'s 5-masted schooner out of Pensacola, Florida, in late March 1917 with lumber for Genoa. On May 25 the schooner was stopped, burned, and shelled by *U-47* about 50 miles southwest of Cape Vincent, Portugal. There were no casualties in the crew of 12.

McIntyre Lumber purchased the vessel in May 1916 for just over $100,000 from the Benedict-Manson Marine Co., about $75,000 more than Benedict paid to have it built at Bath, Maine, in 1904. McIntyre Lumber normally chartered the ship to carry coal from Norfolk to South America and pick up a return cargo of log wood at Jamaica.

John D. Archbold—In March 1917 Capt. Hans B. Thompson's armed Standard Oil of New Jersey tanker loaded 2,500,000 gallons of gas for the French army and departed New York on May 20. The cargo was discharged at Rouen and en route home met the *UC-48* off Ushant Light, Quessant Island, France. John Wilson, of Astoria, New York, was chief engineer. The Armed Guard commander was Chief Boatswain's Mate Timothy O'Donnell, of New York. Capt. Thompson stated:

> I received at Brest, June 16, about 7 a.m. my last instructions to proceed to New York. the instructions were to steer a course from Bertheaulme and I left Bertheaulme about 10:40 a.m. At about 11:30 a fog set in, moderate wind and sea. About 3 p.m. we observed a boat about five points on starboard bow about

> seven miles away in a westerly direction. It looked more like a fishing trawler as it had a sail set, but we kept watching it all the time and suddenly it submerged at 4:16 p.m.
>
> I changed my course to south and making about 11 miles an hour and keeping a zigzag course and at about 4:46 p.m. we were struck by a torpedo that we saw coming through the water about 45 degrees on the starboard quarter, entering the engine room with a terrific explosion, wrecking the engine room and starboard lifeboat and killing three men in the engine room.
>
> We did not see any submarine or periscope at the time the torpedo was fired and my belief is that we were not torpedoed by the same submarine we saw a half hour before as he was too far away and the first submarine had a straight bow and the one that made the attack had a round bow.
>
> We stood by the ship a few minutes after the explosion to see if we could see the submarine. As we could not see anything I gave orders to go to the lifeboats quickly as I was afraid the submarine would fire another torpedo.
>
> The three lifeboats and the crews got away from the ship in good order. The third officer took charge of No. 3 boat, first officer took charge of No. 2 boat and I myself went into No. 2 boat with the armed guard and a few members of the crew.
>
> Lifeboats No. 1 and No. 3 were rowed away in an easterly direction while my boat stood by the wreck for about 20 minutes at a safe distance. The submarine came to the surface and started to shell the wreck until it sank at about 5:50 p.m.
>
> Just about the time of the sinking we saw a steamship coming up towards us from a westerly direction. About half an hour later it passed close to the mate's boat, asking him if he would come on board. The first mate refused, saying that he had better get away quickly as the submarine was not far away.
>
> The captain of the steamer said that he would send a wireless and that the patrol boats would come pick us up.
>
> At about 9 o'clock we saw a submarine about two miles away in a southerly direction, keeping up with us.
>
> At about 10:30 p.m. we saw a patrol boat about three miles away to the northeast. We started to burn distress signal as he observed and about 11 p.m. we were picked up by the French patrol boat Engangennte. As we were climbing over the rails of the patrol boat the latter was firing at the submarine on the port side and had hardly time to stop and pick us up.

The *Archbold* was the first tanker built for Standard Oil of New Jersey and the first ship covered by a government war risk insurance policy. The ship was initially valued at $770,000 and the policy was for $150,000. It was later reinsured prior to the loss for a total value of $2,200,000. It was the largest hull loss paid by the Bureau of War Risk Insurance. There were 12 Americans in the crew of 44. The casualties were:

J.M. Johanesen, 1st engineer, American, wife Jeanne, 157 Bergen Ave., Jersey City. Died in the engine room.
J.R. Martinez, oiler, Spanish. Suffered from shock.
Lorenza Sosa, fireman, Spanish, Villa Hermosa, Spain. Died in the engine room.
Domingo Lago, wiper, Spanish, Coruña, Spain. Died in the engine room.

Two unidentified sailors were reported drowned in the engine room.

Childe Harold—Benedict-Mason Marine Co.'s unarmed schooner was possibly chartered by William M. Mills and partners. In May 1917, Capt. Thomas Byrne and a crew of 8 took aboard 4,000 barrels of refined oil at New York for France. On June 21 the schooner was sunk by gunfire from *UC-17* off Ushant Light, Quessant Island, France. A naval report stated:

> At daybreak an object reported by the mate was believed to be the square sails of a ship hull down. The glasses showed it to be the superstructure of a submarine one mile away. The submarine opened fire and the shells ripped through the schooner's hull and sails. The master ordered the crew into the motor launch

and lowered away. This was plainly visible to the submarine, which had come much nearer. She continued to fire, however, evidently at the boat in the water, which escaped being hit. The captain and his crew were ordered aboard the submarine and noticed that the cap ribbons of the sailors bore the letters "*U-10*" and "*U-17*." Two of the officers spoke excellent English, and Captain Byrne, of the schooner informed them that the firing was totally unnecessary and that he had not expected such dastardly treatment from any white men.

The conversation developed the fact that the commander of the submarine knew the date of sailing of the *Childe Harold* from an American port and was on the lookout for her. He had also received information of the departure of the four-master *Alicia B. Crosby* and the three-master *A.V. Sherman* on the same date, all three vessels having passed out to sea together. He wanted to know where the other schooners were. According to his schedule they should have shown up by this time.

The crew was rescued and taken to Montevideo.

Galena—Galena–Signal Oil Co.'s 3-masted bark was en route from New York to Rouen, France, on June 25, 1917, when they were stopped and sunk by *UC-17* about 70 miles southwest of Ushant Light, Quessant Island, France. There were no casualties in the crew of 15.

Orleans—Oriental Navigation's armed freighter was torpedoed by *UC-71* southwest of Belle-Ile, France, on July 3, 1917, and sank in 10 minutes. The crew of the *Sultana* picked up 45 survivors and landed them at Brest the next day. They returned home on a British steamer. The casualties, all presumed drowned, were:

A.B. Tamarantes, seaman, Greek, mother Basilici, Dimitsaena, Greece.
T. Cleary, fireman, Irish, father Thomas, Kildare, Ireland.
A. Murillo, coal passer, Chilean, father in Talcuano, Chile.
J.A. Nogot, messman, Filipino, brother Harry Delino, 47 Sand St., Brooklyn.

Massapequa—Capt. Alex H. Strumm's New York & Porto Rico freighter was under charter to the French government when they left New York on June 18, 1917, with general merchandise. On July 7 the ship was sunk by gunfire from *UC-72* in the Bay of Biscay. The entire crew of 32 landed on the small island of Sein, 28 miles southwest of Brest on the 9th. In the crew were ten American citizens. The French paid $940,000 for the loss.

Mary W. Bowen—Capt. Stephen McDonough, his mate William Lowe, of Winterport, and nine seamen left New York on the 5-masted schooner with refined oil for Le Havre. On July 8 the boat was sunk by *UC-72* in the Bay of Biscay.

Joseph Bowen, of Fall River, Massachusetts, was in the coal business and lived with his wife Mary and their sons Whitney, Harold, and Edward at 168 Belmont St.

Kansan—American-Hawaiian's 490' armed freighter *Kansan* was chartered by the French & Canada Steamship Co. to carry $3,000,000 in goods consigned to the National War Work Bureau of the YMCA in France. The cargo included 5 tons of milk chocolate from Dorchester, Massachusetts; 200,000 letterhead sheets and 100,000 envelopes from Springfield, Massachusetts; 10,000 pounds of sugar; 600 one-dozen packages of biscuits; a truck, some large tents; 500 folding chairs; 10 Bilhorn portable folding organs; and 35 cases of athletic equipment worth $6,500.

Capt. Enoch A. Forsyth left New York on June 28, 1917, with a crew of 50 and was expected in France on July 8 or 9. On July 10 the ship was sunk without warning by *U-69* in the Bay of Biscay 3 miles east of Kerdonis Point, Belle-Ile, and sank in six minutes with the loss of four:

Jeremiah M. Murphy, 1st engineer, American, mother Mary, 217 Seventh Ave., San Francisco.

Florentine Aguirre, oiler, English, mother Camila, 13 Gibson St., Liverpool.

Alexander P. Kua, fireman, Hawaiian, father, 1501 Beretania St., Honolulu.

Charlie Hanan (aka Chung Mun Tin), cook, Chinese, mother Chung Ching Shi, Hong Shan, China.

John Twohy—Ship broker Albert D. Cummins, of Haddonfield, New Jersey, owned the big 4-masted schooner. On June 18, 1917, Capt. John B. Morehouse left Philadelphia for Algiers. At 6 a.m. on July 21 they were stopped off St. Maria, Azores, by *U-155*. All nine crew got into lifeboats and bombs were placed aboard. The sub towed the lifeboats for a few miles toward shore but it took the survivors a few more days to get to St. Maria. There were no casualties. Three of the crew were Americans: chief mate E.H. Ashley, of Massachusetts; 2nd mate Albert White, of Philadelphia; and Benjamin F. Jones, the cook, of Berlin, New Jersey. All arrived home on August 6.

Augustus Welt—Reed, Welt & Co. owned the 4-masted schooner homeported at Boston. Capt. Hugh Quinlan. The boat was sunk on July 21, 1917, by *UC-17* in the Bay of Biscay en route from New York for Le Havre with refined oil. There were no casualties among the 11 on board. Capt. Sydney G. Hupper was also reported aboard.

Campana—On August 6, 1917, Capt. Alfred Oliver's 335' Standard Oil of New Jersey tanker met Kapitänleutnant Victor Dieckmann's *U-61* about 143 miles west of Ile de Re in the Bay of Biscay. Chief Gunner's Mate James Delaney commenced firing. His Armed Guard crew had 180 shells. After three hours, Capt. Oliver wanted to stop to avoid casualties among the 53 people on board since he could clearly see they were out-gunned, although the sub had been hit three or four times. Delaney objected, but after another hour he ran out of ammunition, fires were burning below decks, and the ship was sinking, Capt. Oliver then had no choice but to heave to.

Dieckmann came alongside and took Capt. Oliver, Delaney, Fred S. Jacob, S1c; Charles Levan Kline, GM3c; William A. Miller, S1c; and Ray Roop, BM2c, as prisoners. The rest of the tanker's crew and eight Armed Guard sailors got into two boats. The French naval vessel *Audacieuse* arrived shortly after and the merchant seamen were landed at La Pallice. The Navymen were taken to Bordeaux.

Armed Guard investigators said there was low visibility, the tanker was slow at 10 knots, the sub presented a much smaller target, the sub's guns outranged the tanker's two 3" guns, and Dieckman had 350 shells. After this, an effort was made to provide ships with more ammunition.

Those taken off were the first American prisoners captured during the war. They went to Heligoland Island, from there to Wilhelmshaven, and after four days to the large prison camp at Brandenburg where conditions were very bad. All survived and left Germany on December 8, 1918. Capt. Oliver lived at 225 E. 123rd St. in New York.

This engagement would prove to be the longest gun battle ever fought between a merchant ship and a submarine.

Christiane—The Bennett Trading Co., of Portland, Maine, bought the Finnish bark *Christiane* and obtained American registry on March 17, 1917. On August 7, 1917, the unarmed bark was homeward bound from London under Capt. Cyrus M. Crooks when they were stopped by *U-155* about 200 miles east of the Azores. Kapitänleutnant Karl Meusel invited Capt. Crooks and three his crew aboard the sub "in excellent English" and they enjoyed a nice lunch with wine while Meusel told his guests how much he hated to sink American ships and the Germans looted the bark. There were no casualties and the survivors arrived home on an Italian liner on August 27. Meusel gave Capt. Crooks a receipt but put an "x" in place of the U-boat's number.

Joseph Cudahy—In August 1918, Capt. Thomas A. Hillgrove's armed American Italian Commercial Co. tanker was en route from France to New York in ballast under War Department operation. At 12:30 a.m. on the 17th they were torpedoed by *U-90* about 700 miles from England. Chief steward Harry Wilson drowned and ordinary seaman Evar William Soderstrom died after contracting tuberculosis from exposure in a lifeboat. He was survived by his stepmother Alva Soderstrom, 812 Jefferson St., Jamestown, New York.

Carl F. Cressy—On August 22, 1917, France & Canada's 4-masted schooner was about 180 miles northwest of Cape Finisterre, Spain, bound from La Pallice to Newport News in heavy weather when the *U-93* spotted the boat and commenced firing. The sub continued firing all night as the opportunity arose. The next day the schooner was abandoned and sank that afternoon. The crew of seven arrived at La Rochelle on September 1.

William H. Clifford—Atlantic Shipping Co.'s 4-masted schooner from Bath, Maine, was en route home from Le Havre on September 8, 1917, when they were met by *UB-50* off Nantes, France. There were no casualties.

Ann J. Trainer—Ernest L. Wasson's unarmed 3-masted schooner, of Calais, Maine, under Capt. Arthur Morisson was en route from New York to Le Havre with case oil. On September 16, 1917. the schooner was sunk by *UC-21* about 30 miles off Ushant Island, France. There were no casualties in the crew of seven.

Henry Lippitt—Adrien E. Cooper's 4-masted schooner from Philadelphia was possibly under charter to Henry Talmadge & Co. The boat was sunk on September 23, 1917, by *U-60* off Brest, France. There were no casualties.

Paolina—Robert Babcock's 3-masted bark under Capt. Hans Mortensen with a crew of 15 was sunk on September 25, 1917, by *U-65* about 140 miles off Brest, France, with a cargo of case oil and steel wire, possibly under charter to Whitney & Boden. The boat was looted of stores then bombs were placed aboard. The entire crew was taken aboard a French destroyer.

Lewis Luckenbach—Capt. James B. Watkins' freighter sailed for France in a 22-ship convoy. On October 11 they encountered *U-53* about 10 miles west of La Vierge Lighthouse on the northwest coast of France, apparently in heavy seas. There were 52 on board. One Armed Guard sailor, J.E. McNeil, S2c, drowned along with crew members when a lifeboat smashed against the side of the ship:

James Watkins, master, American, wife Jennie, 474 Fifth St., New York.
William J. Reichmanberg, 1st mate, American, sister Hannah, Kresma, Russia.
George Benjamin Hassell, quartermaster, British, mother Katherine, St. John Saba, Dutch West Indies.
Irving J. Nissenson, wireless operator, American, father Jacob, 14 E. 106th St., New York.
Wallace Peterson, boatswain, Dutch, mother Judith Ann, St. John Saba, Dutch West Indies.
Lennart Anderson, seaman, Finnish, father Justus, Hitas, Finland.

Laurence S. McCants, 1st engineer, American, parents Paul W. and Susan McCants, Round, South Carolina.

John Erickson, 3rd engineer, American, father Gustaf, 49 Cottage St., New Bedford, Mass.

St. Helens—The Tiltro, Ltd., tanker from New York to Algiers was sunk on October 15, 1917, by *U-22* off Cabo Vilan, Spain, with the loss of 24 of the 26 on board. The American casualties were Frank Golden, oiler, mother Helen, of San Francisco; Bridges Carroll, fireman, mother Fannie, Jacksonville, Florida; and James Roberts, cook.

Antilles—The War Department chartered Southern Pacific's passenger liner *Antilles* for operation by Matson. On the morning of October 15, 1917, the *Antilles*, troopship *Henderson*, and the storeship *Willehad* were anchored in Quiberon Bay waiting for the *City of Atlanta* to arrive from Saint-Nazaire so they could all leave for New York. *Corsair*, *Aphrodite*, *Wakiva*, *Alcedo*, and *Kanawha II* were scheduled as escorts. Word was received that the *City of Atlanta* wouldn't leave until the next morning and then the *Wakiva* developed boiler trouble.

The *Aphrodite* was designated as escort for the *City of Atlanta* and the convoy left at 3:52 p.m. and commenced a zigzag course. The weather worsened and the next morning the *Kanawha II* began taking on water, fell behind, and then the *Willehead* straggled and the convoy slowed to nine knots. At 5 p.m. the *Kanawha II* was forced to turn back.

The morning of the 17th was "thick until about daylight when it cleared with a moderate sea from the southwestward." At about 6:43 a.m. the *Henderson* hoisted a signal and the *Antilles* was observed to sheer away to starboard. The *Corsair* was 1,800 yards ahead of the *Antilles* and turned about at full speed toward the *Antilles*, which seemed to be settling by the stern. At 6:52 the ship "upended" and sank in five minutes.

The *Henderson* turned away to starboard and the *Willehead* to port while escorts searched for survivors. No one saw any sign of the submarine. *Alcedo* and *Corsair* rescued 118 survivors. One Navy sailor, Morris Miller, of New York, was found four days later by the *Alcedo*.

It was later determined the *Antilles* was torpedoed by *U-62*. Admiral Wilson stated:

> There was nothing unusual about the torpedoing and sinking of this vessel, except that the inadequacy of the available escort at our disposal was shown by the fact that three ships were at sea escorted through force of circumstances by only two yachts.

Of 234 aboard, four Armed Guard sailors, 16 Army, 52 merchant seamen, and a civilian ambulance driver were lost. *Alcedo* rescued 117, *Corsair* 50.

The sinking of the *Antilles* was the catalyst for the Navy taking over operation of all Army transports. Because a merchant ship has a continually changing crew, it was felt that sailors assigned permanently to a ship, particularly in the Engine Department, would be far more efficient in keeping the ship's speed up. The maritime unions naturally said it was a convenient way to save on labor costs. In December, the Navy General Board asked Congress for 80,000 additional sailors and on January 9, 1918, the Naval Overseas Transportation Service came into being.

Jennie E. Righter—Shepherd & Morse Lumber Co.'s 3-masted schooner left New York on September 12 under Capt. Stephen Black for Le Havre carrying 8,500 barrels of lubricating oil shipped by the Vacuum Oil Co. The boat was sunk on October 16, 1917, by gunfire from *U-22* off Cabo Vilan, Spain. The crew of 9 was landed at Majia. The Americans aboard were the mate, Dennis O'Neill, of Boston; the cook, Charles E. Blatchford, of Gloucester, Massachusetts, and seaman Albert T. Young, of East Boston, and naturalized citizens John Grant, of Nova Scotia, and Alfred J. Rowe, of England. The boat was valued at $60,000.

Finland—The big liner was under War Department operation when they left Quiberon Bay with two other ships, four destroyers, and three yachts on the afternoon of October 25, 1917. When they were about 150 miles out the next day a torpedo from *U-93* struck the starboard side just forward of the bridge. The escort vessels had fallen somewhat behind their normal position just forward of the beam. Some of the lifeboats swamped but the escorts rescued all but five merchant sailors and three Armed Guard sailors. The ship proceeded to Brest and was repaired.

Merchant sailors lost were M. Cardoza, fireman; Jose Ceuvas, messboy from Havana; Charles H. Maxwell (colored) of Concord, North Carolina; W.F. Phillips, a waiter from New Orleans; and J. Santelo, a barber. Naval gun crew members lost were James W. Henry, of Harrison, New Jersey, and Newton R. Head, of Cleveland, Georgia, and Porter Hilton, from Toccoa, Georgia, was reported missing. Army private Lester Hickey, of Chicago, was also reported lost.

D.N. Luckenbach—The freighter left New York on October 13, 1917, with 29 people aboard and was sunk on October 27 by *U-93* in the Bay of Biscay in rough weather with wind and snow. The crew of a Danish ship rescued the survivors two days later. Five engineers died from the force of the explosion:

Fredrik Trygve Olsen, 2nd engineer, Norwegian, brother Henry, Bjorndalayden, Fredrikstad, Norway.
Albert Pennsa, 3rd engineer, American, mother Marcella, 863 E. 35th St., Brooklyn.
Dennis Evangelacos, oiler, Greek, brother John, 87 Cherry St., New York.
Vallianos D. Panagatos (aka Peter Vallianos), oiler, Greek, father Dionisios Panagatos, Anninata, Greece.
Demvis Chronia, fireman, Greek.

Actaeon—Capt. W.J. Johnson's Shipping Board steamer left New York for Bordeaux with general cargo, a crew of 58, including 31 Americans, and 19 Armed Guard sailors. The ship was sunk on November 24, 1917, by *U-84* in the Bay of Biscay about 150 miles northwest of Cape Finisterre near Coruña, Spain. One lifeboat with 19 Armed Guard sailors and 6 merchant crew became separated from the other boats.

As they rowed and drifted, four survivors died of exposure: John Fitzgerald, an Irish baker, and Armed Guard sailors Seamen 2c Chester Theodore Berry and Stephen John Downey. Thomas John Moran died later in a hospital.

Eleven days later, 21 survivors reached Port Camarinas, Spain, a town about 43 miles southeast of Coruña. Among the crew were 26 Americans and 5 naturalized citizens.

The *Actaeon* was the seized German ship *Adamstrum* and the first seized ship lost in the war.

Harry Luckenbach—The 300' freighter under Capt. Fred S. Jones and 31 crew left Philadelphia on October 17, 1917, for France. On arrival the unarmed ship was requisitioned by Gen. Pershing for use as an Army supply ship.

In January 1918 the ship was assigned to a 15-ship, two-column coastwise convoy from Brest to Quiberon. They left on the afternoon of January 5 escorted by *Wanderer, Kanawha II*, and two small French vessels. The *Luckenbach* was at the head of the right-hand column and the *Henri Le Cour* was on the left. Their speed was 8½ knots and all ships were blacked out except for a dimly-lit stern light. That night was very dark with a rough sea and "Penmarch Pete" had been active of late.

At about 11:10 p.m., a lookout on Lt. Cdr. Wilson's *Wanderer* sighted a "suspicious object" off their starboard beam. The escort flashed a recognition signal, which was not returned, and the object disappeared. They were about 7 or 8 miles off Penmarch.

Twenty minutes later a lookout on the *Henri Le Cour* saw a flash of something in the water and two seconds later an explosion occurred at No. 4 hold. The ship sank without being noticed by any other ship or escort in the convoy and the crew managed to get into three lifeboats. Two reached the village of Penhors and the third containing the master, was picked up at 2:30 a.m. the next morning by the French destroyer *Temeraire*.

At 12:10 a.m. the *Wanderer* heard a loud explosion off their port bow and proceeded at full speed in that direction. They soon found the *Luckenbach* and several sailors in the water. Despite the obvious danger, *Wanderer* turned on searchlights and proceeded to rescue 26 survivors. All were hypothermic and some injured. Two other ships, the *Dagney* and *Kanaris* were also torpedoed later that morning. The sub was the *U-93*.

At that time it was the policy for coastwise convoys to sail at night in order to be less visible to U-boats, but the loss of these four ships prompted a change in thinking. Many already had advocated for a daytime sailing believing the escorts had half a chance of seeing the sub, which more often than not attacked on the surface. From then on coastwise convoys sailed by day. The casualties from the *Luckenbach* were:

George L. Breeden, 1st mate, American, 1227 N. Bond St., Baltimore. Seriously injured.
Peter A. Platteel, 3rd mate, American, 66 Dracutt St., Dorchester, Massachusetts.
Alfred B. Robinson, wireless operator, English, Three Mile Bay, New York.
Harry Carter, seaman, British, Grand Command, British West Indies.
Patrick J. Hunt, chief engineer, Canadian, 1287 E. 8th St., Brooklyn. Injured.
Harry Crouch, oiler, American, 46 W. 53rd St., New York. Injured.
Domingo Paiz, fireman, Venezuelan, 133 Columbia St., Brooklyn.
Jose A. Simoes, fireman, Portuguese, father Florendo, 511 Greenwich St., New York.
Nikolas Zakako, fireman, Greek, 57 Van Brunt St., Brooklyn.
James Townsend, steward, American, 23 Sumpter St., Brooklyn. Injured.

Julia Frances—Victor R. Moderno's 3-masted schooner was sunk on January 27, 1918, by gunfire from *U-152* about 100 miles off Lisbon, Portugal. There were no casualties.

A.E. Whyland—A. Frank Clark's schooner, of New Bedford, was sunk on March 3, 1918, by *U-152* about 55 miles off Tenerife, Canary Islands. There were no casualties in the 11 on board.

Florence H—On April 17, 1918, Capt. Fred J. Butterfield's Shipping Board freighter *Florence H* arrived at Quiberon Bay, France, with 2,200 tons of smokeless powder, a crew of 56 merchant seamen and 23 Navy Armed Guard sailors. They were directed to an anchorage along with other ships and escorts and wait to join another convoy going south.

At about 10:45 p.m. a searchlight was seen flashing from the bridge of the *Florence H* and at 11 p.m. the ship suddenly burst into flame and exploded. Flames reached a hundred feet and the water was covered with burning and smoldering powder containers. In about ten minutes, the hull split apart and the ship sank.

The cries of survivors could be heard above the noise. Destroyer *Stewart*, Lt. Cdr. Francis A. La Roche's escort *Sultana*, Lt. Cdr. Pierre L. Wilson's *Wanderer*, and Lt. Cdr. Millington B. McComb's *Christabel* got underway to search for survivors. Every ship in the vicinity, including those of another convoy that just arrived, lowered boats to help search but the debris was so thick it was impossible to row and the captains of the wood-hulled converted yachts quickly realized their boats would catch fire if they ventured too far into the burning debris.

Lt. Cdr. Harry J. Abbett's destroyer *Whipple* and Lt. Cdr. James G. Ware's *Truxton* went in to open channels for the other boats and threw lines to survivors. Some were seen clinging to unexploded boxes of powder. At one point the *Stewart* got stuck but they picked up three survivors from a lifeboat that was on fire. Several *Stewart* sailors jumped into the water to rescue survivors but in the end only 34 were picked up and several of those were badly burned.

Chief Engineer John B. Watson stated, "She just burned up and melted in about 30 minutes." Of 77 aboard, 32 were rescued by the Navy and 42 were lost or injured. Sixteen out of 23 Armed Guard sailors were lost. The American merchant casualties were:

Fred J. Butterfield, master, wife Abbie, 7 Eleventh Ave., Haverhill, Massachusetts. Died in the explosion.
John S. Beans, 1st mate, wife Alexandra, 67 Linien Ave., Brooklyn. Died in the explosion.
Russell Overington, 2nd mate, wife Marie, 618 W. 148th St., New York. Died in the explosion.
Edward L. Cudahy, 3rd mate, mother Catherine, Multnomah Hotel, Portland, Oregon. Died in the explosion.
Joseph Andrew Washko, reserve officer, sister Mary Wasnak. Died in the explosion.
Leonard Brown Geldart, seaman, father Howard, Beacon St., Everett, Massachusetts. Died in the explosion.
Percy D. West, seaman, Edgartown, Massachusetts. Lost the use of one hand.
John Dawson Blackley, 2nd engineer, 58 Jane St., Brooklyn. Seriously burned.
A.G. Ford, 3rd engineer. Seriously burned.
Joseph L. Lancreaux, Jr., reserve engineer, mother Lilly, 6724 Champlain St., Chicago. Died in the explosion.
Martin L. Collins, cadet engineer, mother Mary Goss, 213 Colfax Ave., Benton Harbor, Michigan. Died from burns.
Carl I. Goodwin, watertender, brother Aubrey, 190 Warwick Rd., Melrose, Massachusetts. Died in the explosion.
John A. Keating, coal passer, 114 Granite St., Malden, Massachusetts. Injured. {end list}

The Armed Guard members who died from explosion: Russell William Clark, S2c; Edward Patrick Connaughton, GM3c; Howard Curtis, S2c; Louis Goetz, CGM; Charles George Gunderson, S2c; Ernest K. Jobes, S2c; Edward Klaphs, S2c; William E. Lennon, S2c; Joseph F. McVeigh, S2c; Bryan B. Long, EN3c; Walter E. Mers, Seaman; Ernest C. Myers, BM1c; George T. Shea, Seaman; Charley A. Tarr, S2c; Horace C. Woodworth, EN3c; Michael Younger, S2c.

John G. McCullough—On May 18, 1918, the 6-knot United States Steamship Co. freighter was under Army orders when it straggled from the convoy and was torpedoed by

UB-74 about 8 miles south of Ile d'Yeu. Survivors from the 32 crew were rescued by *Emeline*. Casualties were:

Frederick C. Keniston, seaman, British, mother, 66 Swancombe St., Swancombe, England. Injured by explosion.
F.W. Doughtry, 2nd engineer, British, mother, 106 Endenbridge Rd., Bush Hill Park, London. Drowned.

Tippecanoe—On July 25, 1918, the Shipping Board steamer was under Army operation en route from New York to St. Nazaire when it was torpedoed by *U-91* about 550 miles from France. An American oiler, William R. Detterer, drowned and was the only casualty out of the 83 aboard. He was survived by his sister Anna, 2537 Orknay St., Philadelphia.

Berwind—On August 3, 1918, Capt. John Hughes' New York & Porto Rico freighter was en route from Nantes to Brest under Army operation with a crew of 32. At 3:20 p.m. they were 3 miles south of Audierne, France, when a torpedo from *UB-88* hit the ship. The casualties, mostly drowned, were:

John Hughes, master, brother Arthur H., Brynderu, Dunpath, Llanelly, Wales.
O. Watts, 1st mate, English, injured.
Lawrence Baines, 3rd engineer, English, wife Lillian, 2 Albert Rd., Woolston, Southampton.
Louis Mardagan, oiler, 49½ Olive St., North Shields, England.
George Edison, messman, English, suffered from shock.
Marcel Janaco, messman, mother Mary, 32 Patissia St., Athens, Greece.

Lake Portage—On August 3, 1918, the Shipping Board freighter was under Army control when it was torpedoed by *UB-88* off Ushant, France, with 48 on board. The casualties were:

Charles McIllsworth, seaman, American. Slightly burned.
Julius Mathew Banks, engineer, American, J.M. Banks, 69 Downing Rd., Bootle, Liverpool, England.
William Sutton, oiler, American, Ann Sutton, 162 Van Brant St., Brooklyn.
Alfred McDonald, fireman, English, wife, 15 Vesuvius St., Liverpool.
Patrick King, fireman, Irish, 2 Hogan Ave., Dublin, Ireland. Severely wounded.
I. Wilkins, coal passer, English. Badly scalded.
John Skogen, chief cook. Badly scalded.
Henry Wellman, assistant cook. Badly scalded.

Cubore—On August 15, 1918, the Ore Steamship freighter was under Army operation in the Bay of Biscay when they were torpedoed by *U-107*. Nine were lost out of 68 on board, most from drowning: Albert Denison Dorey, 1st mate; Elton W. Riley, 1st engineer, apparently the only American casualty, died of exposure and left a wife, Hattie, in Ottawa, Illinois; Antonio Lima, fireman; Joseph Prikol, fireman; John Tisitsiliones, coal passer; Antonio Ancharuco, coal passer; George Mostoropolis, coal passer; Stanitos Tiliotis, coal passer; Bridgewater, 2nd cook; and Pufracio Isberg, messman.

Montanan—On August 16, 1918, the American-Hawaiian passenger ship was under Army operation when it was torpedoed by *U-90* in the North Atlantic. The lost were all

American citizens: L. St. Thoms Duncan, Oliver Haines, Chester O. Lamollate, and Philip Smith drowned. J.W. Johnson died after being struck by the lifeboat tackle. Two Armed Guard sailors were lost.

West Bridge—Torpedoed and damaged by *U-90* in Bay of Biscay on August 16, 1918, en route home with loss of firemen Jens P. Neilsen, of Denmark; William I. Porter, of 114 SE 66th, Portland, Oregon; and Robert J. McConkey, of Mount Vernon, Washington. The ship was abandoned and later towed to Brest.

Dora—The Shipping Board freighter was torpedoed September 4, 1918, by *U-82* about 400 miles off the coast of France. Of 86 on board there was one injury.

The British Isles and English Channel

In early 1917 it was estimated that between 30 and 40 submarines were operating around the British Isles and the coast of France.

Woodward Abrahams—On April 22, 1917, Winfield S. Pendleton's 3-masted schooner loaded cargo at Pensacola, Florida, for Jacksonville and left there on February 28 under Capt. John Vannamee, 62. When they were about 407 miles west of the Fastnet they were met by *U-43*. Capt. Vannamee lowered his sails and hoisted his flag. Three Germans came aboard and took instruments and stores and put two bombs aboard. There were no casualties in the crew of 6 and they arrived at New York on the steamer *Charybdis* on May 31, from Hamilton, Bermuda. Capt. Vannamee lived at 47 Lowell St. in Malden, Massachusetts.

Vacuum—The Vacuum Oil tanker under Capt. Samuel S. Harris left New York on March 30, 1917, for Birkenhead, England, with a crew of 34. Nine of the crew were native-born, 15 naturalized, the remainder foreign-born. April 28 the ship was en route home in heavy seas and was torpedoed by *U-21* about 140 miles west of Barra Island, Hebrides Islands, Scotland.

Capt. Harris, a number of his crew, and five Armed Guard sailors got off in the first boat and spent several days before being picked up and taken to Liverpool. Chief mate Oscar Gailes, a naturalized Russian, a few other crewmen, and three Navy gunners were in the second boat and also made Liverpool. Those lost were mostly from drowning:

Harris, Samuel, master, American, Roanoke, Virginia. Slightly injured.
Frank J. Yerney, 2nd mate, American, 360 2nd Ave., Detroit, Michigan.
E.D. Husted, 3rd mate, American, Jane E. Husted, 112 W. Sidney Ave., Mount Vernon, New York.
Samuel Michaelsen, quartermaster, Danish, 250 Van Brunt St., Brooklyn. Died from exposure.
Olaf Nelson, boatswain, Norwegian. Slightly injured.
Ben Olsen (aka Alten Oltne), seaman, Norwegian, Hittalo, Norway. Died from exposure.
L.K. Rasmussen, seaman, Danish, Maroo, Denmark.
John Kirk, chief engineer, American, 321 Jay St., Brooklyn.
Francis I. Davidson, 2nd engineer, American, 577 W. 130th St., New York.
Jose Casales, fireman, Chilean, Valparaiso, Chile.
Ramon Munoz, fireman, Porto Rican.
Andrew Ortiz, fireman, Chilean, Lota, Chile.

M. Roos, fireman, Porto Rican, San Juan, P.R.
Ramon Torres, fireman, Chilean.
Pedro Villalon, fireman, Chilean, Carrizal Bajo, Chile.
Avelardo Costerla, coal passer, Chilean.
Luis Gistern, coal passer, Chilean.
W.B. Crane, steward, American, 411-A 18th St., Brooklyn.
Jose Miller, 2nd cook, Porto Rican, 39 St. Felix St., Brooklyn. Died from exposure.
Thomas Ellis, messboy, American, mother Bridget, 90 Franklin St., Albany, New York.
William Garvey, workaway, American, 1425 E. McMillan St., Cincinnati. Drowned when lifeboat capsized.

The Armed Guard commander, Lt. Clarence C. Thomas, of Grass Valley, California, was the first American military officer lost in the war. The Armed Guard sailors presumed drowned were Charles J. Fischer, Coxswain; Alexander Donald, S2c; and Seamen Charles F. Inckan and Frank H. Loree.

The *Vacuum* was built at Ecorse, Michigan, in 1912 as the 247' laker *Bayamon* and converted to a tank ship.

A.B. Sherman—Jonathan May & Sons' schooner, of Philadelphia, was shelled, probably by *U-62*, on June 26, 1917, off the Isles of Scilly. Seaman George Tull, 1420 Bainbridge St., Philadelphia, was seriously wounded. The vessel was towed to port.

Hildegard—Around October 23, 1916, Capt. Craig McKeen, 34, left Perth Amboy, New Jersey, in his barkentine for Brunswick, Georgia. Shortly after leaving he became ill and they put in at Newport News where he died in a hospital on Thursday the 26th.

The vessel was put up for sale and purchased by Alfred H. Clement, of New Orleans, and he incorporated the Hildegard Shipping Co. on February 7, 1917. Clement also owned the schooner *General E.S. Greeley*. The *Hildegard* was sunk on July 10, 1917, by *UB-31* in the English Channel about ten miles southeast of Start Point, Devon. There were no casualties.

John Hays Hammond—David W. Simpson's schooner, of Boston, en route from England to Iceland was sunk by gunfire from *U-44* off Ireland on July 27, 1917. There were no casualties. Survivors were taken aboard a British destroyer about 24 hours later 340 miles northeast of Ireland and arrived home on August 12.

Carmela—Capt. John A. Johnson, his mate Ernest Dobert, of Seattle, and a crew of 19, including six Americans, left on June 29, 1917, for France on Arthur A. Miller's 3-masted auxiliary bark. At around 6 a.m. on the morning of July 28 they were fired at "by a German submarine of small type" in the English Channel about 25 miles southwest of Lizard, England. After four shots, the crew got into their boats. The sub came up and the crew was ordered aboard and held under guard while the German's looted the stores of coffee and bacon. A patrol boat appeared during their second trip to the bark. Bombs were hastily placed aboard and the vessel sunk. The sub was the *UC-62*. There were no casualties and the crew was landed at Plymouth.

Motano—On the night of July 31, 1917, the Standard Oil of New Jersey tanker under Capt. Ernest Stratton, of West Philadelphia, and 47 crew was torpedoed by *UC-47* about 20 miles southeast of Start Point, England. C.R. Garrison, CBM, of Cincinnati, was the Armed Guard commander. The ship sank in less than a minute. Wireless operator Samuel C. Tennery told of his coworker, Russell Williams, of Chicago:

It was 7:10 p.m., ship time, and I was on duty. There was no warning. The noise of the explosion at the stern was the first thing we heard. The ship settled at once. I grabbed a life belt and asked Russell to run to the bridge and see if the captain had any orders for me. He put on his belt and with a cheery reply of assurance sprang through the port doorway. That was the last any one saw of him. The ship was awash then and the suction must have got him before he reached the bridge.

Survivors were Capt. Stratton; chief mate Arthur B. Hansen; 2nd mate Fred Stone, of New York; S. Carison, pumpman, of Chicago; L.M. Flynn, seaman, of Philadelphia; 1st engineer B.H. Post, of New York; 3rd engineer Charles McHeron, of Brooklyn; chief cook Alfred J. Quinn, of Osceola, Nebraska; 3rd cook W. Davis, of Binghamton, New York; Alfred G. Deforce, messman, of Jamestown, New York; and Samuel Tennery, of Beecher City, Illinois. The chief engineer, John T. McDonald, of Staten Island, New York, had previously died on board at sea from natural causes. Those lost were all presumed drowned:

Julius G. Haugaard, 3rd mate, Norwegian, brother Harold in the Royal Navy, Norway.
Russell Alexander Williams, wireless operator, American, father J.A., 4634 N. Hermitage Ave., Chicago.
Marius P.M. Madsen, boatswain, Danish, mother Anna, 16 Adelgade, Faaborg, Denmark.
Karl Larson, seaman, Danish, sister Larina, Copenhagen.
Walter Roslund, seaman, Russian, mother Alexandria, Nystad, Finland.
Renaldo R. Lohse, 2nd engineer, American, wife Anna, 120 W. 61st St., New York.
Harry Gregory, oiler, Welsh, mother Mary Alice, 27 Edward St., Cardiff, Wales.
John Nordgren, oiler, Swedish.
Vernon Thorne, oiler, Canadian, mother Isabella, 1 Pitt St., St. John, New Brunswick.
Edward Gilow, fireman, Russian, mother Anna, 60 Nicholas Rd., Rega, Russia.
W. Harwood, fireman, American, sister Minnie, Alderson, West Virginia.
David Lundquist, fireman, Swedish, mother Matilda, Westeros, Sweden.
Robert Tuber, baker, American, mother Alice, 49 Montgomery St., Springfield, Mass.
Lannie Gleason, messman, American, father W.M., Carter, Oklahoma.
Rubin Williams, messman, American, father Hyman, 662 Snediker Ave., Brooklyn.

The Armed Guard sailors lost by drowning were Don Walton Barmore Seaman; Joseph Frank Carlin, Coxswain; and Seamen 2c Sidney Norwig, Andrew Lawrence, James Anthony Sokel, Cornelius Oliver Torjusen, Albert Henry Vogel, and William Whelan.

Laura C. Anderson—William J. Quinlan and partners owned the 4-masted schooner, of Philadelphia. Capt. Charles Davis left New Orleans on July 15, 1917, with general cargo for France. On August 29 they were stopped in the English Channel about 15 miles east of Barfleur by *UC-65*, "stripped of all the stores," and sunk with bombs. There were no casualties. Survivors were taken to Southampton.

Annie F. Conlon—Freeman R. Garrett's schooner, of Portsmouth, New Hampshire, was attacked by *UC-47* on October 3, 1917, about 15 miles southeast of St. Mary's, Isles of Scilly. There were no casualties. The vessel was taken under tow but sank. Garrett owned a retail grocery business at 97 Market St. as Henry L. Garrett & Son.

Rochester—Capt. Erik Kokeritz's Kerr Line freighter departed New York in company with the *Orleans* on February 10, 1917, for Bordeaux, France, and the tanker *Gold Shell* left from Wilmington, Delaware, the first three ships to sail for France after the diplomatic break with Germany. The *Orleans* reached the Gironde River on February 26 in heavy fog and the *Rochester* on March 1. Both carried non-contraband cargo and had large American flags painted on both sides of their hulls. Foreign Secretary Zimmermann said the *Orleans* was lucky and that "neutral ships will be sunk if they enter the barred zone."

On November 2, 1917, the *Rochester* was not a neutral vessel when a torpedoed fired by *U-95* off the western coast of Ireland with 49 on board sank the ship. No sub was seen until the survivors were off the ship in three boats. The sub surfaced and fired ten shots at the ship then two other subs were seen on the surface signaling to each other.

Seventeen of the merchant crew were lost from the explosion, drowning, or exposure in the lifeboats. Capt. Kokeritz, 16 merchant crew, and five Navymen were adrift in one boat for five days before being picked up by the British. The Armed Guard sailors in the boat were Edward McCausland, CBM; William Fouils, GM2c; and Seamen 2c W.B. Thompson, Thaddeus Fellows, and William Eisenhardt. Six Armed Guard sailors died after being rescued. The other two boats reached shore safely. The crew casualties were:

Capt. Kokeritz, master, (naturalized American), sister Ruth Blomberg, 20 Kongsgatan, Gerfle, Sweden, died of wounds.

Christian Kinset, 2nd mate, Norwegian, sister Bessie Jensen, 2107 E. Kedvale Ave., Chicago, probably drowned.

Ragnar Hellstrom, seaman, Swedish, father in Stockholm. Probably drowned.

Axel Pedersen, seaman, Danish, mother Johanna Marie, Copenhagen, Denmark, probably drowned.

Victor Pedersen, seaman, Swedish, sister Alma, Berfendalen, Sweden, died from exposure.

Rex S. Wheeler, seaman, American, father W.T., 735 St. Clair St., Akron, Ohio, died from exposure.

Oscar Sandat, 2nd engineer, Russian, died in the explosion.

E.A. Bekker, 3rd engineer, Russian, possibly drowned.

Thorolf Madsen, oiler, Norwegian, father Kristian Madsen Nostdal, Sarpsberg, Sweden, died in the explosion.

Antonio Avendano Anpuero, fireman, Chilean, father Salvador Avendano, Chiloe, Chile, died from exposure.

George Bulgaria, fireman, Romanian, probably drowned.

Juan Gonzales, coal passer, Panamanian, father in Canal Zone.

A. Manrigues, coal passer, Chilean, mother Flora Dias, Valparaiso, Chile, died at sea.

Benjamin Margoli, coal passer, Spanish, brother Clements in Teruel, Spain, probably drowned.

Ernest Ohman, coal passer, Swedish, father Nils Fritz, Hoganas, Sweden, drowned.

George W. Wehr, chief steward, American, Washington, D.C., injured slightly.

Verner L. Helms, messboy, American, father Thomas A., of Blacksburg, Virginia, died from exposure.

Harry Yorks, messboy, British, 25 South St., New York, injured slightly.

William B. Hinnan, cabin boy, American, father at 23 Greenough Ave., Cambridge, Massachusetts, died of exposure.

Armed Guard sailor Ernest H. Gragg, S2c, died of exposure. Other sailors, all Seamen 2c, drowned: Marshall Underwood Corun, Mearl Ralph Cox, James Crowley, Bernard Joseph Donavon, and George Franklin Wheeler.

Alamance—The Garland armed freighter under Capt. E.E. Johnson was torpedoed by *UB-73*, four miles north of The Maidens, Ireland, on February 5, 1918. Those lost were Spanish wiper, Pio M. Carrasco; British quartermaster Laurie A. Fralic, of Bridgewater, Nova Scotia; Japanese messman Shintaro Fujiwara, of Hiroshima-ken; a Japanese cabin boy; and a Japanese 2nd cook, Magoichiro Uteunomiya, of Chimoken, Japan. American oiler George C. Davis, of 427 Montgomery St., San Francisco, was slightly injured. The ship had a crew of 55 and 23 Armed Guard sailors.

Santa Maria—The Sun Co. tanker, of Philadelphia, was sunk on February 25, 1918, by *U-19* about one mile off the northwest coast of Ireland near Loch Swilly. There were no casualties in the crew of 61. It was reported that Chief Gunner's Mate John Weber "and his armed guardsmen stood at the guns until the water swept around them."

A.A. Raven—Girard Trust Co.'s freighter, operated by the War Department under Capt. Theodore C. Hecker, of Edgemoor, Delaware, and a crew of 37 was sunk March 14, 1918, by *UB-55* about 16 miles southwest of Wolf Rock Light, Cornwall. Capt. Hecker was injured. The reported drowned were:

Dirk Stimis (aka D.P. Stines), chief engineer, American, 25 South St., New York.
James Alfred Stiles, 1st engineer, British, wife Catherine, 46 Stacey Rd., Cardiff, Wales.
Louis Collins, oiler, Belgian, 432 W. 19th St., New York.
John Maseneck, fireman, Russian, 25 South St., New York.
Charles Sousis, coal passer, Russian, father David, Riga, Russia.
Bida Shosaburo, British (naturalized), steward, wife Ada, 30 Court Rd., Grange, Cardiff, Wales.
Jesse George Guy, unk., English, 47 Grange Gardens, Cardiff, Wales.
Joseph Martin, unk., Cuban, father Joseph, 33 Teniente Rey, Havana.

Atlantic Sun—The Sun Co. tanker, of Philadelphia, en route from Philadelphia to Liverpool was torpedoed on March 19, 1918, by *U-46* about 19 miles off Orsay Island, Scotland, with 56 on board. American messman Charles T. Daniels, Jr., and the Belgian 2nd mate Henry Clement Hertoge drowned and three others were slightly injured. Daniels' mother Nellie lived at 507 E. 6th St. in Camden, New Jersey. The American chief mate, David Johnson, was taken prisoner.

Chattahoochee—The Shipping Board steamer was sunk March 23, 1918, by *UB-55* in the English Channel about 28 miles south of Penzance, Cornwall, England. There were no casualties among the 81 on board.

Rockingham—The Garland freighter, sistership of the *Alamance,* was torpedoed on May 1, 1918, about 150 miles northwest of Ireland en route from Norfolk to Liverpool. Survivors were picked up by the Booth Line steamer *Hubert,* en route to New York. All three casualties were American: Harry Margay, 2nd engineer, father 726 Shotwell St., San Francisco, died in the engine room; John Smith, oiler, Honolulu, Hawaii, injured; and William Wond, oiler, Honolulu, Hawaii, died in the engine room.

Dirigo—The big steel steam bark was reported to be the first steel sailing vessel built in the United States. Axim Transportation bought the bark in November 1916 and homeported it at Pensacola, Florida. On May 31, 1918, Capt. John A. Urquhart and his crew of 20 were about 6 miles southwest of Eddystone Light, England, when they were sunk by a U-boat. One American, 3rd mate John M. Ray, of Eastport, Maine, drowned while trying to get into a small boat. The survivors arrived home on June 11 as passengers aboard the steamer *St. Paul.*

Argonaut—The Jason Navigation cargo liner was torpedoed on June 5, 1918, by *U-82* off the Isles of Scilly. There were no casualties among the 73 on board. The ship was insured for $516,500 on the hull and $3,950,836 on the cargo, the largest total loss paid by the Marine & Seamen's Division of the Bureau of War Risk Insurance.

Lake Edon—Capt. George D. Bruce's Shipping Board freighter was under Army charter with a crew of 32 on August 21, 1918, when the ship was torpedoed by *U-107* in the Bristol Channel about 4 miles northeast of Newquay, Cornwall, England. Armed Guard sailor Ralph Eugene Hooten was lost. The crew casualties drowned:

George D. Bruce, master, American, Mrs. Blake, 900 Powell St., San Francisco.
Constantine Cousentis, seaman, Greek, father Dimitri, Samos, Greece.
Vaesilios G. Danielopolos, seaman, Greek, father Georges, Salonica, Greece.
Harold Jensen, seaman, Danish.
Walter Musler, seaman, mother, Russian consul, New York.
Charles W.U. Maller, seaman, wife Margaret. George Stone, 41 W. 82nd St., New York.
Anders Martinson, seaman, Hans Martinson, 289 80th St., Brooklyn.
Charles A. Crass, 1st engineer, Charles A. Crass, Jr., 914 S. Ellis St., Cape Girardeau, Missouri.
Tom Victor Jopling, 2nd engineer, English, wife Violet, 21 W. Road, Southampton.
Paul C. Dunham, 3rd engineer, mother Ora, 1053 Lee Ave., Cleveland, Ohio.
John E. Harms, oiler, American, grandfather, J.M. Harms, Port Clinton, Ohio.
Patrick McKenna, oiler, Irish, father Michael, 49 W. St., Drogheda, Ireland.
W. Dadak, fireman, sister, Theodosia, 28 Rue de la Palaise, Paris, France.
Michael Kelly, fireman, Irish, wife Mary, Dundalk, Ireland.
Patrick Donnelly, fireman, British, wife Sarah, 1 Thornton Pl., Dingle, Liverpool.
Fernando F. Ferros, coal passer, Brazilian, mother, 71 Afonkapell Rua, Rio de Janeiro.
Manuel Lauriano, coal passer, Peruvian, Maria Lauriano, Lima, Peru.
John Weston, coal passer, Swedish, 213 10th Ave., New York.
Meari Ricardi, 2nd steward, Italian, Gasparina Albino, Torino, Italy.
J.W. Stanber, cook, wife, 7 Maitland Place, Grange, Cardiff, Wales.
Alan Mosher, 2nd cook, 71 Spring Garden Rd., Halifax, Nova Scotia.
A. Archeopoulos, unk.

Onega—On August 30, 1918, the Barber freighter was torpedoed by *UB-125* in the Bristol Channel off Newquay, Cornwall. Of the 55 on board, 29 were rescued and 26 were reported lost, including the master whose body was recovered. Lt. Frederic T. Vanauken, the Armed Guard commander, survived.

Lake Owens—The Shipping Board freighter, en route from Nantes, France, to Barry Roads, was sunk on September 3, 1918, by gunfire from *UB-125* in the Bristol Channel about 3 miles off Trevose Head, Cornwall. Out of 49 crew, five were lost and eight wounded.

The Mediterranean

A story was told about the master of a Standard Oil tanker that had been sunk by a sub in the Mediterranean and who enlisted in the Naval Reserve as soon as he got home. The recruiting officer asked him, "Why don't you go and get another ship? Such men as you are needed. Besides, you'll get big money and your life will be insured?" The captain replied, "To hell with the big money. What I want is to get a crack at those damned baby killers." He was probably referring to the several children who were lost on the *Lusitania*.

During the entire war only one convoy sailed from the United States to Gibraltar but it broke up in heavy weather and each ship came in on its own. All the other one million tons of trans-Atlantic shipping to Gibraltar came on ships sailing independently with Armed Guard crews.

A total of 4,245 ships were convoyed through the Mediterranean Sea to supply American forces at Marseilles, French forces in North Africa, allied forces at Salonika, British forces in Egypt and Palestine, and all of Italy.

Seward—On April 7, 1917, Capt. Philip V. Johnson's Alpha Steamship Co. freighter with general cargo was sunk without warning by *U-52* in the Gulf of Lyons about 25 miles east of Port Vendres, France, and 20 miles northeast of Banyuis, Spain. There were no casualties among the 38 crew. Fifteen sailors were picked up by a bark and 16 others, including Capt. Johnson, made landfall at 11:30 that night at Banyuis in a lifeboat.

Edwin R. Hunt—Foreigh & Domestic Transportation's schooner under was under charter to Archibald M. Ostrum when Capt. Nicholas Miller was en route from Crete to New York in ballast. At about 4 a.m., on April 7, 1917, around 40 miles off Cape de Grata, Spain, *UC-25* commenced firing. The nine sailors abandoned ship and were picked up by the crew of the Danish schooner *Konso* and brought to Almeria. There were no casualties.

British Sun—In 1907 the Sun Co., of Philadelphia, got into the overseas market and had the Swan-Hunter yard at Newcastle-upon-Tyne build the world's first tanker from scratch, the *British Sun*. On July 26, 1909, Sun organized a subsidiary, the British Sun Co., Ltd., at Bush House, Aldwych, London, with William Ernest Smellie as operations manager.

The ship's normal routine was to load crude from the Sun Pipe Line Co. at Sabine, Texas, and take it to Italy. During the war the ship operated under charter to the British government until it was torpedoed on May 1, 1917, by *UB-43* in the Mediterranean and lost. The tanker was the only ship Sun ever had in their fleet that was built overseas and the only one in the company's history to fly a foreign flag. The ship's value was $2.5 million.

Hilonian—Universal Transportation's unarmed ex–Matson freighter left New York for Genoa under Capt. H.H. Williams, of Brooklyn, with $2.5 million in general merchandise. On May 16, 1917, the ship was sunk by *UC-35* about 30 miles from Genoa. The ship was valued at $1 million.

Eighteen of the 39 crew were American citizens. Capt. Williams and chief engineer Fred Schmidt, a naturalized Norwegian, were among the rescued. Williams was in a lifeboat with 19 others when the ship rolled over on top of them. One of the sailors panicked and grabbed onto him and dragged him under as three other sailors fought valiantly to free him. The captain "fought his way to the surface."

Four crew drowned: seamen Frank Bach, American, and John Willebrande, Dutch, both of New York; 2nd engineer Robert Hackett, of New York; and Dil Mahomad, 2nd cook, Hindu, of Calcutta. Three others were slightly burned.

Margaret B. Rouss—John Swan's schooner, of New York, left St. Andrews Bay, Florida, on February 4, 1917, under Capt. Fred L. Foot with pine lumber for Genoa. On May 27 the boat was stopped by *U-33* about 42 miles south of Monaco. The crew was robbed and the boat sunk. There were no casualties in the crew of nine.

Petrolite—Capt. Thomas H. McKellum's armed Standard Oil of New York tanker left New York on April 30, 1917, for Savona and Leghorn. There were ten Americans in the crew of 38. On June 10, en route home in ballast, the ship was sunk by *U-39* off the coast of Egypt. There were no casualties. One boat with 20 survivors, including 3rd mate Olaf Mattson, of Brooklyn, made landfall on June 12.

Capt. McKellum, chief mate Philip Richter, 2nd mate Bergen Frederickson, and chief engineer John Torrence, 83 Noble St., Linwood, New York, made landfall in another boat. The survivors left Gibraltar on the steamer *Pancras* and arrived at New York on July 2, 1917.

The *Petrolite* was a German ship purchased from the German-American Petroleum Co. at the beginning of the war and was involved in the diplomatic and Admiralty law controversy when an Austrian sub commander looted their stores.

Moreni—The 354' armed Standard Oil of New York tanker left New York on May 3, 1917, under Capt. Thomas Thompson for Beaumont, Texas. On the 10th the gunners test-fired ten shots from each gun. They arrived at Beaumont on the 12th and left on May 14 with a cargo of gasoline for Savona Italy, in company with the armed steamers *Polarine* and *Harry Luckenbach*.

On May 31 they picked up four survivors in a lifeboat from the schooner *Gipsum Express* and arrived at Gibraltar on June 9. They left Gibraltar the same day and early on June 12 the three ships were off the coast of Spain when the *Moreni* started having engine trouble. At 1 a.m. Capt. Thompson was advised by the engineers to stop. Chief Boatswain's Mate Andrew Copassaki, commander of the Armed Guard crew, was awakened and asked Capt. Thompson what the trouble was. The other two vessels continued on, unaware of an enemy submarine lurking just offshore. After about 45 minutes the engines were restarted and they continued on. Copassaki filed a report about what happened next:

> We were attacked by a submarine at 4:05 a.m. on June 12, 1917. She was off the port quarter about 9,000 yards away. She fired four or five shots before we located her. We swung around until our stern faced the submarine and returned the fire at a range of about 7,000 yards. The submarine stayed about 8,000 to 9,000 yards away, and for that reason our shots fell short. We sent out a radio call for assistance.
>
> After a half hour fight we were hit in the gasoline tank aft and a fire started. It was reported to me that the ammunition aft was running low. Immediately I lined up the forward gun crew with the merchant crew to pass ammunition from forward to aft. About an hour later fire broke out all over the ship and it became impossible for the men to pass any more ammunition aft. When I saw such was the case, I happened to be abaft the mainmast, and I called for one of the gun's crew to go forward with me. I reached the bridge, being burned on the way. The man was unable to follow. About this time the steering gear was shot away and the ship started to go around in circles.
>
> Coming down off the bridge, I saw the captain and the boatswain, who was holding the falls ready to lower the lifeboat. The captain said to come and get in the lifeboat as it was starting to burn. I told him to wait. He said he would hold the boat as long as possible for me.
>
> I went and got the machine gun and ammunition for it to use against the submarine and some clothes in which the radicode book was wrapped and put them all in the boat.
>
> I then noticed two of the gun's crew in the lifeboat. I ordered them to get out and come with me. We went forward and manned the forward gun with which we fired four shots before the firing pin went out of commission.
>
> When the submarine saw our fire was coming closer, she steamed full speed astern so that we could not see her. Then, as we could fire no more, and as the captain called that the lifeboat was burning, we got into the boat.

Meanwhile the engines of the ship had not been stopped and she was going around in circles, which she continued to do for about half an hour after we left.

The lifeboat upset when it hit the water and two of the merchant crew were lost. We swam around until we were able to get to the capsized boat, turn it over, and climb into it.

The after gun's crew and merchant crew left in their boat as the fire was becoming too hot and would have burned the boat. I also found out that the first mate of the ship left much earlier, before any of the other boats, with three men, leaving behind the men that were supposed to go in that boat.

The submarine called the boats alongside and congratulated us, shook hands with the captain and told us it was the best fight he had ever seen any merchantman put up. The lieutenant of the submarine who acted as an interpreter also said our shots were good line shots but fell short (guns were given full elevation) and that he could not have stayed with us if we had had bigger guns. I felt her deck, which was about two inches thick. She carried two guns, one 4-inch and one 3-inch. The lieutenant of the submarine asked me where the gunners were and I told him they had been killed by gunfire. He also asked if there were any Englishmen in the boat. I told him no. The machine gun was still in the boat but it could not be seen as it was covered with water, the ammunition being lost when the boat upset.

The Germans treated two men who had been wounded and returned to the other boats. The commander of the submarine said he would have towed us toward the beach but for the fact that we had called for assistance.

The submarine then left us and commenced firing again into the *Moreni*. She finally hit the *Moreni*'s engine room, which stopped the engines. We then lost sight of the submarine. About an hour after the submarine left us we were picked up by the Spanish steamer *Valbanera*, (about 8 a.m. June 12, 1917). The Captain of the *Valbanera* told us we were about 17 miles off Cape de Palos, Spain.

Shortly afterward the submarine came up near the *Valbanera*. The officer of the ship got in his small boat and went over to the submarine. He returned to the ship, the submarine went around the ship, the submarine's crew cheering and *Valbanera*'s crew also cheering, and then steamed away.

While on the *Valbanera* we were given every consideration, dry clothing, good food, and good sleeping accommodations.

We were landed in Barcelona, Spain, at 8 a.m. June 13, 1917, and reported to the American Consul on arriving, who fitted the crews with clothing.

During the fight, which lasted a little over two hours, we fired about 150 shots, but as the steering gear was shot away and the ship was going around in circles, we could get no range. The submarine fired about 200 shots while we were on the *Moreni*, of which about 45 found their mark.

The *Moreni* was the first armed American ship lost in a surface battle with a sub. Over 250 shells were fired over a period of two hours. The sub was identified as the *U-64*. Out of a total crew of 47 on board the tanker, four were lost:

James J. Curran, Jr., wireless operator, American, 25 Harrison St., New York. Died of heart failure in a lifeboat.

Emil Moustner, seaman, Russian, father John, Verborg, Russia. Treated on the sub but died of wounds.

Edward Cunningham, oiler, English, mother Margaret, 40 Olivia St., Liverpool. Drowned getting into a lifeboat.

A. Giesen, wiper, Dutch, father Arie, Dordrecht, Holland. Treated on the sub but died of wounds.

Six others were injured and three of those were hospitalized. On June 25 the survivors left Barcelona on the steamer *Antonio Lopez* and arrived at New York on July 13.

Grace—The Standard Commercial Steamship Corp. armed freighter left New York on June 1 under Capt. Carl Erickson with a cargo of general merchandise and petroleum for Saloniki, Greece. The ship was under charter to C.B. Richard & Co. The crew and gunners numbered 45 total.

On July 12, 1917, the ship was torpedoed and sunk by *U-23* about 5 miles north of Cape Phesso, Andros Isle, Greece, and sank in 12 minutes. The merchant casualties were A. Anderson, seaman, Finnish; S.J. Van Wyk, seaman, Dutch, 507 W. 28th St., New York, died in the explosion; and E.J. Farrell, messman, of New Jersey, was reported drowned.

Those injured were C.M.A. Sorenson, 1st engineer, seriously burned; Frank Chao, fireman, Spanish, slightly injured; L. Carro, coal passer, Russian, seriously burned; A. Cenarabus, coal passer, Russian, slightly injured; and Felipe Thiao, coal passer, Spanish, seriously injured.

Armed Guard sailors Hugh P. Donnelly, S2c, and George T. Wilson, S2c, were injured. The ship was valued at $900,000.

Wilmore—The Berwind-White Coal Co. collier under Capt. Charles Avery was sunk on September 12, 1917, without warning by *U-64* about 4 miles northeast of Hormigas Light, Cabo de Palos, Spain. Of 43 on board, three were reported lost, including Armed Guard sailor Roscoe Earle Wagstaff, S2c. Survivors reached Newport News on October 1.

Platuria—In August Capt. John Leslie's Standard Oil of New Jersey tanker was en route home from Sweden with 54 on board when they were stopped by Germans and escorted into Swinemunde for examination then released. On September 15 the *Platuria* was sunk at 2:45 p.m. by *U-63* off Tangier, Morocco. One Armed Guard sailor, Roscoe Earle Wagstaff, S2c, was presumed drowned. Merchant crew casualties, most presumed drowned, were:

John Leslie, master, American, brother James, Greenock, Scotland.
Gustav F. Hellstrom, able seaman, Swedish, contracted tuberculosis from exposure in a lifeboat and was transferred to the U.S. Public Health Service Sanitarium at Fort Stanton, New Mexico.
Harold W. Whittier, oiler, American, mother Rita Haines, New Germany, Nova Scotia.
Silveria Pereria, fireman, Portuguese, 1823 38th St., Norfolk, Virginia.
Alberto Gargie, coal passer, Spanish.
Juan Rojas, coal passer, Chilean.
James Jones, chief steward, English, cousin Mrs. Erunie Roberts, 231 W. 21st St., New York.
Lawrence Halpern, messboy, American, mother Fannie, 73 Stanley St., Houston, Texas.
Samuel Wilson, messboy, British West Indies.

Fannie Prescott—Atlantic Shipping Co.'s schooner was sunk on October 25, 1917, by *U-35* off Cape Cantin, Morocco, possibly under charter to the Mandingo Development Corp. There were no casualties in the crew of 8.

Villemer—The Shipping Board's ex–Philippine freighter was en route from Barcelona on November 7, 1917, when the *UC-38* attacked in the Mediterranean with the loss of two.

Rizal—The Tabacalera Steamship Co.'s Philippine freighter was sunk by *UB-50* on November 9, 1917, in the Mediterranean 9 miles off Cape Cavallo. There were no casualties.

Margaret L. Roberts—William S. Roberts' 3-masted schooner, of Crisfield, Maryland, was sunk on November 16, 1917, by gunfire from *U-151* off Madeira Island. There were no casualties in the crew of 7.

Schuylkill—Schuylkill Steamship Corp.'s freighter, of New York, was sunk on November 21, 1917, by *U-39* in the Mediterranean about 15 miles east of Cape Tenes Light, Algeria. There were no casualties in the 37 on board.

Albert Watts—The brand-new American Italian Commercial Corp. armed tanker took on 12,000 cases of gasoline and 187,734 gallons of gas in bulk and left New York with a crew of 36 on October 29, 1917. They joined a convoy for Genoa, Italy, and sailed next to the tanker *Westoil*. In the approaches to Italy a periscope was sighted about 300 yards away and both ships opened fire, commencing a four-hour running battle, both ships firing every time the periscope appeared.

At about 9:55 a.m. on November 28, while in an Italian convoy near Villefranche, about 45 miles from Genoa, an explosion occurred at No. 2 tank, fatally burning John Zwart, a Dutch sailor. The blast opened a large hole above the waterline and gas spewed everywhere. Witnesses said a torpedo leaped out of the water just before striking.

A fire started under the boilers so the engines were stopped while firefighting commenced. The fires were put out and the ship proceeded back to the west about 3 miles and anchored behind a small breakwater and settled on the bottom. The ship was refloated with the assistance of two Italian tugboat crews and around 12:30 p.m. the ship proceeded to Genoa and anchored off the harbor. An Italian destroyer came up and the commander ordered the master to turn on a blue light, the signal for requesting a pilot. A spark from this light reportedly ignited gas fumes on the surface of the water but the tanker got clear of the fire by going full ahead and zigzagging. The ship was declared unsafe to enter the harbor and at about 8 p.m. anchored at the outer entrance. A pilot came aboard but refused to take the captain ashore to consult with authorities, not having been granted pratique.

Early the next morning he went ashore to consult with the Captain of the Port and while there he was told his ship was on fire. Arthur C. Frost, the American consul, investigated and determined:

> A match lighted by a man in a bumboat contrary to orders given him to stay away from said ship by the officer of the deck and that thereupon the flames from the burning match ignited the gasolene upon the surface of the water, rapidly spreading to the leaking gasolene upon the ship's side, which immediately spread to the decks and holds of the vessel.

The bumboatman burned up along with the ship. Chief Boatswain's Mate Reuel V. Hanberry and his Armed Guard crew were commended for trying to save the ship. Government tugs beached the ship, which was declared a total constructive loss, and efforts to salvage the cargo began. Chief mate Herbert C. Goodwin and eight sailors arrived home in December.

Owasco—The Shipping Board cargo liner, en route from Norfolk to Genoa, was sunk on December 11, 1917, by *U-64* in the Mediterranean off Villajoyosa, Spain. Armed Guard commander, Thorwald Smith, CBM, was reportedly the last one off the ship. Three merchant crew were injured and two were presumed drowned: Johan A. Jakobeson, seaman, Swedish, father Johan Petter, Klinteham, Sweden; and Jesus Garcia, coal passer, Spanish, father Jose Gutierres, Vallaiolid, Spain.

Suruga—The freighter was torpedoed on December 20, 1917, by *UB-49* about 1 mile off Santa Stefano, Italy. There were no casualties and the ship was beached and salvaged.

Chincha—On January 18, 1918, at 7:25 a.m., the Nafra-operated Shipping Board freighter was 2 or 3 miles off Sabinal Point, Spain, when an "object like an enormous whale" was sighted. Chief Gunner's Mate E.E. Nordquist decided it was one of the larger Hun subs and commenced firing at 2,000 yards. After ten shots the sub disappeared. At 8:15 it surfaced again and with the fifth shot an explosion and black smoke was seen. The sub appeared to be heading for the beach about 1½ miles away so firing ceased.

On March 21 the ship was shelled by *U-154* off Gibraltar with the loss of six: Karl Victor

Falt, carpenter, Finnish; two Japanese mess attendants, K. Kurackichi and B. Yamanoto; and Allen Seth Edwards, S2c, of the Armed Guard. The ship's gunners believed they sank a second sub that fired a torpedo and then charged them.

City of Pensacola—Pensacola Steamship Corp.'s auxiliary barkentine, of Port Arthur, Texas, was sunk on April 29, 1918, by *UB-105* in the Mediterranean near Garrucha, Almeria, Spain. There were no casualties in the crew of 9.

Tyler—Old Dominion's armed freighter departed Genoa, Italy, in a convoy on April 30, 1918, in ballast en route to New York. The ship was sunk on May 2, 1918, by *UB-48* in the Mediterranean off the French coast with the loss of six merchant seamen and five Armed Guard sailors out of 57 on board. The merchant seamen lost were:

Edward W. Mears, 3rd engineer, American.
Clarence Knowlton, oiler, American, sister Hazel Halstead, 401 35th St., Norfolk, Virginia.
Fernando Posse, fireman, Spanish, father Joaquin, Coruña, Spain.
Jose Rodriguez, fireman, Spanish.
Gregario Carro, coal passer, Spanish, wife Antonio Carro Lourido, Coruña, Spain.
F. Kamamoto, messman, Japanese, Kobe, Japan.

Five Armed Guard sailors were lost: Howard Richer Smith, GM2c; Lee Martin Carey, EN3c; Seamen Arthur Edward Johnson; Henry Christian Mihalik; and Jesse Clifford Sampson.

Mopang—A.H. Bull's freighter, en route from Constantinople to New York, hit a mine on July 1, 1921, off Bourgaas in the Black Sea at the entrance to the Gulf of Bougas and sank. There were no casualties.

West from 40° Longitude to Atlantic Coast and South to Caribbean Waters

Southern Pacific Co.'s Morgan Line shipped goods from Atlantic ports to Galveston, New Orleans, and California via their Sunset Gulf Route but by April 1917 merchandise had piled up in warehouses because the company did not have enough ships.

The British freighters American shippers had used for shipments to Europe were no longer available, goods, were piling up on docks, and every available American ocean freighter was taking advantage of the high freight rates to Europe.

On April 9 the company announced they would underwrite all war risk policies on westbound cargo shipped coastwise to Galveston, New Orleans, or California. The company wanted to express confidence in the Navy's ability to protect coastal shipping and to stabilize freight rates.

By early 1918, Germany's land war had ground to a halt and the U-boat war against American troop and supply ships had failed due to new anti-submarine warfare tactics and technology and the many new merchant and naval vessels coming into service. German naval planners felt there was nothing left to do but send U-boats into American coastal waters. Their goal was to create a panic, destroy freight and troop ships, take the pressure off attacks on Germany itself, and force the United States to sue for peace. The first goal was accomplished. It was generally thought the Germans must have had a base somewhere in North America where the subs could refuel or even have a "mother ship" large enough to refuel and restock the attack subs.

British intelligence suspected a big U-boat had left Germany on April 14, 1918, and was probably heading for the Atlantic coast of America. They thought it might be the *U-151*, a *Deutschland*-class boat that carried two 5.9" deck guns, two 22-pounders, and one machine-gun. The complement was 8 officers and 65 sailors and they thought the captain was probably Korvettenkapitän Waldemar Kophamel.

On April 14 the *U-151* under Korvettenkapitän Heinrich von Nostitz left Kiel and on May 2, the British notified Adm. Sims in London that a "large-type" submarine had left Germany for American waters. The same day the Office of Naval Operations received a message from Kingston, Jamaica, reporting a U.S. steamer had engaged an enemy submarine about 400 miles north of the Azores.

On June 4 Sims informed the Navy Department that there was one submarine off the Atlantic Coast, probably the *U-151*. Sims continued to receive updates including minelaying operations on the East Coast and relayed all to Washington.

On May 15 the British steamer *Huntress* reported they had escaped a torpedo attack about 1,000 miles off Cape Hatteras. The next day, May 16, the Navy Department sent a coded message to all Section Bases:

> *Most Secret:*—From information gained by contact with enemy submarines, one may be encountered anywhere west of 40 degrees. No lights should be carried, except as may be necessary to avoid collision, and paravanes should be used when practicable and feasible. Acknowledge, Commander-in-Chief Atlantic Fleet; Cruiser Force, Commander Patrol Squadron, Flag San Domingo, Governor Virgin Islands, Commandants 1st to 8th, inclusive, and 15th Naval Districts. 13016.
>
> OPNAV.

At 12:14 p.m. on May 19 the first definite information about a German sub off the Atlantic Coast was received at Atlantic City, New Jersey, in the form of an intercepted S.O.S. from the American steamer *Nyanza* about 300 miles off Maryland saying they were being fired on by a submarine heading west. They managed to escape.

The same day, the tanker *J.C. Donnell* intercepted a radio message from the American steamer *Jonancy* saying they were being "gunned" 150 miles off Winter Quarter Shoals. The *Donnell* arrived at Lewes, Delaware, the next day and the master reported the message.

On the 20th, the British steamer *Crenelia* reported to the Cape May radio station they had sighted a sub about 50 miles off Maryland.

Hattie Dunn—On May 25 the *U-151* was about 30 miles off the coast of Virginia when it encountered George Elliott's schooner *Hattie Dunn*, of Rockland, Maine, under Capt. Charles E. Holbrook. Holbrook gave a statement to the Secret Service in New York on June 4:

> We left New York for Charleston in ballast on May 23. Two days later we were about 15 miles south of Winter Quarter lightship, bowling along under an 8-knot breeze. At about 10:10 a.m. I heard a shell pass near the vessel, then another shell, which fell perhaps a quarter of a mile away. I was not taking much notice because I believed the vessel, which I saw about two miles away, was an American submarine at target practice. A third shell exploded close by us on the weather quarter and I knew that whoever it was wanted us to stop. I brought the vessel up into the wind.
>
> The submarine, with her superstructure and conning tower showing plainly above the water, came within two hundred yards, and I saw that she was flying the two code letters "A.B.," meaning "Stop immediately."

From a small staff at the rear end of the superstructure fluttered a small flag of the Imperial German Navy. An officer and three men came over in a small boat, not over 12 feet long, and in perfect English the officer told us to get into our boats and that we had ten minutes allotted to us to get clear of our vessel.

They placed bombs along the sides of our vessel and blew her up immediately, in the meantime putting an armed German sailor on board the small boat, in which were seven men and myself. This did not give me time to rescue my personal effects and nautical instruments and so I lost them all. My men saved only what they stood in.

Perhaps I would have been given more time if the commander of the submarine had not seen the *Hauppauge* under full sail about four or five miles away. Like us, the *Hauppauge* was light and, I understand, bound from Portland to Newport News. The U-boat destroyed Captain Sweeney's fine new schooner after ordering him and his crew to take to their boats and within a half hour both crews were on board the submarine and both the small boats had been placed on the submarine's deck and lashed down.

We were kept below for several hours until the submarine picked up Captain Gilmore and the *Edna* at 4 o'clock in the afternoon. Then, I guess, the commander thought he had done a good day's work for he was in excellent humor and told us we could go on deck for a smoke. He did not extend liberty to the others that day but later they got their chance once in awhile.

The boats from the *Hattie Dunn* were lashed to the sub and subsequently bashed to splinters by the sea. Capt. Holbrook gave another account:

I started my ship full speed to the westward. He fired again and finally came alongside and said, "Do you want me to kill you?" I told him I thought his was an American boat. He told me to give him the papers and get some foodstuff. He then wanted me to get into his small boat but I was anxious to get ashore so I immediately got into one of my own boats and shoved off. He halted me because he did not want me to get ashore. He then put a man into my boat so that I would come back to the submarine.

There were no casualties. The weather was fine and clear, the sea was calm. We kept aboard the submarine until the morning of June 2. While we were aboard, the second officer and others of the submarine crew wrote some letters and gave them to me to mail. I told them I would not mail the letters if there was anything in them detrimental to my country. I handed them to the first naval officer I came to.

There were times when I could hear them using their wireless. One night the sputtering was so loud it woke me up. They were sending messages either to another ship or to some shore station. Every night the operators listened to press bulletins sent out from America and one of them told me of the battle drive now on in France. They also said that a Whitehead torpedo factory had been blown up in Austria.

Thus was the American coastal fleet introduced to von Nostitz in the *U-151*.

Hauppauge—The 4-masted schooner was, as Capt. Holbrook stated, about 5 miles from the *Hattie Dunn*. The boat was owned by R. Lawrence Smith, in business as the East Coast Transportation Corp. in New York. The schooner left Portland, Maine, in ballast on Friday, May 17, for Norfolk, Virginia, under Capt. John Sweeney. As was typical, the crew was taken off, interrogated, and bombs placed aboard their vessel and sunk. Capt. Sweeney was given a receipt for his ship:

Am 25 ten Mai 1918, 11 Uhr 10 Min. Vormittags ist auf 37° 27' N. und 75° 09' W. der amerkanische 4 Mast Schooner Hauppauge L T Q H von S.M. Unterseeboot vernichtel worden.

Aug See, den 25 ten Mai 1918.

Der Kommandant,

v. Nostitz,
Korvettenkapitän.

The mate, M.H. Sanders, later said the sub had no external identification but that he saw "*U-151*" markings on bunks and blankets and another sailor said he saw the same thing on hammocks and machineguns.

The survivors had seen a strange looking apparatus on the foredeck of the sub and said they were off New York Harbor on May 28 when it took a sudden lurch. At the same time the sub lurched, one trans–Atlantic cable to Europe and one to Central America stopped transmitting. It was then surmised that the device they saw was for cutting cables.

Sanders later gave a statement regarding their captivity:

> The food was good. In the morning we had rolls and fresh butter. The butter was fine. The bread was black and came in loaves about 3 feet long. We had cognac nearly all the time.
>
> They had three graphophones on board. The members of the crew were cheerful and joked with us, especially after indulging in cognac. They were apparently young fellows and frequently talked of their mothers. The crew expressed great surprise when Capt. Sweeney told them we had shipped two million men overseas and had ten million more as reserves.
>
> None of the Germans would give us any information as to the number of submarines over here. We were told that the *U-151* left Kiel on April 14, 1918. The bread wrappers bore the stamp of April 9. The commanding officer said he expected to remain out about eight weeks.
>
> At 5:30 a.m., June 2, word was passed to prisoners by an officer, who said, "Get ready, there's a sailing vessel alongside we are going to put you aboard of." All the 21 men were ordered up on deck. A little later the schooner *Isabel B. Wiley* and the steamship *Winneconne* were sighted. Instead of putting us aboard either of those vessels, they sank them and transferred us to the four boats—one from the *Wiley* and three from the *Winneconne*—with the survivors.

The bombs on the *Hauppauge* failed and the vessel stayed afloat bottom up and was later towed to port and salvaged.

Edna—Von Nostitz next came upon C.A. Small's 3-masted, 62-year-old schooner *Edna*, of Machias, Maine. Capt. Charles W. Gilmore left Philadelphia on May 17 with case oil for Santiago, Cuba. Captain Gilmore stated:

> The submarine came right up to us. The small boat was lowered and an officer came aboard telling us, "You have ten minutes in which to abandon ship." When I was telling the men how to get the boat, which was lashed on the deck, clear, the lieutenant told me to come below.
>
> I suppose I acted as if I was in a hurry to get away from the ship but when we got below the lieutenant said, "Don't get excited, captain. Take your time, we'll be around here an hour and a half." So I picked up everything I could think of that belonged to me and when I got over to the submarine I found I'd left my new silk umbrella. After they blew up the schooner the Germans rowed back to the submarine and I found that besides the few things they had picked up for themselves, they brought me my umbrella. They took a few cases of oil as, of course, they had electric lights on the submarine.
>
> We captains had plenty to eat and they told us we might go where we pleased on the submarine. We could go into any room except when we were submerging. At that time we had to stay in whatever compartment we found ourselves. It seemed to me this submarine was about 300 feet long. There were 76 men in the crew and there were two 6-inch guns mounted—one forward and one aft, the guns being about 32 feet long each.
>
> Of course, we prisoners, although we were treated well, made a pretty big ship's company and the commander said he was looking for a sailing vessel to put us on board as he didn't want to take us to Germany. Although we were below Sunday morning when the submarine overhauled the *Winneconne*, we knew at once when she was sunk and 26 men were set adrift that something would have to be done with us. They couldn't keep us all on board. For after the steamer was sunk we went into her boats, while Captain Sweeney and his crew set out in the launch belonging to the *Isabel Wiley*, which sunk in a few minutes before the *Winneconne*.

These bombs also failed and the next day, May 28, the *Edna* was found adrift near Winter Quarter Shoals Lightship by the Clyde Line steamer *Mohawk*. No sub was sighted so the liner commenced towing and was relieved by the tug *Arabian* and the schooner arrived at Philadelphia on the 29th. Investigation discovered two holes in the hull and a time fuse.

Isabel B. Wiley—The Atlas Shipping Corp.'s schooner under Capt. Thom I. Thomassen, of Philadelphia, left Princess Bay anchorage at 3 p.m. on Saturday, June 1, en route to Newport News, Virginia, to pick up a load of coal for Montevideo.

The same day, Capt. Waldemar Knudsen and a crew of 26 took the American Transatlantic Co.-operated Shipping Board freighter *Winneconne* out of Newport News, Virginia, for Providence, Rhode Island, with 1,819 tons of coal. Chief Mate Henry Walsh later stated:

> At that time we were about 60 miles east of Cape May and looking about I saw a 3-masted schooner with all sails set lying nearly becalmed. There was enough wind to fill her sails and I thought it strange she lay there doing nothing. Then I saw what looked like a destroyer come out from behind her. When we got about two miles away I could see it was a submarine and that the men on deck were not dressed in white as they would be at this time of the year on an American vessel.
>
> Then I saw the German flag and the submarine sent a solid shot across our bow. They lowered a small boat and a lieutenant came with some of the crew alongside, telling us to get into the boats and to row over and lay alongside the submarine.
>
> Just about the time we were leaving our boat the men, on what proved to be the *Isabel B. Wiley*, were going over the side into their gasoline yawl. Of course we obeyed and all went to the side of the submarine and the German crew first sank the *Winneconne* and then the *Isabel B. Wiley* with small bombs. The *Winneconne* went down in five minutes. They gave us officers time enough to save most of our things.

The survivors from the *Hattie Dunn, Hauppauge, Edna, Isabel B. Wiley*, and *Winneconne* were given a large keg of water from the sub. The *Wiley* had a motor whaleboat and it was agreed it would head for shore and try to spot a rescue vessel while the other boats stayed put. At 5:30 they sighted the Ward Line southbound steamer *Mexico* and were picked up. At 5:30 on the evening of June 2 the *Mexico* radioed that the schooner *Isabel B. Wiley* had been torpedoed and they had rescued 50 people in three lifeboats from the *Wiley* and other destroyed schooners.

At about 7:30 the *Mexico* hailed the northbound *Santiago*. The survivors were put aboard the *Santiago* and they proceeded to the location of the other lifeboats. All 48 survivors were picked up and landed at New York at 12 noon on Tuesday, June 4.

Jacob M. Haskell—Crowell & Thurlow's schooner under Capt. William H. Davis, Jr., en was en route from Norfolk to Boston with 2,547 tons of bituminous coal for the Boston & Maine Railroad was sunk. The crew was picked up by the Merchant & Miners steamer *Grecian*.

Samuel H. Hathaway—Crowell & Thurlow's 4-masted schooner *Samuel H. Hathaway* under Capt. Emil Bonamo and a crew of ten was reported lost en route from Norfolk to Para with coal.

Edwin H. Cole—The first evidence of a second submarine was reported by Capt. Humphrey G. Newcombe on Crowell & Thurlow's 4-masted schooner, en route from Newport News to Portland, Maine, with 2,516 tons of coal. On Sunday afternoon, June 2, a submarine suddenly "popped up" on his port side, as he related:

> We couldn't tell at first what nationality she was so that when the commander, through his megaphone, told us to lay to, we were sore. "We're Americans" I signaled back and hoisted our flag. We certainly were surprised when they answered us by breaking out the flag of the Imperial German Navy. The commander then shouted to us that we had ten minutes' time to leave the boat, which was to be sunk.
>
> I got all my men on deck and lost no time getting a boat ready for launching. The U-boat meanwhile stayed about 150 feet away with 15 of its crew on deck. It was about 200 feet long and had two guns mounted fore and aft manned by sailors.
>
> While we waited, the commander and three or four of his marines got into a boat and came on board.

The commander was a handsome chap of about 35, clean shaven, and went through his business of sinking the *Cole* in a businesslike but thoroughly courteous manner.

I'll give you ten minutes to leave the ship" he said in perfect English, without ceremony. Then he asked me where the *Cole* was from, where bound, and what her cargo was. After I had answered these questions he asked for the ship's papers and I took him down to my cabin and handed them over.

You have only seven and a half minutes left now," he then said. I told him we wouldn't wait for the half minute and got on deck as fast as I could. One of the crew started to grumble but the commander turned to him like a flash and said menacingly, "You cut that out and be quick about it."

The German sailors meanwhile had been busy placing bombs about the schooner. They put two on the port side, two on the starboard side, and to more on deck. The German sailors rowed back to the U-boat and we pulled as hard as we could before the detonation.

We were a half mile off and it was just 16 minutes after we had been hailed when all the bombs exploded together and the *Cole* seemed to fall in—crumple up—and disappear at once.

The survivors headed for the New Jersey shore and about four hours later saw smoke on the horizon and realized it was a steamer heading right for the last location of the sub that sank the *Cole*. Captain Newcombe continued:

We thought she would be the next victim and rowed with all our might to cut across her course and head her off in time. We tried to attract the attention of the commander by shouting altogether and by waving our coats. Our signals were finally seen.

"Don't stop to pick us up! Beat it quick or that submarine will get you!" we shouted to the commander of the vessel, which turned out to be the collier *Bristol* of the Coastwise Transportation Company, but Captain Frederick Hart refused to look out for his own safety before taking us on board.

We had only one U-boat up to this time but when we got on the *Bristol* and started heading toward port away from the submarine we ran plump into another one. The one which sank the *Cole* was busy firing at another ship and the newcomer was only 500 yards away.

"There's a nest of 'em," was my first thought. We were all prepared to be shelled if we kept up our flight but decided to make a try at it. We were going to give Fritz a run for his money.

Six of my men went down to help the nine firemen and coalers of the *Bristol* with orders to get every darned knot of speed out of the ship. They did. While the speedy U-boat, with only its periscope visible, came plowing along toward us, the *Bristol* actually lept forward.

The *Bristol* had never done better than nine knots. It did 17 then. The men worked like devils in the stoke-hole and slowly we drew away from the submarine. For half an hour we could see her hanging on grimly, however. Then we lost sight of her.

But for several hours afterward, while we were running full speed for port, we heard the sound of shellings and explosions and I'm sure a number of other boats were sunk.

I've been across three times to France in command of vessels and I've been a master for 12 years and this is the first submarine I've ever met, right here near our shore. This was to be my last trip on the *Cole* for I was going to Rockland to get a new ship, just built.

I'm going home to see the kids and rest up for a week or two then I'm going to get that new boat and try my luck again. I'm not scared and Fritz won't frighten any other American skipper either.

The ship had no war risk insurance and the big 4-masted schooners were valued at around $150,000 each. Capt. Newcombe, 43, lived in Somerville, Massachusetts, with his wife Margaret, 32, son Charles, 12, and daughters Sadie, 10, and Madeline, 8, at 39 Whitman. Humphrey and Margaret were Canadians.

Capt. Frederick Hart reported:

The *Bristol* left Boston on June 1 en route to Norfolk, Virginia, in ballast. On June 2, about 8 p.m., we sighted a lifeboat with 11 men, crew of the American schooner *Edward H. Cole*, schooner having been sunk by submarine at 3:30 p.m. about 50 miles southeast of Barnegat Light. The crew of the *Edward H. Cole* was picked up.

At about 8:20 p.m., when about 38 miles southeast of Barnegat Buoy, bearing 120° true, we sighted a submarine about 5,000 yards off the starboard quarter, heading directly toward the ship. By traveling at maximum speed of 12 knots, we managed to outdistance him and arrived at New York at 5:37 a.m., June 3.

The submarine appeared to be about 200 feet long and to be armed with two rifles which appeared to be either 5 or 6 inch.

Texel—Capt. Kenneth B. Lowry's Shipping Board freighter left Ponce, Puerto Rico, on May 27 for New York with a cargo of sugar. At about 4:21 p.m. the *U-151* stopped the ship with several close shots and one hit. Everyone got off and about an hour later bombs were put aboard and the ship sunk. Capt. Lowry, described Korvettenkapitän von Nostitz:

He was about 5 feet 8 inches, probably weighed about 200 pounds, stocky build. He had light hair, wore a mustache and Van Dyke beard. He was about 40 years of age and wore a Navy uniform with long overcoat. His rank was indicated by two gold stripes on his sleeve, slightly above the wrist. He spoke English.

Carolina—On Sunday, June 2, 1918, Capt. Thomas R.D. Barbour's New York & Porto Rico passenger liner was en route from Puerto Rico to New York with 333 passengers and crew. That afternoon they were about 60 miles off Tucker Beach Light at Atlantic City, New Jersey. Erwin W. Vogel, 20, of Paterson, New Jersey, was the chief wireless operator. His assistant was H.W. Werner. Vogel later stated:

I was awakened at 5:35 o'clock on Sunday afternoon for supper. The other fellow came running in while I was dressing and yelled, "Come up here quick!" I rushed into the wireless room and there was an S.O.S. from the schooner *Isabel B. Wiley*. I clamped the receivers to my ears and heard it myself—"S.O.S. *Isabel B. Wiley* being shelled by armed German submarine latitude 38.07 longitude 75.10."

I told Captain Barbour he better make quick headway for, judging by the strength of the signals, the sub must be right on top of us. You're right the skipper said and he bawled out to change our position and we began to zigzag and speed up. At about 6 o'clock I was down for supper and had just ordered my meal when I heard something smack alongside. I had a hunch what was going on but I went out on deck without attracting any more attention than possible. I hurried up to the wireless house and sent the other operator up top to get our position. He returned shortly saying someone else was figuring out where we were and meanwhile I was setting up and forthwith shot out our first S.O.S.—"S.O.S. steamship *Carolina* being gunned by German submarine." I sent a couple of times then I set my instrument for receiving. Within two minutes Cape May station had picked up and queried for our location. But just then the German operator butted in and said, "You don't use wireless—we don't shoot." Sounds German doesn't it? By the way, I could tell from the faintness of the vibrations that the Hun sender was geared to be heard not over a half-mile away.

On order then from my captain I replied that we were laying to but I followed that by repeating the S.O.S. call.

I forgot to say that the first shot went right across our bow. When I repeated the call for help the Huns sent over another, which whizzed some distance over our top. A third went to the rear and the fourth and last fell short. It seemed as though it was designed to. I guess they intended only to scare us but I thought they were a bit careless where they shot.

By this time everyone had rushed from the dining room and was trying to find shelter behind the deckhouse or below. I listened in again and caught repeated calls from Cape May and one from the Brooklyn Navy Yard asking where we were. I hadn't our bearings from above yet but knew approximately where we were. Just when I was rigging up to give this, Captain Barbour ordered "No."

I went out of the wireless room and watched the passengers and crew going over the side in boats then went back to the operating room and tore up a secret navy code book which the Huns might have found useful. All other papers of any importance that I found were put in a silver water pitcher and thrown overboard. It sank. Then the Captain came in and with him I took to a boat. Everyone else had left.

Capt. Barbour had immediately blacked out his ship, increased his speed, and commenced a zigzag course due west. Just as the navigator had established their base course, a lookout sighted a submarine on the surface about two miles away. As fate would have it, he ran right into the *U-151*, the first meeting between an American passenger ship and a U-boat.

Korvettenkapitän von Nostitz used flag signals to order everyone off. About 40 or 50 sailors were seen gathered on the sub's deck. At first, Capt. Barbour thought the German might try to seize his liner as a commerce raider but then changed his mind due to the heavy wireless traffic and he knew von Nostitz was aware that American destroyers were in hot pursuit of his sub. Von Nostitz didn't even take the time to send a boarding party aboard to loot the ship's stores for food.

As the passengers and crew got into the ten lifeboats they began singing The Star Spangled Banner and all were off within 15 minutes. Von Nostitz then fired three shells into the port side and three shells into the starboard side from about 1,500 yards and the ship sank around 8 p.m. Before leaving, he used his big searchlight to indicate a course toward shore for the lifeboats.

The survivors headed west in smooth seas until around midnight when the weather turned rough. The boats became separated and conditions in the open boats were bad, especially for the numerous children.

At about 8 a.m. on Monday the 3rd, the occupants in No. 3 lifeboat spotted a 4-masted schooner through a light fog. The occupants of the other nine lifeboats were notified and the six oarsmen in each boat rowed toward the schooner. At about 10:30 they came up to Capt. Lauro's schooner *Eva B. Douglas*. It took over an hour for 160 passengers and 94 crewmembers to get aboard the *Douglas*. Capt. Lauro headed for New York in very light winds and at noon on Tuesday the 4th they anchored outside the gas buoy off Barnegat Light, New Jersey.

Capt. Barbour went ashore to send messages and the chief mate went to request a tug. He returned around 2:30 p.m. just as the Navy's converted yacht *Machigonne* arrived. A towline was rigged and when they arrived at the submarine net at New York a tugboat took over the tow.

M.R. Spellman, the Alien Property Custodian for Puerto Rico, and his wife, who had become very ill, were taken aboard the Navy boat. When the *Douglas* arrived at the Quarantine Station, they were met by tugs with Red Cross supplies and were directed to Pier 35 in Brooklyn where they tied up at 6 p.m. The National League for Women's Service gave the survivors food, supplies, and cash. On the 5th, 254 survivors reached New York.

Survivors reported that a lifeboat with seven passengers and six crew capsized in the heavy seas and all were lost. On Monday, June 3, the 29 survivors in Lifeboat No. 5 rowed into Atlantic City, New Jersey, through the surf to the astonishment of thousands of Boardwalk visitors.

The next day, Tuesday the 4th, 19 more survivors were found and taken to Vineyard Haven and 18 and more were taken aboard the British steamer *Appleby* and brought to Lewes, Delaware. Of the 333 aboard, 13 were reported lost, including five merchant crew:

Nathaniel George Mussenden, purser, wife Helen, 811 St. John's Place, Brooklyn.
C. Johansen, 1st engineer, 62 First Place, Brooklyn.
Alejandro Cordero, waiter. Mayaguez, Porto Rico.
Ethel Margaret McHarvey, stewardess, mother Isabel, 295 President St., Brooklyn.
Carlos Munoz, messman, wife in Mayaguez, Porto Rico.

Sam C. Mengel—at 6 p.m. on June 3 the Axim Transportation Co. schooner from Pensacola, Florida, was sunk by *U-151* in the Atlantic at 38°08'N/73°35'W. There were no casualties in the crew of 9. Chief mate John W. Wilkins reported that one of the Germans shook his hand and said, "Send Wilson out here and we will finish him in ten minutes. Wilson is the only one prolonging the war."

Pinar Del Rio—Capt. Mackenzie's American & Cuban freighter was en route to New York with sugar. On June 8 they were off the coast of Maryland when they met the *U-151*. Everyone was ordered off and the ship was sunk with gunfire. There were no casualties from the 35 crew. Chief mate Arkes and 15 others landed at the Life Saving station at Manteo, North Carolina. The master and 17 others were in the second boat and apparently arrived at Norfolk on the 11th.

On about July 1, Von Nostitz headed for home and reached Kiel on August 1, having sunk 23 vessels. On June 15, Kapitänleutnant Richard Feldt left Germany for America in the *U-156* and arrived on July 1. On July 19 he sank the cruiser *San Diego*.

On June 22, Korvettenkapitän Waldemar Kophamel left Kiel in the *U-140* for the American coast via north of the Shetland Islands. While the *156* operated to the north, the *140* went south. On July 14 the sub was spotted by Cdr. Wallace Bertholf's troop transport *Harrisburg* at 45°33'N/41°W. The ship sent an "A.L.L.O." to report the sighting.

On July 20, 1918, the cruiser *Galveston* picked up a radio message, supposedly from the tanker *Joseph Cudahy*, which contained confidential information. Because of that, the cruiser's captain thought the Germans had seized the ship and had the ship's confidential papers and were setting up a decoy. It turned out the message was legitimate.

Kapitänleutnant Feldt in the *U-156* had a very good day on July 21. The tugboat *Perth Amboy* was towing four barges about 3 miles off Orleans, Massachusetts, south of Cape Cod when the *U-156* fired reportedly three torpedoes at the tug, which seems unlikely unless the tug and tow looked like a large, low-slung vessel and worth a torpedo. Feldt surfaced and commenced firing. One shell came through the wheelhouse, cut the helmsman's hand, and set the tug on fire. He then fired on the barges. Three women and five children were on one barge. Thirteen crew were injured by shrapnel.

The episode was witnessed by Robert F. Pierce, keeper of the Coast Guard Station #40 at Orleans. Pierce ordered the motor surfboat out to assist. The sub submerged and Pierce and his crew treated the wounded. One of the barges, the Pennsylvania Railroad Co.'s *P.R.R. 403*, of New York, was apparently sunk. Seaplanes from the Chatham Naval Air Station were sent out and would probably have sunk the sub if the bombs dropped had exploded.

About 3 miles off Orleans, Massachusetts, he sank Lehigh Coal & Navigation Co.'s schooner *Lansford*, of New York, with gunfire and wounded three. The barge *Barge No. 403* was next. There were no casualties.

Lehigh Valley Transportation Co.'s schooner barge *Barge No. 740* and schooner barge *Barge No. 766* of Perth Amboy, New Jersey, were sunk by gunfire. There were no casualties.

Robert & Richard—On July 22 Feldt sank the Chisholm Corp. schooner with gunfire about 60 miles southeast of Cape Porpoise, Maine. There were no casualties. One member of the schooner's crew said he recognized the sub's commander and the officer "boasted" that he had a home in Maine for many years.

Dornfontein—Feldt headed north and at 11 a.m. on August 2 surfaced on sighting Capt. Charles E. Dagwell's brand-new lumber schooner about 25 miles southwest of Brier

Island en route to South Africa. The boat was looted then set on fire. Dagwell and his crew of eight landed on Grand Manan Island in the Bay of Fundy at 6:30 a.m. on the 3rd. Dagwell said the sub commander told him he had been on the coast for six months and was leaving in October and he was fairly certain it was the same sub that sank the tug and coal barges off Cape Cod.

These sinkings caused widespread panic and big newspaper headlines. It was known that the merchant sub *Deutschland* had been armed and was in the naval service in American waters. Just prior to the *Pratt* sinking it was reported that a sub fired at the British tanker *Arco* but they were able to escape, being in ballast. Navy ships in the area fired on what they believed was a submarine but nothing came of it. The Bridgeport Telegram stated, "Never since the time the British fired upon Lewes has this town been so stirred as it is tonight, and for the greater part of the populous sleep is out of the question."

All sailings from the ports of New York, Boston, and Philadelphia were cancelled. Secretary of the Navy Daniels reported that within 24 hours the Navy Department received 5,000 telegrams, radio messages, and phone calls demanding action and asking questions. Many wanted destroyers recalled from Europe. As he later stated:

> The halls and offices of the Department were thronged with anxious people, shippers and shipowners, friends and relatives or captains and crews. And everybody wanted information.
>
> There was alarm along the coast, from Cape Cod to Key West. If one U-boat was over here, two might be or three or more. That was the general feeling.
>
> One of the most persistent questions, which came from the country, as well as the press, was whether we were going to recall our destroyers from Europe—and in many cases this was not put as an inquiry but a demand.
>
> We could not tell the public what we were doing, what ships were being sent out, and where. That was just what the Germans wanted to know. Most of our destroyers and the best of our patrol craft were in European waters 3,000 miles away, performing vital duty against the enemy in England, Ireland, France and Italy. We had no idea of recalling them.
>
> Thousands of vessels would have been required to patrol every mile of our long coast-line, and guard all the boats off our shores. Our duty was clear. The Germans had sent their U-boats across the sea mainly to interrupt the transportation of troops and supplies. If they did not succeed in that, their coming would have no real military effect.

The *U-156* and the Seal Island Fishing Fleet

On August 3 the *U-156* came upon a group of fishing schooners in the vicinity of Seal Island, Nova Scotia. The 1907 Hague Conference produced a number of conventions regarding merchant vessels. Convention XI, Chapter II, Article 3 stated:

> Vessels used exclusively for fishing along the coast or small boats employed in the local trade are exempt from capture, as well as their appliances, rigging, tackle, and cargo. They cease to be exempt as soon as they take any part whatever in hostilities. The Contracting Powers agree not to take advantage of the harmless character of the said vessels in order to use them for military purposes while preserving their peaceful appearance.

Germany participated in the Conference and ratified this agreement. The sole purpose in attacking these boats could only have been a desperate attempt to cause panic and capitulation. They were certainly no threat to anyone militarily.

Annie Perry—116-ton schooner, crew of 20. Philip P. Manta, owner. Boston. Capt. James Goodwin. Sunk with bombs.

Muriel—120-ton schooner, crew of 22. Atlantic Maritime Co., Boston. Capt. Eldridge Nickerson. Escaped.

Rob Roy—111-ton auxiliary sword schooner, crew of 18. Gorton-Pew Vessels Co., Gloucester. Capt. Freeman Crowell. Sunk with bombs.

Sydney B. Atwood—100-ton. Reported sunk with bombs. This vessel does not appear in any official register nor was its loss reported by the Commerce Department.

The *Muriel* was about 45 miles northwest of Seal Island; the *Sydney B. Atwood* was about 30 miles east; the *Rob Roy* was about 35 miles east; and the *Annie Perry* was about 35 miles east. Manta was in the retail grocery business with Jozo N. Souza as Manta, Souza & Co., 221 Commercial St. He lived at 61 St. Andrew Rd. in East Boston.

Charles Swain was a sailor aboard the *Annie Perry* and gave a statement to a notary:

> Mr. Charles H. Swain states that they left Boston, Friday noon, August 2, 1918, and that on Saturday about 2:30 p.m. they sighted a boat about 5 miles due north. They did not know it was a submarine till it was about 4 miles away. The submarine then fired a shot across the bow when about 3 miles away and they hove to. Mr. Swain states that they saw the *Rob Roy* sunk. The crew then put off in four dories with 10 men. They were 35 miles from land off the Nova Scotia coast, W. half south from Seal Island. The sub came close to the dory and the officer of the sub called for the captain to go aboard. Our captain and four men went aboard. The German captain was a young man, German, about 23 years old. He said, "Don't be afraid, we won't hurt you, but we're going to sink your ship." He further stated that we were fortunate to be near land and told us that he sank one ship 400 miles from land and that he was the man who had sunk the *San Diego*. He said that he had supplies to remain out three months. The sub crew consisted of 60 or 70 men and about 30 or 35 were on deck. The submarine was battleship gray and had two large guns fore and aft. The submarine was clean. They took all the supplies from our ship and the sub captain gave the crew in [the] dory brandy and cigarettes. The sub captain then asked for newspapers.

Several crewmen from these boats who went aboard the subs claimed that the commanders had so much knowledge of coastal waters "from Wood's Hole, Massachusetts, to Nova Scotia" that they must have been engaged or shipping or fishing at one time. One man even thought he recognized one of the German officers even with the beard he had grown.

O.B. Jennings—On March 24, 1918, Capt. George W. Nordstrom's Standard Oil of New Jersey tanker collided with the British freighter *War Knight* in the English Channel, which resulted in the deaths of seaman James J. Shea and Armed Guard sailor Charles La Rue Phillips, S2c. Shea was survived by his mother Josephine of 77 Eaton Place, East Orange, New Jersey.

On July 20 Capt. Nordstrom left for home with 50 total on board. The *U-140* spotted the ship off Cape Henry, Virginia. Capt. Nordstrom later reported:

> While on a voyage from Plymouth, England, to Newport News, Virginia, in ballast, on August 4, 1918, in position 36° 40' N., 73° 58' W., a torpedo was sighted about 1,000 yards, four points on port bow. By maneuvering ship, torpedo missed and passed 3 to 4 feet astern. At 9:30 a.m. the submarine opened fire from a distance of about 8 miles and we immediately hoisted our flag and opened fire after we broke out our smoke boxes and made a smoke screen, changing course to hide ship behind smoke screen. At this time the submarine was observed five points abaft port beam. The submarine followed us and kept shelling until 11:40 a.m. having our range finely all the time, several shells bursting so close to ship that ship's side was punctured in several places. At 11:40 a.m. a direct shot hit the engine room through the counter, smashing port engine and wrecking main steam line. Several men were wounded. at the same time another shell hit magazine and exploded, destroying all ammunition. Previous to this one man was killed by gunfire and some minor wounded on deck. I pulled my colors down and sent wireless calls. I surrendered at 11:45 a.m., ordered all hands in the boats, and abandoned ship. Pulling away from the ship, the submarine came up

and interrogated boat crews, took second officer prisoner, asking the boat crew where the captain was, and the crew answered that the captain was killed. Then the submarine returned to ship and commenced to shell her. We pulled away to northward until out of sight and then hauled around to westward. At 3 p.m. all boats together, sails set, course given by captain, compass regulated and commenced to sail for land. During the night, the first officer's boat and chief engineer's boat dropped out of sight. At 8:30 a.m. we were picked up by Italian steamship *Umbria* and given food and relief. This ship took us down off Currituck beach, put the boat into water, and we sailed for shore. We landed at 3:30 p.m. August 5, 1918.

At 9:30 a.m. sent out allo call and at 9:50 approximately sent out S.O.S. at irregular intervals thereafter. The surrender call was sent out on emergency set as dynamo was smashed.

Breech locks for our guns were thrown overboard upon the approach of the submarine to the life boats. The telescope sights were left with the chief engineer's boat. Moving pictures were made of the *O.B. Jennings* and her lifeboats from the deck of the submarine.

Survivors were rescued by destroyer *Hull,* whose commander forwarded a report to the Bureau of Operations:

At 9 a.m. (ATS), August 4, 1918, while proceeding at a speed of 10 knots, a torpedo wake was sighted four points off port bow. Helm was put hard astarboard and quarters was sounded and full speed ahead ordered. *Jennings* had been zigzagging and continued to do so, holding the same course. About 9:30 a.m. several shots were fired by the submarine, which was then sighted. The *Jennings* opened fire on the submarine, which was then apparently 7 miles away. Shots from *Jennings* fell short. *Jennings* continued at full speed, zigzagging and uses smoke boxes. Submarine continued firing, discharging about 40 or 50 shots at two-minute intervals. At 11:40 a.m. shot from submarine exploded in the engine room of the ship, disabling the engine and wrecking the main steam line. A few minutes later another shell hit the magazine, exploding and destroying all ammunition remaining. Prior to this time the *Jennings* had fired 60 rounds at irregular intervals, causing submarine to submerge twice. During the engagement submarine fired 150 rounds at approximately two-minute intervals. At 11:45 ship's flag was struck and wireless call sent out that the *Jennings* had been captured. In this connection it had been the intention to use the word surrendered instead of captured.

Preparations were immediately made for abandoning the ship, and at 12:20 p.m. all survivors had left the ship in three small boats. Position of *Jennings* at this time was latitude 36° 40' N., 74° W. Submarine then approached the *Jennings* and continued shelling her. The boats drew away from the vicinity and at 2:20 p.m. the *Jennings* turned on port side and sank slowly, disappearing from view 15 minutes later. At this time the lifeboat in charge of the first officer, W.J. Manning, was about half a mile away from the Jennings. The lifeboat in charge of the chief engineer, Albert Lacy, was 2 miles ahead of the first officer's boat and the boat in charge of the captain was ahead of the chief engineer's boat, thus the actual sinking of the Jennings was observed only by the first officer's boat. After the crew abandoned the ship and before they were permitted to leave the vicinity the three boats were called to the side of the submarine and a conversation between the second officer of the submarine and the men in charge of the three lifeboats took place as follows:

Q. We got you at last. I knew we would. What damage did the shell in the engine room do?

A. Put the engine out of commission.

Q. Where is the captain?

A. He is dead.

Q. Where is the chief wireless operator?

A. I don't know, he must be dead too.

The above conversation was carried on by one of the men in the captain's boat, during which time the captain and the chief wireless operator were both present, but the captain's clothes had been placed on the body of the second steward, who had been killed and left on the deck of the *Jennings*.

The chief engineer was in charge of one of the lifeboats and in this lifeboat was also Rene Bastin, second officer of the *O.B. Jennings,* who had joined the ship at Southampton, England. Bastin insisted that he be permitted to speak with the officer on the submarine and, despite the fact that he was slightly wounded, jumped from the lifeboat to the deck of the submarine and began speaking rapidly in German to the officer and men on the deck, finally shaking hands with them and without further conversation with his companions in the lifeboat went below decks of the submarine and never returned. The men and officers on the *O.B.*

> *Jennings* had been suspicious of this man during the entire voyage. At the time he joined the ship in Southampton he claimed to be a Belgian and produced proper credentials to substantiate his claim. He spoke French and German fluently. The captain of the *O.B. Jennings* feels confident that this man could have been carrying no confidential documents to the officers aboard the submarine.
>
> The submarine was about 300 to 325 feet in length, and the top of the conning tower was about 20 feet from the surface of the water, very rusty looking, with a guard around the propeller. It was armed with two guns, 6-inch caliber, barrel 20 feet long. They were placed fore and aft and about 6 feet from the conning tower. There were two periscopes about 5 inches in diameter on the conning tower. No masts or wireless visible. Three officers and 30 men were observed on the deck of the submarine, all dressed in regulation blue uniforms. The hat of the men had the inscription" "Untersee Boat Deutschland Untersee Hamburg."
>
> At 2:20 a.m., August 5, two boats in charge of the first officer and chief engineer were picked up by the U.S.S. *Hull*. About 30 minutes before this occurred a submarine was sighted moving slowly along the surface. Submarine passed about 50 yards from first officer's boat and about 300 yards from the captain's boat. Submarine was about 200 feet long, conning tower about 20 feet from the surface of the water.
>
> The U.S.S. *Hull* searched the vicinity for 30 minutes, endeavoring to locate the captain's boat, but without success, as the captain refused to show any lights or answer any signals, believing, as he stated later, that the U.S.S. *Hull* was the supposed submarine sighted about 30 minutes earlier. The only fatality was that of the second steward. Several of the crew sustained minor injuries from shrapnel.

Inquiries were made concerning Rene Bastin and sometime later a reply arrived from the American consul at Le Havre, France. The office reported that Rene Henry Bastin presented himself at the consulate on January 4, 1919, and asked for transportation back to the United States. Bastin claimed he was held aboard the sub *UK-140* as a prisoner of war along with four or five other survivors until the ship arrived back at Kiel on October 25, 1918, after sinking several other ships and survived a depth charge attack. He said that on October 29 a mutiny occurred in the German navy's surface and undersea fleets, and on November 2 he was transferred to a POW camp at Wilhelmshaven and then "interned" on the cruiser *Hamburg*. On the 9th of November the prisoners were told an armistice had been signed, mutinies had broken out in the British and German navies, and "peace conditions would be favorable to Germany." On November 11, when the terms of the treaty were made known, everyone became depressed. On the 21st, the prisoners were transferred to an American officers' POW camp at Karlsruhe where there were about 20 American Flying Corps pilots. Bastin was released on the 29th and went to Villingen, about 20 miles from Switzerland, where he was cared for by the Red Cross. From there he said he was sent to the American Base Hospital No. 26, APO 785, arriving on December 1 and from there sent by rail to John Ball Osborne, the American consul at Le Havre, as a destitute American seaman. He told the consul he was born in Ostend, Belgium, on June 21, 1889, came to New York in June 1918, and took out citizenship papers but that he was not then a citizen. Because of that, the consul said he could not issue him a destitute seaman's certificate. But Osborne arranged for his passage to New York on the NOTS freighter *Newton*, then under charter to the Army. The captain, Lt. Cdr. Richard C. White, took him aboard as a consular passenger. On January 11, 1919, the *Newton* left Le Havre for Newport News, Virginia, via Plymouth, England.

Bastin gave his address as 24 John St., Southampton, England. The casualties on the *Jennings* were Thomas McCarthy, 2nd engineer, English, 61 Northumberland Rd., Southampton, seriously wounded, and James H. Scott, 2nd steward, English, wife Alice Marie, 2 Alderly Terrace, Southampton, who died from gunfire.

Stanley M. Seaman—Lewis K. Thurlow's 4-masted schooner was en route from Newport News, Virginia, to Porto Plata, San Domingo, with coal when they were stopped by a shot from *U-140* at 11:30 a.m. on August 5 about 100 miles off Cape Hatteras. The crew immediately got into their boats and rowed toward the sub with the ship's papers. The sub then towed them back and they were ordered to tie the schooner's stern to the bow of the sub. The sub's crew then looted the schooner of provisions and supplies. The crew was allowed to leave and the vessel was sunk. There were no casualties and they were put aboard a steamer three days later.

Agnes G. Holland—The fishing trawler was sunk on August 5 by *U-156* in the Atlantic and then the British-flag London & Pacific Petroleum Co. tanker *Luz Blanca* was torpedoed 40 miles off Halifax with two lost.

Merak—On August 6 Capt. Searborough's unarmed Shipping Board freighter was steaming at 8 knots through thick haze off the coast of North Carolina. At about 1:40 p.m. they were about 4 miles west of the Department of Commerce Light Vessel *No. 71* at Diamond Shoals when they heard a report and saw a flash through the mist then a column of water off their port bow. The freighter turned about and headed for shore on a zigzag course. More firing ensued at intervals of about one minute. After about 30 shots had been fired, the ship ran aground and the crew took to the lifeboats. The *U-140* then came alongside and a boarding party went aboard, set charges, and detonated them. The lifeboats were ordered alongside the sub and the master was interrogated. The sub commander checked the master's answers in a copy of Lloyds Register of Ships. Visibility was excellent there and the Lightship, the British steamer *Bencleuch*, and Capt. Louis Hansen's New York & Porto Rico Steamship Co. freighter *Mariners Harbor* were clearly in sight. The lightship was sunk by gunfire and the sub chased and fired on the British ship, which escaped along with the *Mariners Harbor*.

The freighter had a crew of 40. Twenty were picked by the Coast Guard and arrived at Elizabeth City on the 8th.

The Norwegian steamer *Sydland* was chartered by the Allies to bring relief supplies to Belgium. Capt. Alexander N. Larson left Bagen, Norway, for Hampton Roads, Virginia, and on August 8 was stopped by the *U-156*. The chief mate struck up a conversation with one of the sub's sailors who told him he had sailed with the Hamburg-American Line as a chief officer and was qualified to be an officer in the German Naval Reserve but had to sail as an ordinary seaman because he had gotten married in New York and had a brother in the U.S. Navy.

On the morning of August 10 an S.O.S. was heard from the Brazilian steamer *Uberaba*. Destroyer *Stringham* responded but found nothing. Three days later the destroyer *Pastores* sighted a sub on the surface about 6 or 7 miles away. The sub then commenced firing at the destroyer, which was returned. After firing 15 rounds the sub took off and disappeared about five minutes later. The destroyer dropped 15 depth charges but got no results. The next week was quiet until August 21 when the British steamer *Diomed* was sunk far to the north of the original sighting and it was later learned to be the *U-140*.

The *U-117* and the Georges Bank Fishing Fleet

In early July 1918, Kapitänleutnant Otto Dröscher left Kiel and headed his cruiser-minelayer *U-117* toward the East Coast of North America. On the way across, on July 26, he attacked the British steamer *Baron Napier* and on August 5 sank the Standard Oil tanker *Louis Blanchet* off Halifax, Nova Scotia, after a 3-hour gun battle. The chief cook and chief steward were lost when the torpedo hit the stern.

Around this time, the torpedo boat *Tingey* reported a submarine 5 miles south of Sable Island, which was later identified as the *U-117*.

On Saturday, August 10, around 30 or 40 auxiliary fishing schooners were spread out over a wide area 140–150 miles off Massachusetts on the Georges Bank and Dröscher ran right into one group. The boats he sighted on surfacing were:

Albert W. Black—54-ton auxiliary sword boat, crew of 12. Portland, Maine. Capt. Granville Johnson. Escaped.

Aleda May—31-ton auxiliary schooner, crew of 9. Leandro J. De Costa, Jr., East Boston. Capt. Frank Lynch. Sunk with bombs.

Alice M. Doughty—33-ton auxiliary sword boat, crew of 10. Portland, Maine. Capt. Reuben Doughty. Escaped.

Cruiser—28-ton auxiliary cod schooner, crew of 6. Bridget A. Barnwell. Sunk with bombs. Bridget owned a restaurant in Boston at 18 Howard St. and lived at 31 Temple with Richard J. Barnwell.

Earl and Nettie—24-ton general fishing, crew of 6. Anton Hendrikson, New York. Sunk with bombs.

Gleaner—44-ton auxiliary sword boat, crew of 7. New York. Capt. Edward A. Proctor. Escaped.

Katie L. Palmer—31-ton sword boat, crew of 8. Frank S. Willard, Portland, Maine. Capt. Edward Russell. Sunk with bombs.

Mary E. Sennett—27-ton sword boat, crew of 8. Gloucester. Capt. Manuel Dias. Sunk with bombs.

On Time—18-ton auxiliary cod boat, crew of 4. Gloucester. Escaped.

Progress—34-ton sword boat, crew of 7. Capt. Robert L. Jackson, owner and master, Vineyard Haven, Massachusetts. Sunk with bombs.

Reliance—19-ton cod boat, crew of 6. Commonwealth Fisheries Co., Boston. Capt. James Nickerson. Sunk with bombs.

William H. Starbuck—53-ton sword boat, crew of 8. Robert A. Sanchez, Jr., Newport, Rhode Island. Capt. Albert Sanchez. Sunk with bombs.

On seeing the sub or being fired on the *Black*, *Doughty*, *Gleaner*, and *On Time* immediately gunned their engines and put on every square inch of sail they had and left. The *Black* was fired at around 9 a.m. The *Doughty* was able to make 12 knots with motor and sail and landed $5,000 worth of swordfish at Gloucester on the 13th.

The other crews took to their dories, uncertain of the intention of the submarine. Dröscher brought his sub up to each boat he caught, took off all the provisions, stores, and supplies and then put a bomb aboard and detonated it. He rounded up all the dories he could, firing at some to stop them, and took some of the fisherman aboard for questioning. More than anything, Dröscher wanted to know where any minefields were. If they had no provisions in their dories, he gave them supplies and sent them on their way.

Andrew St. Croix, the mate on the *Katie L. Palmer* said there were 30 or 40 boats of various flags fishing in the area and he believed the sub got most of them before the day was over.

Robert Sanchez said the sub fired on the dories before coming up to them and stated, "As many as 20 of the crew were standing in the conning tower laughing as the *Starbuck* was being rifled of all her provisions by some of their mates."

At around 1 p.m. the *Mary E. Sennett* was fishing with the *On Time*. Capt. Dias was up on the mast looking for sword fish when he spotted two submarines to the southeast. One left and wasn't seen again but the other began circling the boats. The boats took off to the northwest but when each one was fired on, the crews got into their dories. Capt. Dias believed the *Cruiser* and *On Time* had been sunk with no survivors before his boat was boarded. He reported that they sailed northwest for home and at 2 p.m. on Sunday the *Goodspeed*, "a knockabout trawler on Charles Bank" picked them up, took them back to the fishing grounds, and transferred them to a U.S. subchaser. From there they went aboard the steamer *Rijndijk* at Island Light and were brought into port where they reported to the immigration office.

Capt. Russell of the *Katie L. Palmer* and his crew left Boston on August 6. He said they heard gunshots at around 10 a.m. but continued fishing. At 3:30 p.m. they sighted a sub about 8 miles away. Louis Amirault was in the crew and the engineer was French Canadian Fred Doucette, 33. Doucette said when he heard the first shot he gunned the engine but the second shot hit the mainsail and at that point the crew got into the dories. The sub was about a quarter mile off when firing then came up and took Capt. Russell and Doucette aboard the sub. A boarding party came aboard and took off all the stores and supplies of all kinds. They put a bomb aboard, "painted red, about the size of a tomato can" and about 10 minutes later it exploded. They could see the sub had a cutting device of some sort on the bow and saw the magazine of the forward gun, which Russell thought held about 200 shells. Doucette said he was surprised when Dröscher told him nobody would win the war. Russell later stated:

> I was questioned as to whether or not I was armed. Four of my men boarded the submarine, but the other three did not. I was told to go below. The crew were sociable and said, "Don't hurry" and asked if we had water and provisions. I replied that we had. The submarine crew then took our stuff out of the dory. I had a drink with some of the members of the crew and was given a bottle of cheap rum, unsealed. An officer of the crew asked about the mines and tried to get us to talk. We said we were not permitted to know anything about these matters.

Before they left, all of the goods removed from the dory were put back aboard and they left the sub about 8 p.m. They headed toward the South Shore Lightship in the hope of meeting a vessel and at around 10 or 11 they were picked up by the *Helen E. Murley* and landed at New Bedford at 3 a.m.

All the fishermen were greatly alarmed by this submarine activity and requested naval protection. All who were taken aboard the subs reported that the Germans were very afraid of mines in the coastal waters. The officers wanted to know what the Americans thought about the war. They also seemed to be very ignorant of the land war and had no idea of how many U.S. troops were in France, greatly underestimating the number. No one seemed to express virulent anti–American feelings or ill will toward the United States. They seemed to believe that war was war and sinking enemy vessels was their job. More than one survivor said some sub officers expressed concern over injuring or killing anyone. Others who went aboard the subs reported the crews were dirty and had a yellowish cast to their complexions or described them as the picture of health. Some boats were cleaner than others and reports indicated the officers and crews wore a variety of uniforms, combinations of uniforms, or no uniforms. Some survivors said the Germans were drunk and abusive and desecrated the American flag from one boat.

Reports of a second sub were also made. It took about a week for naval authorities to get all the facts regarding this raid and put the whole story together as the survivors were all spread out and arrived ashore at various times and on different vessels.

Dröscher continued his patrol after looting and sinking nine boats. On August 12 he torpedoed Capt. George Hansen's Norwegian steamer *Sommerstadt*.

On Sunday, August 11, at about 2 p.m. the *U-117* met the trawler *Acushla* and 14 crew from the *Aldela May*, *Progress*, and *Starbuck* were transferred off and they arrived at Boston at 9 a.m. on Tuesday, August 13.

Frederick R. Kellogg—Capt. C.H. White's Pan American Petroleum & Transport Co. tanker was 30 miles south of the Ambrose Channel Lightship, 12 miles north of Barnegat Light, and 5 miles offshore at 5:10 p.m. on August 13 when Dröscher's *U-117* fired without warning. The ship was hit in the engine room and sank stern first in 15 seconds. Capt. White stated:

> The torpedo must have been close to the water for the reason that it blew up two steel decks and a wooden deck and a lifeboat on the port quarter clean into the air. There was one engineer, the third assistant, one fireman, and one oiler on watch, and all three were killed. Four others were killed or drowned. These men were in the vicinity of the engine room. They were the second engineer, one mess boy, and two cadets.
>
> The ship was unarmed. However, she was measured last trip for gun platforms. It was impossible to do anything for the defense of the ship for she was torpedoed without warning and the sinking was instantaneous. Even if the ship had been armed nothing could have been done.

The ship sank in shallow water and was later salvaged. Chester Curtis Cubberly, QM3, and William T. Stillman, QM3, from the Naval Auxiliary Reserve office in New York were aboard for instruction and were lost. The merchant crew casualties were:

Leo Volgemuth, seaman, American, Riverdale, New York, injured slightly.
Samuel L. Johnson (aka Albert Myron Hamilton), 2nd engineer, American, sister-in-law Mrs. C.A. Hamilton, Racine, Wisconsin.
James Kramer, 3rd engineer, Norwegian, mother Marie Hansen, 425 61st St., Brooklyn.
John K. Carlsen, oiler, Norwegian, mother Julia, Christiania, Norway.
Axel Albert Jorgensen, fireman, Norwegian, mother Anna Caroline, Copenhagen.
Francisco de Sousa, messboy, Portuguese, sister Elvira, 165 Bemformosa St., Lisbon.

Dorothy B. Barrett—Gardiner G. Deering's big schooner under Capt. William Merritt with a crew of 10, encountered the *U-117* on August 14 and then Lt. (jg) Charles L. Greene's minesweeper *Kingfisher*. Greene reported later to naval authorities:

> The schooner *Dorothy Barrett* was at 9 o'clock about 6 miles NE of Five Fathom Bank Lightship heading NNE when from the WNW a submarine on the surface about 4 miles away ESE fired one shot. The *Dorothy Barrett* hove to and lowered her jib and abandoned ship. The submarine came within 2 miles and submerged. I started my boat NW to try to get some assistance. After getting within 1½ miles of the submarine I stopped the boat for awhile and then thought I had better try to secure some help.
>
> I was to the north and west of the submarine. Then he made a short circle and then I had to go on the outside. I then had him on a line between me and the ship. Now I was between him and the shore. He waited awhile to see what I would do. I thought if he wanted that vessel he would take it and in the meantime I would try to get some help. I then gave our boat a NW course then headed WNW and ran in on that until I found the *Kingfisher*. The submarine followed me for a while and then stopped and showed his conning tower. I kept going and bye and bye I got away. He was going to the ship again and worked on to a line between me and the ship with his conning tower still up. Then for about 10 minutes after that I saw the *Kingfisher* coming from the northward. I laid my course for the *Kingfisher* to see if he could not give us some assistance. He swung around as soon as I signaled him and I went aboard.

I asked the captain what he had to fight with. He said he had three-inch guns and I told him that the submarine was too big for him to tackle with that kind of gun. I then went up on the bridge and asked him what he was going to do. He said he was going to run for shoal water. We had been running about west only a short time when we sighted the oil tanker. I said to the captain, "He is going to get that fellow also and to shift our course to warn the tanker." The submarine was on our starboard quarter and we commenced firing. We were then heading about west and that submarine was about 2½ miles on our stern, just so we could fire by the pilot house. I did not see the conning tower then but the gunner said he saw it and he fired a couple of shots.

When I saw the tanker she was headed SE. by this time she had also seen the submarine, I guess. We did not have but little time to go down toward him and started to fire on this submarine and then we had to watch the fellows. We kept right on down that way toward him and started to fire and put on all the steam we could but, as I said before, we wanted to get in shoal water.

Then the submarine came up and went right ahead of us and we fired about three shots. From the time we abandoned our ship to the time the shots were fired, I judge it was about three quarters of an hour.

Some of them came pretty near. Just before we shot the last few times a big black smoke cloud came up out of the water. It was coming up just like steam. Then, about five minutes later, another big black steam cloud seemed to rise from the water from near the vicinity of the submarine and kept coming up for about 10 minutes. I called the captain's attention to it, but he said he could not tell what it was. I do not believe it was over 5 fathoms deep where that submarine was. The last thing I did before I left my vessel was to heave the lead and it showed 15 fathoms. If we only had one of your fast submarine chasing boats there we could have gotten that fellow because he was in shallow water and could not turn very fast.

Madrugada—The 5-masted Gravenhorst & Co. auxiliary schooner was sunk by gunfire on August 15 about 35 miles off Norfolk, Virginia. There were no casualties among the 22 on board.

On August 20 the Italian steamer *Ansaldo* fought a three-hour gun duel with the *U-117* where over 200 shots were fired. The Italians quit only after they ran out of ammunition and put on speed and fled. The freighter was hit several times and had three crewmen wounded.

The Raider *Triumph*

In August 1918, Kapitänleutnant Richard Feldt's *U-156* was outfitted with equipment to cut trans-Atlantic cables. One terminated at Canso, Nova Scotia, so he was dispatched there.

On August 12, Arthur Boutilier's 125' steel-hulled Canadian steam trawler *Triumph*, of Halifax, left Portland, Maine, under Capt. Gjert Myhre for the Western Banks fishing grounds around Sable Island. He only had a 10-day coal supply so grueling work was carried on 24 hours a day in all weather to fill his hold with 300,000 pounds of cod, halibut, and haddock. The *Triumph* was one of three steam trawlers that fished out of Halifax. Everything iced up in winter so the boats had to maintain a rigid schedule to meet the year-round demands of buyers and processors.

Around noon on the 20th, the crew saw a sub surfacing astern of them and around 2 p.m. Feldt ordered a shot fired. Capt. Myhre heaved to and the sub came up. The 11 fishermen were brought aboard and photographed and given refreshments. When the festivities were over they were put into their dory and commenced rowing toward Canso.

Feldt put 16 sailors aboard the boat with arms and ammunition, including two rapid-fire guns, and the new raider set out to destroy as many fishing boats as they could.

A. Piatt Andrew—At around 2:45 p.m. the raider came upon the Chisholm Corp.'s Gloucester fishing schooner under Capt. Wallace Bruce. The 24 fishermen aboard were stunned when shots were fired at them. Bruce hove to and when the stranger came alongside they realized they were Germans. Capt. Bruce was ordered to bring his papers over then three Germans got into his dory with revolvers and bombs and went back to the *Andrew*. The rest of the crew were ordered off with very little time to gather personal effects. About 15 minutes later the bombs exploded while Feldt waited 4 or 5 miles away. The boat sank at about 3:30 p.m.

Sylvania—Jeffry F. Thomas's Gloucester schooner was sunk on August 21 about 90 miles southeast of Canso. There were no casualties from the 22 on board.

Francis J. O'Hara, Jr.—Manuel P. Domingues' Gloucester schooner under Capt. Jose P. Mesquita was sunk about 55 miles off Canso with bombs. Eighty survivors from the fishing smacks arrived at Canso late on the 21st.

J.J. Flaherty—Gorton-Pew's fishing schooner *J.J. Flaherty* under Capt. Charles T. Gregory was sunk on August 25 by *U-156* about 30 miles from Saint-Pierre and Miquelon Islands off Canada. There were no casualties in the crew of 25. The boat was valued at $50,000.

Rush—The Commonwealth Fisheries steam trawler *Rush* left Boston on August 20 under Capt. Alvro P. Quadros to fish at Quero Banks. At 5:45 a.m. on August 26, about 135 miles southeast of Cape Canso, Nova Scotia, they sighted a submarine that came up and crossed their stern within 50 feet. The trawler hove to and crewmen Joseph Golart, Joseph Telles, and Joseph Tasida were taken aboard the sub. Three hours later they came back outside and saw the wreckage of their trawler. They said they heard no explosion or noises. Golart said the captain showed him a list of the boats he had sunk including the *Progress, On Time, Katie L. Palmer*, and *Mary Sennett*. The sub took 1,200 gallons of fuel oil from the trawler. They got off the sub at 9 a.m. and at 11 were picked up by the *John J. Fallon* and taken to Canso. There were no casualties.

The *Triumph* ran out of coal in three days after sinking 7 vessels in all so Feldt ordered it scuttled. No cables were cut and the *U-156* eventually headed for home. At around 44°20'N/51°09'W encountered the Shipping Board steamer *West Haven*. The sub opened fire and the steamer sent a radio message:

> 8:25 a.m. Attacked by an enemy submarine which opened fire upon us. Immediately brought guns into action. At 8:46 a.m. enemy submarine ceased firing and turned broadside to.

Richard Feldt's sub hit a mine in the Northern Mine Barrage on the way home and sank leaving 21 survivors who were landed in Norway after the Armistice. The rest of the crew was presumed lost. Feldt is remembered in Canada for treating survivors well and never taking a life.

The subs were off the East Coast for about four months but none were able to attack a trans–Atlantic convoy or do any real damage to the Allied war effort.

The *U-117* headed for home but on August 30 they sank the Canadian fishing schooners *Elsie Porter* and *Potentate*. Most of their crews were picked up by the steamer *Solberg* on September 1. W.M. Rheinhard, captain of the *Elsie Porter*, was taken aboard the sub. He described the commander as "about 35 years old, medium height, thin face, sharp nose, dark complexion, black hair, and appearance dirty." This was the last adventure for the *U-117* on the American coast.

On September 9 the *U-117* came to assistance of the *U-140*, which had sprung a leak, apparently from some previous damage, and the two proceeded to Kiel. The *140* arrived on

October 25 but the *117* apparently came in on fumes and had to be towed in by two destroyers.

Kingfisher—On September 20, 1918, the East Coast Fisheries Co.'s steam trawler was stopped by *U-155*. The captain stated:

> The *Kingfisher* was out of Boston and at the time she went down was at latitude 43.31 north and longitude 61.53 west. At this position a submarine was sighted on the starboard quarter about two-thirds of a mile away. The submarine fired a shot across the stern of the trawler and as soon as the shot was fired the trawler stopped and I ordered the small boats launched. A second shot was then fired by the submarine which passed directly over the smokestack and we then immediately abandoned the ship. We started to row towards the submarine which had in the meantime come up nearer the trawler, motioned us to come alongside, which we did.
>
> When first seen the submarine was apparently coming to the surface and the first shot was fired within a few seconds thereafter. The submarine was flying the German flag which was hoisted on the corner of the bridge. There was no number on the submarine, which I estimated was about 250 feet long. The conning tower seemed to be lower than that usually seen on submarines and also was apparently of larger size. The submarine carried two six-inch guns with portable mount. The two large guns were permanently fixed. I did not notice a searchlight and saw only one periscope which was in the conning tower. The submarine was equipped with wireless and had what looked like a large pair of steel scissors over her stern and rudder. She was camouflaged on both sides and apparently freshly painted with grey, blue and white. The submarine carried a boat which I saw and which probably was 30 feet in length and was raised up out of the hold just aft of the gun.
>
> Most of the officers and crew word corduroy jackets. The trousers looked old and worn and some of them were torn. Some wore wooden shoes, others were barefoot and a few had long leather boots. The officer who was on the bridge was probably 25 years of age and wore side whiskers. Most of them were unshaven and looked rather ill kempt and none of them wore uniforms. The three officers that I saw all wore a large number of medals, probably 10 or 15. We counted 23 members of the crew including the three officers and as the engines of the submarine were working, a portion of the engine crew probably was below. The officer who interrogated me spoke good English without any foreign accent. The questions he asked were as follows: "Where are you from? When did you leave? Where are you bound for? What is the name of the captain of the trawler? Have you seen any large ships? He then said, "Captain, I am going to sink your ship," and motioned us to row away.
>
> After we had rowed a short distance the boat from the submarine rowed over to the trawler carrying 10 to 12 of the submarine crew, who placed bombs on board the trawler as we heard three loud explosions at about 11:35 a.m. The bombs which I saw on the deck of the submarine were large brass bombs. When I last saw the trawler, which was before we heard the explosions, we were five miles away and the submarine was then alongside of her.
>
> The crew of the trawler, including officers and men, numbered 26 in all. We landed at West Quaddy, Nova Scotia, Sept. 21, at about 10 a.m. and were driven over by farmers to Fort Dufferin, which place we left Sept. 23 on the steamship *Dufferin* for Halifax, arriving on the morning of Sept. 24.

One crewman stated:

> The commanding officer was about 6 feet tall, thin, ruddy complexion, sandy hair, wearing short side whiskers, and appeared to be between 25 and 30 years of age. Another officer of the submarine was about 5 feet 8 inches tall, light complexion, clean shaven, and weighed about 180 pounds. About 53 men were counted on the deck of the submarine wearing various kinds of uniforms. The crew were all young men and apparently healthy and contented.

San Saba—At 5 p.m. on October 3, 1918, Capt. Bergen G. Birdsall's Mallory freighter *San Saba*, under operation of the U.S. Railroad Administration, left New York for Tampa and Mobile with general merchandise. The next day at 12:45 a.m. about 15 miles southeast of Barnegat Light, New Jersey, they hit a mine set out by *U-117* and sank in five minutes with the loss of 30. The only known survivors were 2nd mate Adolph Beer; seaman Edwardo Simona; and coal passer Pedro Aceredo. They got ahold of debris and were picked up between 4 and 4:30 p.m. by the Norwegian steamer *Brieford*. The survivors left Norfolk, Virginia, for New York on the 6 p.m. New York, Philadelphia & Norfolk train. Of the 30 lost, the American casualties, all presumed from drowning or exposure, were:

Bergen Birdsall, master, wife Alice, 202 Pearl St., Somerville, Massachusetts.
John L. Downie, quartermaster, brother J. Vale, Beaver Falls, Pennsylvania.
William J. McBride, chief engineer, wife Sarah, 290 Vanderbilt Ave., Brooklyn.
Frank Miller, 2nd engineer, wife Annie, 502 E. 73rd St., New York.
Ralph Richards, 3rd engineer, mother Julia, 151 Davis St., Jacksonville, Florida.
Harry Cobb, junior engineer, wife Laura, 714 1st St., Brooklyn.
Arthur Ray, oiler, father John E., Marine Operating Co., 29 Broadway, New York.
Joseph Robinson, messman.
Fred West, messman, mother Julia, 915 Stonewall St., Brunswick, Georgia.

Amphion—In October 1918 the Shipping Board armed steamer was en route from Bordeaux to New York in ballast under War Department operation. On October 12 they were attacked near Bermuda by *U-155*. The sub reportedly fired over 200 shots and the freighter 72. The freighter's superstructure and lifeboats were badly damaged but the Navy gunners believed their last shot hit the sub, which then fell off astern and submerged. The *Amphion* was the ex–German ship *Koln*. An American seaman, Diamond Atkinson, died from wounds. His father was Diamond C., of Buford, Georgia. A Russian able seaman, Karl Denau, was wounded by shrapnel and transferred to the Marine Hospital at Stapleton, New York.

Lucia—Capt. Charles Frederic Leary's Shipping Board freighter was under charter to the Army and had the distinction of being the last American merchant ship sunk in the war. Ironically, the ship was sunk by the first submarine to cross the Atlantic, the *Deutschland*, now the *U-155*.

The *Lucia* was en route from New York to Marseilles, France, in an unescorted convoy. At 5:30 p.m. on October 17, at 38°05'N/50°50'W a torpedo from *U-155* hit the engine room on the port side with the loss of all four watch engineers. The Armed Guard trained their guns in the direction from which the torpedo came while the deck gang got the lifeboats swung out.

The freighter was fitted with "buoyancy boxes"—rubber-bound barrels—which were intended to keep a ship afloat after a torpedo hit allowing the crew plenty of time to get off. The gunners were forced off at 1:30 p.m. the next afternoon when waves started breaking over their gun platforms.

The steamer *Hawaiian* was the "guide ship" about 1,000 yards ahead and got their S.O.S. Survivors got off and the ship sank at 3:20 p.m. the next day. A very heavy sea was running and on October 19 at 9:26 p.m. they sighted Lt. Cdr. Stanford C. Hooper's destroyer *Fairfax* and 86 survivors were brought aboard.

Of the 90 on board, four were lost from drowning: James Kenny; E.C. Young; 3rd engineer Michael B. Gaynor; and assistant engineer Daniel Edward Moynihan who was survived by his mother Margaret of St. Helena, Maryland.

The *U-155* then headed for the Azores and was reported at Kiel on November 12. The *U-155* was later put on display in London and scrapped in 1921.

The West Coast and Pacific

German influence and intrigue in Mexico was worrisome for two reasons: The Germans could establish bases for raiders or U-boats or they could stop the flow of oil from Tampico. Navy Secretary Daniels declared after the war:

> Long before 1914, Great Britain had felt dependence upon Mexican oil for its increasing oil-burning navy, and had made provision for securing it through acquisition of Mexican oil fields. American captains of industry had likewise large oil fields in Mexico. From the minute war was declared in 1914, Allied dependence was upon Mexican and American oil. Tampico and Port Arthur were strategic points in all Allied plans of campaigns on sea or land. If this supply of oil had been interrupted, the war might have gone on much longer.
>
> From the day the first German raider sank a British ship or a submarine fired at an Allied vessel, the British and French were zealous to protect the oil supplies in Mexico. They maintained patrol vessels in that region and kept ceaseless vigil of sea routes to protect this priceless agency of war. However great their need of ships on their own coast, they knew that if the oil supply failed at Tampico they would lose the only adequate available source of oil for all their operations.
>
> No U-boat bases were ever discovered in Mexico but a plot to fit out a raider was. Those waters were constantly patrolled by the Navy. The only problem American shipmasters encountered in the Pacific were the raiders *Wolf* and *Seeadler*. The effort to put another raider into service in the Pacific was broken up and made spectacular headlines at the time.

The Raider *Seeadler*

The Imperial German Navy decided the big sailing vessel *Pass of Balmaha* they had captured and condemned as a prize of war in December 1915 could serve as an armed merchant cruiser in the Pacific. To get through the British North Sea blockade it was decided to disguise the *Balmaha* as a Norwegian vessel taking lumber to Melbourne, Australia, which at that time was a very common voyage. A more pressing problem was finding an experienced naval officer who could handle the boat and who spoke enough Norse to pull it off.

The only officer with experience under sail was Kapitänleutnant Graf Felix von Luckner, a real count who had run away to sea as a cabin boy, breaking his father's heart and his royal family's long tradition as cavalry officers. He concealed his true identity—a fairly common practice among sailors—and signed on as Phelax Luedige. He had many adventures before wearing the Emperor's uniform, although not the one his long-suffering father had desired.

When asked, von Luckner jumped at the chance to get back under sail and serve the Emperor. Much work then began on the ship. Two new "ultra-modern" 500-hp motors were installed along with heavy guns in concealed locations in addition to rifles, grenades, bombs, and other armaments. Bunks for 400 prisoners were installed. Tanks for 480 tons of fuel oil and 480 tons of fresh water were added. A cozy dining room, spacious, modern, living quarters, and a well-appointed recreational lounge were constructed for officer prisoners equipped with a gramophone and records, books, and magazines in French and English.

Secret passageways were constructed so the movements of the crew could not be observed and so British prize crews could be captured while on board, which they did on numerous occasions.

A deck load of lumber was put aboard and the Norwegian flag painted on the hull. According to international law, a blockade runner manned by navymen was allowed to disguise itself as a neutral and the sailors could wear any clothes they wished. By contrast, a soldier on land engaging enemy forces was always required to be in his country's uniform. However, when a blockade runner was confronted by an enemy vessel the runner was required to show its true colors, which, of course, von Luckner had no intention of doing.

The tricky part was finding a name for the ship. The ship's official German name was *Seeadler* (Sea Eagle), but they needed something that would pass muster with a British commander. They searched the Lloyds Register for a very similar-looking Norwegian ship and found the *Maleta* but after expending considerable effort to create an elaborate disguise, including von Luckner actually stealing the logbook from the *Maleta*, the plan fell apart when the *Maleta* sailed sooner than expected. When a second name, the *Carmoe*, also failed, von Luckner chose "*Irma*," the name of a girl he had fallen in love with on a cruise as a young naval officer.

They left the Weser on December 21, 1916, and headed north toward Iceland. On Christmas Day they were stopped by Capt. Arthur L. Ashby's British armed merchant cruiser *Avenger*. Their ruse worked and Ashby let them go.

On his way south, Von Luckner sank 11 ships, continued on around the Horn to the Pacific then sank three 4-masted American schooners. All were reported overdue in October 1917.

A.B. Johnson—James Rolph, Jr.'s lumber schooner left Raymond, Washington, on May 14 under Capt. Andrew B. Petersen for Newcastle, Australia. On June 14 they were captured at 1°33'S/151°15'W. The crew of nine was taken aboard the raider and the next day von Luckner burned the schooner.

R.C. Slade—Capt. Haldor Smith took Pacific Freighter's schooner out of Sydney, Australia, with a cargo of copra for San Francisco under charter to Comyn-Mackall & Co. Capt. Smith stated:

> I left Sydney on April 24, 1917, and proceeded without any incident until the evening of June 17 when I was in latitude about 2 north and longitude 150 west. On the evening of June 17, about 5 o'clock, the second mate reported to me that a ship was firing on us. I went on deck and looked aft, and instantly, as I came on deck, they fired again and I saw the shell fall short about two miles. She was about eight miles off. There was a heavy squall starting to eastward, wind favorable at this time, and I thought it possible to get away and kept holding on. But she kept firing on me at intervals of about 5 to 10 minutes and was coming upon me fast.
>
> The ninth shot fired, about 6 o'clock, struck very close, passing the poop and splashing water on the ship. Then I concluded that there wasn't any use, and I lowered down spanker, clewed down topsail, hoisted American flag and hove to. About 7 o'clock the raider was up alongside and asked what ship. I told him what it was and he told me to lower down sails and stand by and he would send an officer aboard me. Shortly after, the prize officer came aboard, and a doctor and about 10 men. The officers were in uniform. They told me to leave the ship and to go on board the raider and they would give me time in the morning to pack my clothes.
>
> They took all our men aboard the raider except the cook. Next morning I went back on board with all my men and packed up.
>
> We left the ship with our belongings June 18. We were put on board the "raider again. Shortly after I saw from the raider that they cut holes in the masts and placed dynamite bombs in each mast and put fire to both ends of the ship and left her. I saw the masts go over the side and the ship was burning from end to end, and the raider steamed away.

The crew from the *A.B. Johnson* was aboard the raider. Capt. Smith was 72 and had 58 years at sea when he took command of the Liberty ship *William J. Gray* in July 1944 at San Francisco.

Manila—Capt. Fred E. Southard's Pacific Freighter schooner, under charter to Comyn-Mackall & Co., loaded lumber at Raymond, Washington, for New Castle, Australia, and after an uneventful passage loaded coal and left for Honolulu on May 25. On July 8 they were captured by the *Seeadler.* Among the crew was 1st mate Hans Hansen, a naturalized American, 2nd mate Frederick John Williams and able seaman Charles Thompson.

Von Luckner took them aboard and cruised around for about three more weeks looking for ships without success and on Tuesday, July 31, sighted Mopeha Island, an atoll in the Leeward group of the Society Islands. He decided to enter the lagoon but wasn't able to get in and instead anchored on the lee side. The wind freshened, the anchor dragged, and on August 2 the ship was blown ashore. Von Luckner asked Capt. Southard for his opinion, and Southard replied that it was impossible to get an iron ship off a coral reef. Von Luckner then took off all the beer and wine, usable equipment, arms, tents, two boats, and a powerful wireless set and dynamited the wreck. The wireless set was installed between two coconut trees. There were three Tahitian men and a boy already on the island making copra for Grand, Miller & Co., of Papeete, who said a boat would be along soon to pick them up and load the copra.

On August 23 von Luckner and four officers put hand grenades, 6 rifles, a machinegun, and 6 months' provisions, a fuel into a motor launch. He said they were going to the Cook Islands with the intention of capturing an American ship then return to Mopeha, pick up his remaining crew and prisoners, and continue on their merry way.

At daybreak on September 5, the 2-masted French trading schooner *Lutece* from Papeete the Tahitians were expecting arrived. A German officer named Kling immediately saw an opportunity to escape and ten men set out in a launch and captured the schooner. The two French owners, Pedro Miller and M. Fain, were on board with the native captain and 14 native men.

At 9 a.m. Kling brought two boatloads of prisoners ashore and the Germans prepared to leave the island. The schooner had 4 tons of water, 450 sacks of flour, 150 cases of salmon, and 150 cases of beef. They destroyed the wireless plant and gave the castaways some bad flour, sugar, canned mutton, butter, syrup, and molasses and the schooner left at 9 p.m. that evening, leaving all the prisoners on the island.

The Germans left a small, damaged boat and the captain of the *Lutece* said it would be possible to reach Tahiti in it. Repairs commenced and on September 7 the boat was declared seaworthy. Capt. Smith was against making the attempt but the castaways, some armed, turned on him. At that point, Capt. Southard, the "old man" at 60, decided to make an attempt. His plan was to sail to Maupiti Island, around 90 miles to the north, and hire a motor schooner to Papeete. Supplies were put into the boat and he and John Johansson, Frederick Williams, and Charles Thompson set out at 3 a.m. on the 8th. At 8 p.m. on the 16th they arrived back at Mopeha, thoroughly exhausted, having failed in the attempt due to bad weather.

Capt. Smith then said he would make an attempt. Johansson, Williams, and Thompson agreed to go again and the four left at 8 a.m. on September 19, leaving 44 people total on the island. The days later they arrived at Pago Pago and reported to Governor John M. Poyer, who sent the steamer *Tiare Taporo* to rescue the castaways. The boat arrived on October 6 and left on the 11th.

In the meantime, unbeknownst to anyone on the island, von Luckner and five of his crew were captured on September 21 off Fiji. The castaways reached San Francisco on December 12.

The Raider *Wolf*

"Kaiser Bill" wanted another raider in the Pacific but Germany had no coaling stations so it was imperative that the ship carry a sufficient supply for a long voyage. Hamburg-American had a subsidiary, the Hansa Line, whose freighter *Wachtfels* was outfitted to carry 7,500 tons of Westphalian coal. Fregattenkapitän Karl A. Nerger took command of the *Wolf* and left Kiel on November 21, 1916, with a total crew of 375 including two surgeons, 500 mines, and 2,500 tons of food and ammunition. They were overloaded two feet below the ship's Winter North Atlantic marks but by the spring of 1917 they were in the Pacific. Nerger sank four American ships.

Winslow—Capt. Robert Davie Trudgett took a load of lumber in George E. Billings' schooner to Sydney, Australia, then loaded 325 tons of coal, benzene, and firebrick for Samoa. On June 16, 1917, they were off Raoul (Sunday Island) in the Kermadec Group, about 500 miles south of New Zealand. At around 3:30 p.m. a seaplane came over and gave Capt. Trudgett a message to "steer southeast to a German cruiser, or else the cruiser would come up and shell you."

At 6 that evening an innocent looking freighter appeared. Kapt. Nerger identified himself and said he had intercepted a wireless message and had been waiting for the schooner for several days. A prize crew was put aboard with orders to sail to the raider's anchorage at Sunday Island. The sea turned rough so a towline was rigged and the cargo was transferred to the *Wolf*. The firebrick was just what they needed for the boilers and the benzene was fuel for the seaplane that flew over. On the 22nd the schooner was set on fire but the raider had to fire 39 rounds to sink it.

Capt. Trudgett, the chief mate Carl Teigland, a naturalized American; 2nd mate C.A. Anderson, also naturalized; William Alford, an American able seaman; Jacob Evanoff, a Russian able seaman; George McMahon, an Irish able seaman; A. Nelson, a Swedish able seaman; Charles Nelson, a Norwegian able seaman, and the Japanese cook Y. Murakama were held aboard the raider for over ten months and then interned in Germany.

Beluga—Capt. John Stanley Cameron had quit the sea after he married his Australian wife Mary and their daughter Juanita was born in 1911. By 1917 Mary hadn't seen her family in ten years. Over the years Capt. Cameron had promised her several times they would make the trip but their plans always fell through. Now with the war on and his services in great demand and a nice, safe voyage to Australia possible, it seemed like a good time to fulfill his promise.

The Pacific Steam Whaling Co. had purchased the bark *Beluga* in February 1917 for $27,000 and he agreed to take 15,000 cases of benzene from San Francisco to Sydney, Australia. At some point he realized he had a crew of ten, and with himself, Mary, and Juanita, that made 13. He, like many sailors, was superstitious, but he tried to brush it off as silly. After all, this was 1917, not the Dark Ages! Yet he couldn't shake the nagging, uneasy feeling he secretly held.

Everything went well until July 9 when Fred "Fritz" Jensen, a Norwegian seaman, was working aloft and spotted smoke on the horizon off their port beam. Since it was the was the first vessel they had seen since leaving port, they were all interested in the stranger and they might get some news.

When Capt. Cameron looked through his glasses, the vessel appeared to be coming toward them. As he said later, "To all appearances she was the ordinary black-painted, dingy-looking tramp." His chief mate, Matthew Buckard, took the glasses and said, "By God, Captain, I don't know her nationality but she carries the largest crew I have ever seen." Capt. Cameron immediately became tense and just then the stranger hoisted the Imperial German Navy ensign and signaled "G-T-E" for "Heave to and I will send a boat on board." Two minutes later, the Germans dropped their bulwarks and fired a shot across the bow. Capt. Cameron was incredulous. They were 15,000 miles from the war zone, or so he thought, and said later:

> In a moment there flashed through my mind all the accounts I had read in the papers of the German atrocities towards women and children in Belgium and barbarisms practiced along the Russian front and the thought of my wife and child being at the mercy of these people nearly drove me crazy. On walking aft I saw my wife leaning against the wheelhouse, her face absolutely bloodless and a look of horror in her eyes that fairly chilled my blood. When she saw me she came over and took my hand in hers, looking all the time into my eyes and not saying a word. We stood there for what seemed a century. Presently I called to Juanita to us and the three of us went down below to the cabin. We sat on the settee, never saying a word and poor little Nita started to sob, feeling something sinister in the air, which she did not understand. In a minute the mate came to the cabin skylight and sang out that the launch would be alongside in a minute. I answered, "All right." My wife got up and walked over to the bed and took one of my revolvers—I had two—from under the mattress and handed it to me.
>
> Suddenly she threw both her arms around my neck and drew my head into such a position that she could look into my eyes, and said, "Stanley, I want you to promise me that they will never get Juanita." I threw both my arms round her, hugging her tight to myself and said, "Mamie, I promise, but you must leave it to me," and with a sob I left her and started on deck.

A young officer, Lt. Zelasco, came aboard and in excellent English asked Capt. Cameron the usual questions. He said he had sailed as an able seaman on the American steamer *Roanoke* some years before, a ship Capt. Cameron had also been on, and assured him that no harm would come to his family. He said he was taking his vessel and ordered the Stars and Stripes hauled down. The Germans had brought aboard 20 pairs of handcuffs and three large bombs to sink the bark with. The crew was lined up and searched and when Zelasco learned they had benzene, Capt. Cameron was ordered to sail 60 miles due east and wait there. They wanted the benzene as fuel for their float plane but had gotten word of an approaching steamer they wanted to capture. In the meantime, the German prize crew took to little Juanita immediately and she enjoyed playing games with them.

On the 11th the *Wolf* returned after an unsuccessful attempt at capturing the steamer. They took 300 cases of benzene and all the provisions and stores and 1:30 p.m. everyone got off into boats and on approaching the raider Capt. Cameron could see all the German sailors smiling and grinning with the anticipation of sinking his ship. Once on board, he was met by the officer of the deck who said, "Tell your wife and little girl that they have nothing to fear, that we are not the Huns you probably think we are." At around 4:30 Capt. Cameron was told they were going to sink his ship with gunfire from 2½ miles distance and fired 19 shots from a 6" gun that exploded the benzene into a spectacular fireball and set the sea ablaze. Cameron and his family were on the raider for ten months and treated very well, unlike the common sailors who lived in miserable conditions. At one point Nerger told Capt. Cameron he would have taken the benzene and let him go if he had known his wife and child were aboard.

In the course of time, Capt. Cameron met other members of the raider's crew who had actually held American masters' licenses before the war and stated later, "I am led to believe that at the outbreak of hostilities the German consuls at the port where their vessels hailed from ordered these men to resign and go home to the Fatherland. I also believe that their fare and expenses were paid."

His crew taken prisoner were Matthew Buckard, Canadian 1st mate; Axel Johnson, Norwegian 2nd mate; Fred Jensen, Norwegian seaman; N.M. Bennett, American seaman; Charles Anderson, Swedish seaman; Bertil Torsten Lindholm, Russian seaman; Johnan Martin, Swedish seaman; J. Morris, Scottish seaman; Daniel Parra, Chilean seaman; and R.L. Weatherwood, English seaman. On October 10 Axel Johnson died aboard the raider from heart failure.

All were imprisoned in Germany. Capt. Cameron and his family were released in Denmark on February 24, 1918, left Christiania, Norway, on the steamer *Bergensfjord* on April 11, and arrived at New York on the 28th. After repatriation, the crewmembers wanted to know if the owners intended to pay wages lost during their detention. On December 24, 1918, the Department of Commerce ruled there was no liability "that these seamen were taken as American citizens who were liable with the rest of their countrymen to the hardships and hazards of war."

Encore—The Simpson Lumber Co.'s 4-masted schooner loaded 630,000 feet of Pope & Talbot lumber and 100,000 laths at Port Gamble, Washington, and left the Columbia River for Sydney, Australia, under Capt. Antone Olsen and a crew of nine. The normal voyage was 75 days but on November 7, 165 days out, no word of their fate had been received and they were considered lost. They had been captured on July 15 and the crew taken prisoner. Two days later the schooner was burned. Capt. Olsen, of Oakland; Chief mate Hillamer Lennardsen Richardsen, of Willapa, and 2nd mate Colstad were held at Camp Karlsruhe.

John H. Kirby—Gaston, Williams & Wigmore's 3-masted bark was en route from New York to Port Elizabeth, Africa, under Capt. John Arnold Blom with general merchandise and automobiles. They were captured on November 15, 1917, about 300 miles south of Port Elizabeth, Africa. Blom and his crew of 15 were taken prisoner aboard the raider.

A German Consul's Foiled Plan

In late 1916, the Regents of the University of California decided to sell their nine-year-old auxiliary schooner *Alexander Agassiz* because the 70' marine research vessel no longer suited the needs of the scientists who used it. The price was set at $6,000 and was sold to Frank C. Wheeler, Maude M. Lochrane, and William Taylor.

Maude was a striking 38-year old red-head who lived in Los Angeles and worked as a reporter. Frank was a Los Angeles businessman from Michigan whose family had adopted Maude when she was six. William Taylor was an African American who had business experience in Mexico but unbeknownst to the others, was under indictment by the U.S. Justice Department for dealing in counterfeit Mexican money.

None of the three had the money for the boat but Maude knew Joseph Mesmer, president of the St. Louis Fire Brick & Clay Co., with whom she had worked on various civic projects. She told him their plan was to got to Mazatlán, Mexico, and take cargo and passengers between San Pedro and Mexican ports. Several other American boats were known to be operating out of Mazatlán at the time under the American flag and regular passenger and freight service was provided to San Francisco by steamship companies.

Mesmer agreed to finance the purchase, having full faith in her character and credibility. On January 2, 1917, she made a $600 down payment on the boat and Mesmer executed a $5,000 bond in favor of the University.

Maude, Frank, and William formed the Pacific Coast Trading & Shipping Co. and rented office space at 207 S. Broadway, Room 411, in Los Angeles. Maude was secretary of the company and it was agreed that she would run the office and pay off the debt to Mesmer while Frank and William ran the boat. Neither of them had a navigator's license, were mechanics, or had experience operating a trading vessel so a local captain and mechanics would have to be hired.

On January 27, 1917, the boat was issued a Temporary Register at Los Angeles and the same day Wheeler and Taylor left San Diego for Mazatlán.

The primary exports from Mazatlán were gold and silver, spilt peas, tomatoes, various ores, ixtle fiber, sugar, fish, wax, pearls, and coconuts.

There were no wharves in town so vessels were obliged to anchor out anywhere from one to two miles from shore and hire stevedores and lighters to transfer cargo. A shallow-draft vessel like the *Agassiz* was ideal for Mexican waters. Lighterage charges were $3.00 per ton and stevedores got 75 cents an hour with double time for nights and holidays. The stevedores had to carry goods on their backs from the customer's warehouse to the lighter and at low tide they had to walk from 100 to 200 yards across the mudflats. The lighterage service could generally handle 1,000 tons of goods during a typical 12-hour day.

Launch services took passengers between ship and shore for $1.00 each way per passenger plus .25 cents for a suitcase and .50 cents for a trunk. All port charges and fees were based on meters of draft in Mexican currency. Ten Mexican dollars equaled $4.98 U.S. Mexico used the old Spanish system of weights and measures. The use of a pilot was compulsory and the pilotage charge was $12.00 per meter of draft to enter and clear and that applied at each port of entry for all vessels trading on the Mexican coast. Sanitary dues were 2 cents per net ton, a bill of health was $5.00, and a clearance charge from the Captain of the Port was $8.00.

At precisely the same time the *Agassiz* arrived at Mazatlán, a 27-year-old mechanical engineer from Los Angeles named Cornelius Heintz left home for Guaymus, Mexico, and thus, by sheer chance, a fateful chain of events was set in motion.

Normally, any American arriving at, or doing trading business in a foreign port, would call on the American consul at their first opportunity to inform him of their plans and to insure they receive any pertinent information but Wheeler and Taylor never did. The American consul at Mazatlán was William Chapman.

They hired an agent and a captain and the boat found work right away, including transporting soldiers for the Mexican government, but their happy situation would not last long.

On February 5, 1917, President Wilson proclaimed that no American vessel could be transferred to foreign registry or sold or chartered to a foreigner. This was not a problem until Mexico passed a law shortly after prohibiting foreign-flag vessels from operating coastwise between Mexican ports.

To Frank and William the obvious solution was to change the boat to Mexican registry. They called at the American consulate but Chapman was away and they were referred to the French consul, who was acting in Chapman's stead. The French consul was unaware of President Wilson's proclamation and was unable to help anyway so Taylor and Wheeler consulted an attorney and the Captain of the Port at Mazatlán. Whether anyone involved at that point knew of the new American law is unknown and at that time communication with the outside world was very difficult. All mail was opened and examined, trains were held up by bandits, passengers robbed and murdered, and telegraph messages intercepted. Messages sent by diplomatic cable were somewhat safer but not guaranteed. Mail brought on steamers by State Department couriers was about the safest mode of communication.

Frank and William changed the registration to Mexican and posted a $4,500 gold bond in favor of the Mexican government. But Taylor and Wheeler represented themselves as the sole owners of the boat and the deal was made without Maude's knowledge or consent. As an officer in the company her signature was required on all official documents. Under the terms of the agreement, the boat could be seized for unpaid port or pilotage fees or for any debt owed to the Mexican government until the bond was paid off. Any unpaid charges would also prohibit the boat from clearing for a foreign port.

Since the boat was still under foreign ownership, they now had to wait in line behind Mexican-owned vessels for passengers and cargo. They were also charged higher port and pilotage fees and duties and Frank and William soon found themselves without operating capital.

Just then, at the point of financial collapse, they met a tomato buyer named H.A. Macintoch who had his own transportation problems. He saw the *Agassiz* as the ideal solution but his first impression of Frank and William was unfavorable and he was leery of getting involved with. When they offered Macintoch at-cost transportation for his goods in exchange for his business and his connections, he put aside his reservations and advanced them money for operating costs, which was to be deducted from his freight charges. As a prudent businessman, Macintoch consulted an attorney who advised him to put a lien on the boat so it couldn't be sold without his consent. All of this was done without consulting William Chapman or Maude, whom they had come to regard as a "girl Friday" instead of an officer in the company.

Some trouble started over the tomato contract and Taylor accused Wheeler and Macintoch of trying to cut him and Maude out of the business. Taylor apparently threatened to sail the boat as far south as he could and sell it, along with other unpleasant consequences. All of this was too much for Frank and in mid–May 1917 he went back to Los Angeles, reported to Maude, and quit the venture.

When Maude heard the boat was operating under the Mexican flag she immediately contacted the collector of customs at Los Angeles and he told her to get the flag changed back to American as quickly as possible. When she reported all this to Joseph Mesmer he decided to send her down to Mazatlán and take over the operation of the boat. During her time in Los Angeles she had obtained a commercial boat operator's ticket and Mesmer had resolved Taylor's problem over the counterfeit money. Taylor wrote him a very eloquent letter of appreciation.

When she arrived by steamer on June 4, the boat was not in port. She soon learned that Taylor was on a second voyage to Manzanillo so while she waited, she called on the boat's original agent. The agent told her that Macintoch was a "bad fellow" and also learned that William Taylor was well-liked and highly regarded by his business acquaintances and customers. She was relieved when Taylor arrived at Mazatlán a few days later and she told him of Mesmer's wishes. Being grateful to Mesmer, he graciously bowed out of the business and set off to parts unknown.

When Maude called on Chapman she initially found him unhelpful, possibly because he regarded her partners as shady characters. Chapman said her only options were to legally change the flag to Mexican or sell the boat under Mexican registry. Either action would require the permission of the U.S. Shipping Board.

Maude moved onto the boat to save expenses but soon learned the engines were a continual source of trouble and that local mechanics were of little use.

Just then, Cornelius Heintz arrived in Mazatlán. His employment in Guaymus had not worked out and he heard about a salvage prospect at Mazatlán. He immediately reported to Chapman and in July met Maude at the consulate. After learning of her mechanical troubles he agreed to help with the engines and Chapman eventually got the flag changed back to American.

Since arriving in Mexico, Heintz had acquainted himself with the various German businesses in the area and through them the German consul in Mazatlán, Friedrich Unger, who was also the head Melchers in Mexico. Melchers specialized in banking, dry goods, and fancy goods and ran the steamers *Pinotepa, Josefina,* and the motor vessel *San Cosmo*. When war came in 1914, the Germans in Mexico shifted from a commercial to a war footing, directing all of their efforts and profits toward aiding the homeland.

Heintz improved the engines but Maude could not trade coastwise in Mexico and she had heard, no doubt from Chapman, that vessels of all kinds were being requisitioned by the Shipping Board and in July 1917 exports from the United States were severely restricted. At that point she should have sailed the boat back to San Diego and faced whatever came her way, but she didn't and by the end of 1917 her situation had become desperate.

Friedrich Unger had spies everywhere and when he heard about Maude's troubles he saw a golden opportunity. Twelve big German sailing vessels were holed up at Santa Rosalia just across the Gulf and most of their crews were still aboard. He concocted a scheme to use the *Agassiz* to capture a larger ship that could operate as a raider in the Pacific. He reportedly had his eye on the steamer *San Juan* and was then in port.

Unger's first step was to recruit a crew from Santa Rosalia. Fritz Baumann was the chief mate on Capt. August Kohnke's *Reinbek*. Kohnke later said Baumann had "aspirations toward command and indulged his mind in fancy." At one point in 1914 he left the ship, with no opposition from Capt. Kohnke, visited relatives in California, and was briefly interned on Angel Island in San Francisco Bay as an enemy alien then released to return to Mexico. He might even have been employed as captain on the *Agassiz,* learned of Maude's troubles, and reported this to Unger.

Unger's plan was for Baumann to convince her that she could make a lot of money in the South Sea Island trade, keep her boat, and pay off all her debts. Because the boat still owed port fees, she knew she could not legally clear for a foreign voyage but she nevertheless agreed to go forward with the notion. That was Maude's first mistake. Her second mistake was failing to deliver the ship's papers to Chapman after a voyage as the law required. In order for the scheme to work, the ship's papers had to remain aboard. The boat had carried a load of corn in December and when they ran into heavy weather, it swelled and caused some damage to the boat. Maude used that as an excuse and repeatedly told Chapman the papers had gotten wet and were in a drawer that had swelled shut and couldn't be opened. Chapman then became suspicious of her.

In the meantime, Baumann brought charts, a sextant, a German flag, and other equipment from Santa Rosalia and got them aboard the *Agassiz* as unobtrusively as possible. When questioned later, sailors from Santa Rosalia adamantly denied that weapons of any kind were aboard their vessels so it appears that Unger somehow procured a machinegun, small arms, and ammunition. The machinegun was brought aboard concealed in a wooden box.

The boat had to leave the harbor legally so a scheme was concocted whereby Maude would get permission from the Captain of the Port to test the engines after drydocking for repairs. Cornelius Heintz would go as engineer and his wife Cora would go as a passenger. A few Germans would be concealed aboard and six Mexicans would be hired as the sailing crew. Everything would appear normal. The boat would go north and pick up Baumann on Venados Island who would be waiting with 200 pounds of supplies, stores, and provisions for a six months' voyage, all furnished by Melchers. The Mexicans would be put ashore and the boat would sail to Santa Rosalia to pick up additional crew and arms.

Consul Unger's big mistake was assuming that Cornelius Heintz was a loyal German and thus included him in the scheme but when he had learned enough details he reported everything to Chapman. Chapman then reported the scheme to the Navy and the *Agassiz* was easily captured by Lt. Charles Edwin Reordan's gunboat *Vicksburg*. His report stated, in part:

> On March 17th Mr. Heintz stated several dates that the *Agassiz* might attempt to escape and one of these was Sunday morning, March 17. Word was received that the *Brutus* and Submarine Chaser *302* would arrive from the south that afternoon. The *Vicksburg* therefore stood to the northward and anchored in Viejo Bay, hoping thus to lure the *Agassiz* out of port to be captured by the Submarine Chaser if he went south, or by the *Vicksburg* if he went north. At 12:25 pm., he was sighted standing out of the harbor. *Vicksburg* immediately gave chase, hoisting signal to "Stop." *Agassiz* turned and attempted to escape but was overhauled and stopped by a shot across his bow. He was seen to be throwing articles overboard before he stopped. Sent Lieutenant (jg) F.W. Dorr, N.N.V., as boarding officer to search him. Found aboard a German Flag, a secret cipher, two rifles, three pistols, ammunition and the following persons:
>
> Mr. Cornelius A. Heintz,
> Mrs. Cornelius A. Heintz,
> Miss Maud M. Lochrane
> Frank Volpert,
> Richard Charles Lewis Brandt,
> Charles F. Boston,
> Arthur Martens,
> Hendrik Koppalla,
>
> and six Mexicans (names unknown). There were no proper ship's papers aboard. The vessel was therefore taken as a prize for the following reasons—(1) The vessel's papers were not complete and what papers were aboard were concealed; (2) The vessel attempted to escape after being summoned by International flag signal to stop; (3) The vessel was carrying armed forces of the enemy; (4) The affidavit of a member of the crew to prove that the vessel was proceeding to engage in belligerent operations against the United States and its allies. A prize crew was put aboard, and the personnel taken aboard the *Vicksburg* as prisoners. At first only one man admitted that he was a German. This man was found hiding in the double bottom of the *Agassiz* with a pistol in his possession. The others claimed American, Scandinavian, and Dutch nationality, but later more of them admitted being German. A certificate was found on one of them (Volpert), stating under the seal of the German consulate, that he was a German Reservist. All of them disclaimed any knowledge of how the arms and ammunition got aboard.

Lt. Reordan convened a meeting aboard the *Vicksburg* attended by the commanding officers of the *Brutus, SC-302,* the American and British consuls, and the Captain of the Port. The port captain demanded the vessel be released to him on the grounds it left without clear-

Top: The *Agassiz* after being captured by the *Vicksburg*. *Bottom:* Unidentified prisoners being held under guard aboard the *Vicksburg* (both photographs, National Archives).

ance and several hundred pesos were owed in port charges. Reordan offered to pay the fees if the Mexican government would drop all charges relating to the capture. The Captain of the Port refused the offer so Reordan kept the schooner and released the six Mexicans as he felt they had no knowledge of the scheme. Three hours later the Captain of the Port left after telling Reordan he understood his position but was very afraid of what the German's would do to him for not getting the schooner back.

Since Reordan anticipated a prize hearing, he took a sworn statement from Cornelius Heintz on board the *Vicksburg* on March 18. In his statement he referred to Arthur Martens as "Madden":

Cornelius A. Heintz, being first duly sworn, on oath deposes and says: My full name is Cornelius Adolph Heintz. I was born in Los Angeles, California on April 4th, 1890. My father was a naturalized German and my mother an American. My regular occupation is mechanical engineer. I left the United States in January 1917, going to Guaymas, Mexico, and returning to Los Angeles the same month. At the instance of W. Loyaza & Co., of San Francisco and Mazatlan, I went to Guaymas to inspect the motor ship *Anvil*, then anchored in the harbor at Guaymas, and on my return to Los Angeles I took a contract with H. Hilbing [Bernard Hilbing] and J.B. (?) Franklin [John B. Franklin] to put the *Anvil* in shape for sea. About the 1st of April 1917, I returned to Guaymas and commenced work on the *Anvil*, continuing until June 1, 1917. I was to stay on the ship for 6 months as chief engineer at $240 per month, but I broke my contract and left the ship because I was informed that Mr. Hilbing had been taken into custody by the U.S. Immigration authorities for his pro-German activities. This information came from Mr. Franklin, father-in-law of Mr. Hilbing. Another reason for breaking my contract was that Mr. Hilbing, who spent all of his time in Guaymas, had numerous immoral parties on board the vessel which interfered with my work. I reported the facts in full to Dr. McDougall who was then acting American Consul at Guaymas.

After leaving the *Anvil* I remained in Guaymas for about three weeks and then left for Mazatlán to buy the old gunboat *Morelos* then sunk in the harbor. Through Mr. Hilbing I had met some Germans at Guaymas and they were backing me financially putting up $8,000 bond so that I could bid on the ship at public sale. The names of these Germans were Mr. Unger, head of the house of Melcher's and Mr. Haüs, of the same firm. Melcher's signed the bond. I consulted with the American Consul at Mazatlan concerning all of my dealings with the Germans, and later on October 6th, 1917, I severed all business connections with them. In the meantime I had started to raise the *Morelos*, but the work was not completed and is not yet fully completed. This ship was to be sold when raised and the venture was for speculation only. Louis Arzac, a Mexican residing in Mazatlán was in partnership with me on this deal and advanced me the money.

After severing business connections in October 1917, I retained my social connections with the Germans on the advice of the American Consul. I had a card to the German club, went out to the brewery and was familiar with all of their plans. I was looked upon as a true German, knew all of their secrets and associated with them constantly.

My living expenses came from money advanced by Arzac. I furnished the American Consul with much confidential information from time to time, but never received any pay for this, nor asked for any. I kept the American Consul fully advised of everything that went on during the past four months. I would say that I know a hundred Germans in Mazatlan and vicinity and know the heads of all the mercantile houses there.

I first became acquainted with Miss Maude Lochrane at Mazatlan in July 1917. I met her at the American Consuls. She was then part owner of the auxiliary ketch *Alexander Agassiz* and was having trouble with her co-owner, Taylor, who had placed the vessel under Mexican registry illegally. I helped her to get the vessel out of Mexican registry and back under the American flag. We held court before the American Consul who heard the facts and declared the American flag to be again placed on the vessel. This boat was hauling freight and passengers up and down the coast between Mexican ports and I saw it frequently at Mazatlan.

About the first part of February 1918, I was informed by Mr. Unger of the house of Melchers and also German Consul at Mazatlan, that they were going to try to send the *Agassiz* to sea as a raider with a German crew, their intentions being to use her to capture a larger boat on which to operate. On this occasion Mr. Unger sent for me and discussed freely their plans. I was to put the boat into commission, to go into the deal as a partner, and to go to sea in her as chief engineer. As a partner I put into the deal about $300 gold to show my good faith and place the boat in commission. I never intended to go to sea in the vessel. I informed the Consul immediately of their plans and he told me to keep both eyes open and keep him posted.

About that time Maude Lochrane negotiated with Captain Baumann and First Mate Madden for a trip to the South Sea Islands. Captain Baumann was formerly interned in San Francisco but was allowed to return to Mexico and is now in Mazatlan. He was formerly a captain in the German navy. Madden was formerly in the German navy and came to Mazatlan from one of the interned German sailing vessels at Santa Rosalia. Baumann and Madden were acting for Unger. She agreed to make the trip and commenced to get the boat ready. After my talk with Unger in February, he sent Captain Baumann to Santa Rosalia to get charts of the Pacific Ocean, navigational instruments, books and equipment. He also brought back one machinegun of a German make and one box of ammunition. These articles were left at the German Consul's and I looked them over, except the machinegun which was in a box. I saw the box of ammunition and examined the cartridges. I notified the American Consul as soon as I saw them.

On the night of March 15th all of the above articles were taken aboard the *Alexander Agassiz*. About 11:30 a.m. on March 16th they were taken off because of information the Germans had received that they were being watched. They were taken off by order of the German Consul who knew the USS *Vicksburg* had arrived and he was afraid that the *Alexander Agassiz* would be boarded. Captain Baumann and Mate Madden made up a list of stores and supplies which were furnished by Melchers and were taken aboard early on the morning of March 16th under my supervision. I had orders from Unger to get the stores aboard without exciting suspicion. There was gasoline on board sufficient to cruise 500 miles, also water and wood for cooking. We figured on using the sails most of the time. The night of March 16th, Mate Madden said he was game to make a dash and would take a chance. The charts and instruments were taken aboard and all except 200 pounds of supplies which we were to pick up with Captain Baumann on the farthest island to the northward outside of the harbor. The machinegun was also to be taken aboard.

Miss Lochrane was not present at the conference where the use of the boat was discussed and so far as I know did not know that the Germans intended to seize any ships. The boat was in debt and could not be cleared and she agreed to slip out without clearing. They were to take the boat down for her without pay and after getting down to the South Seas they would act as crew, run the boat and furnish the provisions. They represented to her that the boat would run between the islands trading. She knew of everything that was being taken aboard except possibly the machinegun.

The whole crew was picked by Captain Baumann. Madden was second in command. Madden gave me the names of the crew that I could expect and I gave them to the American Consul. I knew one of the crew for about two months, he worked for me for about two months, the tall fellow who claimed to you that he was a Hollander but afterwards admitted that he was a German. His name is Hendrik Koppalla. He was in on the plans, to my knowledge, about one week before we left. All of the crew knew that she was going out as a raider and all except one had taken an oath before the German Consul that they would stand by the captain. The one who claims he is an American, Charley, did not take an oath before the German Consul but agreed to the plans in the presence of all the others. At the time of taking the oath, which was about two days before we left, the German Consul read to us the orders of Captain Baumann, and in substance as follows: "He was to pick up the first suitable ship, take possession of it, destroy all commerce that he could and do all that was possible to return to Germany. He was entrusted with letters in cipher code to be delivered to the German government."

The orders were written out by the German Consul and signed by him.

During the preparation, Miss Lochrane was living aboard the boat. We decided to leave on the morning of the 19th at 7:00 a.m. but were delayed on account of the tide and also were waiting for instructions from the German Consul in regard to picking up Captain Baumann. We got those instructions about 11:00 a.m. and got underway shortly afterwards. We were to get Baumann off the north island and were to land some Mexicans which we had taken aboard as a blind. They were to return in a large canoe which was used by Baumann to take over the rest of the supplies.

In order to get the boat out without clearing, Miss Lochrane obtained permission from the Captain of the Port to take the boat out for a trial run. I obtained similar permission from the American Consul. We stood out of the harbor and headed direct toward the north island. We saw the *Vicksburg* when we went outside but we had orders from the German Consul not to go more than one mile from shore and figured that she could not touch us. We had a Mexican pilot going over the bar. Madden was in charge and I acted as chief engineer.

When we noticed that the *Vicksburg* was coming after us we attempted to dodge and return to the harbor. Madden immediately grabbed several packages, including a sextant, and threw them overboard. We also had a full set of German flags on board, including a battle flag, which he must have also thrown overboard. As soon as I saw the stop signal I slowed the engines down by flooding the carburetors to keep the boat from getting back into the harbor. When the shot was fired I immediately jumped below to shut off both engines. I was below when the boarding officer came aboard.

On the night of the 15th I was called before the German Consul and was told that if I didn't shoot square they would see that I wouldn't be walking around in the morning. They were suspicious because of the actions of Gardiner, the American Vice Consul.

I am a married man and my wife knew that we might be stopped by a gunboat and I told her not to worry. She didn't know anything about the plans and I took her along on the advice of the American Consul. I also discussed it with the captain of the *Vicksburg*.

The foregoing statement contains all of the facts which I know relating to the seizure of the *Alexander Agassiz* and the events leading up to it.

Those arrested on the boat were transferred to the custody of the Justice Department's Bureau of Investigation for interrogation and reports were filed on each one.

Richard Charles Lewis Brandt was described as 5'-10", 147 pounds, dark brown hair, blue eyes, and ruddy complexion. He was born in Mecklenburg, Germany, on April 27, 1885, and arrived at New York in 1901 at age 17. After about six months he went to San Francisco and in 1903 was a clerk at the Sterling Furniture Co., a mattress manufacturer. Two years later he was a cashier at Franz Walter and George Kronert's restaurant.

He survived the big earthquake and fire in April 1906—Kronert might not have or he left town—as the next year Brandt partnered with Walter as the Walter & Brandt restaurant. The venture was short-lived because in 1908 Brandt was on his way to Chihuahua, Mexico.

In 1909 he was a lumber inspector for the Madero Co. in Chihuahua. In 1913 he returned to the United States and went to Doucette, Texas, to work for the Long-Bell Lumber Co. In 1914 he was working for the Standard Lumber Co. in Sonora, California, and in 1915 the restless soul was a ranch hand at Imperial, California. In 1917 he went to El Paso, Texas, to manage a Madero Co. mill there when it opened. In April 1917 he went to El Oro on the west coast of Mexico to see about employment with the Aloha Co. He stated that on or about March 16 he was approached by Cornelius Heintz and asked if he would like to go on an excursion aboard the *Alexander Agassiz* while waiting for employment and that he had no knowledge of anything, anyone, or the rest of the crew. He was found to have 8 pesos and 71 centavos on his person and gave as a reference Samuel G. Thomas, a wood carver and neighbor who lived in San Francisco. In 1920 he was 35, single, and a lumber inspector living in a boarding house in Los Angeles.

Charles Franklin Boston was described as 5'-9", 160 pounds, dark brown hair, brown eyes, and ruddy complexion. He was born on December 7, 1889, in Fristoe, Missouri.

In 1911 he lived in Canada and in 1912 went to Mexicali on a prospecting tour. On June 5, 1917, he registered for the draft at Bisbee, Arizona, while working on a farm. On October 1 he left for Mazatlan and arrived on January 25, 1918. On February 26 he registered at William Chapman's office to work as a seaman on the *Alexander Agassiz* to earn enough money to return to the United States in about a month. He was employed on the *Agassiz* as a cook around February 5.

On March 1 he was called up for service in the Army but failed to report. He said he attempted to enlist but was rejected for service by "Dr. Fly at 5th and Broadway in San Diego."

He had 50 centavos in his possession when he was arrested. In 1930 he was working on a farm in Silsbee, California. He died on October 28, 1961, in La Habra, California.

Hendrik Koppalla was described as 5'-4", 125 pounds, light brown hair, brown eyes, and ruddy complexion with several tattoos and scars. He was born on April 29, 1895, in Amsterdam, Holland, came to America in July 1916, and worked as a marine fireman. He was arrested and incarcerated at the immigration station at Angel Island in San Francisco Bay after having been in Mexico and on October 25, 1917, he was deported to Mexico as an undesirable alien. He said he entered the United States illegally by swimming across the river at Laredo, Texas. He was working on a farm outside Mazatlán when he was "invited" to go on a trip on the *Alexander Agassiz* "by a party that did not make the trip." He knew nothing about anything.

Arthur Martens was described as about 5'-10", 170 pounds, age 36, ruddy complexion, light brown hair, blue eyes, slender build with a "morose expression to face." He was born in Lübeck, Germany, on November 5, 1880, and entered a maritime academy but when he was on a training cruise he deserted the ship in Chile. He was known to speak German, French, Spanish, and English. In 1900 he went to Mexico and in 1914 came to the United States and worked as a timekeeper and storekeeper for the Imperial Irrigation District at Imperial, California, for two and a half years. He went back to Mexico in 1916 and said he was recruited by Cornelius Heintz to work on the *Alexander Agassiz* and knew nothing about anything. He had 3 pesos and 91 centavos in his possession.

Franz Volpert was described 5'-11", 138 pounds, dark brown hair, blue eyes, and ruddy complexion. He was born in Hoppecke, Germany, on July 8, 1887, came to America in 1910, and became a brick layer as Frank Volpert. He entered Mexico through Laredo, Texas, on April 4, 1917, and went to Monterey. He said he was in Mazatlán waiting for passage to Manzanillo when Charles Boston asked him to go on the *Alexander Agassiz* and that Maude agreed take him to Manzanillo. He said he always carried a gun and was hiding in the double bottom in the boat with his pistol when they were captured but he didn't know anything about any flags or weapons. He had a letter written in German in his suitcase along with 13 pesos and 50 centavos. He was single and said he intended to go to Cedar Rapids, Iowa.

Cornelius Heintz was born in California on April 4, 1890. In early 1918 married Cora Soto, a Spanish national. In 1920, Cornelius and Cora lived in Los Angeles with their son new Carl and Cora's younger sister Lucy. Heintz later married Adella Metzener, of Idaho, 21 years his junior, and they had four children. Cornelius manufactured air conditioning equipment and died in Los Angeles on June 20, 1958.

Francis Clark "Frank" Wheeler was born on November 27, 1876, in Wheeler, Michigan, to farmers James B., of New York, and Ida Wheeler, of New Jersey.

In 1899 Frank married Minnie Elliott, a Canadian who came to the United States in 1895. They lived in Aberdeen, Washington, where Frank was a grocer. By 1910 Frank was a real estate agent in Aberdeen.

In 1915 Frank came down to Los Angeles and partnered with Harold S. Cross as debt collectors. The business dissolved in January 1917 when Frank became a co-owner of the *Alexander Agassiz*. After the whole *Agassiz* deal had blown up Frank worked as a cashier with the Davis Manufacturing Co. Frank remarried and became a manager at an auto factory. In

1929 he married an 18-year-old Mexican girl named Josephine and was the manager of a collection agency. Frank died on July 7, 1964, in West Covina, California.

Lt. Charles Edwin Reordan was born on July 11, 1888, in Junction City, Kansas, and appointed to the Naval Academy on July 28, 1905, from Dallas, Texas. He took command of the *Vicksburg* on April 13, 1917, and served in that capacity throughout the war.

Reordan was well known in the service as the skipper who captured a German raider during the World War. He made captain on September 1, 1937, and after serving as commanding officer of the transport *Henderson* and director of the naval reserve in the 12th Naval District in San Francisco, he took command of the battleship *Tennessee* on December 30, 1940. A year later his ship was at Pearl Harbor as flagship for Rear Adm. David W. Bagley. The ship was moored inboard of the *West Virginia* on December 7, 1941, and became pinned against the dock when the *West Virginia* sank.

Capt. Reordan was in his street clothes and wearing a straw hat when the attack began. He was out on deck when a projectile hit Turret II, throwing shrapnel everywhere. He escaped injury but the skipper of the *West Virginia*, Capt. Mervyn S. Bennion, was mortally wounded in the blast while standing on his bridge. The *Tennessee* took aboard many survivors from the *Arizona*, moored directly astern. Capt. Reordan later stated:

> I thought we might be able to move forward a little and get into a safer position and I ordered slow speed ahead. We didn't move. I ordered the speed increased until the engines were turning over at full speed. Still we didn't move, because the *West Virginia* had us jammed. But about that time I noticed the wash from the propellers was holding back the flames from the *Arizona*. I felt dumb for not thinking of it before.

The *Tennessee* was freed on the 16th. He served on the *Tennessee* until June 20, 1942, then became CO of the Key West Naval Operating Base. He died on March 22, 1947, and was buried at Arlington National Cemetery.

Maude was born on April 16, 1878, in Michigan, adopted at age six by James B. and Ida Wheeler. In 1903 at age 25 she was a stenographer for Almon C. Varney's architectural company in Detroit. The next year she was an architectural draftsman living with Michael J. Carr.

In 1906 she moved out to San Bernardino, California, and worked a bookkeeper for Hiram and George Roach's Southwest Electrical Co. at 323 D St. The company installed electric power plants and wiring, provided local electric service, manufactured electric fixtures, provided electrical engineering services, and repaired equipment.

By 1908 she was a manager at the company and in 1910 she moved to El Paso, Texas, as the company's Southwest sales representative. In 1912 she owned the Raymond Hotel in El Paso.

By 1915 she had moved back to California and was working as a reporter for the New & Greater Los Angeles, a weekly newspaper, and it was around this time that she became acquainted with Joseph Mesmer.

Maude was described by Bureau of Investigation agent Dave Gershon as Irish, 5–4, 110 pounds, slight build, auburn hair, light blue eyes, with a ruddy complexion and freckles. Her nickname was "Geranium." One acquaintance stated, "She is quite an entertaining talker and will immediately impress one as being an adventuress. She knows navigation thoroughly, in fact she possesses a navigator's license."

The *Agassiz* was towed to San Diego and Maude and the other prisoners were lodged in the county jail under federal custody.

The prize hearing was held in Judge Benjamin F. Bledsoe's district court in San Diego. This was the first admiralty prize case ever held on the Pacific Coast and the first since America entered the war. District Attorney Robert O'Connor represented the crew of the *Vicksburg* and Maude hired attorney Alfred L. Black to present her case. The hearing began on May 14.

Maude and the others appeared in court on a charge of conspiring to destroy shipping but none of them testified. On July 15 a verdict was returned in favor of Maude on the grounds that she knew nothing of the plot to seize shipping. Koppalla and Boston were released and the others jailed.

On October 4, 1918, Maude obtained a Permanent Register for the boat at Los Angeles and resumed her business but things did not work out. In 1919 the boat was libeled for unpaid bills and crew wages and in October was sold at San Pedro.

The boat was enrolled at Los Angeles by new owners on May 28, 1920. Early on Friday, November 19, 1920, it went aground in dense fog on an island in Drake's Bay near Point Reyes with seven people aboard. There were no injuries but the boat was a total loss.

On December 25, 1944, an article appeared in *The Bakersfield Californian*:

> Police today found the body of Mrs. Maude Lochrane, 50, in her lodgings after an unidentified woman gave the coroner this message by telephone: "Go to 111 G South Lake Street and you will find the body of a woman who has been poisoned and is dead." Detective Lieutenants E.B. McCreadle and Charles L. Dietrich said that almost a year ago they received an anonymous call reporting the death of another woman, Mrs. Grace Irwin, by poison at the same address. They said Mrs. Lochrane's body was found upright in a chair.

17. The Final Tally

Compiling accurate statistics for a very complicated event like a war is an almost impossible task. Figures from the various reporting government agencies, bureaus, and professional associations vary, often widely, probably because the methods of calculation and what was tallied up or included in any given report varied, especially when there were overlapping functions.

The Treasury Department estimated the net cost of the war to the United States at $41,873,948,225 after credits totaling $2.3 billion were deducted, mostly from the repayment of loans. This was slightly larger than Great Britain's costs over four years.

Pre-war loss of life claims totaling $205,346.74 were filed on 15 persons from belligerent causes. Personal injury claims for pre-war losses were filed by 46 Americans totaling $1,761,316.41 and 40 belligerent claims totaling $2,495,553.64. Beginning on June 12, 1917, every merchant seaman who signed on a Shipping Board steamer was required to carry government war risk insurance.

The Marine & Seamen's Division of the Bureau of War Risk Insurance identified 2,769 merchant seamen who died or were slightly or seriously injured from all causes or taken prisoner during the war. Between June 30, 1915, and June 30, 1919, 338 deaths were reported on American merchant vessels; 72 sailors were seriously injured, and 46 were detained aboard enemy ships.

Between June 30, 1915, and June 30, 1919, a total of 2,493 American merchant seamen reportedly died in marine hospitals and relief stations from disease or injuries, including 245 from the flu pandemic of 1918.

According to the Bureau of Marine Inspection & Navigation, 372 seafarers were unaccounted for by war's end, either from desertion, murdered in a foreign port, kidnapped, dying unidentified in a foreign hospital, etc. They never rejoined their ship and simply disappeared.

The Marine & Seaman's Division reported collecting $843,172.89 in premiums on $321,946,263.02 worth of policies and paying out $367,096 in losses on 561 deaths "directly or indirectly due to war."

The Bureau of War Risk Insurance paid out $303,807.81 on claims for injuries and the detention of American merchant seamen after April 6, 1917. A number of merchant seamen who had been taken prisoner filed claims for back wages.

American-owned insurance companies reported paying $3,280,342 on death and injury claims due to submarine warfare.

By March 1921, American citizens had filed 451 claims for damages against Germany totaling $110,254,058 for "loss of life, personal injuries, loss of hulls, cargoes, and personal effects, war risk insurance, losses due to submarines, raiders, and mines."

During four years of war, Great Britain lost 15,361 merchant seamen on 3,145 merchant ships, far greater losses than any other nation. A total of 1,519 Merchant Navy officers and sailors were added to the Roll of Honour.

Prior to the United States entering the war, 21 merchant ships were reported lost from probable enemy action, including the *Orleanian*, which disappeared. During the war 125 ships were reported lost and 13 ships inducted into the Naval Overseas Transportation Service were lost for a total of 159 ships.

Private vessel owners filed 11 claims for pre-war sinkings totaling $6,604,487.96 and 89 claims for wartime losses totaling $23,807,276.17.

Government policies were taken out and paid on 164 American vessels lost, many of those small fishing craft. Receipts on premiums from the Bureau of War Risk Insurance Marine and Seaman's sections totaled over $17 million and, combined with the $50 million sinking fund allocated by Congress, over $67 million was returned to the Treasury Department.

The Army and Navy reported in 1920 that between August 1, 1914, and November 11, 1918, 12,852,000 gross tons of Allied and neutral shipping were lost to enemy action. Of that, 11,154,000, or 87 percent, was lost to U-boats; 1,121,000 tons was lost to mines; 569,000 tons to surface ships; and 8,039 tons, representing 5 ships, were sunk by aircraft—49 vessels were attacked by aircraft, 44 escaped, and 5 were sunk.

It was reported that 1,880 British vessels escaped attacks and 45 of those were attacked by aircraft. Four ships were lost to aircraft.

Norway lost 703 ships, France 697, Italy 621, Denmark 241, Sweden 185, the United States 154, Greece 168, Holland 134, Russia 126, Belgium 60, and Japan 29. Holland lost the only ship sunk by aircraft.

A total of 48 ships either owned or chartered by the Shipping Board were lost to submarines. Of those, 19 were owned, including six former German vessels, and 29 were privately-owned ships under charter. Some of the ships were being operating by the Army or Navy at the time they were lost.

Approximately one-third of the world's total merchant shipping that existed in July 1914 was lost to enemy action, including mines. The normal percentage was lost to collision, stranding, and the ordinary perils of the sea.

Approximately $60 million worth of cargo was lost during the war to submarines. More than 70 insurance companies paid more than $53 million on war risk policies. The Bureau of War Risk Insurance and the Treasury Department paid another $4,272,164,01 to cargo owners. Premiums on war risk policies were very high, as can be imagined, and one shipping company reported spending $1,212,000 just between April 1917 and November 1918.

Admiral Sims estimated that because the Navy Department failed to "enter the war immediately and whole-heartedly" it cost 2,500,000 tons of Allied shipping to be lost unnecessarily. He further stated:

> The loss of this amount of shipping can also be translated into a definite prolongation of the war and unnecessary sacrifice of blood and treasure in accomplishing the victory. As General Pershing clearly shows

in his report to the Secretary of War, the primary consideration limiting the number of American troops that could be sent to France was that of tonnage. The tonnage losses of 1917 made it impossible at the time to transport any considerable American Army and, and, at the same time, continue the absolutely essential military supplies and food for the civil populations of the allied countries. It therefore became necessary to limit the number of American troops that could be sent abroad during the first year to an average of approximately 25,000 men per month. If the additional million and a half tons sunk unnecessarily in 1917 had been saved by prompt cooperation of our Navy, the number of American soldiers sent to France could have been doubled or trebled. If the tonnage had been available and the additional American troops had been sent to France, and the new drafts called more promptly in this country, America could have had a million men in France by March, 1918, instead of 300,000.

The loss unnecessarily of the two and a half million tons of shipping, therefore, in all probability postponed the end of the war at least four months. The average loss of life per day to the Allies during the war was 3,000 men. This prolongation of the war, therefore, cost half a million lives. Similarly, as the war cost the Allies $100,000,000 a day on the average, this prolongation resulted in the unnecessary expenditure of $15,000,000,000, of which at least one-third was expended by the United States directly or loaned to the Allies.

The December 29, 1918, newspapers carried a wire service article:

Amsterdam, Dec. 28.—A gloomy picture of Germany's future was painted by Grand Admiral von Tirpitz in his recent speech in Essen, according to the Hamburg News. Regarding a successful economic war against Germany after the war, he said:

"Imagine the position if we simultaneously have to bear the burden of taxation which must fall on every German and, despite the fallen value of German money, we still have to buy the most necessary food and raw materials from abroad. Can anyone in his heart of hearts really believe that in these circumstances, without an increase in power, without indemnity and without security we could avoid Germany's ruin?

Not only has England taken our colonies and Mesopotamia, but everywhere she has made deeper and firmer bases for her maritime and colonial supremacy. She has tarnished and trodden down the prestige and honor of Germany by unprecedented calumnies. In the whole trans–Atlantic world we are considered as conquered and done for."

Russia announced they would lift the White Sea blockade pursuant to the Brest-Litovak negotiations "for resumption of commercial relations" with Germany. A German ship was expected to arrive shortly at Revel with supplies and German commercial agents were taking orders for January deliveries.

Conclusion: Toward the Future

The Great War never really ended. Deep-seated conflicts, turmoil in the vanquished states, new nationalistic ambitions, and national memories that lasted much longer than the flight of bullets set the stage for future events.

The big winner in the World War was Japan. Japan expended approximately $481,818,000 toward the Allied cause, about one-tenth of one percent of the United States, and ended up with huge territorial acquisitions that left a credit indemnity of only $231,818,000.

Japan's gain went back to March 1898 when Baron Edmund Friedrich Gustav von Heyking of Germany signed a 99-year lease with China in the Bay of Kiao-chau for a naval base, fortifications, railroads, mines, and commercial enterprises in the Prefecture of Tsao-chau-foo in Shantung. After the war started, Japan saw an opportunity to expand its empire by aligning itself with Great Britain and the Allies, whom they believed would win. They immediately demanded that Germany surrender its leased property, to be held in trust by them for China for the duration of the war. Germany refused and on August 23, 1914, Tokyo date, Japan declared war on Germany and took possession of the entire province. Shantung by then had a population of 30 million, 308 miles of railroad, two iron mines, two gold mines, and coal mines producing 814,000 tons a year. No objection was made by the Allied nations.

Germany also had possessions in the Pacific including German New Guinea, the Bismarck Archipelago, Bougainville, Buka, and Nauru Islands, German Samoa, and the Caroline, Mariana, and Marshall Islands. German New Guinea and German Samoa were seized by Australian troops in 1914.

Japan actively participated in the war alongside British forces as an Ally and on May 6, 1919, the League of Nations awarded 1,040 square miles of former German islands to Japan. It was felt in Washington that these islands should have gone to the United States, but the United States was not formally an Allied nation during the war. These so-called Mandate Islands were quickly fortified and formed the basis of Japan's ultimate plan to control the Pacific. In 1924, in anticipation of almost certain war with Japan, the U.S. Navy's General Board adopted War Plan Orange.

Bibliography

Abbot, Willis J. *The Blue Jackets of 1918*. New York: Dodd, Mead & Co., 1921.

American Battle Monuments Commission. *American Armies and Battlefields in Europe*. Washington, D.C.: Government Printing Office, 1938.

Belknap, Capt. Reginald R. *The Yankee Mining Squadron, or the Laying of the North Sea Mine Barrage*. Annapolis, Maryland: U.S. Naval Institute, 1920.

Bureau of the Census. *Statistical Abstract*. Foreign and Domestic Commerce. Washington, D.C.: Government Printing Office, 1923.

Cameron, Capt. John S., as told to Cyril Brown. *Ten Months in a German Raider*. New York: George H. Doran Co., 1918.

Clephane, Lewis P. *History of the Naval Overseas Transportation Service in World War I*. Washington, D.C.: Naval History Division, 1969.

Daniels, Josephus. *Annual Report of the Secretary of the Navy for the Fiscal Year*. Washington, D.C.: Government Printing Office, 1917, 1918, 1919.

Daniels, Josephus. *Our Navy at War*. New York: George H. Doran Co., 1922.

Department of Justice, Bureau of Investigation Case File #78249.

Exporters' Enclyclopaedia Co. *Exporters' Encyclopaedia, 1917–18*. New York.

Fasset, F.G., Jr., ed. *The Shipbuilding Business in the United States*. The Society of Naval Architects and Marine Engineers. New York, 1948.

Ferris, Theodore E. *Douglas Fir Ship. Specifications for the Construction of a Standard Wood Steamship*. Washington, D.C.: Government Printing Office, 1917.

Hook, Frederick Arthur. *Merchant Adventurers, 1914–1918*. London: A. & C. Black. Ltd, 1920.

International Seamen's Union of America. *Coast Seamen's Journal*, September 12, 1917, to September 4, 1918.

Jackson, Claude V., III, U.S. Army Corps of Engineers, and State Historic Preservation Office. *A Maritime History and Survey of the Cape Fear and Northeast Cape Fear Rivers, Wilmington Harbor, North Carolina*. Wilmington: North Carolina Dept. of Cultural Resources, 1966.

James, Marquis. *Biography of a Business, 1792–1942; The Insurance Company of North America*. New York. The Bobbs-Merrill Co., 1942.

Jenkins, Capt. John Wilbur, USN, et al. *The World War and Historic Deeds of Valor, Vol. VI*. Chicago: National Historic Publishing Assoc., 1919.

Jones, John Price. *The German Secret Service in America*. Boston: Small, Maynard & Co., 1918.

Jones, John Price. *The German Spy in America*. London: Hutchinson & Co., 1917.

Kennedy, Philip B. *Annals of the American Academy of Political and Social Science*, Vol. 63, January 1916.

King, W.C., ed. *King's Complete History of the World War*. Springfield, Massachusetts: The History Associates, 1922.

Marquis, Albert Nelson, ed. *Who's Who in New England*. Chicago: A.N. Marquis & Co., 1916.

McMaster, John Bach. *The United States in the World War*. New York: D. Appleton & Co., 1918.

McNairn, Jack, and Jerry McMullen. *Ships of the Redwood Coast*. Palo Alto, California: Stanford University Press, 1945.

National Archives, Record Group 41, Entry 251. Records of the Bureau of Marine Inspection, and Navigation. Lists of Merchant Seamen Lost in World War I, 1914–1919.

Navy Department. *American Ship Casualties of the World War Including Naval Vessels, Merchant Ships, Sailing Vessels, and Fishing Craft*. Washington, D.C.: Government Printing Office, 1923.

Navy Department. *German Submarine Activities on the Atlantic Coast of the United States and Canada*. Washington, D.C.: Government Printing Office, 1920.

Navy Department. *Ships' Data. U.S Naval Vessels, January 1, 1914*. Washington, D.C.: Government Printing Office, 1914.

Newell, Gordon. *Pacific Lumber Ships*. New York: Bonanza Books, 1960.

Nutt, John L. *Newburgh, Her Institutions, Industries and Leading Citizens*. Newburgh, New York: Ritchie & Hull, 1891.

Palmer, A. Mitchell. *Annual Report of the Alien Property Custodian*. Washington, D.C.: Government Printing Office, 1919.

Perry, Ralph Barton. *The Plattsburgh Movement: A Chapter of America's Participation in the Great War*. New York: E.P. Dutton & Co., 1921.

Potter, Frank Hunter. *The Naval Reserve*. New York: Henry Holt and Company, 1919.

Pratt, Fletcher. *The Navy, a History: The Story of a Service in Action*. Garden City, New York: Garden City Publishing Co., 1938.

Rathbone, Alfred D, IV. *Romance of the Sea Lanes*. New York: National Travel Club, 1944.

Redfield, William C. *Report of the Secretary of Commerce and Reports of Bureaus*. Washington, D.C.: Government Printing Office, 1919.

Society of Naval Architects, and Marine Engineers. *Transactions, Volume 15*, "Submarines of Battleship Speed" by Mason S. Chace, page 43.

Thomas, Lowell. *Count Luckner, the Sea Devil*. Garden City, New York: Garden City Publishing Co., 1927.

Thomas, Lowell. *The Wreck of the Dumaru*. New York: Doubleday, Doran & Co., 1930.

United States Army. *United States Army in the World War, 1917–1919: Reports of the Commanders-in-Chief, Staff Sections and Services, Vol. 15*. Washington D.C.: Center of Military History, U.S. Army, 1991.

U.S. House of Representatives, 66th Congress, 2nd Session. *Claims of Wooden-Ship Builders*. Hearings before the Committee on the Merchant Marine and Fisheries, January 14 and 15, 1920, Vol. 1. Washington, D.C.: Government Printing Office, 1920.

U.S. Senate. Naval Investigation Hearings Before the Subcommittee on Naval Affairs, 66th Congress, 2d Session, Vol. 1. Washington, D.C.: Government Printing Office, 1921.

United States Shipping Board Emergency Fleet Corporation: Hearings on Concrete Ships Before the Committee on Commerce, United States Senate, Sixty-Fifth Congress, commencing on January 26, 1918, at Washington, D.C.

United States Shipping Board Emergency Fleet Corporation. *Contract and Requisitioned Steamships Taken Over on August 3, 1917, Under Authority of the President's Executive Order of July 11, 1917, Up To and Including May 31, 1919*. Washington, D.C.: Government Printing Office, 1919.

United States Shipping Board Emergency Fleet Corporation. *List of American and Foreign Ships*. Washington, D.C.: Government Printing Office, October 16, 1919.

Weiss, George. *America's Maritime Progress*. New York: The New York Marine News Co., 1920.

Wilson, Vice Admiral Henry B. *An Account of the Operations of the American Navy in France during the War with Germany*. Washington, D.C.: Navy Department, 1919.

Index

Note: All vessels are steel or wood steam freighters unless otherwise noted. Other types are sch.–schooner; shp.–ship-rigged sailing vessel; sail–sailing vessel of unknown rig. A "laker" is a freighter designed for service on the Great Lakes.

Foreign vessels are identified as Braz.–Brazilian; Br.–British; Can.–Canadian; Cub.–Cuban; Dut.–Dutch; Fin.–Finnish; Fr.–French; Ger.–German; Gr.–Greek; It.–Italian; Mex.–Mexican; Nor.–Norwegian; Per.–Peruvian; Sp.–Spanish; Sw.–Swedish; Ur.–Uruguayan.

www.ingramcontent.com/pod-product-compliance
Ingram Content Group UK Ltd.
Pitfield, Milton Keynes, MK11 3LW, UK
UKHW051928100926
8556IPUK00022B/1445

9 781476 667034